SPACE SCIENCE SERIES
Tom Gehrels, General Editor

Planets, Stars and Nebulae, Studied with Photopolarimetry
Tom Gehrels, editor, 1974, 1133 pages

Jupiter
Tom Gehrels, editor, 1976, 1254 pages

Planetary Satellites
Joseph A. Burns, editor, 1977, 598 pages

Protostars and Planets
Tom Gehrels, editor, 1978, 756 pages

Asteroids
Tom Gehrels, editor, 1979, 1181 pages

Comets
Laurel L. Wilkening, editor, 1982, 766 pages

Satellites of Jupiter
David Morrison, editor, 1982, 972 pages

Venus
Donald M. Hunten et al., editors, 1983, 1143 pages

Saturn
Tom Gehrels and Mildred S. Matthews,
editors, 1984, 968 pages

Planetary Rings
Richard Greenberg and André Brahic,
editors, 1984, 784 pages

Protostars & Planets II
David C. Black and Mildred S. Matthews,
editors, 1985, 1293 pages

Satellites
Joseph A. Burns and Mildred S. Matthews,
editors, 1986, 1021 pages

The Galaxy and the Solar System
Roman Smoluchowski, John N. Bahcall,
and Mildred S. Matthews,
editors, 1986, 485 pages

THE GALAXY
AND THE
SOLAR SYSTEM

THE GALAXY AND THE SOLAR SYSTEM

Roman Smoluchowski
John N. Bahcall
Mildred S. Matthews

Editors

With 34 collaborating authors

THE UNIVERSITY OF ARIZONA PRESS
TUCSON

About the cover:

Painting by William K. Hartmann. A range of astronomical phenomena from galactic to planetary is symbolized in the cover painting. In the background are the galactic hub and various stars and nebulae of the spiral arms. In the foreground is part of a planetary system, illuminated by an off stage star. The back cover shows a comparison of the size of the Oort cloud surrounding our Sun with the distances to a few of the nearest stars.

THE UNIVERSITY OF ARIZONA PRESS

Copyright © 1986
The Arizona Board of Regents
All Rights Reserved

This book was set in 10/12 Linotron Times Roman
Manufactured in the U.S.A.

Library of Congress Cataloging-in-Publication Data

The Galaxy and the solar system.

  Bibliography: p.
  Includes index.
  1. Galaxies. 2. Solar system. I. Smoluchowski, Roman. II. Bahcall, John N. III. Matthews, Mildred Shapley.
QB857.G4   1986   523.1'12   86-14665

ISBN 0-8165-0982-4 (alk. paper)

British Library Cataloguing in publication data are available.

CONTENTS

COLLABORATING AUTHORS	ix
PREFACE	xi

Part I—THE SOLAR-GALACTIC NEIGHBORHOOD: GALACTIC GRAVITATIONAL FIELDS

THE GALACTIC ENVIRONMENT OF THE SOLAR SYSTEM *J. N. Bahcall*	3
STARS WITHIN 25 PARSECS OF THE SUN *W. Gliẹse, H. Jahreiss and A. R. Upgren*	13
PRESENT, PAST AND FUTURE VELOCITY OF NEARBY STARS: THE PATH OF THE SUN IN 10^8 YEARS *F. Bash*	35
THE LOCAL VELOCITY FIELD IN THE LAST BILLION YEARS *J. Palouš*	47

Part II—MASSIVE INTERSTELLAR CLOUDS

MOLECULAR CLOUDS AND PERIODIC EVENTS IN THE GEOLOGIC PAST *P. Thaddeus*	61
OBSERVATIONAL CONSTRAINTS ON THE INTERACTION OF GIANT MOLECULAR CLOUDS WITH THE SOLAR SYSTEM *N. Z. Scoville and D. B. Sanders*	69
INTERSTELLAR CLOUDS NEAR THE SUN *P. C. Frisch and D. G. York*	83

Part III—OTHER GALACTIC FEATURES

THE CHEMICAL COMPOSITION OF COMETS AND POSSIBLE CONTRIBUTION TO PLANET COMPOSITION AND EVOLUTION *J. M. Greenberg*	103

TEMPORAL VARIATIONS OF COSMIC RAYS OVER A
    VARIETY OF TIME SCALES   116
J. R. Jokipii and K. Marti

A LOCAL RECENT SUPERNOVA: EVIDENCE FROM X
    RAYS, ^{26}AL RADIOACTIVITY AND COSMIC RAYS   129
D. D. Clayton, D. P. Cox and F. C. Michel

Part IV—THE OORT CLOUD

DYNAMICAL INFLUENCE OF GALACTIC TIDES AND
    MOLECULAR CLOUDS ON THE OORT CLOUD OF
    COMETS   147
M. V. Torbett

COMETARY EVIDENCE FOR A SOLAR COMPANION?   173
A. H. Delsemme

THE OORT CLOUD AND THE GALAXY: DYNAMICAL
    INTERACTIONS   204
P. R. Weissman

Part V—PERTURBATIONS OF THE SOLAR SYSTEM

GEOLOGIC PERIODICITIES AND THE GALAXY   241
M. R. Rampino and R. B. Stothers

GIANT COMETS AND THE GALAXY: IMPLICATIONS OF
    THE TERRESTRIAL RECORD   260
S. V. M. Clube and W. M. Napier

DYNAMICAL EVIDENCE FOR PLANET X   286
J. D. Anderson and E. M. Standish, Jr.

PLANET X AS THE SOURCE OF THE PERIODIC AND
    STEADY-STATE FLUX OF SHORT PERIOD COMETS   297
J. J. Matese and D. P. Whitmire

Part VI—EXISTENCE AND STABILITY OF A SOLAR COMPANION STAR

EVOLUTION OF THE SOLAR SYSTEM IN THE PRESENCE
    OF A SOLAR COMPANION STAR   313
P. Hut

MASS EXTINCTIONS, CRATER AGES AND COMET
    SHOWERS   338
E. M. Shoemaker and R. F. Wolfe

EVIDENCE FOR NEMESIS: A SOLAR COMPANION STAR   387
R. A. Muller

DEFLECTION OF COMETS AND OTHER LONG-PERIOD SOLAR COMPANIONS INTO THE PLANETARY SYSTEM BY PASSING STARS *J. Hills*	397
IS THERE EVIDENCE FOR A SOLAR COMPANION? *S. Tremaine*	409
COLOR SECTION	417
GLOSSARY *M. Magisos*	423
BIBLIOGRAPHY *M. Magisos*	439
LIST OF CONTRIBUTORS WITH ACKNOWLEDGMENTS TO FUNDING AGENCIES	469
INDEX	473

COLLABORATING AUTHORS

J. D. Anderson, *286*
J. N. Bahcall, *ix, 3*
F. Bash, *35*
D. D. Clayton, *129*
S. V. M. Clube, *260*
D. P. Cox, *129*
A. H. Delsemme, *173*
P. C. Frisch, *83*
W. Gliese, *13*
J. M. Greenberg, *103*
J. Hills, *397*
P. Hut, *313*
H. Jahreiss, *13*
J. R. Jokipii, *116*
K. Marti, *116*
J. J. Matese, *297*
F. C. Michel, *129*

R. A. Muller, *387*
W. M. Napier, *260*
J. Palouš, *47*
M. R. Rampino, *241*
D. B. Sanders, *69*
N. Z. Scoville, *69*
E. M. Shoemaker, *338*
E. M. Standish, Jr., *286*
R. B. Stothers, *241*
P. Thaddeus, *61*
M. V. Torbett, *147*
S. Tremaine, *409*
A. R. Upgren, *13*
P. R. Weissman, *204*
D. P. Whitmire, *297*
R. F. Wolfe, *338*
D. G. York, *38*

PREFACE

While two broad fields of astronomical research—our galaxy and our solar system—have been progressing rapidly, areas of concern common to both have remained until recently relatively inactive. In the last decade there has been intense interest in a more detailed study of the galactic neighborhood of the Sun, of the outer limits of the solar system and in topics such as the velocity and mass functions of neighboring stars, the location and motion of the Sun in the Galaxy, the penetration of interstellar and interplanetary dust, the distribution of molecular clouds and of galactic radiation and gravitational fields, of galactic cosmic rays and the role of nearby supernovae. The general question of the influence of the galactic neighborhood on the solar system has received particular attention recently because of the expected perturbations of the outer reaches of the solar system by passing stars, by molecular clouds or by gravitational fields and because of the proposal that the Sun may have a companion star. The latter possibility aroused special interest because of its possible relation to the suggested periodic extinctions of species on the Earth. The many papers being published in the general area of the effect of the Galaxy on the solar system brought to mind the need for a topical conference and book.

The Space Science Series of the University of Arizona Press, which has as its prime goal the encouragement of new fields, was actively involved in this project from the start. As expected, the conference stimulated new ideas on the Galaxy and the solar system. The book is a text for advanced students and a source book for scientific workers in the field; it is not a conference proceedings.

The conference was held in Tucson, Arizona, in January 1985 with the purpose of emphasizing the influence of the Galaxy on the solar system rather than to discuss the Galaxy and the solar system *per se*. The meeting was apparently the first get-together of the galactic and the solar system scientific communities and, as a result, the discussions were lively and productive—the more so because of the strong opinions expressed on controversial subjects. As a result of dealing with such a relatively new and controversial topic, it has taken somewhat longer than usual to get the chapters completed.

We are indebted to the National Science Foundation and the National Aeronautics and Space Agency for their generous financial support and to the American Astronomical Society, and particularly its Division of Planetary Science and Division of Dynamical Astronomy, for sponsorship.

>Roman Smoluchowski
>John N. Bahcall
>Mildred S. Matthews

PART I
The Solar-Galactic Neighborhood: Galactic Gravitational Fields

THE GALACTIC ENVIRONMENT OF THE SOLAR SYSTEM

JOHN N. BAHCALL
Institute for Advanced Study

The general features of the galactic environment of the solar system are described. The principal components of the distribution of matter in the solar vicinity are discussed. Special emphasis is placed upon the question of determining the total amount of matter in the vicinity of the Sun.

My task is to describe the galactic context in which the interaction between the solar system and the Galaxy occurs. For most of this chapter, I shall focus on that part of the Galaxy in which the solar system resides. Before doing that, however, I want to remind the reader of the overall nature of our galactic system. Table I gives some of the essential numbers for considering this problem, culled from a variety of sources.

In this chapter, I concentrate on the problem of determining the total volume density and the total column density of matter at the solar position. In the course of doing this, I review some of the important characteristics that are known about the solar neighborhood. In conclusion, I use the results of our discussion of the mass distribution near the Sun to evaluate the period and the phase jitter of the solar motion perpendicular to the galactic plane (following the analysis of Bahcall and Bahcall [1985]). As discussed in other chapters, the value of the solar period is similar to the apparent periods that have been inferred from the cratering and extinction data (see Rampino and Stothers 1984*a*; Schwartz and James 1984), although the similarity may be accidental.

I. WHAT IS NEW?

Some of the early studies determining the total amount of matter in the solar vicinity by Oort (1932,1960) led to one of the first astronomical sug-

TABLE I
The Galaxy Context

Parameter	Approximate Value
Total disk mass at solar position	$0.2\ M_\odot\ pc^{-3}$
Total unseen halo mass at solar position	$0.01\ M_\odot\ pc^{-3}$
Typical exponential scale height, young stars and gas	80 pc
Typical exponential scale height, old disk stars	0.3 kpc
Exponential scale length of disk stars	2–4 kpc
Distance to galactic center	7–10 kpc
Total number of stars ($M_V \leq 16.5$ mag)	5×10^{10}
Circular velocity at the Sun	2×10^2 km s^{-1}

gestions of a large "missing mass." The method of weighing the matter in the local neighborhood can be summarized as follows. The density distribution and velocity dispersion of a set of tracer stars (e.g., K giants or F dwarfs) is measured perpendicular to the galactic plane (typically out to distances of order a few hundred parsecs). Theoretical models are then computed for the expected distribution of tracer stars in different gravitational potentials (mass distributions). The amount of matter actually present in the Galaxy is determined by comparing the observed and computed distributions. The problem is similar to computing the distribution of an isothermal atmosphere (since for the tracer stars of interest, the velocity dispersion changes much more slowly with height above the plane than does the density). Clearly, the more matter there is close to the plane, the more quickly the density will fall off with height above the plane.

The availability of modern computers has made possible significant advances in the theoretical analysis of this problem at the same time that better observational samples of tracer stars have been obtained. I have taken advantage of these developments to improve and strengthen the determinations of the total amount of matter in the solar vicinity, using more realistic Galaxy models and more accurate theoretical solutions. I have solved numerically the combined Poisson-Vlasov equation for the gravitational potential of Galaxy models consisting of realistically large numbers of individual isothermal disk components in the presence of a massive unseen halo. The calculations were carried out with different assumptions about the unseen matter and compared with the observed number densities of F dwarfs and K giants versus height above the plane.

The principal result of this work (Bahcall 1984a,b) is that about half of the matter in the vicinity of the Sun is in the form of unseen disk material which has an exponential scale height < 0.7 kpc. The unseen material that is inferred from galactic rotation curves at large galactocentric distances and from applying the virial theorem to groups and clusters of galaxies may not be the same as the unobserved disk matter. The unseen material inferred on large scales is often discussed by particle physicists in terms of dissipationalless particles (various 'inos') while the unseen disk material is presumably dissipational. It is possible that the unseen material in the disk consists of stars or planets that are not massive enough to burn hydrogen and hence have a luminosity too low to have been detected by searches carried out thus far.

II. GALAXY MODELS

Table II summarizes the relative amounts of the observed mass components, and their velocity dispersions (i.e., temperatures) that were derived (using data from many sources) by Bahcall and Soneira (1980) and by Hill, Hilditch, and Barnes (1979), often referred to as the B&S and the HHB Gal-

TABLE II
The Galaxy Model for Observed Components[a]

Component	B & S Mass Fraction (A_i)[b]	$\langle v_z^2 \rangle^{1/2}$ (km s^{-1})	HHB Mass Fraction (A_i)[c]
Main sequence stars			
$M_V < 2.5$ mag	0.021	4	0.038
2.5 mag $\leq M_V \leq$ 3.2 mag	0.015	8	0.019
3.2 mag $\leq M_V \leq$ 4.2 mag	0.031	11	0.033
4.2 mag $\leq M_V \leq$ 5.1 mag	0.035	21	0.034
5.1 mag $\leq M_V \leq$ 5.7 mag	0.025	20	0.023
5.7 mag $\leq M_V \leq$ 6.8 mag	0.037	17	0.036
	0.0358	8	
	0.0626	13	
$M_V \geq 6.8$ mag	0.0536	15	0.0262
	0.0626	20	
	0.0834	24	
Subgiants and giants	0.016	~20	
White dwarfs	0.052	21	0.185
Atomic H and He and	0.469	4	0.287
Molecular H and dust			0.083
Spheroid	0.001	~100 km s^{-1}	
Total	0.0958 M$_\odot$ pc^{-3}		0.108 M$_\odot$ pc^{-3}

[a]Disk luminosity functions and velocity dispersions from Wielen (1974).
[b]Bahcall and Soneira (1980) model.
[c]Hill, Hilditch and Barnes (1979) model.

axy models. The models contain many observed disk components (typically 14) whose characteristics are determined by local measurements, a population 2 spheroid inferred from faint star counts, different models for the unobserved disk components, and an unseen massive halo whose normalization is fixed by the solar rotation velocity. The mass fractions are defined in terms of the total *observed* mass density (in stars, gas and dust), i.e.,

$$A_i = \frac{\rho_i(0)}{\rho_{\text{obs}}(0)} \qquad (1)$$

Here the densities are evaluated at the galactic plane $z = 0$ indicated by the a "0" in Eq. (1).

I use the difference between the results obtained with the B&S and the HHB Galaxy models as one measure of the uncertainty. The two models are similar because the luminosity function of the disk stars is reasonably well determined (see Wielen 1974) over much of its range. The B&S and HHB models differ mainly in the mass density assigned to white dwarfs and to interstellar matter; in both cases, I have made use of more recent determinations. For example, fewer white dwarfs are observed at faint absolute magnitudes than had been expected on the basis of earlier theoretical estimates. I have used in the B&S model the observed number density (Green 1980; Liebert et al. 1979) down to $M_V = 17.2$ and a white dwarf mass of 0.6 $M_\odot$. I have also adopted the value for interstellar matter density that has been estimated by Spitzer (1978), which is consistent with the recent value inferred by Sanders et al. (1984). This value is rather larger than the interstellar matter density that was used by HHB.

Previous theoretical studies of the total amount of matter in the vicinity of the Sun have been limited mainly to simplified Galaxy models with one or, at most, a few disk components and no spherical component. The previous solutions were also limited either by what was tractable analytically or by assuming a numerical form for the total matter density that was independent of the potential. I have calculated numerical models with many different sets of input data and several assumptions about how the unseen material is distributed. I estimated the major uncertainties in the determination of the distribution of unseen matter by comparing an extensive collection of theoretical models with the available data.

III. THE EQUATIONS

The basic equation used is the combined Poisson-Vlasov equation for the potential. This equation describes how the gravitational potential at a given height above the plane can be calculated from the mass densities and velocity dispersions that are specified in the plane of the disk for any number of isothermal disk components (some observed as stars, dust or gas and some unobserved) plus a halo mass density (constant, to first approximation, with height above the plane). The dimensionless form of the combined Poisson-Vlasov equation is:

$$\frac{d^2\phi}{dx^2} = 2\left(\sum_{i=1}^{N_{obs}} A_i e^{-\alpha_i \phi} + \sum_{j=1}^{N_{unobs}} B_j e^{-\beta_j \phi} + \epsilon\right) \quad (2)$$

with $\phi(0) = (d\phi/dx)_o = 0$. The gravitational potential has been divided by the square of a velocity dispersion which is taken here to be $(10 \text{ km s}^{-1})^2$ for numerical convenience. The quantities $\alpha_i = [(10 \text{ km s}^{-1})^2/<v_z^2>_i]$, with a similar definition for the unobserved β_j. The height z above the plane is taken to be $z = z_o x$, where the unit of length is $z_o = [(10 \text{ km s}^{-1})^2/2\pi G \rho_{obs}(0)]^{1/2}$. The quantity N_{obs} is the total number of observed mass components. The unobserved mass fractions B_j are defined, by analogy with Eq. (1), as the ratio of the mass density in component j to the total observed mass density. Finally ϵ is defined as the ratio of $\rho_{halo}^{eff}(0)$ to $\rho_{obs}(0)$. The effective halo mass density is equal to the total halo mass density for a constant rotation curve but is slightly different if the rotation curve is not exactly flat.

IV. MISSING MATTER

An illustrative model in which the unobserved disk mass is proportional to the observed disk mass has some especially simple features. In this model, the unobserved mass density in every component i is proportional to the observed mass density in the same component,

$$B_i \equiv P \times A_i \quad (3)$$

and the unobserved and observed velocity dispersions for the ith component are equal, $b_i \equiv a_i$. Of course, this is only one of the many different models that have been explored.

Figure 1 compares the observed star counts taken from the work of Oort (1960) with four different but illustrative models for the distribution of unseen material. All of the models shown give satisfactory fits to the data.

Table III gives the ratio of unobserved to observed mass density for 28 detailed models (see Bahcall [1984b] for a description of these models) that fit the observed distribution of K giants. The models represent numerical solutions of the combined Poisson-Vlasov equation for different input parameters, as well as for several assumptions about the distribution of the unobserved disk material. There are separate columns referring to the observed K giant samples of Oort (1960) and to the Upgren (1962) K giant density distributions. For both the volume and the column density, the typical best-fit model has, for the Oort densities, about equal amounts of unobserved and observed

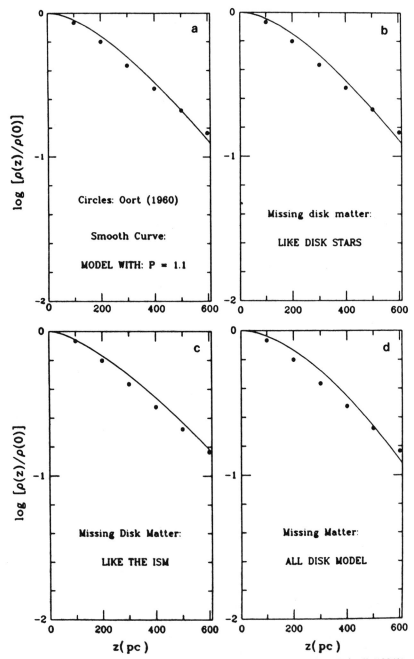

Fig. 1. Comparison of computed and observed K giant densities, taken from Bahcall (1984b). The circles represent the relative densities of the observed K giants, as reported by Oort (1960). The theoretical models are different in each comparison: (a) the proportional model is described by Eq. (3); (b) the unseen material is distributed like the average population of disk stars; (c) all of the unseen material is in a small scale height population like that of the interstellar medium (ISM); and (d) all of the unseen material is in a disk population with an exponential scale height of 0.65 kpc.

TABLE III
Ratio of Unobserved to Observed Disk Material

Row[a]	$\frac{\rho_{unobs}(0)}{\rho_{obs}(0)}$	$\frac{\sigma_{unobs}}{\sigma_{obs}}$	$\frac{\rho_{unobs}(0)}{\rho_{obs}(0)}$	$\frac{\sigma_{unobs}}{\sigma_{obs}}$
	Oort Densities		Upgren Densities	
1	1.1	1.1	1.5	1.6
2	1.6	1.6	2.1	2.1
3	0.6	0.6	1.0	1.0
4	0.9	0.9	1.3	1.4
5	1.3	1.3	1.8	1.8
6	1.3	1.1	1.8	1.6
7	0.6	1.2	0.8	1.6
8	0.7	0.7	1.0	1.0
9	0.4	0.7	0.5	1.0
10	2.4	0.5	2.6	0.5
11	1.5	0.3	2.2	0.5
12	0.6	2.5	0.7	3.2
13	1.1	1.1	1.5	1.6
14	1.5	1.5	2.0	2.0
Average	1.1	1.1	1.5	1.5

[a] The models used here are described in the corresponding rows of Tables 5 and 6 of Bahcall (1984b).

material. For the Upgren densities, the typical best-fit model has about 40% more unobserved than observed matter. These averages are illustrative since at most only one of the models considered for the distribution of unseen matter can be correct. Similar results are obtained by comparing theoretical models to the observed sample of F dwarfs (Bahcall 1984a).

I conclude that a typical best-fit model implies that about half of the disk material at the solar position has not yet been observed, a conclusion in qualitative agreement with previous major studies (see, e.g., Oort 1932,1960; Hill 1960; Woolley and Stewart 1967; Lacarrieu 1971; Hill et al. 1979), although I find a larger ratio of unobserved to observed matter than in some of the earlier analyses. The present investigation establishes more firmly and specifically the existence of unobserved disk material. The added confidence in the results arise because: (1) more realistic Galaxy models are used; (2) the Poisson and Vlasov equations are solved self-consistently; (3) improved (and more homogeneous) observational data are utilized; and (4) many theoretical models are compared with the observations in order to estimate the uncertainties. However, one should not be too confident. There is no modern data sample of K giants; the samples that I have been forced to use are a quarter of a century old. The stars are very bright (apparent magnitudes < 10) so that it

TABLE IV
Individual Disk Column Densities in Some Galaxy Models[a]

Component	$\langle v_z^2 \rangle^{1/2}$ (km s^{-1})	Column Density $M_\odot pc^{-2}(L_\odot pc^{-2})$ Row 1, Table 5[b] $P = 1.06$	Column Density $M_\odot pc^{-2}(L_\odot pc^{-2})$ Row 9, Table 5[b] Like K Giants	Column Density $M_\odot pc^{-2}(L_\odot pc^{-2})$ Row 12, Table 5[b] All Disk	Column Density $M_\odot pc^{-2}(L_\odot pc^{-2})$ Row 10, Table 5[b] ISM
(1)	(2)	(3)	(4)	(5)	(6)
Disk main sequence stars					
$M_V < 2.5$ mag	4	0.4 (2.3)	0.2 (2.5)	0.2 (2.7)	0.2 (1.8)
2.5 mag $\leq M_V \leq$ 3.2 mag	8	0.6 (0.9)	0.3 (1.0)	0.3 (1.1)	0.3 (0.8)
3.2 mag $\leq M_V \leq$ 4.2 mag	11	1.9 (1.7)	1.0 (1.8)	1.0 (1.8)	0.9 (1.6)
4.2 mag $\leq M_V \leq$ 5.1 mag	21	4.9 (2.3)	2.4 (2.3)	2.4 (2.3)	2.5 (2.4)
5.1 mag $\leq M_V \leq$ 5.7 mag	20	3.3 (0.9)	1.6 (0.9)	1.6 (0.9)	1.7 (0.9)
5.7 mag $\leq M_V \leq$ 6.8 mag	17	3.9 (0.6)	1.9 (0.6)	2.0 (0.6)	2.0 (0.6)
$M_V \geq 6.8$ mag	8	1.5 (0.0)	0.8 (0.0)	0.8 (0.0)	0.6 (0.0)
	13	4.7 (0.1)	2.4 (0.1)	2.5 (0.1)	2.2 (0.1)
	15	4.8 (0.1)	2.4 (0.1)	2.5 (0.1)	2.4 (0.1)
	20	8.2 (0.1)	4.1 (0.1)	4.1 (0.1)	4.2 (0.1)
	24	14.0 (0.2)	6.8 (0.2)	6.8 (0.2)	7.4 (0.2)
OTHER		—	37.5 (0.0)	86 (0.0)	—
Subgiants and giants	20	2.1 (13.1)	1.0 (13.3)	1.0 (13.5)	1.1 (13.9)
White dwarfs	21	7.3 (0.0)	3.6 (0.0)	3.6 (0.0)	3.8 (0.0)
Atomic H and He / Molecular H and dust	4	8.3 (0.0)	4.5 (0.0)	4.8 (0.0)	19.6 (0.0)
Total column density of mass (luminosity)		66 (22)	71 (23)	120 (23)	49 (22)

[a] Table includes both observed and unobserved material. The entries without parentheses refer to mass column densities (in $M_\odot pc^{-2}$); the entries in parentheses refer to luminosity column densities (in $L_\odot pc^{-2}$, visual band). The spheroid column density of mass (at the solar position) is (cf. Bahcall et al. 1983) about 3 $M_\odot pc^{-2}$ and about 1.5 $L_\odot pc^{-2}$ (in the visual band).

[b] See Bahcall 1984b, Table 5.

would be easy to get a much improved sample with modern techniques, using spectroscopic observations to assure that the population was homogeneous with height above the plane. The velocity dispersions of both the K giants and the F dwarfs could be improved with modern radial velocity techniques. Finally, the absolute magnitude of the K giants should be redetermined using Hipparcos as well as the soon-to-be-published Yale parallax catalogue.

The largest identifiable source of uncertainty in the Oort limit is the unknown form of the distribution of unseen matter (see Bahcall 1984b, Table 9, last row). In the future, it should be possible to constrain sharply the distribution of unseen matter by requiring consistency with observations of several carefully selected samples of tracer stars with different scale heights.

To illustrate the contributions to the mass (and light) of various components, Table IV lists the individual column densities of mass and of luminosity for four very different models. The column densities are given in columns 3 through 6 of Table IV. In column 3, I give the individual densities for the standard proportional model, defined by Eq. (3) above. Column 4 lists the individual densities for the model in which all of the unseen material is assumed to have a distribution like that of the K giants themselves. In column 5, I show the densities for the extreme model in which *all* of the unseen material is in the disk. In this model, the material that holds up the rotation curve is also assumed to be in a flattened disk. The final values listed in Table IV refer to a model in which the unseen disk material has a small scale height like that of the interstellar medium (i.e., an ISM model). For all of these models, about half of the disk light comes from the giants even though they only provide at most a few percent of the mass density.

If the missing material is in the form of stars that are not massive enough to burn hydrogen ($M < 0.1 M_\odot$), then the nearest such brown dwarf is probably less than a parsec away and has a proper motion of more than an arcsec per year. Brown dwarfs of the required number density might be detected in future dedicated large-area surveys for very red, high proper motion objects. If the unseen material has a typical mass like that of Jupiter, the nearest such object would be about 0.2 pc from the Sun, moving with a proper motion of order 5 arcsec yr^{-1}. Such remarkable objects might be discoverable with the Infrared Astronomical Satellite (IRAS).

The unseen material must be mostly in a disk form, i.e., be dissipational. If all of the material were in a relatively round halo, then the rotation velocity at the solar position would have to be as large as 500 km s^{-1}. For a given local volume density of unseen mass, the total amount of mass required in a round halo is larger than the amount of mass needed in a disk by about the ratio of the galactocentric distance of the Sun to the disk scale height, i.e., by more than an order of magnitude. The largest scale height of the unseen disk material that is consistent with the solar rotation velocity is 0.7 kpc (see Bahcall 1984b, Tables 5 and 6, row 12).

V. THE SOLAR OSCILLATION PERIOD

Finally, I wish to discuss some recent work that S. Bahcall and I have done (Bahcall and Bahcall 1985), which relates the above analysis to the central topic of this book. We have used the exact numerical solutions to the Poisson-Vlasov equation with realistic Galaxy models to evaluate the period of the Sun's motion perpendicular to the galactic disk. We have *not* made the usual assumption that the potential in which the Sun moves is quadratic. We considered a variety of theoretical models for the unseen matter, from a very squashed distribution (like that of the interstellar matter) to the most extended distribution that is consistent with the Galaxy's rotation curve.

We find half-periods for the vertical oscillation that range from 26 to 37 Myr (bracketing the range of periods that have been inferred from the terrestrial records on mass extinctions and on cratering) and maximum heights above the plane from 49 to 93 pc. For the simplest model (see Eq. 3) in which the amount of unseen material is everywhere proportional to the amount of observed material, the computed half-period is 31 Myr. For all the models we considered, the most recent passage of the Sun through the galactic plane occurred in the last 3 Myr provided only that the present position of the Sun is between 0 and 20 pc above the plane. We also calculated the fluctuations in the time of crossing the galactic plane produced by interactions with whatever objects cause the velocity dispersion of observed stars to increase with galactic age. We found an average phase jitter per half-period of order 6 to 9%. The largest uncertainty in all these calculations is again contributed by the unknown distribution of the unseen mass that must be postulated to explain the distribution of observed stars.

In addition, we (Bahcall and Bahcall 1985) applied the argument of Thaddeus and Chanan (1985) to show that the apparent periodicity in the mass extinction and cratering records cannot be caused by any population of objects (observed or unobserved) that contributes a major fraction of the total mass density at the solar vicinity. The argument of Thaddeus and Chanan applies very generally because all of the major components listed in Tables II and IV have exponential scale heights larger than, or comparable to, the maximum height that the Sun reaches above the plane (with its present orbital parameters).

Acknowledgments. This work was supported in part by grants from the National Aeronautics and Space Administration and the National Science Foundation.

STARS WITHIN 25 PARSECS OF THE SUN

WILHELM GLIESE and HARTMUT JAHREISS
Astronomisches Rechen-Institut, Heidelberg

and

ARTHUR R. UPGREN
Van Vleck Observatory

The stars within 25 pc of the Sun constitute the best known and most complete stellar sample in the Universe. This sample includes more than 800 stars for which trigonometric parallaxes and B-V photometry yield absolute magnitudes with errors < 0.3 mag. The majority of them define the field star main sequence. They provide an excellent sample for calibrating mean spectral type and color-luminosity relations. The stars can be divided into two groups; those brighter than M = +9 are statistically complete to the limit of 25 pc. This completeness allows predictions of the total numbers of solar-type stars most likely to possess planetary systems to be made with some confidence. It also provides a reliable estimate of their contribution to the total stellar and mass densities in the solar neighborhood. The newly revised stellar luminosity function of Wielen et al. is most valuable for making estimates of these properties for stars fainter than +9, the M dwarfs and degenerate stars.

I. INTRODUCTION

The solar neighborhood should be defined as a volume which is large enough so that stars of different kinds are sufficiently numerous for investigations relevant to many problems of interest, but small enough so that the stars within it are as statistically complete as possible, and whose distances can be determined with greatest possible accuracy.

A distance of 25 pc from the Sun seems to be an acceptable limit for these purposes. Outside this volume the only direct method for deriving distances, the measurements of trigonometric parallaxes, has yielded reliable individual distances for only a very few objects but inside it, precise distances are plentiful. Furthermore, interstellar matter does not noticeably contaminate magnitudes and colors within 25 pc of the Sun. Within this volume, we find a few giants and subgiants, many main sequence members from type A to type M, some subdwarfs and the class of degenerates (white dwarfs). Not present in the immediate solar neighborhood are high-luminosity objects of luminosity classes I and II, main sequence stars with spectral types earlier than B8 V and even hot subdwarfs. On the other hand, there is an abundance of low-luminosity dwarf stars, very red dwarfs, brown dwarfs, cool degenerates, and stars with astrometric (unseen) companions which can be observed and studied only in the very nearest regions of the solar neighborhood. Only here may objects be selected and identified for a search for planetary companions.

At present, a new edition of the *General Catalogue of Trigonometric Stellar Parallaxes* (YPC) is being compiled by W. F. van Altena at the Yale Observatory. Comparisons between parallax series observed at various stations with various instruments show small systematic differences which were eliminated in former editions of the YPC by "observatory corrections" although explanations for their appearance were not understood. Van Altena no longer followed these rules. Therefore, the new catalogue is a heterogeneous collection of trigonometric parallaxes. However, the restriction to parallaxes $> 0\rlap{.}''040$ keeps effects due to systematic differences between parallaxes very small in our sample.

The distribution of nearby stars with known trigonometric parallaxes is shown in Table I. The decrease of our knowledge of parallaxes with increasing distance is obvious. This is mainly due to the incompleteness of fainter stars and, to a lesser degree, to the increasing incompleteness of known compan-

TABLE I
Distances of Nearby Stars with Trigonometric Parallaxes

Distance Limit (pc)	Number of Stars (N)	Star Density (pc^{-3})	Density Relative to 0–5 pc
0– 5	61	0.116	1.00
5–10	239	0.065	0.56
10–20	1271	0.043	0.37
20–25	802	0.025	0.22
0–25	2373	0.036	0.32

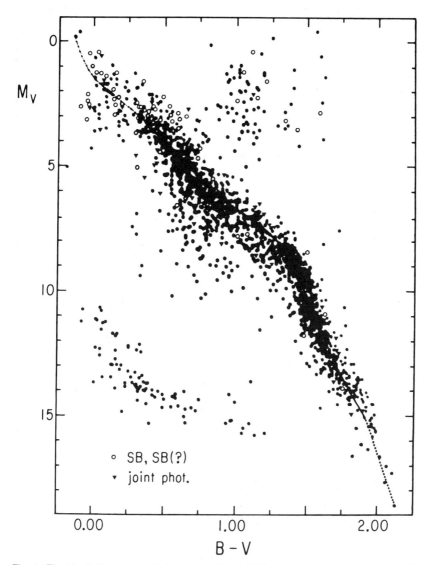

Fig. 1. The M_v, B-V color-magnitude diagram for all 1919 stars with photometry in B and V within 25 pc. Open circles indicate spectroscopic binaries (SB) or suspected spectroscopic binaries (SB?) and triangles indicate two or more stars with joint photometry.

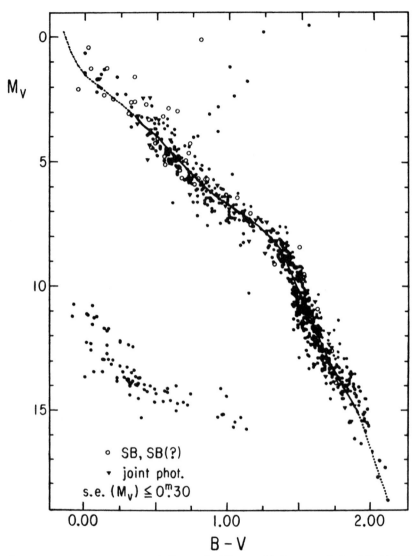

Fig. 2. The M_v, $B-V$ color-magnitude diagram for 861 stars with photometry in B and V within 25 pc and for which the trigonometric parallax yields a standard error in $M_v < 0.3$ mag. The symbols have the same meaning as in Fig. 1.

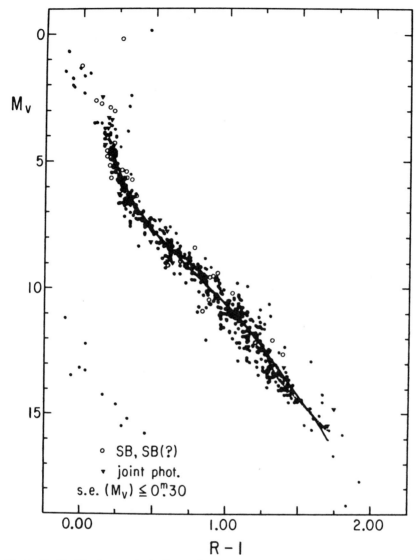

Fig. 3. The M_v, $R-I$ color-magnitude diagram for 596 stars within 25 pc with photometry in R and I on the Kron system and for which the trigonometric parallax yields a standard error in $M_v <$ 0.3 mag. The symbols have the same meaning as in Fig. 1.

TABLE II
Distribution of Stars within 25 pc in Visual Magnitude

M_v range	<6	6–8	8–10	10–12	12–14	14–16	>16
Number of stars	474	411	496	478	313	177	20

ions. The 0.11 stars per cubic parsec within 5 pc of the Sun certainly represents only a lower limit to the local stellar number density. Probably there are still some unknown stars within 5 pc. For example, only a few years ago LHS 292 was discovered.

The representation of our nearby-star sample by an HR diagram including spectral classes vs. absolute magnitude is unrealistic because no single classification system covers a majority of the stars in it. In fact, the majority of the very faint red stars and the white dwarfs have no spectral types on any system. However, for more than 80% of the stars, including many of even the faintest objects, V magnitudes and B-V colors have been determined photoelectrically. Therefore a (B-V)-luminosity diagram will be the most appropriate way to present a reasonably complete picture of our knowledge of the solar neighborhood. Figures 1 and 2 show such diagrams for the stars nearer than 25 pc, and Fig. 3 shows the M_v,R-I diagram for the stars with R-I colors (all references to R and I are on the Kron system unless otherwise indicated), an index which is better for revealing the faintest part of the main sequence. Figure 2 has been confined to only those stars with very accurate trigonometric parallaxes.

We can see from these diagrams, especially as shown in Fig. 2, that the restriction to stars with accurately determined absolute magnitudes shows the main sequence as a fairly narrow band with a certain real or cosmic dispersion but not as a single line. Incomplete knowledge affects especially the lower part of the dwarf sequence. From the luminosity function (see Sec. VII below), we suppose the maximum frequency of stars nearer than 25 pc to appear between +14 and +16 apparent visual magnitude. Table II showing the observed numbers available up to 1985, demonstrates again the incompleteness of our knowledge from trigonometric distance measurements. The trigonometric parallax data can be augmented with about 400 stars with photometrically and/or spectroscopically determined distances within 25 pc. Because these stars are mostly faint, the above situation will be slightly improved.

II. BASIC DATA

A new edition of the *Catalogue of Nearby Stars* (see Gliese and Jahreiss 1979) with an extended distance limit of 25 pc is planned for the near future.

TABLE III
Distribution of the Standard Errors in Absolute Magnitude

s.e. (M_v)	N	Sum
0.1	274	274
0.2	386	660
0.3	369	1029
0.4	303	1332
0.5	268	1600
0.6	226	1826
0.7	203	2029
0.8	136	2165
0.9	101	2266
1.0	48	2314
>1.0	55	2369

Recent compilations of nearby-star data can be found in Lippincott (1978) and Gliese (1982) for the stars within 5.3 pc; in Gliese (1969), and Gliese and Jahreiss (1979) for the stars within 22 pc; and in Woolley et al. (1970) for stars within 25 pc.

Trigonometric Parallaxes

The numbers in Table I demonstrate the incompleteness of our knowledge of trigonometric parallaxes, showing a rapid decrease with increasing distance from the Sun. This situation is even more unsatisfactory if we consider the relative parallax errors defined as σ_π/π which, of course, are larger on average with increasing distances. They produce standard errors (s.e.) in absolute magnitudes, s.e. (M_v) = 2.1715 σ_π/π as shown in Table III. A remarkable percentage of the parallaxes between 0″.040 and 0″.050 have measures which are too large; the true values of many stars in our lists will be smaller than our distance limit of 0″.040. The number of stars with trigonometric parallaxes below our limit but with true distances nearer than 25 pc can be expected to be smaller than the number of stars erroneously included in our sample. In other words, for a volume-limited sample the positive parallax errors predominate over the negative ones due to the Lutz-Kelker effect (see Sec. III).

In spite of the revival of trigonometric parallax programs in the 1960s, the numbers in Table I increase only very slowly. We expect that in the early 1990s the results of the astrometric satellite Hipparcos will determine the parallaxes for all nearby stars brighter than apparent magnitude +8 since its observing program is intended to include all stars in the sky brighter than this limit. In addition to this program, several groundbased observatories continue to include fainter stars. These are red and white dwarf stars, mostly with large proper motions. Most are objects in the Luyten half-second (LHS) catalogue (Luyten 1979a) which contains 527 stars with annual proper motions, $\mu >$ 1".0 and more than 3000 stars with 1".00 $> \mu >$ 0".50. However, systematic parallax observations of all new Luyten two-tenth (NLTT) stars (Luyten 1979b, 1980) numbering nearly 60,000 with $\mu >$ 0".18 per year, are far beyond present capabilities. Selection is necessary and supported by spectroscopic and/or photometric data. VRI photometry by Weis (1984) for 413 NLTT stars of color class m with $m_R <$ 13.5 (subscript R, red magnitude) revealed about 90 nearby stars. Geneva photometry (Golay 1973) of 737 NLTT stars in the declination zone 0° to $-10°$ with $m_R <$ 12.5 yielded 45 new stars within 25 pc (Grenon, personal communication, 1984). Similar programs involving objective-prism surveys and searches for degenerate objects are under way.

Proper Motion

Proper motions are available for nearly all of the known nearby stars, but with very different accuracies. In addition, some of them are given as absolute data, and others as relative values only. However, even at the distance limit of 25 pc, an error of 0".01 in the yearly proper motion corresponds to only 1.2 km s^{-1} in tangential velocities. This is practically insignificant when compared to the uncertainties in parallax data and measured radial velocities.

The most extensive source of proper motions for the fainter stars extending to photographic magnitude $m_{pg} =$ 21 are the NLTT catalogues which are compilations of all known stars with proper motions exceeding 0".175 per year. At the outer limit of 25 pc, this corresponds to a tangential velocity of 21 km s^{-1}; thus most of the stars within 25 pc with tangential velocities larger than this amount are listed in Luyten's catalogues. From objective-prism plates, Vyssotsky (1963) has detected a sample of late K and early M dwarfs, which is not biased towards large proper motions. Nevertheless, among Vyssotsky's K and M dwarfs nearer than 25 pc, about 80% have proper motions exceeding 0".175 per year. Taking these stars as a representative sample, we can guess that about 80% of all nearby stars are already contained in Luyten's catalogues.

Observations and measurements of trigonometric parallaxes of stars include the determination of their proper motions as well. Therefore, present-day proper motions are unknown only for a very few faint stars assumed to be nearby objects on the basis of their spectra and colors.

Spectral Types

One essential condition for the use of spectral types as distance indicators is the knowledge of a reliable luminosity class. In some cases there are gradual transitions between dwarfs and subdwarfs or dwarfs and subgiants. Among the stars within 25 pc, various systems of spectral classification are of importance not only for characterizing an intrinsic stellar quantity but also as luminosity indicators which allow the derivation of stellar distances. The most significant of these systems are:

1. *The MK system.* Data are available for nearly all nearby stars of classes A through K. The MK data are of varying quality and are not in a uniform system. This means that one must calibrate each of the various sources of spectral classifications. Compared to slit spectra, objective prism classifications of K and M dwarfs yield mean luminosities with considerable scatter, and often the given luminosity classes must be taken with care. For fainter M dwarfs, only the work of Boeshaar (1976) achieved accurate classifications using wide image-tube spectrograms. Yet up to now, these reliable spectra constitute only a small sample.
2. *The Mt. Wilson types.* These are mostly found in the *General Catalogue of Stellar Radial Velocities* (Wilson 1953) with additional M dwarf star classifications given by Joy and Abt (1974).
3. *The Kuiper types.* Between 1938 and 1944, Kuiper observed about 3500 stars with large proper motion, brighter than $m_{pg} = 18$ mag. Most of these data were not published, but a new compilation has been made and published by Bidelman (1985). The HR diagram of Kuiper's K and M dwarfs with relative parallax errors $< 10\%$ is displayed in Fig. 4. It shows that these types are still valuable in the search for nearby stars.

The various systems of spectral classifications for M dwarfs are discussed by Wing and Yorka (1978), using eight-color photometry to test the usefulness of the different types as temperature indicators. Despite the usefulness of these spectral systems, only the color class estimates by Luyten given in his NLTT catalogues are known for most of the nearby stars fainter than 12 mag.

Photometric Data

As mentioned above, Figs. 1 through 3 show color-luminosity diagrams with broadband colors *B-V* and *R-I* (Kron's system). Such photometry is already available in *B-V* for about 81% and in *R-I* for about 45% of the members of our nearby-star sample. This covers objects of all spectral classes including even the faintest red main sequence stars, the most numerous type of star in the solar neighborhood. We observe that the dispersion along the mean main sequence appears somewhat smaller in the $(M_v, R\text{-}I)$ diagram than in a $(M_v, B\text{-}V)$ diagram. But as stated above, the *R* and *I* colors have not yet been measured for as many stars as have *B* and *V* colors. Furthermore, in addition to Kron's system for red magnitudes, other *R* and *I* systems are in use, including Johnson's *R-I* and more recently the Kron-Cousins system.

Narrowband photometry proves to be more suitable for luminosity determinations than broadband colors. However, such data are available at present only for certain spectral classes and for limited magnitude regions, and these do not yet include large numbers of low-luminosity stars. The *General Cata-*

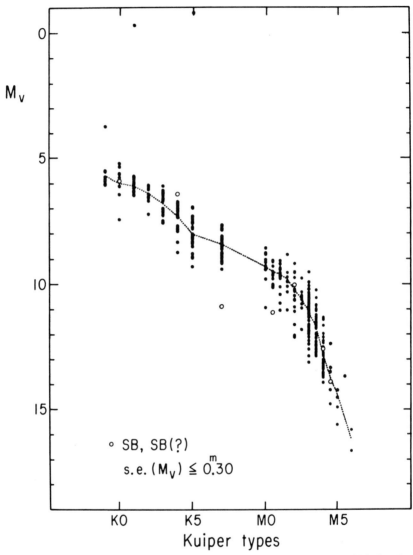

Fig. 4. The HR diagram for 383 stars with spectral types by Kuiper and for which the trigonometric parallax yields a standard error in $M_v < 0.3$ mag. The symbols have the same meaning as in Fig. 1.

TABLE IV
Percentage of Faint Nearby Stars with Measured Radial Velocities

Visual Magnitude	Total Number of Stars	Stars with Radial Velocities	
		Number	%
10–11	215	154	72
11–12	139	71	51
12–13	127	45	35
13–14	97	27	28
14–15	77	9	12
15–	51	3	1

logue of Photometric Data underway at the Geneva Observatory offers information on the number of different photometric systems which can apply to a nearby star (Hauck and Mermilliod 1983).

Radial Velocities

The radial velocities of stars brighter than $M_v = 8$ in our sample are almost completely known, whereas for the fainter stars we have to deal with incompleteness along with other astrophysical data. At present, only 28 out of 71 white dwarfs have measured radial velocities, and for the faint red dwarfs the existence of radial velocity data rapidly decreases with fainter apparent (and therefore also absolute) magnitude. This is demonstrated in Table IV. Furthermore, many of the measurements for the fainter stars have standard errors of 10 to 20 km s^{-1}, leading to large errors in space velocities, unlike the errors in proper motion, which result in velocity errors of < 2 km s^{-1}. Often only a single plate was obtained for a radial velocity determination for a faint star, which gives us no information on possible variation in radial velocity.

For late-type stars brighter than about apparent magnitude 11, the situation will probably improve considerably in the near future. This is due to an extensive photoelectric observing program with CORAVEL (correlation radial velocity) (Mayor, personal communication, 1985) and similar programs elsewhere. First results increased the percentage from 4 to 15%, of detected spectroscopic binaries among the nearby K dwarfs north of $-10°$ dec (Halbwachs 1985). In the forthcoming new *Catalogue of Nearby Stars* we estimate the percentage of stars with measured radial velocities at 65% (Jahreiss and Gliese 1985).

III. LUMINOSITY CALIBRATIONS

About seventy years ago it was realized that the position of a star in the HR diagram is determined approximately by its spectral type. Today, colors are often preferred over spectral types (see Figs. 1–3) for deriving the absolute magnitude of a star and its distance modulus, from which its spectroscopic or photometric parallax can be found. Spectral types and colors prove to be valuable secondary distance indicators, increasing significantly the number of recognized nearby stars. Figures 1 through 3 show that the derivation and calibration of mean spectral types and color-luminosity relations for the objects in the solar neighborhood is recommended only for main sequence types from F5 to M; white dwarfs must be considered separately.

From a comparison between Figs. 1 and 2, it is evident that large relative accidental errors of trigonometric parallax measurements in a sample of objects with a lower parallax limit, yield more values with positive parallax errors than with negative deviations from the true values. Therefore, mean relations derived from all available observations are systematically biased and the resulting mean absolute magnitudes will be too faint. To eliminate such effects, Lutz and Kelker (1973) developed and evaluated corrections which should be applied in such calibrations. The Lutz-Kelker correction shifts the mean M_v curve to somewhat brighter luminosities. We must emphasize once more that the application of such statistical corrections to the measured parallax of an individual star is meaningless. Because the correction has only been evaluated for cases where the error $\sigma_\pi/\pi < 0.175$, the material used for calibrations should be restricted to stars with errors smaller than this limit. In this chapter, we have restricted this ratio to < 0.14 which means that the standard error in the absolute magnitudes will not exceed 0.30 mag.

Any group of stars is limited in magnitude but the faint limit in apparent magnitude may or may not be well defined. Main sequence stars show a certain scatter along a mean line, the dispersion resulting in part from different chemical compositions and from different ages. Near the faint limit of the sample, the high-luminosity objects of a certain specific spectral type or color are still included in the sample whereas the stars with lower luminosities are not. The resulting mean M_v will be brighter than true mean absolute magnitudes, an effect occurring in magnitude-limited samples known as the Malmquist bias (Malmquist 1936). In theory, a statistical correction seems to be as necessary as for a distance-limited sample, but in practice it proves to be difficult to apply because in many cases the faint magnitude limit cannot be defined as a fixed value of apparent magnitude. Furthermore, in searching for nearby stars in such a magnitude-limited sample, the uncorrected mean luminosity relation must be used, namely the relation derived from the stars of this special sample with trigonometrically determined M_v. The true mean values of this color region which could be derived only by inclusion of fainter or more distant objects of lower luminosities not contained in the observed sample, would not represent the mean relation for this special material.

The absolute magnitude of a main sequence star may be affected by its chemical composition, rotation, chromospheric activity or star spots. It may also be affected by its age, which is statistically correlated with space velocity

TABLE V
Color-Luminosity Relations M_v, $B-V$ and M_v, $R-I$ (Kron) for Main Sequence Stars

B-V (mag)	M_v (mag)	R-I (mag)	M_v (mag)
+0.40	+3.4	+0.20	+4.4
0.45	3.7	0.25	5.2
0.50	4.0	0.30	5.9
0.55	4.3	0.35	6.5
0.60	4.6	0.40	7.0
0.65	4.9	0.45	7.4
0.70	5.3	0.50	7.7
0.75	5.6	0.55	8.0
0.80	5.9	0.60	8.3
0.85	6.1	0.65	8.6
0.90	6.3	0.70	8.9
0.95	6.5	0.75	9.2
1.00	6.7	0.80	9.5
1.05	6.9	0.85	9.8
1.10	7.1	0.90	10.1
1.15	7.3	0.95	10.4
1.20	7.5	1.00	10.7
1.25	7.8	1.05	11.0
1.30	8.1	1.10	11.3
1.35	8.4	1.15	11.6
1.40	9.0	1.20	12.0
1.45	9.7	1.25	12.3
1.50	10.4	1.30	12.7
1.55	11.0	1.35	13.1
1.60	11.6	1.40	13.5
1.65	12.3	1.45	13.9
1.70	13.0	1.50	14.3
1.75	13.5	1.55	14.7
1.80	14.0	1.60	15.1
1.85	14.5	1.65	15.5
1.90	15.0	1.70	16.0

(see Sec. IV). These effects produce a cosmic dispersion which can be observed along the main sequence in an HR diagram. Therefore the use of a spectroscopic and/or photometric parallax derived from a mean relation suppresses the dispersion existing among objects of the same spectral type or color.

This is not the case for stars with accurate trigonometric parallaxes. Only for a very small percentage of objects can a trigonometric parallax determination be erroneously shifted by the motion of the photocenter of an undetected astrometric binary. Thus it can be assumed that trigonometric parallaxes are superior and preferable to results from secondary distance indicators, providing only that the parallaxes are significantly larger than their errors.

Since the nearby main sequence stars are a mixture of such different objects, the mean dwarf sequence from F5 V to M5 V cannot be determined more accurately than 0.2 or even 0.3 mag. As an example of the uncertainties of a mean luminosity relation, we point to the M_v-$(R-I)$ diagram (Fig. 3) in the region $+1.25 < R-I < 1.50$ where the recently measured data by Weis (1984) are, on the average, 0.1 to 0.3 mag fainter than the mean relation derived from luminosities formerly available for our calibration. His calibration is indicated by a dotted line in Fig. 3. Individual luminosities depending on these mean relations have errors of the order of 0.4 mag, and even somewhat larger for high-velocity stars. Table V gives mean color-luminosity relations for the preferentially used broadband photometry of nearby stars.

IV. AGES AND VELOCITY DISTRIBUTION

It follows from the discussions in Sec. II that for almost all stars within a distance limit of about 20 pc and brighter than $M_v = +8$ mag, space velocities (U, V, W) can be computed. The accuracy of these space velocities is such that a typical mean error in one component is of the order of 3 to 4 km s^{-1}. This underlines the importance of the nearby-star sample for kinematical studies. The main results have already been published elsewhere by Jahreiss (1974), Wielen (1974), and Jahreiss and Wielen (1983). Here we give only a brief discussion of their conclusions.

Due to the completeness of the brighter ($M_v < +8$) stars, this sample of nearby stars should be relatively free of selection effects. The fainter red dwarfs and the white dwarfs are mostly found in the course of proper motion surveys, and this results in a predominance of stars with high tangential velocities. Therefore, the subsample of McCormick K and M dwarf stars is often chosen as a representative group for the study of kinematical behaviour of the common stars in the solar vicinity; this is because they have been selected by means of an objective-prism survey (Vyssotsky 1963) and should be free of kinematical bias as a consequence. A further consideration stems from the fact that our sample of nearby stars lies in a small volume near the galactic plane. The older stars are thus underrepresented due to their larger velocity dispersion perpendicular to the plane. To reduce this bias each star can be weighted according to its W velocity component perpendicular to the galactic plane. The analyses mentioned above also eliminated the white dwarfs and old giants after identifying them in the $M_v(B-V)$ diagram. Then, for the remaining stars on or near the main sequence, groups were formed according to their average $B-V$ values in color-magnitude diagrams and designated "CM-groups." For each CM-group, a mean age was determined. This method gives a reliable age separation only for the stars brighter than $M_v = +5$, since the fainter main sequence consists of a mixture of stars of all ages. To obtain an age separation in this lower part of the main sequence, the emission intensity in the H and K lines of Ca II were used, which were estimated for a large number of McCormick K and M dwarfs by Wilson and Woolley (1970). The stars were grouped by their intensity and called "HK groups."

The resulting dispersions in the different velocity components as well as the total velocity dispersion are plotted in Fig. 5 against age. In addition, groups of nearby stars are plotted with individually determined ages making use of available Strömgren (1966) photometry and appropriate calibrations. All of these groups reveal a rather monotonic increase with age τ. This increase can best be modeled by a diffusion process in velocity space. A good fit of the observations was obtained by the relation $(\sigma_o^2 + C\tau)^{1/2}$ plotted in the upper part of Fig. 5, where σ_o is the initial velocity dispersion and C is a constant diffusion coefficient (Wielen 1977). Regarding the McCormick K and M dwarfs as representative for the nearby stars, we get typical velocity dispersions of 48, 29 and 25 km s^{-1} in the U, V and W directions, respectively, and for the solar motion with respect to the local standard of rest (LSR), we find $U_\odot = 8$ km s^{-1}, $V_\odot = 12$ km s^{-1}, and $W_\odot = 6$ km s^{-1}.

There is some evidence indicating that a single solar motion is not representative for these stars. Upgren (1978) and more recently Poveda and Allen (1985) find an outward motion away from the galactic center of 5 to 10 km s^{-1} for the nearby McCormick stars belonging to the young-disk population relative to those of the old-disk population and possibly the dynamical LSR for circular orbital motion as well.

Matching the CM-groups against the HK-groups, there seems to be no significant dependence of the velocity distribution on the stellar mass. Therefore we may conclude that the results for the relatively bright McCormick stars are also valid for the stars with lower absolute luminosity. However, recently Poveda and Allen (1985) postulate that the very faint stars ($M_v > 13$ mag) beyond the maximum of the luminosity function should be, on the average, significantly younger than the brighter ones; they present five tests in favor of this hypothesis. They argue that the relative youth of the fainter stars is due to their failure to reach a hydrogen-burning state, and so they shine as detectable stars for a relatively short period of time.

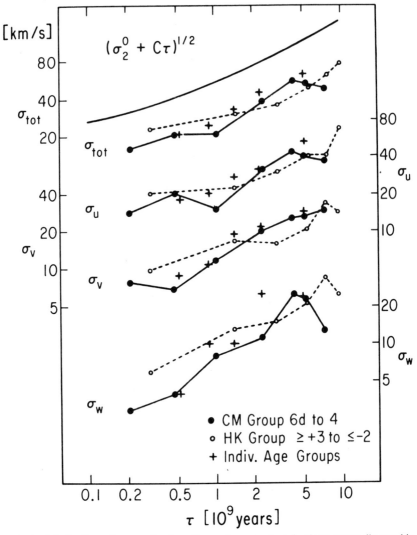

Fig. 5. Velocity dispersions as a function of age τ for variously defined age groups discussed in the text.

V. FREQUENCY OF BINARIES AND MULTIPLE SYSTEMS

Observations of the nearest stars reveal binaries and multiple systems of various categories: visual double stars with well-known orbits, wide pairs, and also very close systems detected by interferometric means and, in recent years, also by speckle observations. In addition to single and double-lined spectroscopic binaries (SBs), we recognize many astrometric binaries and unseen companions, some of which have been later detected visually. Suspected unseen companions (UC) often are still unconfirmed, their existence doubtful. The search for planetary companions continues with the additional help of new nonastrometric techniques (Black 1980). For example, the existence of possible companions to Barnard's star and van Biesbroeck 8 (McCarthy et al. 1984) are not yet fully confirmed.

It is not easy to make a reliable estimate of the relative binary frequency, defined as number of components in double or multiple systems divided by the total number of stars considered. At present, we know of 27 single stars, 14 double stars (including 5 UCs and 1 SB) and 5 triple systems (including 1 UC and 2 SBs) within 5 pc of the Sun. This yields a relative binary frequency of 61% for stars within 5 pc, but only 55% if 4 dubious UCs are omitted. This frequency decreases to 49% for the stars within 5 to 10 pc of the Sun.

In the region between 5 and 10 pc, we find altogether 258 individual stars divided as follows: 131 single stars, 48 doubles (including 12 SBs and 1 UC), and 10 multiple systems (including 7 SBs and 1 UC). The above figures, which can only be regarded as lower limits, compare closely with other findings which indicate that 60% of the brightest stars ($M_v < 1.65$), 53% of solar type F3–G2 main sequence stars, and 50% of the main sequence stars from B to M appear to be binaries or multiple systems. A detailed discussion is given by Abt (1979) and by Herczeg (1984). It should be emphasized here that, due to the lack of the necessary information for fainter stars, our knowledge of the frequency of nearby SBs is still poor. Research on this subject is in progress (see Sec. II).

For the phenomenon of wide pairs among nearby stars with parallel space motions implying a common origin, we refer to Upgren and Chabotte (1983). They found no statistical evidence for wide pairs or triples among 171 late-type dwarf stars. Earlier, Vandervoort (1968) found five pairs and one triple system among 20 nearby A type stars, indicating a relatively high frequency of wide pairs. This difference between the dA stars and the dK-M stars can be interpreted to mean that the motions of the much younger (on average) A stars still reflect the presence of the Ursa Major and Hyades star streams in the solar vicinity, and also perhaps that the motions of young stars are not yet fully relaxed.

From the 61 stars within 5 pc of the Sun, we get a mean distance of 2.2 pc between a star and its nearest neighbor. Among the nearby stars, the Alpha and Proxima Centauri triple system and Barnard's star are closer to the Sun than this distance. Although no stars are known that are unusually close to the Sun at present, we can look for possible past and future close approaches to the Sun which may occur, and we can detect some of these if we search for stars with proper motions much smaller than their parallaxes. There is one object in our sample, the red dwarf Gl 710 = BD $-1°3474$, with a present distance of 15 pc and a radial velocity of -23 km s^{-1}. Its proper motion is very small, being of the order of 0.″01 per year. About 650,000 years from now, Gl 710 will pass by the Sun with a probable minimum distance of 1 or even 0.5 pc (Gliese 1981).

VI. COMPLETENESS OF THE SAMPLES

The space distribution of the nearby stars can be analyzed to determine whether or not they are statistically complete. Two methods have been used for this purpose. The first makes use of the vol/vol(max) ratio defined by Schmidt (1975). This volume ratio has a value of 0.5 for a uniform distribution of stars in space, but would be smaller if the observed stellar density decreases with distance. The second is due to the fact that for uniform density, the distribution of parallaxes should increase with the inverse fourth power in parallax. Any variation less extreme than this indicates an increasing incompleteness with distance.

It has long been known that the sample taken as a whole is increasingly incomplete with distance. This is mainly because the fainter stars in it have been discovered from proper motion surveys which miss increasingly greater numbers of stars with distance. Most of the stars closer than 25 pc will have

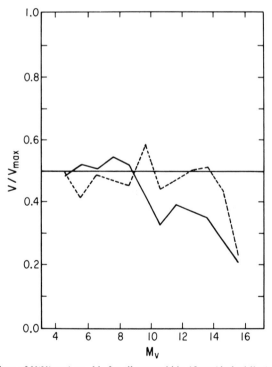

Fig. 6. The values of $V/V(max)$ vs. M_v for all stars within 10 pc (dashed line) and 20 pc (solid line). The straight line at $V/V(max) = 0.5$ represents the case for uniform distribution.

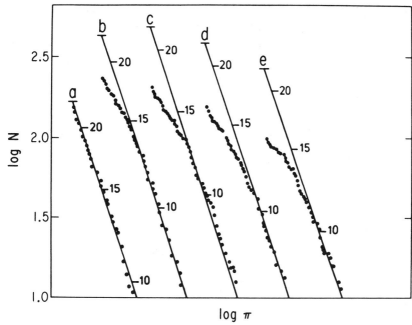

Fig. 7. The cumulative values of log N vs. log π for five ranges in B-V (and M_v): (a) 1.30–1.40 (8.0–9.0); (b) 1.40–1.50 (9.0–10.2); (c) 1.50–1.60 (10.2–11.5); (d) 1.60–1.70 (11.5–12.8); (e) 1.70–1.80 (12.8–14.0). The straight lines have slopes of -3 with distances of 10, 15 and 20 pc indicated. The values of log π for each line are -1.00, -1.18 and -1.30 for these three distances, respectively. The symbol π refers to parallax.

been recognized because of their relative brightness. Both tests confirm that the absolute visual magnitude brighter than that of the nearby star sample is statistically complete, and is very close to $+9$, as demonstrated by Upgren and Armandroff (1981). This corresponds closely to $B - V = 1.40$ and a spectral class of M0V on the MK system. Incompleteness becomes immediately apparent for stars only slightly fainter than this. The situation is shown in Fig. 6 for the vol/vol(max) test and in Fig. 7 for the inverse power law, where the cumulative distribution in parallax has been shown. The data for the red dwarf stars grouped by absolute magnitude (or by B-V color) indicates that for stars fainter than $+9$ but brighter than $+13$ or even $+14$, the nearby stars are essentially complete to a distance of about 13 pc, but far from complete at larger distances. The near-constancy of this distance across this range of luminosity strongly suggests that almost all nearby stars have motions large enough to be detected by motion surveys but beyond 13 pc, substantial numbers of slower-moving stars are missed.

One implication of the study by Upgren and Armandroff (1981) and the earlier one by Jahreiss (1974) is that the luminosity function derived from the nearby stars by Wielen (1974) yields a true picture of the composition of the galactic disk near the Sun, at least for $M_v < +9$. The dip in the function for stars slightly brighter than this, first found by van Rhijn (1936) and confirmed by Wielen, is real whereas the monotonically increasing function of Luyten (1968) is in need of correction. The luminosity function is discussed more completely in Sec. VII below.

VII. LUMINOSITY FUNCTION AND TOTAL MASS DENSITY

Counting the number of known stars within a certain distance limit is a straightforward method for the determination of the luminosity function of nearby stars. The preceding section illustrates the rapidly increasing incompleteness of our stellar sample with increasing distance and with fainter absolute magnitudes. In order to derive a representative luminosity function, a restriction is necessary to samples which are presumably complete. The evaluation of sampling volumes treated as "complete" are described by Jahreiss (1974), Wielen (1974) and Wielen et al. (1983). The resulting luminosity function of 1983 is displayed in Fig. 8, where the units are the numbers of stars within 20 pc per interval of visual absolute magnitude.

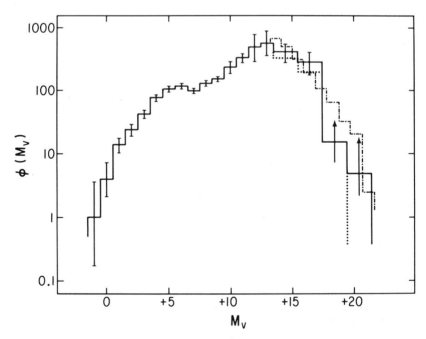

Fig. 8. Luminosity functions for nearby stars for which Φ (M_v) is the number of stars within 20 pc of the Sun. The solid line is from Wielen et al. (1983), the dotted line from Wielen (1974) and the dash-dotted line from Luyten (1968).

The reliability of the luminosity function depends on the basic assumption of a constant true space density of stars within a specific distance from the Sun. This is probably a valid assumption even for relatively young stars since their motions are well mixed. The dotted curve (Fig. 8) shows the luminosity function of 1974 which is based almost exclusively on data included in the catalogue of Gliese (1969). The revised function of Wielen et al. (1983) differs from the earlier result only at fainter magnitudes, $M_v > 14$ mag, due to recently detected new nearby stars within 10 pc. A preliminary study based on the data of Gliese and Jahreiss (1979) also suggests a slight revision for the brighter magnitudes. For example, due to many new trigonometric parallaxes for McCormick K and M dwarfs, the obviously existing dip at about $M_v = 7$ mag becomes slightly less pronounced.

Although the maximum of the luminosity function at $M_v = 13$ mag is probably real (but see Sec. V above), the values for fainter absolute magnitudes $M_v > 14$ represent most likely only lower limits because of the still possible incompleteness within 5 pc of the Sun. In addition, one has to bear in mind that there are still unresolved binaries in our samples which could increase the function at the faint end. Nevertheless we see the luminosity function, determined by mere counts of known nearby stars, approaching the faint end of Luyten's (1968) function, which is represented by the dash-dotted line in Fig. 6. The latter one, depending on Luyten's extensive proper motion survey, was determined by a completely different method of mean absolute magnitudes, which requires additional assumptions about the velocity distribution of stars used. The luminosity function predicts about 7800 stars within 25 pc of the Sun, compared with our estimate of about 2800 known stars within this distance limit. It follows, then, that 64% of the nearby stars, most of which must be very faint, are still unknown.

TABLE VI
Total Mass Density in the Solar Vicinity

Source		Mass Density $M_\odot/pc^{-3}$	
Stars on or near the main sequence	$M_v \leq 9$	0.020	
	$10 \leq M_v \leq 13$	0.014	
	$14 \leq M_v$	0.004	
25 giants within 20 pc		0.001	
5 wide dwarfs within 5 pc		0.007	
Sum of stars			0.046
Interstellar hydrogen		0.028	
Dust, other elements		0.011	
Sum of interstellar matter			0.039
Total observed material			0.085

The resulting local stellar mass density (see Table VI) is 0.046 $M_\odot$ pc^{-3}. The use of the faint end of Luyten's luminosity function would increase this value to 0.049 $M_\odot$ pc^{-3}. A conservative estimate of the percentage of unresolved binaries would give an additional contribution of at least 0.004 $M_\odot$ pc^{-3} (Mezzetti et al. 1983). The total mass density of all observed matter near the Sun is about 0.092 $M_\odot$ pc^{-3}, using an estimate of the mass density of interstellar matter by Wielen (1982). Three extremely high-velocity stars and six stars classified as subdwarfs within 10 pc of the Sun yield a local mass density of 0.94 10^{-3} $M_\odot$ pc^{-3} for the halo stars (Jahreiss and Wielen 1975).

In a recent investigation Bahcall (1984) concludes, in agreement with other authors, "that about half of the disk material at the solar position has not yet been observed." He estimates the mass density of the unobserved matter to be about 0.1 $M_\odot$ pc^{-3}, and discusses the possible nature of the unseen material. If this material is in the form of brown dwarfs, the nearest one would probably be $<$ 1 pc from the Sun.

THE PRESENT, PAST AND FUTURE VELOCITY OF NEARBY STARS: THE PATH OF THE SUN IN 10^8 YEARS

FRANK BASH
University of Texas

The velocities of stars near the Sun are discussed in terms of their velocity ellipsoid as a function of stellar spectral type. The direction showing the largest velocity dispersion (the vertex) points toward the galactic center except for early-type stars; for them, the vertex deviates from the center by a significant amount. This deviation is interpreted as the effect of the star's initial velocity, and it is consistent with stars being born at the post-shock velocity in the spiral arms near the Sun. Older stars have suffered more two-body interactions, and this relaxed system shows no deviation. In addition, the older stars show larger velocity dispersions than the younger ones which may have been caused by interactions with giant molecular clouds. Finally, the galactic orbit of the Sun is integrated over 10^8 yr with gravitational perturbations by the spiral arms taken into account. This integration is meaningful only so long as no significant two-body interactions have occurred during the period covered by the integration. If major two-body interactions are those with giant molecular clouds, and if those clouds are concentrated in spiral arms, then the Sun's path can be traced up to the next spiral arm crossing if that time is less than the relaxation time with respect to two-body stellar interactions. The vertical (Z-motion) path of the Sun is treated as a simple harmonic oscillation because that model is probably adequate in view of the state of our present knowledge.

The material in this review chapter has been taken from Delhaye (1965), Mihalas and Binney (1981), Mihalas and Routley (1968), and Goulet (1984), and from various papers cited in these works. A portion of Goulet's results can be found in Goulet and Shuter (1984). In addition, Eggen's work on very near stars (see, e.g., Eggen 1983a,b,c), should be mentioned although in what follows we take a somewhat larger view.

We adopt a velocity coordinate system (u, v, w) in which: u is the velocity positive outward away from the galactic center with respect to the local standard of rest (LSR), v is the velocity in the galactic plane positive in the direction of galactic rotation with respect to the LSR, and w is the velocity in the Z direction positive toward the north galactic pole (NGP). In this convention $(u, v, w) = 0, 0, 0$ for an object at R_o (the Sun's distance from the galactic center) moving only in the galactic plane on a circular orbit.

I. THE SOLAR MOTION

From Delhaye (1965), the *standard solar motion* is $u_\odot, v_\odot, w_\odot = 10.4$ km s^{-1}, 14.8 km s^{-1}, and 7.3 km s^{-1}, respectively, and is the average of the Sun's velocity with respect to stars in the galactic disk averaged over all those objects whose radial velocities and proper motions are catalogued.

Its physical interpretation is not clear because, especially in the v component, the average of such a stellar population does not move on a circular orbit, and the deviation from a circular orbit depends on the details of the stars' galactic radial density distribution, assuming random orbital eccentricities.

However, the *basic solar motion* derived from the most commonly measured velocities for stars near the Sun (A stars, giant K stars and dwarf M stars) can be used to estimate the Sun's velocity with respect to an LSR moving in the plane on a circular orbit with $R = R_o$. From the basic solar motion, Delhaye (1965) gets

$$u_\odot = -9 \text{ km s}^{-1}$$
$$v_\odot = +12 \text{ km s}^{-1}$$
$$w_\odot = +7 \text{ km s}^{-1} \qquad (1)$$

for the Sun's motion with respect to an LSR on a circular orbit in the plane with $R = R_o$. We shall call this the Sun's motion with respect to the local *dynamical standard of rest* (DSR). The Sun is moving inward toward the galactic center at 9 km s^{-1}; it is moving faster than the local circular velocity by 12 km s^{-1} and is moving upwards toward the NGP at 7 km s^{-1}. We shall adopt this value.

Goulet (1984) reanalyzed the velocities of groups of nearby stars using a full least-squares analysis which also allowed for local velocity gradients and carefully computed the formal uncertainties. If we take the average values of the solar motion with respect to his sample of 1061 AV stars, 981 K III stars and 237 FV stars, we get (standard deviation σ indicated)

$$\bar{u}_\odot = -12.3 \pm 2.3 \text{ km s}^{-1}$$
$$\bar{v}_\odot = +14.5 \pm 3.6 \text{ km s}^{-1}$$
$$\bar{w}_\odot = +7.9 \pm 2.0 \text{ km s}^{-1}. \qquad (2)$$

Comparing these values with our adopted values of the Sun's motion with respect to the DSR, we see that the largest difference is in $\bar{u}_\odot$ where the difference is 3.3 km s^{-1}. Since it is uncertain that the average velocity of the stars used for the DSR defines on a circular orbit with $R = R_o$, the uncertainties in the velocity of the Sun with respect to the local dynamical standard of rest are probably at least $\sigma \approx 3$ km s^{-1}.

II. MEAN VELOCITIES OF NEARBY STARS

We can now take Goulet's samples of B, AV, K III, and FV stars and ask if any of them have significant mean velocities with respect to the DSR by taking the solar motion with respect to each group and subtracting it from the Sun's motion with respect to the DSR. We write the mean velocity of a group of stars with respect to the DSR as $(u^*_{DSR}, v^*_{DSR}, w^*_{DSR})$. We take our adopted uncertainty of 3 km s^{-1} in the Sun's motion with respect to the DSR and the average value of the uncertainty determined by Goulet of the Sun's motion with respect to each group of stars and find that the standard deviation of the difference is ~ 4 km s^{-1}. The only groups of stars where the difference is greater than 4 km s^{-1} are the following:

Sample	Velocity	
1301 B stars	$v^*_{DSR} = -4.2$ km s^{-1}	
451 gB stars	$v^*_{DSR} = -5.4$ km s^{-1}	
330 B stars $d < 200$ pc	$v^*_{DSR} = -6.2$ km s^{-1}	
233 late AV stars	$\begin{cases} u^*_{DSR} = +7.1 \text{ km s}^{-1} \\ v^*_{DSR} = -5.4 \text{ km s}^{-1} \end{cases}$	
981 K III stars	$v^*_{DSR} = -6.6$ km s^{-1}	
237 FV stars	$u^*_{DSR} = +5.9$ km s^{-1}.	(3)

The late AV and FV stars seem to be moving outward from the galactic center and the 3 B star samples plus the late AV stars and K III stars seem to lag behind the DSR.

It is important to point out that none of these velocities is very significant, and no sample shows a significant (compared to 4 km s^{-1}) mean motion in the Z direction (w component).

III. VELOCITY DISPERSION OF DISK STARS

Through the invention of Schwarzschild (1907, 1908), the velocity dispersions of disk stars are described in terms of the velocity ellipsoid whose three axes measure the velocity dispersions of stars in three orthogonal directions. In general, the principal axis of the ellipsoid, the one in which the velocity dispersion is largest, points approximately toward the galactic center and another axis is perpendicular to the galactic plane.

The velocity ellipsoid is usually described by the velocity dispersions in the three directions (u, v, w) plus the longitude toward which the longest axis of the ellipsoid points. This axis is found always to lie in the plane of the Galaxy, and the distribution of velocities in the u, v, w directions looks Gaussian.

As summarized by Mihalas and Binney (1981), the main results are:

1. $\sigma_u > \sigma_v \geq \sigma_w$ with σ_v only slightly greater than σ_w.
2. The ratio of the maximum velocity dispersion to the other in-plane axis dispersion is 1.3 to about 2.
3. When we look at later spectral types, all three velocity dispersions increase by about a factor of 3 from B0 to G stars and are about constant from G to M stars. This suggests that something stirs up the stars as they age. The ages of stars from GV to MV are about the same and set by the age of the Galaxy so long as their relative star formation rates, over the history of the Galaxy, have been similar. This discontinuity in the trend of velocity dispersions at late F or early G was noticed by Parenago (1950), and is called Parenago's discontinuity.
4. The longest axis of the velocity ellipsoid (the vertex) points at $\ell \approx 310°$–$320°$ for early B stars, toward $\ell \approx 20°$ for early A stars, and apparently toward $\ell \approx 0°$ for F stars and later types. This vertex deviation of the longest axis of the ellipsoid away from $\ell = 0°$ is discussed in Sec. IV below.
5. Early spectral types (O and B stars) have nearly circular velocity ellipsoids as far as the in-plane axes are concerned; that is, $\sigma_u \approx \sigma_v$ but σ_u and σ_v are still about 40% larger than σ_w.

In his elaborate mathematical analysis of the velocities of nearby stars, Goulet included velocity gradient terms as well as the ellipsoidal model. The classical analyses do not include these gradient terms. Goulet finds general results for the velocity ellipsoids which are in agreement with the classical ones implying that ignoring the gradient terms has not confused the previous results.

Clube (1985) has suggested that a revival of Kapteyn's star-stream hypothesis, which preceeded Schwarzschild's velocity ellipsoid description, might better describe the local stellar velocity field. He identifies two star streams. Stream I is the young population from the nearby spiral arms contaminated with some older stars in the solar neighborhood. Stream II is a mostly late spectral-type population made up of stars with well-mixed orbits corresponding to the older galactic disk. Clube argues that stream I dominates in the solar vicinity, and so the solar motion and velocity ellipsoid describe that stream, but, farther from the Sun stream II is more dominant and that those more distant stars define a different solar motion and velocity ellipsoid. For example, he finds that stream II stars produce a less deviated velocity ellipsoid. As pointed out below, Hilton and Bash (1982) have argued that the cause of the vertex deviation of the velocity ellipsoid is the peculiar velocity of young stars recently born in the spiral arms.

IV. THE CAUSE OF THE VERTEX DEVIATION

In an axisymmetric system with stars distributed randomly in their orbits, we should have no vertex deviation. Wielen (1974) argues that the vertex deviation decrease from young to old stars may be due to a selection effect in choosing specific old stars. He suggests that the vertex deviation may not actually decrease for old stars. However, this idea remains unproven, so here we shall take the data literally and assume that the deviation does increase. In general, the explanations for the vertex deviation divide into two general classes:

1. The vertex deviation is caused by a perturbation (possibly a local one) in the flow of nearby stars around the Galaxy. Lindblad ($1927a$) and Oort (1928) argued that the deviation is evidence for a third integral of motion. Heckmann and Strassl (1934) claimed that the deviation is caused by random local irregularities in the velocity distribution of nearby stars. Mayor (1970) believed that the deviation is caused by the local spiral structure. Rholfs (1972) suggested that the deviation is caused by the gravitational response to a local spiral arm and that the Sun is on the outside edge of the arm. House and Innanen (1975) showed that the vertex deviation as a function of stellar age could be caused by random perturbations in the velocities of stars; however, their model shows no significant deviations for the first 10^8 yr.

All of these theories have the same flaw: they do not explain why the youngest stars should show the greatest vertex deviation and why the deviation becomes smaller with time. The earliest spectral-type stars are so young that they have not had time to respond to a local potential corrugation caused by the spiral arms and their vertex deviations must be mostly due to their peculiar initial velocities.

2. The vertex deviation is caused by peculiar velocities possessed by the stars at their birth and subsequent random perturbations later reduce the deviation. Woolley (1970) suggested that the vertex deviation is caused by stellar initial velocities at their birth. Yuan (1971) showed how the vertex deviation changes with time by integrating stellar orbits with spiral perturbations included. Hilton and Bash (1982) showed that the initial vertex deviation of early B stars can be quantitatively understood if stars are born in molecular clouds moving initially on post-shock trajections from the spiral arms.

In his recent analysis, Goulet (1984) also argues that the velocity field depends on the ages (spectral types) of the stars in his samples, and thus must be an evolutionary rather than a steady-state process.

V. INCREASE OF VELOCITY DISPERSION WITH TIME

The classical work on this subject was done by Spitzer and Schwarzschild (1951, 1953). They argued that the growth of stellar velocity dispersions with time could be understood if disk stars encounter interstellar clouds whose masses are $\sim 10^5$ $M_\odot$. When, about 20 yr later, molecular clouds were discovered, it seemed that the problem had been solved. However, Barbanis and Woltjer (1967) have shown alternatively that the growth of velocity dispersions could be understood if disk stars periodically encounter spiral arms of stars. This works only if the spiral potential over the mean radial potential was ~ 3 times its present value "a few billion years ago," or the spiral arms dissolve and reform on a time less than a stellar orbital period near the Sun.

Much recent work has been done on this subject. In general, the work divides into papers by authors who examine the effects of random encounters with massive disk clouds and claim that such an interaction is a sufficient explanation, or papers by workers who argue that random encounters with giant molecular clouds (GMCs) are not sufficient when the detailed velocity dispersions are examined and that encounters with spiral arms must be added to the theory. Unfortunately, the papers which deal with spiral arms do not treat the observed growth of the Z-component (w) of the velocity dispersion.

Wielen (1974, 1977) uses a diffusion picture to examine the growth of the velocity dispersions (all three axes) and the functional dependence of the rate of growth on time. He concludes that such a picture works well. On the other hand, Villumsen (1983) examines the growth of the disk scale height and velocity dispersion with time using numerical models, and predicts that the velocity dispersions grow as $t^{1/2}$. He also finds that the velocity ellipsoids grow rounder with time (in contrast to what is observed), and that in order to explain the hottest disk populations, there must have been more GMCs in the past than are present now.

Lacey (1984) argues that in his model of stars scattering from GMCs, the ratio of the Z velocity dispersion to the tangential dispersion σ_w/σ_v is > 1, whereas it is observed to be < 1. Furthermore, he finds that σ_u/σ_v is too large compared with the observed value. He concludes that large-scale perturbations from, for example, spiral density waves must be important. Sellwood and Carlberg (1984) look at the growth of spiral waves in a stellar disk which is cooled by star formation occurring in gas in the disk. Although their model is patterned on an Sc spiral galaxy, they feel that it is capable of explaining the growth of the in-plane velocity dispersions. They do not treat σ_w.

It may be that the growth of the velocity dispersions in the disk is a result of stellar encounters with GMCs, which have been more numerous in the past, and with spiral arms. In the latter case, the heating may be due to the heating of the stars in a disk which is unstable to the growth of spiral modes. As far as we know, no one has treated the picture (possibly correct) for the distribution of the largest GMCs, namely that they lie in a spiral pattern and move with a peculiar, noncircular velocity (Bash 1979).

VI. GOULD'S BELT

Gould (1879) studied the distribution in the sky of bright B stars with the general conclusion that between $\ell = 120°$ and $230°$ bright B stars lie below the galactic plane, while from $\ell = 10°$ to $210°$, they lie above it. The system of bright (nearby) B stars apparently lies in a plane inclined by $\sim 16°$ to the galactic plane: it is called Gould's Belt. Lesh (1968) found that stars of type B5 or earlier, $\delta \geq -20°$ and $m_v \geq 6.5$ define Gould's Belt, and that this system apparently is expanding. She discovered that faint B stars $m_v > 6.25$, lie in the galactic plane.

Olano (1982) has studied the H I gas in Gould's Belt. He finds an expanding ellipse of H I, coincident with Gould's Belt, of dimensions 364×211 pc centered at 166 pc from the Sun in direction $\ell = 131°$ with an expansion age of 3×10^7 yr. (The H I shell would have passed by the Sun $\sim 10^7$ yr ago.) Weaver (1974) suggested that there was an H I expansion coincident with Gould's Belt possibly caused by H I infall from the direction of the galactic poles.

Goulet (1984) excluded Olano's Gould Belt region from his B star sample. This eliminated 29% of his sample of 1301 B stars. There was no significant change in any parameter which describes the B star velocity field. He concludes that the Gould's Belt B stars are kinematically the same as the non-Gould's Belt stars, or they are so local or so dilute in the B star sample that they have no noticeable effect. The same result even obtains when he eliminates Olano's Gould's Belt region from a sample of nearby H I clouds; i.e., when Gould's Belt is eliminated from this sample, the H I is still found to be an expanding system of clouds.

Westin (1984) has also studied the distribution of local O-A0 stars. Those stars whose ages are $< 6 \times 10^7$ yr are found to be located in a system inclined to the galactic plane; i.e., they are found to define Gould's Belt. The stars younger than 3×10^7 yr were found to be in a 2 to 3 times larger region than that defined by Olano and the system of stars was found to be expanding. But he also found that simple expansion away from a point as well as a density-wave velocity perturbation both fail to describe the complexity of the actual velocity field and the spatial extent of the expanding region. He suggests that a more elaborate expansion model which takes into account the distribution of interstellar matter in the Gould Belt region plus the birth of massive stars in the expanding region, some of which become supernovae and act as secondary expansion centers, may be a better description of the Gould Belt region.

VII. THE SUN'S IN-PLANE GALACTIC ORBIT

Before showing the result of a numerical integration of the Sun's orbit in the galactic plane, it is necessary to discuss the relaxation time for the Sun in order to estimate how long one can safely integrate the orbit.

Spitzer and Schwarzschild (1951,1953) give a charactcristic energy-exchange time of 2×10^8 yr for stars in the disk of the Galaxy. Wielen (1977) gives values of this energy-exchange time from 5×10^7 yr to 2×10^8 yr with the larger values preferred. Clube and Napier (1984a) suggest that the Sun has suffered 10 encounters with major molecular clouds during its lifetime; $\bar{t} = 4.5 \times 10^8$ yr is the mean time between encounters. Napier (1985) in a more detailed investigation argues that during its lifetime, the Sun has had close (impact parameter < 20 pc) encounters with 56 GMCs having $M \geq 3 \times 10^3$ $M_\odot$ ($\bar{t} = 8 \times 10^7$ yr) and 8.2 close encounters with GMCs having $M \geq 10^5$ $M_\odot$ ($\bar{t} = 5.5 \times 10^8$ yr).

In its 4.5×10^9 yr lifetime, the Sun has crossed spiral arms (in a two-armed galaxy) about 17 times. Again, if massive molecular clouds are concentrated in the spiral arms, then their surface density in those arms is much greater than if they were spread evenly around the Galaxy. In that case, the relevant time may be the time between spiral arm passages of the Sun, $\bar{t} \sim 2.6 \times 10^8$ yr, and it may not be safe to integrate the Sun's orbit, using a global potential, past one spiral arm passage. The other interesting feature of spiral arm passages is that, according to Bash (1979), the GMCs in the arms have a large (10 to 20 km s^{-1}) systematic, noncircular velocity. In any case, and erring perhaps on the conservative side, we do not choose to integrate the Sun's orbit past 10^8 yr into the future. Perhaps it is not even safe to do so past the next arm passage about 0.5×10^8 yr from now.

Before showing the results of the integration, we can discuss the Sun's galactic orbit ignoring the spiral perturbations and the encounters with massive GMCs (and stars) by using the Haas-Bottlinger diagram as discussed by Trumpler and Weaver (1953). Since the local force law is not inverse-square, the galactic orbit of the Sun will not be a closed Keplerian ellipse. However, the best-fitting, approximate, Keplerian ellipse to the Sun's current orbit shows $a = 1.07$ R_o and $e = 0.07$ where a and e are the semimajor axis and eccentricity of the Sun's orbit, respectively.

The Sun is currently moving inward toward perigalacticon about 4% faster than the circular velocity at R_o. The Sun reaches perigalacticon at $R = 0.995$ R_o and apogalacticon at $R = 1.145$ R_o. The Sun is presently only about 15 Myr away from perigalacticon which is only 0.005 R_o inward of its present distance from the galactic center.

In order to do the actual numerical integration of the Sun's in-plane galactic orbit, we assume the standard spiral pattern given by Yuan (1969), a 5% spiral perturbation potential (5% of the axisymmetric potential), a spiral pattern speed of 13.5 km s^{-1} kpc^{-1}, and for the Sun's initial velocity, our values adopted above for the Sun's motion with respect to the DSR ($u_\odot$, $v_\odot$, $w_\odot$) = -9 km s^{-1}, $+12$ km s^{-1}, $+7$ km s^{-1}. We have done two different integrations for the Sun's orbit by assuming for the Sun's distance from the galactic center and the local circular velocity either the "standard values" of $R_o = 10$ kpc and $V_o = 250$ km s^{-1} or the "best values" (de Vaucouleurs 1983), of $R_o = 8.5$ kpc and $V_o = 220$ km s^{-1}. In both cases, the adopted rotations curve is flat and has the value of the local circular velocity.

The results of the numerical integration, using the standard values, are shown in Figs. 1 and 2. Figure 1 shows the Sun's orbit in a nonrotating frame with the spiral pattern at $t = 0$. The orbit is integrated over 10^8 yr; the marks on the coordinate axes are spaced 1 kpc apart and the position of the Sun is shown each 5 Myr.

Figure 2 shows the same orbit in a frame rotating at the spiral pattern speed so that the spiral arms are stationary, and encounters with the spiral arms can be seen. Here we see that the Sun crosses a spiral arm about 55 Myr from now. With our smooth spiral perturbation potential of only 5%, the effects of the spiral perturbation are difficult to see. The dominant visible effect is the Sun reaching perigalacticon and then conserving angular momentum as it moves toward apogalacticon.

Figures 3 and 4 show the Sun's orbit for our "best values" of R_o and V_o in the nonrotating and the rotating frame, respectively. From Fig. 4 one can see that the Sun now crosses an arm about 50 Myr from now. To estimate the effect of errors in the Sun's initial velocity, we added 3 km s^{-1} to $u_\odot$ and $v_\odot$ and repeated the integration shown in Fig. 3. After 10^8 yr, the Sun's position and velocity differ from those in Fig. 3 by 400 pc and 3.5 km s^{-1}.

VIII. THE SUN'S VERTICAL MOTION

In view of the uncertainty caused by the perturbation of the Sun by giant molecular clouds which, as Thaddeus and Chanan have pointed out (see their chapter), lie in a layer which is thick compared to the Sun's vertical excursions, it is probably sufficient to treat the Sun's vertical motion as a simple harmonic oscillation as in Mihalas and Routley (1968). In that case,

$$w = w_o \cos \lambda t \quad \text{and} \quad Z = w_o \sin_\lambda \lambda t \tag{4}$$

where $\lambda = 92.9$ km s^{-1} kpc^{-1} and w_o is the Sun's velocity when it crossed the galactic plane ($t = 0$) and Z is the Sun's distance from the plane. The vertical oscillation period is $P_Z = 2\pi/\lambda = 6.6 \times 10^7$ yr, and thus the Sun crosses the plane every 33 Myr.

If the Sun's present position and velocity are $Z_\odot = +15$ pc (Allen 1973) and $w_\odot = +7$ km s^{-1}, then the Sun crossed the galactic plane 21 Myr ago

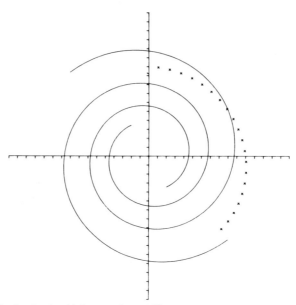

Fig. 1. The Sun's galactic orbit integrated over 10^8 yr shown in a nonrotating frame. The galaxy is viewed from the north galactic pole and rotates clockwise. The assumed spiral pattern is shown at $t = 0$ and the Sun's position is marked each 5×10^6 yr. The tick marks on the coordinate axes are spaced 1 kpc apart and the Sun starts out at $R_o = 10$ kpc from the galactic center with an initial velocity given in the text. A 5% spiral perturbation potential is included in the orbital integration. For this integration, the galactic rotational velocity at the Sun V_o is assumed to be 250 km s^{-1}.

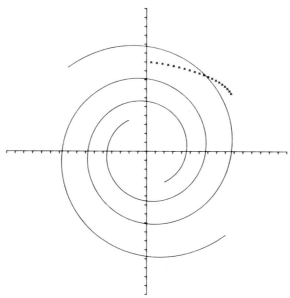

Fig. 2. Same as Fig. 1, except in a frame which rotates at the spiral pattern speed so that the spiral arms are stationary for the period of the integration. The Sun is seen to cross a main spiral arm about 55 Myr from now.

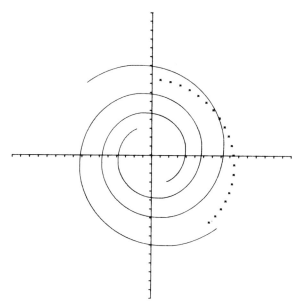

Fig. 3. Same as for Fig. 1, except that $R_o = 8.5$ kpc and $V_o = 220$ km s^{-1}.

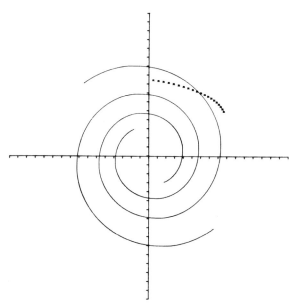

Fig. 4. Same as Fig. 3, but in a frame that rotates at the spiral pattern speed. The Sun is now seen to cross the major spiral arm about 50 Myr from now.

moving at $w_o = 7.1$ km s^{-1}. It will reach its maximum distance from the galactic plane $Z_{max} = 76.8$ pc about 14 Myr from now, at about the same time that it reaches perigalacticon.

The more recent models of the Galaxy's gravitational potential at the Sun in Bahcall (1984a) can be used to compute the Sun's oscillation amplitude. From Bahcall's value, Smoluchowski (personal communication) gets $Z_{max} = 81.8$ pc which is 5% larger than our cruder calculation.

IX. CONCLUSION

In the near future it will be possible to improve our understanding of the velocities of nearby stars. If the dominant interactions, which change the velocities of stars, are those with giant molecular clouds, then as we improve our knowledge of the mass distribution and the galactic location of GMCs, our ability to compute the effect of the interaction will improve. At present, it appears that at least 50%, and possibly as many as 80%, of the GMCs lie in the spiral arms. If only the most massive GMCs are considered, the fraction may be even higher. Further, it now seems clear that the massive GMCs which lie in spiral arms have systematic peculiar velocities. Those peculiar velocities should have a systematic effect on the evolution of the velocity distributions of stars in the galactic disk.

Even though our knowledge of the Sun's location in the Galaxy is imperfect and the circular velocity at the Sun is also uncertain, to first order these quantities only cause a scale change in the size and period of the Sun's galactic orbit. Knowledge of the Sun's galactic orbit depends much more critically on encounters of the Sun with GMCs. Although this interaction has a stochastic effect on the orbit, if GMCs lie in spiral arms, then improved knowledge of the location of the local spiral arms would allow a better estimate of when the Sun was likely to encounter a GMC, or when it might have encountered one in the past. Local spiral arms are the most difficult ones to locate kinematically because, locally, radial velocities caused by differential galactic rotation are small compared to random motions or the systematic peculiar velocities of spiral arm GMCs. It would seem that a fruitful avenue for exploration is to study the stars which are recently born in the local GMCs. Those young stars are very luminous and can be seen up to 2 kpc away. The distances to the stars can be determined by normal photometric techniques and the physical association of a group of young stars with its parent molecular cloud can be established easily (e.g., by their identical radial velocities) making mistaken associations of stars and GMCs quite infrequent.

THE LOCAL VELOCITY FIELD IN THE LAST BILLION YEARS

JAN PALOUŠ
Astronomical Institute of the Czechoslovak Academy of Sciences

We have analyzed the motion of young stars of ages $< 10^9$ yr, which deviates from older stars; they do not conform to the ellipsoidal hypothesis of peculiar stellar velocities. The complicated pattern of the space and velocity distributions and the deviation from the asymmetric drift relation of stars younger than 10^8 yr are explained as the relic of both the protostellar gas cloud motion and the star formation process. The Sirius and Hyades superclusters influence the velocity distribution of stars $< 10^9$ yr old, and we assume that the deviation of the vertex is the result of the sample contamination by these two superclusters.

I. INTRODUCTION

Until the beginning of the 20th century, it was generally assumed that stellar space motions are random, analogous to the motion of molecules in a volume of gas. However, in 1904 Kapteyn announced the discovery of two stellar streams. Relative to the Sun, stars move in two preferred directions, with the observed distribution of proper motions a superposition of the solar motion and of the motion of the two intermingled streams.

Later, in his paper on The Sidereal System, Kapteyn (1922) wrote:

> "Observation has already proved that there really exists a systematic motion of the stars, that it is exactly parallel to the plane of the Milky Way, and that the motion takes place in two exactly opposite directions, the two streams having a relative velocity of about 40 km/s."

As an explanation of this counter-streaming motion within the symmetry plane of his small Galaxy (the distance from Sun to center being only 2000 pc), Kapteyn proposed:

> "Nothing prevents us from assuming that part of the stars circulate one way, while the rest move in opposite direction."

But the center of Kapteyn's Galaxy is about 90° away from the center of our Galaxy and, therefore, the assumed rotation around the center of Kapteyn's Galaxy is actually in our center-anticenter direction.

In the first two decades of this century, it was generally believed that Kapteyn's streams are two independent stellar systems. However, Eddington (1914) wrote:

> ". . . reserving judgments as to whether they are really two independent systems or whether there is some other origin for this curious phenomenon."

He examined whether there was any physical difference between the members of the two drifts and concluded that the complex distribution of motions in different parts of the stellar system and among different classes of stars renders all statistical results inadequate.

An excess of stars moving towards the solar antapex was attributed by Eddington to a small third stream but this stream seems to be more than just a mathematical abstraction since it contains a large amount of early-type bright O and B stars. Eddington (1914) states:

> ". . . it corresponds to some real physical system, and places it on a somewhat different footing from the two older drifts, for which we have as yet failed to find any definite characteristics apart from motion."

An alternative representation of peculiar stellar motions was suggested by K. Schwarzschild (1907), who found that this motion can be described by ellipsoidal distribution, where the largest axis of the velocity ellipsoid points in the direction of greatest stellar mobility. Having compared in detail the two-stream and ellipsoidal hypotheses, Eddington asserts:

> "The distinction is a small one and it is found that the two hypotheses express very nearly the same law of stellar velocities; but by the aid of different mathematical functions."

The era of the "Kapteyn Universe" came to an end with Shapley's (1918) discovery of the space distribution of globular clusters. In 1927, Oort confirmed, by analyzing proper motions and radial velocities, Lindblad's previous hypothesis of the rotation of the galactic system around the center identified with the center of the system of globular clusters. This assumes a much larger Galaxy with a Sun-to-center distance of about 10,000 pc.

The discovery of differential rotation in stellar proper motions and radial velocities led to the general acceptance of Schwarzschild's ellipsoidal distribution. Oort (1928) commented:

> "The differential rotation of the galactic system is seen to be directly tied up with the ellipsoidity of the distribution of peculiar velocities. In a steady system the one cannot exist without the other."

This was also the end of Kapteyn's two-stream hypothesis although it was never disproved statistically.

The approximation of the nearly circular orbits in epicycles of ellipsoidal shape yields a direct connection between the motion of an individual star and the statistical representation of motion around the Sun in the ellipsoidal hypothesis. However, there remain some difficulties and unsolvable problems. The ellipsoidal velocity distribution leads, in steady state, to physically impossible systems with infinite total mass (Perek 1962 and references therein). The velocity dispersion perpendicular to the galactic symmetry plane should be equal, in accordance with Jeans (1919), to that in the radial direction, which is not confirmed in observation. Problems are particularly evident with young stars. We investigate these difficulties for stars younger than 10^9 yr in greater detail below.

II. STARS YOUNGER THAN 10^8 YR

As already noted above, the bright stars of early spectral types O and B form a real physical system separable from the older stars. They are concentrated within a plane different from the galactic symmetry plane. This is referred to as Gould's belt after B. A. Gould, who investigated its properties in the second half of the last century. The long story of the investigation of Gould's belt has been reviewed by Stothers and Frogel (1974) and Frogel and Stothers (1977) (see also references therein).

A system X, Y and Z of spatial coordinates and U, V and W of space velocity components is used throughout this chapter. The X, Y, Z coordinates are centered on the Sun. The X axis points towards the galactic center, the Y axis in the direction of galactic rotation and the Z axis towards the galactic north pole. The U, V, W velocity components are oriented along the X, Y, Z axes, respectively. The galactocentric coordinates (R, ϑ, Z) are sometimes also used: R is the galactocentric distance in the galactic symmetry plane, ϑ is the angular distance in the galactic symmetry plane from the Sun-center line measured clockwise when seen from the galactic north pole.

Space Distribution

The young stars show an uneven distribution in the solar vicinity, which is represented in Figs. 1 and 2. The distinct appearance of the Scorpius-Cen-

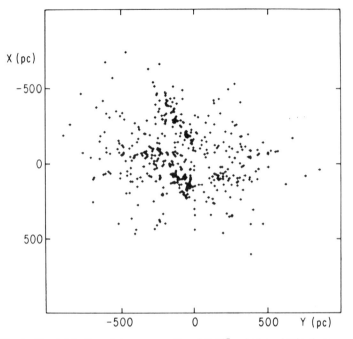

Fig. 1. The distribution of stars younger than 5×10^7 yr in the galactic plane.

taurus and Orion associations can be seen in Fig. 1. These associations and the effects of obscuration by dark clouds of interstellar matter shape the X-Y distribution into a dragonfly pattern, as called by Stothers and Frogel. The above associations form large clumps in the space distribution and they influence the velocity distribution of stars from this region.

The distribution perpendicular to the plane of the Galaxy can be seen in Fig. 2. The orientation and extent of Gould's belt has been discussed by Westin (1985). The tilt of Gould's belt to the galactic plane is some 19° and the line of nodes of these two planes practically coincides with the direction of the galactic rotation. The parts of the belt farther from the center extend to about 500 pc and are below the galactic equator, but the parts closer to the center extend to some 250 pc and are above the galactic equator.

Kinematics of Stars

A proper interpretation of the velocity distribution of stars younger than 10^8 yr is a delicate problem. Their uneven distribution makes it difficult to represent the motion in terms of statistical quantities. However, let us attempt this difficult step from which we obtain a velocity distribution far from ellipsoidal. It is closer to the spherical Maxwellian distribution with the velocity

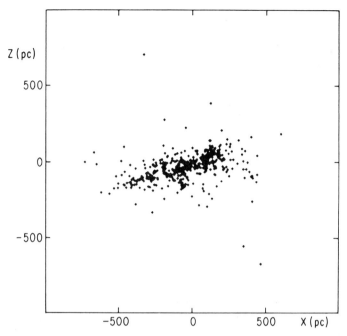

Fig. 2. The distribution of stars younger than 5×10^7 yr perpendicular to the galactic plane.

dispersion near 10 km s^{-1} (see, e.g., Delhaye 1965; Palouš and Piskunov 1985).

The Oort-Lindblad model of galactic differential rotation, with its ellipsoidal epicycles, does not apply to this spherical velocity distribution and we should, therefore, adopt a different model.

Milne (1935) expanded an arbitrary velocity field around the Sun and derived four first-order terms in distance. Their amplitudes, which were later denoted as A, B, C and K, are defined as follows:

$$\begin{aligned} A &= \tfrac{1}{2} (\partial U/\partial Y + \partial V/\partial X)_o \\ B &= \tfrac{1}{2} (\partial V/\partial X - \partial U/\partial Y)_o \\ C &= \tfrac{1}{2} (\partial U/\partial X - \partial V/\partial Y)_o \\ K &= \tfrac{1}{2} (\partial U/\partial X + \partial V/\partial Y)_o. \end{aligned} \quad (1)$$

The subscript o indicates that the quantity refers to the position of the Sun.

The Oort-Lindblad model, which proposes a specific circular streaming around the far center, yields

$$C = K = O \quad (2)$$

which corresponds to the motion of the old disk stars. However, young stars deviate from conditions given by Eq. (2).

Lindblad (1980) reviewed the two possible models: an expansion of the local system from a region near the α Per cluster and the influence of the density wave. Both models yielded specific combinations of the A, B, C and K constants for certain expansion ages, or for certain parameters of the density wave. Lindblad (1980) was able to predict the radial velocity and proper motions as a function of galactic longitude, but, unfortunately, comparison with the kinematical data shows serious discrepancies between these models and observed motions (see also Palouš 1985; Westin 1985).

The velocity field of young stars is quite complicated and cannot be properly represented by Milne's simple model. These complications are connected with (1) the Scorpius-Centaurus and Orion associations, and (2) Gould's belt. The motion of the expanding associations deviates distinctly from the mean and we should isolate their influence. The common motion of Gould's belt stars deviate from the motion of non-Gould's belt stars and, therefore, the distinction between two such young-star subpopulations around the Sun can be important. We should try a combined model where Gould's belt is an expanding system of young stars, which was formed from a giant molecular cloud (GMC) after its transition through a global shock front connected with a density wave. The Scorpius-Centaurus and Orion associations are formed from fragments of a GMC and their own expansion is added to the overall expansion of Gould's belt.

The stars younger than 10^8 yr spend less than one half of the epicycle in orbit after their formation, which leads us to the opinion that their space and velocity distribution are not completely relaxed and that they still hold some information from the era of star formation. Therefore, the complicated pattern of young stars demonstrates the complex nature of the motion of protostellar gas clouds and of star formation process.

III. THE ASYMMETRIC DRIFT AND THE DEVIATION OF THE VERTEX

Strömberg (1924) found a linear relation between the mean motion of certain stellar groups relative to the Sun in the direction of the galactic rotation $\bar{V}$ and the radial component of the velocity dispersion σ_U^2: for small σ_U^2, $\bar{V}$ is also small, which is referred to as the asymmetric drift. This relation is simply explained in the model proposed by Lindblad (1925). In Lindblad's model the Galaxy is divided into subsystems, the orbital motion of each of which decreases with increasing internal velocity dispersion.

In the frame of Lindblad's rotating Galaxy, Oort (1928) analyzed Liouville's equation

$$df/dt = 0 \qquad (3)$$

using the ellipsoidal distribution function f

$$f = f_o \exp(-U'^2/\sigma_1^2 - V'^2/\sigma_2^2 - W'^2/\sigma_3^2) \qquad (4)$$

and assuming that the Galaxy has an axis and a plane of symmetry and that it is in steady state. σ_1, σ_2, σ_3 are the velocity dispersions in the directions of the principal axes of the velocity ellipsoid U', V', W'. He found several conditions for the coefficients of the velocity ellipsoid, some of them of particular interest for this discussion:

1. The velocity ellipsoid must always be so oriented that its principal axes are parallel to the directions U, V and W;
2. $\sigma_V^2/\sigma_U^2 = -B/(A-B)$ (5)
   where A and B are Oort's constants defined in Eq. (1);
3. $\bar{V} = V_o + \sigma_U^2/2(A-B)\ [\partial \ln \rho/\partial R + \partial \ln \sigma_U^2/\partial R + (1 - \sigma_V^2/\sigma_U^2)/R + (1 - \sigma_W^2/\sigma_U^2)/R]$ (6)
   where $-V_o$ is the solar motion in the V direction and ρ is the space density of stars.

Equation (6) yields an explanation for Strömberg's asymmetric drift relation and can be used to determine V_o and/or the radial density and radial velocity dispersion gradients (see, e.g., Mayor 1974).

Mayor (1974) pointed out that the kinematical data from the solar vicinity collected by Delhaye (1965) show the linear relation of $\bar{V}$ versus σ_U^2 except for early-type O and B stars, which deviate significantly. This deviation was also observed for stars $< 10^8$ yr by Palouš and Piskunov (1985).

Mayor (1970, 1972, 1974) analyzed the response of the stellar distribution to the imposed nonaxisymmetric perturbation and found that in the linear approximation, the relative perturbation of the distribution function is inversely proportional to the square of the velocity dispersion. Consequently, the young stars, with their small velocity dispersion, are extremely sensitive to any perturbation. The local mass concentration or the galactic spiral arms are possible perturbers. According to Mayor, the deviation of young stars from Eq. (6) is due to such perturbations.

But the velocity distribution of young stars is far from ellipsoidal and their space distribution also deviates from the galactic symmetry plane, as shown in Sec. II, which means that the assumptions used by Oort (1928) in his analysis of Liouville's equation are not fulfilled. Therefore, his condition expressed in Eq. (6) does not apply in this case. The ratio σ_V^2/σ_U^2 is also far from satisfying the condition in Eq. (5). This is further evidence that Oort's (1928) analysis cannot be applied to stars $< 10^8$ yr.

In our opinion the deviation of the early O and B stars from the linear asymmetric drift relation is the result of the systematic motion of Gould's belt, which has its origin in the motion of the protostellar gas clouds combined with effects of the process of star formation.

TABLE I

Mean Distances from the Sun, Dispersions of X, Y, Z, Mean Values of $\bar{U}$, $\bar{V}$, $\bar{W}$, and their Dispersions for Seven Age Groups of B and A Stars

Age (10^8 yr)	<0.5	0.5–1.0	1.0–2.0	2.0–4.0	4.0–6.0	6.0–8.0	8.0–10.0
No. of stars	210	217	158	141	238	126	86
$\bar{r}$ (pc)	194	192	151	118	98	80	77
σ_X (pc)	130	129	107	79	64	51	55
σ_Y (pc)	138	136	103	79	58	45	37
σ_Z (pc)	64	66	60	59	56	52	54
$\bar{U}$ (km s^{-1})	−10.7	−12.9	−11.6	−8.7	−5.5	−13.1	−12.3
$\bar{V}$ (km s^{-1})	−14.8	−12.4	−9.0	−8.6	−6.0	−8.3	−8.3
$\bar{W}$ (km s^{-1})	−6.7	−8.0	−6.8	−7.0	−7.0	−7.7	−5.9
σ_U (km s^{-1})	9.5	11.0	14.1	18.3	18.0	18.9	18.2
σ_V (km s^{-1})	10.4	12.0	11.2	10.8	12.2	11.6	12.8
σ_W (km s^{-1})	8.1	9.6	9.2	8.7	8.2	9.3	8.3

Let us now discuss slightly older stars, with ages $< 10^9$ yr. The velocity-versus-age relation for B and A stars (Table I) have been discussed by Palouš and Piskunov (1985). The individual stellar ages used in that paper are derived from a comparison of the stellar positions in the HR diagram with those derived from theoretical models of stellar evolution. Their errors are also briefly discussed there. The ratio σ_V/σ_U evolves from 1.09 ± 0.07 for stars $< 5 \times 10^7$ yr to 0.64 ± 0.06 for stars $> 2 \times 10^8$ yr. The latter value is very close to that predicted from Eq. (5): if $A = 15$ km s^{-1}kpc^{-1} and $B = -10$ km s^{-1}kpc^{-1}, $\sigma_V/\sigma_U = 0.63$.

This confirms Mayor's (1974) earlier conclusion that the time required to establish a well-mixed state in statistical equilibrium is close to 2×10^8 yr. This time interval is also plausible from the point of view of stellar dynamics since the epicyclic period is about 2×10^8 yr near the Sun. This is in agreement with the results given by Wielen (1974) and Carlberg et al. (1985), although the velocity-versus-age relation, discussed in these papers, concerned a longer interval of up to 10^{10} yr.

What about the orientation of the principal axes of the velocity ellipsoid? It is well known that stars $< 10^9$ yr old display a deviation of the vertex from the direction to the galactic center of some $20°$ (see Mayor 1972 and references therein; Palouš 1983). Mayor argued that it cannot be a fossil of the initial conditions because the orbital mixing time is about 2×10^8 yr and the mean age of the samples having a considerable deviation of the vertex is larger. He concludes that the deviation of the vertex can be interpreted as the consequence of perturbation by spiral density waves or local mass concentration. However, an interesting question is why σ_V/σ_U remains so undisturbed by this perturbation and why it is so close to the value predicted for an axisymmetric galaxy. An alternative interpretation of the deviation of the vertex will be discussed below (Sec. IV).

IV. THE SUPERCLUSTERS

The existence of moving groups was discovered in the last century by Proctor (1869) and confirmed by Kapteyn, Eddington and others in the first half of this century. These groups contain stars that are widely separated in the sky, but move parallel in space towards a common point of convergence. They played quite an important role in astronomy, because they yielded distances for stars which cannot be reached by trigonometric parallaxes. The idea of moving clusters was extended in this century by Mohr (1931) and later mainly by Eggen (1970,1982,1983a,b,c,1984). Eggen's moving groups or superclusters are stellar streams moving with a small internal velocity dispersion almost parallel in space. Their extent is at least 100 pc. The solar vicinity is penetrated by these superclusters and the Sun itself is surrounded by their members, which are dispersed over the whole celestial sphere.

There are at least three large superclusters near the Sun: Pleiades, Sirius and Hyades. The Pleiades supercluster or the local association, as it is also called by Eggen, contains a large number of early-type B stars $< 5 \times 10^7$ yr.

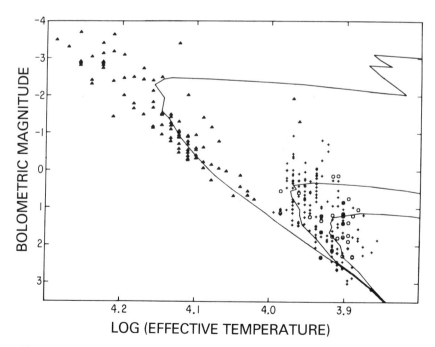

Fig. 3. The HR diagram for the Pleiades ($\triangle$), Sirius (+) and Hyades ($\bigcirc$) superclusters. The solid lines are the isochones for 0.75, 4.5 and 7.5×10^8 yr derived from the theoretical models of stellar evolution.

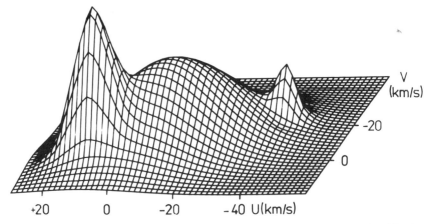

Fig. 4. The predicted number of stars projected into the (U,V) plane as a function of position in this plane. The three ellipsoidal distributions correspond to (from left to right) Sirius supercluster, field stars, Hyades supercluster.

Therefore, the presence of this supercluster near the Sun mainly influences the velocity distribution of young stars discussed in Sec. II. The Sirius and Hyades superclusters are older: the Sirius supercluster is about $4.9 \pm 1.3 \times 10^8$ yr old (Palouš and Hauck 1986) and the Hyades supercluster about $6.3 \pm 0.6 \times 10^8$ yr (Bubeníček et al. 1985). In Fig. 3 we show the HR diagram of these three superclusters; the differences in the ages are remarkable.

The velocity distribution of stars $> 2 \times 10^8$ yr and $< 10^9$ yr is undoubtedly influenced by the Sirius and Hyades superclusters. Their space velocities $(U, V, W) = (11, 3, -8)$ km s^{-1} (Sirius) and $(U, V, W) = (-41, -21, -6)$ km s^{-1} (Hyades) places these superclusters in opposite but peripheral regions of the velocity distribution in the (U,V) plane; according to Palouš and Piskunov (1985) the mean motion for A stars is $(U, V, W) = (-9, -7, -7)$ km s^{-1}.

The velocity distribution should be decomposed into three parts: field stars, Hyades and Sirius superclusters. This is done in Fig. 4, which is a representation of the three velocity ellipsoids produced in a fit of the velocity distribution of a sample of A stars (Bubeníček et al. 1985; Palouš and Hauck 1986). The predicted number of stars projected onto the (U,V) plane according to our fit is plotted on the vertical axis of a three-dimensional diagram in Fig. 4.

How can we now interpret the deviation of the vertex? Let us imagine, for example, that our sample of stars $> 2 \times 10^8$ yr and $< 10^9$ yr only involves members of the Sirius and Hyades superclusters, and that both are represented by the same number of stars. If we fit only one velocity ellipsoid to this bimodal velocity distribution, we find the deviation of the vertex to be about 25°. This is very close to the values found in the literature (Mayor 1972) and will differ for unequal representations of the two superclusters. On the other hand, if we separate the Sirius and Hyades superclusters and if we have a sample composed only of field stars, the deviation of the vertex disappears.

Our interpretation of the deviation of the vertex is close to Lindblad's (1927b), who suggested that it is a result of the existence of the Ursae Major stream. In our opinion, the presence of the Sirius and Hyades supercluster members in the samples investigated in the literature previously (see, e.g., Mayor 1972) resulted in the appearance of the deviation of the vertex. The recognition of the existence of superclusters eliminates this, now rather old, problem.

V. CONCLUSIONS

The early O and B stars $< 10^8$ yr show a rather uneven distribution in space; their velocity distribution is closer to a spherical Maxwellian than to an ellipsoidal distribution. They also deviate from the linear asymmetric drift relation observed for older stars.

Stars $< 10^8$ yr spend only a small fraction of the epicycle in orbit after they were formed; therefore, the orbital mixing in epicycles is not sufficient to relax the initial conditions. The observed pattern is the relic of both motions of protostellar gas clouds and star formation process.

The samples of stars $> 2 \times 10^8$ yr and $< 10^9$ yr are contaminated by superclusters. The superclusters are an astrophysically distinct system and must be eliminated to obtain an unbiased field-star velocity ellipsoid. This also yields a possible explanation of the deviation of the vertex, which is, in our opinion, the result of the contamination of the sample by the Sirius and Hyades superclusters.

The pure velocity ellipsoid does not exist for local stars $< 10^9$ yr. The ellipsoidal hypothesis, which considers the local stellar system as a homogeneous system, leads to erroneous results. Kapteyn's theory of stellar streams is, in a sense, closer to reality because it assumes that intermingled, but astrophysically distinct, stellar systems are present in the solar vicinity at the same time.

We agree with Clube (1983a,1985) that "history was less than fair to Kapteyn" because of the sudden and undiscussed change from the two-stream to the ellipsoidal hypothesis, but, as described by Kuhn (1970), such changes are inevitable in science.

Acknowledgment. The author wishes to thank S. V. M. Clube, J. A. Frogel, L. Perek and an anonymous referee for the critical reading of the original version of the manuscript.

PART II
Massive Interstellar Clouds

MOLECULAR CLOUDS AND PERIODIC EVENTS IN THE GEOLOGIC PAST

P. THADDEUS
Goddard Institute for Space Studies

The suggestion that a claimed 30 Myr period in the geologic past resulted from cometary impacts following encounters with molecular clouds as the solar system oscillates about the Galactic plane poses a well-defined problem in the theory of stochastic processes and, in particular, in the theory of shot noise. All recent CO surveys of the Galaxy clearly indicate that the concentration of molecular clouds in the galactic plane is not sufficient to allow a statistically significant period to be extracted from the small number of dated events (e.g., ~ 9 mass biological extinctions and ~ 40 large impact craters). Of the order of 1000 events is probably required to obtain a credible period.

Rampino and Stothers (1984a; also see their chapter) suggest that encounters with molecular clouds as the solar system oscillates about the plane of the Galaxy are the cause of the 30 Myr periodicity that several investigators claim in the terrestrial record of mass biological extinctions and large impact craters (Raup and Sepkoski 1984; Alvarez and Muller 1984; see chapters by Muller and by Hut). Such encounters, like the passage of the solar companion star Nemesis postulated by Davis et al. (1984) in an alternate explanation, are supposed to perturb gravitationally the Sun's large family of comets, causing many to enter the inner solar system where one or more collide with the Earth. The trouble with this mechanism, as Chanan and I (Thaddeus and Chanan 1985) recently pointed out, is the signal-to-noise ratio: the concentration of molecular clouds in the galactic plane is not sufficient to allow a statistically significant period to be extracted from the small number of well-dated events in the terrestrial record (~ 9 mass extinctions and ~ 40 craters).

The mechanism of Rampino and Stothers (1984a; also see their chapter) is stochastic and poses a straightforward problem in the theory of shot noise. Molecular clouds are concentrated both in the plane and also within the spiral arms (Cohen et al. 1985), but the arms only modulate the solar encounter rate on a fairly long time scale (> 100 Myr) and can be neglected to a first approximation. Let us therefore assume that the clouds are randomly distributed in the plane, with a number density that falls off perpendicular to it with some characteristic scale height. The random rate at which the solar system encounters molecular clouds will then be modulated at *twice* the oscillation frequency, because the number density of clouds is less near the turning points of the oscillation than it is in the plane itself, which the solar system crosses twice per period. Supposedly, this modulation is impressed on the terrestrial record. The degree of modulation obviously depends on the ratio of the scale height of the cloud distribution to the amplitude of the solar oscillation and tends toward zero when the ratio is large. Encounters with stars and with the diffuse clouds of gas observed at 21 cm, which have large scale heights with respect to the solar oscillation, are essentially random and will contribute only noise. We do not attempt here to evaluate this source of noise but evaluate only the signal and noise inherent in the encounter with the molecular clouds themselves.

Chanan and I based our analysis on the assumption that the number density of molecular clouds is Gaussian in z, the distance from the galactic plane, this being the distribution function which best fits the large-scale CO surveys and the one radio astronomers undertaking such surveys have generally adopted in analyzing their data; our conclusions, however, are not sensitive to this assumption and remain essentially unchanged if an exponential or other plausible distribution function is adopted instead. For the half-width at half-maximum, $z_{1/2}$, of the Gaussian distribution, we obtained a value of 85 ± 20 pc, as shown in Fig. 1, by extrapolating slightly outward to the solar circle (taken to be at $R = 10$ kpc from the galactic center) the Massachusetts-Stony Brook (Sanders et al. 1984) and the Columbia CO determinations of $z_{1/2}$ in the inner Galaxy ($R = 4$ to 9.5 kpc) and by extrapolating slightly inward to $R = 10$ kpc the Columbia CO determinations of $z_{1/2}$ in the Carina and Perseus arms ($R = 10.5$ to 13 kpc). This procedure yields a value of $z_{1/2}$ at the solar circle, which is an average over a fairly large part of the Galaxy and therefore likely to be a good estimate of the average thickness of the molecular cloud layer actually encountered by the Sun over the last 250 Myr—roughly speaking, the period of time over which the 30 Myr periodicity has been claimed.

As a check, Dame and I (1985) have recently attempted to determine a purely *local* value of $z_{1/2}$ from the northern Columbia wide-latitude galactic CO survey from New York and from parallel surveys of the southern Galaxy recently completed in Chile with a similar 1.2 m telescope. Distances of local clouds, on which the determination of $z_{1/2}$ depends, have generally been determined from associated optical objects and so are free of any assumption

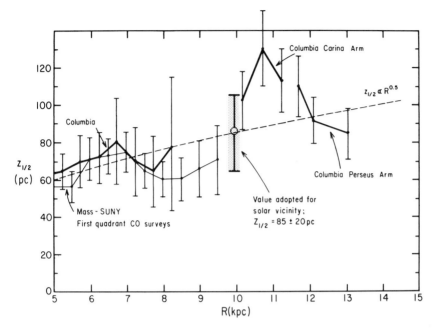

Fig. 1. Gaussian half-thickness at half-maximum of the distribution of molecular clouds with distance from the galactic plane, as a function of distance from the center of the Galaxy, from the Massachusetts-Stony Brook (Sanders et al. 1984) and Columbia (Cohen et al. 1986), CO surveys.

about the distance to the galactic center. From the large system of clouds that comprise the Great Rift and other dark nebulae along an entire quadrant of the northern Milky Way (Fig. 2) and the extensive complex of clouds in Orion and Monoceros (Maddalena et al. 1986), we derived for the rms deviation of molecular gas from the plane a value of 64 pc, which for a Gaussian distribution corresponds to $z_{1/2} = 75$ pc (Dame and Thaddeus 1985). Subsequently, we extended this calculation to include the CO surveys of the well-known dark nebulae in Taurus, Ophiuchus, Cepheus, Lupus, the Coal Sack, Scorpius and Vela. Most of the molecular gas within 500 to 600 pc probably has now been taken into account. The rms deviation from the galactic plane found from this inventory of local molecular clouds is 76 pc, which corresponds for a Gaussian distribution to $z_{1/2} = 89$ pc. At the present time, therefore, we conclude that *no* CO survey of molecular clouds, either local or galactic, furnishes evidence of an appreciably thinner molecular disk than the $z_{1/2} = 85 \pm 20$ pc that Chanan and I used.

The amplitude of the solar oscillation is actually fairly uncertain (on the order of 30 to 40%) owing to uncertainty in the local total mass density of stars and gas and the resultant uncertainty in the gravitational restoring force

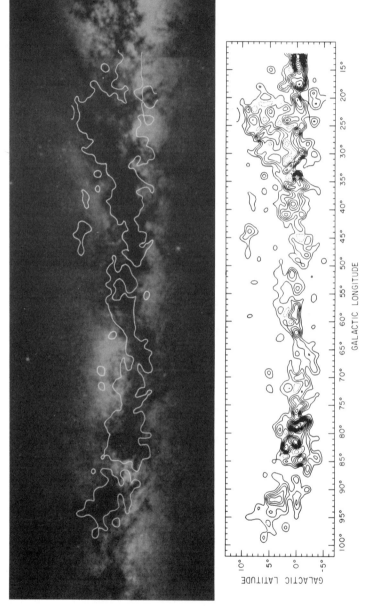

Fig. 2. The Columbia wide-latitude CO survey of the northern Milky Way, superposed on a Mt. Wilson photographic mosaic of the same region (Dame and Thaddeus 1985). The CO survey, at an angular resolution of 1°, has been integrated over a limited range of radial velocity, -10 to $+34$ km s^{-1}, to best exhibit local molecular clouds. The plotted isophotes are linear in integrated line intensity $W = \int T dv$, the contour interval being 4 K km s^{-1}. The conversion factor from W to molecular hydrogen column density N is $N/W = 2.8 \times 10^{20}$ cm^{-2} (K km s^{-1}).

on the Sun when it is displaced from the plane (Bahcall and Bahcall 1985). However, this amplitude cannot be regarded as so uncertain in the present analysis for, *by hypothesis,* the oscillation period is 60 Myr (i.e., twice the putative period in the terrestrial record), and simple harmonic motion requires that $z_o = v_\perp T/2\pi = 72$ pc, where $v_\perp$ is the in-plane z component of the solar motion relative to local stars and interstellar gas, safely approximated by the current z component 7.4 km s^{-1}, the Sun now being quite close to the plane (Mihalas and Binney 1981).

Similarly, to recapitulate the deviation in Thaddeus and Chanan (1985), we may calculate the encounter rate as the solar system oscillates sinusoidally through a Gaussian distribution of clouds by neglecting the small present displacement of the Sun from the galactic plane. The solar-velocity components parallel and perpendicular to the plane then are $v_\parallel = v \cos b = 18.5$ km s^{-1} and $v_\perp = v \sin b \sin \omega t = 7.4$ km s$^{-1} \sin \omega t$, where $\omega_o = 2\pi/60$ Myr, $v = 20$ km s^{-1} is the present solar motion relative to local stars and interstellar matter, and $b = 22°$ is the galactic latitude of the solar apex (Mihalas and Binney 1981). Because the perpendicular velocity is always small relative to the parallel velocity, the solar speed as a function of time is only slightly modulated by the oscillation: $v(t) = v_\parallel (1 + 0.164 \cos^2\omega_o t)^{1/2} \approx v_\parallel (1 + 0.082 \cos^2\omega_o t)$; the solar speed also is always large with respect to the random motion of the clouds, which may therefore be neglected. Letting $\zeta = \ln 2 \, z_o^2/z_{1/2}^2$, the number of encounters per unit time then is

$$r(t) = n(z(t))v(t)\sigma = C (1 + 0.082 \cos^2\omega_o t) \, e^{-\zeta \sin^2\omega_o t} \tag{1}$$

an explicit function of time with no free parameters except the constant of normalization $C \equiv n_o v_\parallel \sigma$, where n_o is the in-plane number density of clouds and σ the encounter cross section. It is implicitly assumed that the encounters are short-range, i.e., that the impact parameter b is small with respect to $z_{1/2}$; the effect of long-range encounters is clearly to reduce the modulation and somewhat tighten our conclusions, but this effect has not yet been treated explicitly.

We now wish to calculate the power spectrum $p(\omega)$ for a sample of N encounters drawn from this distribution with respect to time and, specifically, the signal strength of the leading Fourier component at $2\omega_o$ relative to the purely statistical fluctuations in the absence of modulation, i.e., the shot noise. We define the complex Fourier amplitude as

$$A(\omega) = \sum_{n=1}^{N} e^{i\omega t_n} \equiv \sum_{n=1}^{N} (x_n + iy_n) \tag{2}$$

where the t_n are the encounter times and, as usual, define the power spectrum as $p(\omega) = |A(\omega)|^2$. The first and second moments of $A(2\omega_o)$ are readily calculated. We may take $<y> = 0$ without loss of generality. Then

$$\langle x \rangle = \int r(t) \cos 2\omega_o t\, dt / \int r(t)\, dt \equiv R \tag{3a}$$

$$\langle x^2 \rangle = \int r(t) \cos^2 2\omega_o t\, dt / \int r(t)\, dt \equiv S^2 \tag{3b}$$

$$\langle y^2 \rangle = 1 - S^2. \tag{3c}$$

The variances corresponding to the second moments are $\sigma_x^2 = S^2 - R^2$ and $\sigma_y^2 = 1 - S^2$. The expectation value of the power at $2\omega_o$ then is

$$\langle p(R) \rangle = N^2 \langle x \rangle^2 + N\sigma_x^2 + N^2 \langle y \rangle^2 + N\sigma_y^2 = N(N-1)R^2 + N \tag{4}$$

Care is required in calculating the signal-to-noise ratio S/N, since in the absence of modulation the statistical fluctuations are distributed normally in A but exponentially in p; when the phase of the signal is unknown the expression for this ratio is

$$S/N = \left[\frac{\langle p(R)\rangle - \langle p(0)\rangle}{\langle p(0)\rangle}\right]^{1/2} = \sqrt{N-1}\, R \tag{5}$$

which for a large number of events is proportional to $\sqrt{N}$, as it should be.

When ζ is not too large the normalized Fourier transform R in Eq. (2) is conveniently approximated analytically by expanding the exponential in Eq. (1) in a power series, and at all ζ it is readily evaluated numerically. The signal to noise by both methods for $N = 9$ events is shown in Fig. 3. For the observed thickness of the population of molecular clouds at the solar circle the signal to noise is small, 0.45 ± 0.17; and, clearly, molecular clouds would have to be much more concentrated to the galactic plane (by a factor of at least 3) for a well-defined period to result (i.e., to achieve $S/N \gtrsim 2$).

A graphic demonstration of this conclusion, and a useful check as well of the analytic theory, has been obtained by Monte Carlo simulation. The top of Fig. 4 shows the power spectrum derived numerically from a sample of nine random events drawn from the distribution of Eq. (1): the signal is clearly swamped by the purely statistical fluctuations, i.e., the shot-noise spikes. The expectation value of the peak power can be obtained by averaging together 100 such (independent) spectra, as in the bottom panel of Fig. 4; its smallness, compared with the noise spikes in the top panel, is evident. To express this result another way, scaling from Eq. (5) we find that even a relatively modest signal characterized by $S/N = 3$ would require not nine extinction events but over 300.

Encounters between the solar system and molecular clouds would have been more strongly modulated in the past, of course, were the amplitude of the solar oscillation larger then than now, and this possibility deserves consid-

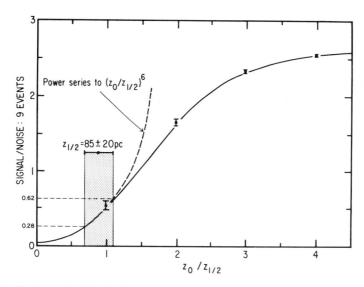

Fig. 3. Signal to noise of the leading Fourier component at $2\omega_o$ for nine events, as a function of the ratio of the amplitude of the solar oscillation to the half-thickness of the molecular cloud disk. The four points with vertical error bars indicate the results of a check on Eq. (5) with the Monte Carlo code used to produce Fig. 4.

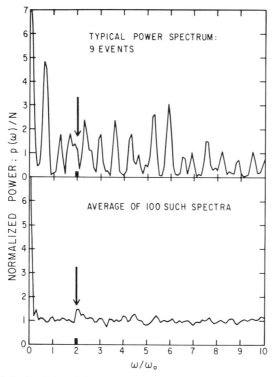

Fig. 4. Monte Carlo simulation of the power spectrum of random events drawn from the distribution given by Eq. (1), for $z_0/z_{1/2} = 1$. *Top*: nine events, the number of mass extinctions analyzed by Rampino and Stothers (1984a); *Bottom*: the average of 100 such spectra, showing the small expectation value $<p(R)> - <p(0)>$ of the nine-event peak (short vertical bar at $\omega/\omega_o = 2$).

eration because it is well known that most stars as old as the Sun possess significantly larger amplitudes of oscillation (Mihalas and Binney 1981). To salvage the mechanism of Rampino and Stothers (1984a; also see their chapter), it is tempting to suggest that the Sun has only recently acquired its present low-amplitude oscillation and that over most of the last 250 Myr its amplitude was typical of that of other G stars: several hundred parsecs. This sudden braking of the Sun's motion, however, would require an extremely improbable close encounter with another star and amounts, therefore, to an arbitrary, *ad hoc* assumption, which destroys the attractive simplicity of the entire scheme.

OBSERVATIONAL CONSTRAINTS ON THE INTERACTION OF GIANT MOLECULAR CLOUDS WITH THE SOLAR SYSTEM

N. Z. SCOVILLE
and
D. B. SANDERS
California Institute of Technology

Over two-thirds of the interstellar gas in the interior of our Galaxy is now believed to be in molecular form. Most of this gas resides in giant molecular clouds with masses 10^5 to 4×10^6 $M_\odot$ and sizes 20 to 100 pc. The properties of the molecular clouds as determined from CO surveys of the Galaxy are reviewed with particular emphasis placed on the characteristics in the solar neighborhood. Near the solar circle the mean volume density of molecular material is 0.033 $M_\odot$ pc^{-3} with a half width at half maximum in Z of 75 pc. Estimates for the frequency and duration of encounters between the solar system and giant molecular clouds are presented. Typical times are $\sim$ 1 Gyr for penetrating encounters, each of which would extend over a period $\sim$ 4 Myr. The mean densities interior to these clouds are sufficiently low ($\bar{n}_{H_2} \sim 250$ cm^{-3}) that tidal effects can be important for comets in the outer Oort cloud (at R = 40,000 AU) only during penetrating encounters of the clouds by the solar system. Such encounters, occurring only every 1 Gyr, cannot therefore be linked with short-term periodicities in the geologic record.

I. INTRODUCTION

Since the mid-1970s, extensive observations of carbon monoxide emission in the disk of our Galaxy have shown that molecular (H_2) gas rather than atomic (H I) gas is the major active component of the interstellar medium. In the inner Galaxy as much as 80% of the interstellar medium is contained in massive molecular clouds; at the solar circle the abundances of H_2 and H I are approximately equal (Sanders et al. 1984). (A somewhat lower abundance of $\sim 50\%$ is estimated by Cohen et al. [1980].)

The extraordinary properties of these clouds were not at all anticipated in the 1960s. Typical masses for the molecular clouds surveyed in CO are in the range 10^5 to 4×10^6 $M_\odot$; their sizes range from 20 to 100 pc (Sanders et al. 1985a). The objects at the top end of this mass range may be more in the form of complexes than individual clouds. It is now clear that the giant molecular clouds (GMC) are the most massive gravitationally bound components in the Galaxy; as such, they are important to both the dynamics and stability of the galactic disk. Even before their discovery, it was hypothesized by Spitzer and Schwarzschild (1951) that objects of this mass could account for the observed high velocity dispersions of disk stars.

More recently it has been suggested that these clouds may also be important to solar system dynamics. Biermann and Lüst (1978), Clube and Napier (1982b), and Bailey (1983a) have pointed out that the tidal forces of GMCs and the substructures within them should disrupt the outer comet cloud at 25,000 AU. And Clube and Napier (1984a) and Rampino and Stothers (1984a,b) have gone further to link the disruption of cometary orbits with periodic, catastrophic extinctions and impact cratering on the Earth.

In this chapter we summarize the properties of the molecular cloud distribution, i.e., size and mass spectrum from 10^3 to 2×10^6 $M_\odot$, with special emphasis on the solar neighborhood. Our discussion is based largely upon a comprehensive survey of the inner Galaxy comprising over 40,000 CO spectra, taken as part of the Massachusetts-Stony Brook survey (Sanders et al. 1985b). This unbiased survey with spectra obtained every 3' in the first galactic quadrant can be used to put into perspective more limited data on the molecular clouds associated with OB associations (Blitz 1979) or small, optically selected clouds in the solar neighborhood (Lynds 1962).

We show that the mass density within molecular clouds is sufficiently low that passing or grazing encounters cannot be significant in the perturbation of cometary orbits at 40,000 AU. The mean time for the Sun between penetrating encounters of GMCs is approximately 1.5 Gyr with a typical duration of 1 Myr. Independent of the exact consequences to the comet cloud when the Sun passes through a molecular cloud, the long time interval between encounters necessarily rules out a link between short-term periodicities in the geologic record and molecular cloud passages. The molecular cloud distribution vis-à-vis the spiral arms in the solar neighborhood is reviewed later (Sec. VI) since the passage of the Sun through the spiral arms has been linked to long-term periodicities (~ 250 Myr).

II. PROPERTIES OF THE MOLECULAR CLOUD DISTRIBUTION

Although CO is only one of over thirty trace molecules detected in molecular clouds, it has been adopted as the most general, widespread probe. CO

TABLE I
Empirical Measures of the Ratio $N(H_2)/\int T(CO)dv$

Value ($\times 10^{20}$ cm^{-2} K km s^{-1})	Source
4	^{12}CO vs. A_v; Young and Scoville 1982
4.8	^{13}CO vs. A_v in Taurus; Frerking et al. 1982
2.9	^{13}CO vs. A_v for 5 dark clouds; Sanders et al. 1984
1.8	^{13}CO vs. A_v in ρ Oph; Frerking et al. 1982
2.5–4.5	Virial masses; Sanders et al. 1985
≤3	γ ray in inner Galaxy; Lebrun et al. 1983
2.6	γ ray in Orion; Bloemen et al. 1985

is the most abundant molecule with a permanent dipole moment and its fundamental transition ($J = 1$-0) at $\lambda = 2.6$ mm is excited easily by collisions with molecular hydrogen, even in clouds with very low kinetic temperature ($T \sim 5$ K). The minimum H$_2$ volume density required to produce excitation appreciably above the microwave background is about 100 to 300 cm^{-3}.

The first surveys of CO emission in the galactic plane (Scoville and Solomon 1975; Burton et al. 1975) revealed widespread emission with typical lines of sight in the galactic plane between longitudes 20° and 40° typically showing five or more distinct emission features, each from a discrete cloud.

CO as a Tracer of H$_2$

The use of CO as a tracer of the molecular hydrogen distribution within clouds and throughout the Galaxy is based on the premise that there exists an approximately linear correlation between the observed CO line flux and the column density of molecular hydrogen. In recent years, there have been several attempts to evaluate the constant proportionality between the CO emission and H$_2$ column density and to test the linearity of the relation (see Table I).

Four general techniques, all pseudoempirical, have been applied: (1) the measurement of ^{13}CO emission which is presumably less saturated than ^{12}CO; (2) the measurement of visual extinctions in nearby dark clouds and their correlation with CO intensities; (3) the estimation of virial masses using the CO and ^{13}CO line widths and cloud sizes; and (4) estimates of the total column density of nucleons derived from the gamma-ray observations in both the galactic plane and the nearby Orion molecular cloud. The results of these attempts to calibrate the number/intensity ratio (N_{H_2}/I_{CO}) are presented in Table I along with references. The total range in estimates for the constant proportionality is 1.8 to 4.8 $\times 10^{20}$ cm^{-2} K km s^{-1}. Although the relatively good agreement of these totally independent methods does not imply that the derived mass estimates must be correct to a factor of 2, it does cast doubt on the basis for claims that molecular cloud masses are uncertain to an order of magnitude.

One point which must be emphasized is the fact that several of these determinations apply in different areas of the Galaxy, thus suggesting that possible gradients in the metallicity or the temperature of molecular clouds do not significantly affect the CO emissivity per unit mass of H_2. In particular, the virial mass estimates were derived for clouds within the molecular cloud ring at galactic radius 5 to 8 kpc while the extinction analysis was based on clouds in the solar neighborhood, generally within 1 kpc of the Sun. Especially noteworthy are the gamma-ray analyses which yield nearly identical calibration constants comparing the Orion molecular cloud with areas in the inner Galaxy. In the following discussion we adopt

$$N_{H_2}/I_{CO} = 3.6 \times 10^{20} \text{ cm}^{-2}/\text{K km s}^{-1}. \tag{1}$$

Molecular Cloud Properties

Several comprehensive surveys of CO emission from the inner Galaxy have now been performed. Although the first surveys contained observations solely at galactic latitude $b = 0°$, this deficiency was remedied in a low-resolution survey done with the Columbia mini-telescope by Cohen et al. (1980) which covered $b = -1$ to $1°$ and $\ell = 12$ to $60°$ at 8 and 16 arcmin resolution. This survey provided a good picture of the large-scale distribution for molecular clouds as a function of R and Z; however, the resolution was inadequate to reveal cloud properties. In order to sample a cloud, there must be at least two sample points within the cloud. For the Columbia survey these points would encompass a minimum of 16 arcmin, corresponding to ~ 45 pc at a typical distance of 10 kpc. Recently a complete high-resolution survey has been done with the Five College Radio Astronomy Observatory (FCRAO) 14 m telescope at a resolution of 45 arcsec and a sample spacing of 3 arcmin over the region $\ell = 8$ to $90°$ and $b = -1$ to $1°$. This survey, comprising over 40,000 CO spectra covering all velocities permitted by galactic rotation, provides a virtually complete sample of all molecular clouds > 12 pc in the northern hemisphere of the inner Galaxy. For clouds within 10 kpc of the Sun, the survey is complete for clouds > 5 pc.

A sample of the data may be seen in Plates 1, 2 and 3 (see Color section) which show the integrated CO emission for the clouds and cloud complexes in the vicinity of M17, W44 and W51. The former may well be a single giant molecular cloud elongated roughly parallel to the galactic plane; the latter two are giant cloud complexes.

Initial studies of the M17 region focused mainly on the hot molecular gas adjacent to the H II region at the lower left of the cloud ($\ell = 15°.0, b = -0°.6$). More extensive mapping by Elmegreen et al. (1979) revealed a much larger extent for the cloud (~ 100 pc) and showed it inclined to the galactic plane.

The FCRAO survey now shows the M17 cloud as one link in a chain of clouds including those associated with M8 and M16 (seen at $\ell = 17°\!.0$, $b = 1°\!.0$).

The W44 region was mapped with an 8 arcmin beam by Dame (1983); the high-resolution data shown in Color Plate 4 now reveal the CO emission in W44 to be made up of numerous individual clouds, all grouped into a fairly spherical cluster. In both W44 and W51 one sees emission with scales ranging between 5 and 100 pc.

The overall distribution of cloud sizes, that is, the cloud size spectrum, has recently been measured by Sanders et al. (1985a). In a sample of 80 clouds with known distances in the first galactic quadrant, they find a distribution which is fit with the counting statistics by a power law,

$$N(D) = 1.3 \times 10^{4.1} D^{-2.3} \text{ clouds pc}^{-1}. \qquad (2)$$

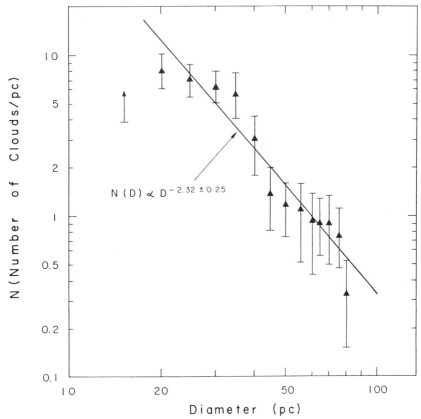

Fig. 1. The number distribution of cloud diameters is shown for a sample of clouds with diameter > 10 pc obtained from CO observations of the inner Galaxy by Sanders et al. (1985a). The 80 clouds were taken from a 10 kpc^2 region at $R = 5$ to 8 kpc and $\ell = 5$ to 41°.

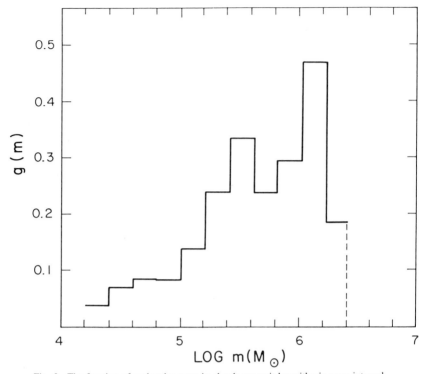

Fig. 2. The fraction of molecular mass in clouds per unit logarithmic mass interval.

The distribution which this fit was derived from is shown in Fig. 1; it is well determined at sizes between 15 and 80 pc. There is good indication that the spectrum may actually turn over at the lower size end. A similar result has been found by Dame (1983) for clouds in the outer Galaxy Perseus arm. Stark (1979) measured clouds near the terminal velocity in three $1° \times 1°$ grids at $\ell = 34°$, $36°$ and $51°$. The size spectrum we deduce from his list of clouds is $N(D) \propto D^{-2.5}$, in good agreement with Eq. (2).

Virial masses were determined for the same 80 clouds shown in Fig. 1 and the derived mass spectrum is shown in Fig. 2 as the function $g(m)$ which is the fractional mass per logarithmic mass interval. (In general, it is believed that the cloud lifetimes are > 30 Myr which is at least five times the internal dynamical crossing time obtained from velocity dispersion and measured size. Thus the clouds should be dynamically relaxed enough for an approximate application of the virial theorem. Similar estimates of cloud masses are obtained from other methods such as from ^{13}CO measurements.) We find that $g(m) \propto m^{0.42}$; the positive exponent indicates that the majority of mass in molecular clouds is at the high-mass end of the spectrum. The integral mass

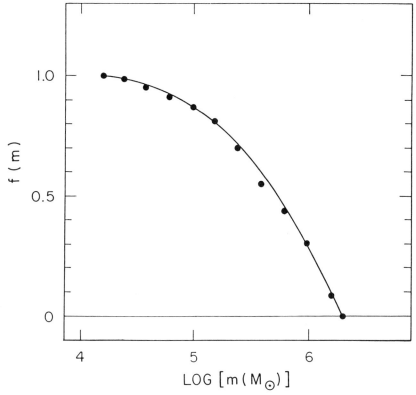

Fig. 3. The fraction of gas in molecular clouds with mass greater than a given mass m. Circles denote the data points derived from the sample shown in Fig. 1, the curve represents the mass fraction function derived based upon the analytic fit $N(m) \propto m^{-1.58}$.

spectrum, that is, the fraction of molecular mass in clouds of mass greater than m, is given by the function $f(m)$ shown in Fig. 3.

The mass spectrum derived from these data is given by

$$N(m) \propto m^{-1.58}. \qquad (3)$$

Lastly, we note that the mean internal density within the clouds can also be described by the power law

$$\bar{n}(H_2) = 290 \, (D/20 \text{ pc})^{-0.75} \text{ cm}^{-3}. \qquad (4)$$

Comparing this cloud sample in the inner Galaxy with the standard H I clouds of mass ~ 500 $M_\odot$ and density ~ 30 cm^{-3}, one finds that the molecular clouds which contain most of the H_2 have masses higher by a factor of 1000 and mean densities a factor of ten higher. In fact, within the molecular

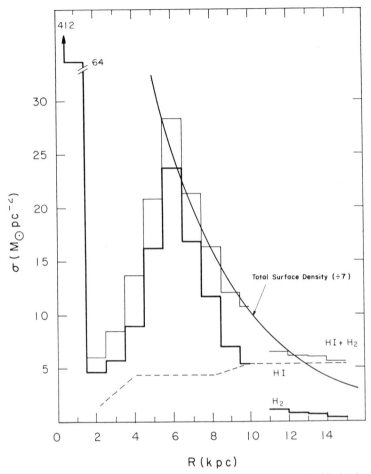

Fig. 4. The gas and total surface densities shown for the Milky Way disk. The H_2 density was derived from CO survey results, the H I surface density from 21 cm surveys. Values for both H_2 and H I include a 1.36 correction factor for He. The total surface density of the galactic disk component is derived from the model of Bahcall et al. (1983). Note that in the vicinity of the Sun, the interstellar matter, H I + H_2 amounts to approximately 14% of the disk surface density and that the mass in H_2 and H I are approximately equal.

clouds there undoubtedly exists considerable clumping of the material. At present, it is difficult to characterize in a simple way the degree of clumping because the fraction of volume filled by clumps, i.e., the filling factor, probably changes from place to place within the clouds. In cloud cores this filling factor may well be considerably higher than in the envelopes (Mundy 1984). Observations of high-excitation molecular transitions with rapid radiative decay rates clearly indicate volume densities in the core regions in excess of 10^5

to 10^6 cm^{-3}. The fraction of the mass contained in the core regions is in most cases insignificant compared to the GMC mass.

III. MOLECULAR CLOUDS AT THE SOLAR CIRCLE

At the position of the Sun (R_o = 10 kpc), the radial distribution of molecular gas is falling off steeply with increasing radius. The overall galactic distribution of H_2 and H I gas is shown in Fig. 4 where the H_2 mass scale has been derived from the CO emissivity using Eq. (1). The total mass of molecular clouds in the inner galaxy (including the helium contribution) is 4.3 × 10^9 M$_\odot$. About 60% of this is the molecular cloud ring at 4.5 to 8.5 kpc and about 15% in the galactic center region. Outside the solar circle, the molecular cloud mass is 0.4 × 10^9 M$_\odot$.

The smoothed-out density of molecular clouds in the vicinity of the Sun is 0.033 M$_\odot$ pc^{-3} at the molecular midplane (Sanders et al. 1984). The comparable number for H I is 0.017 M$_\odot$ pc^{-3}. However, the total masses in H I and H_2 are not as dissimilar as these numbers suggest because the overall scale height for the atomic gas is approximately twice that of the molecular gas. The Z scale height for the clouds, that is the half thickness to half density,

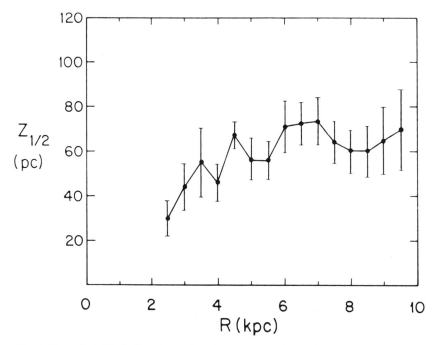

Fig. 5. The scale height (half width at half maximum intensity) $Z_{1/2}$ of CO emission as a function of galactocentric radius. The error bars represent one standard deviation from the mean. The scale height may be fit by the analytic expression: $Z_{1/2}(R) = 26 R^{0.47}$ pc.

is shown in Fig. 5. This half thickness is 65 pc at the peak of the molecular cloud ring, rising to 75 pc at the solar circle. The surface densities of H I and H_2 clouds (including the helium contribution) at the solar circle are approximately equal and amount to 5.2 $M_\odot$ pc^{-2} for each species.

IV. GRAVITATIONAL INTERACTIONS WITH THE COMET CLOUD

A key parameter in discussions of the interactions between clouds and the solar system is the mean time between encounters. For penetrating encounters this quantity is easily estimated from the data given in Secs. II and III. The time τ between penetrating encounters is given by $\tau = \lambda/v$ where λ is the mean free path along the line of sight and v is the rms velocity of the Sun relative to the clouds. For the latter quantity we adopt the quadratic mean of the cloud-cloud velocity dispersion and the solar peculiar velocity. The cloud-cloud velocity dispersion in one dimension has been determined by Clemens (1985) who finds a value of 4 km s^{-1}; the 3-dimensional velocity dispersion of the clouds would therefore be 7 km s^{-1}. For the solar rms velocity, we adopt 18 km s^{-1} (Hut and Tremaine 1985). Adding these two velocities quadratically we find an rms relative velocity of v = 19 km s^{-1}. The mean free path λ may be evaluated from the volume filling factor of the molecular clouds and their size. Thus

$$\lambda = \tfrac{2}{3} f_v^{-1} D \tag{5}$$

where f_v is the volume filling factor for molecular gas and D is the typical cloud diameter. The volume filling factor of molecular gas may be evaluated at the midplane using the estimates presented in Sec. III, that is, $\bar{\rho} = 0.033$ $M_\odot$ pc^{-3} and $\rho_{MC} = 11.6$ $M_\odot$ pc^{-3} which corresponds to the mean density of 175 H_2 cm^{-3} found for a cloud of diameter 40 pc. (The size of 40 pc is used since this is the mean diameter of the clouds weighted by mass.) Using these numbers we find $f_v = 0.28\%$ at the solar circle. The mean free path using these numbers is then $\lambda = 9.5$ kpc, and the mean time between penetrating encounters in the midplane is therefore $\tau = 490$ Myr.

The actual collision time will be ~ 2 times larger (i.e., 1 Gyr) due to the motion of the Sun up out of the galactic midplane and the epicyclic motion of the Sun to a larger galactocentric radius where the cloud density is about 70% lower (cf. Hut and Tremaine 1985).

Given our knowledge of the mean densities within molecular clouds, it is possible to evaluate the relative importance of distant versus penetrating encounters on the orbits of comets at distances of 40,000 AU. Applying the impulse approximation to the perturbation of a cometary orbit with semimajor axis R, it is easily shown that the impact parameter for the solar system relative to the center of mass must satisfy the condition

$$p < (4GM_{GMC}^2 R^3/M_\odot v_{rms}^2)^{1/4} \tag{6}$$

if there is to be a significant perturbation. Here M_{GMC} is the mass of the giant molecular cloud, and v is the rms relative velocity evaluated above. Since the above relation was derived for point masses, it applies explicitly to the case of a distant encounter. Evaluating Eq. (6) for typical cloud masses of 10^5 to 4×10^6 $M_\odot$ and a cometary orbit $<R> = 30{,}000$ AU (assuming a typical semimajor axis of 20,000 to 30,000 AU and $<R^2>^{1/2} = 1.3\ a$), we find that the critical impact parameter is $p < 7$ to 50 pc. Since these are less than the cloud radii for this mass range, it is clear that only penetrating encounters need be considered for disruption of the cometary cloud.

An important factor in evaluating the effects of penetrating encounters is the internal clumping of the clouds. In the limit that all the cloud mass is contained in small clumps, then the disruption of the comet orbits occurs at the maximum rate as given with a half-life equal to the time between encounters with clouds (see Hut and Tremaine 1985); this is because the half-life is independent of the "cloud" mass (see discussion following their Eq. 5). In intermediate cases where the cloud is partially clumped, the half-life is increased by a factor $(f_2)^{-1}$ equal to the mass-weighted surface density in the cloud divided by the critical density $N_{crit} \simeq 3.7 \times 10^{22}$ H cm^{-2} (Hut and Tremaine 1985). Thus f_s is given by

$$f_s = \frac{1}{N_{crit}} \frac{\int N_{H_2}^2 dA}{\int N_{H_2} dA} \qquad (7)$$

where the integral is performed over the surface area A of the cloud. If the mass in the 40 pc GMC is uniformly distributed, then $\bar{N}_{H_2} \simeq 3 \times 10^{22}$ cm^{-2} and $f_s = 0.8$.

The maximum value for the clumping factor is $f_s = 1$, so that the minimum disruption time for the comet cloud is simply the mean time between penetrating encounters, estimated above to be 1 Gyr. For $f_s = 0.8$, the disruption time increases only slightly.

V. NONGRAVITATIONAL EFFECTS OF GMCs

Possible nongravitational effects of molecular cloud encounters with the solar system include an increased luminosity of the Sun due to accretion (McCrea 1975), sweeping back the solar wind due to the ram pressure of the interstellar gas (Begelman and Rees 1976), and deposition of gas and dust in the upper atmosphere. In order for the accretion luminosity to be significant ($\sim 1\%$ $L_\odot$), the cloud density must exceed 10^5 H_2 cm^{-3} and the relative velocity of the Sun through the cloud should be ≤ 10 km s^{-1} (McCrea 1975). Because the mean density in the GMCs is ~ 250 H_2 cm^{-3} and the typical relative velocity is 18 km s^{-1}, this process is unlikely to be significant. In the rare circumstance that the Sun passed into the core region of a cloud where the densities do exceed 10^5 cm^{-3}, then other effects such as the high luminosity of other stars and the higher dust opacities are probably more important.

The choking of the solar wind due to the ram pressure of the dense gas in the cloud is a clear possibility. Begelman and Rees (1976) suggest that the decreased flux of high-energy particles at the Earth would change the chemistry in the upper atmosphere and thus affect the radiative transfer of solar energy. However, a detailed scenario for this has not been worked out.

Finally it is worth noting that the dust column in a GMC is sufficiently large (A_v = 20 mag) that even if a small fraction of this could be retained in either the ecliptic or the upper atmosphere during the 1 Myr passage of the Sun through the cloud, it would surely affect the solar energy flux in the lower atmosphere. For example, if the retention time were as great as 0.5 Myr, then the typical opacity would be A_v = 0.02 mag. Since typical settling times for dust in the upper atmosphere are of order years, it would seem that an additional mechanism to keep the dust from settling must be invoked if this effect were to be significant.

VI. 33 AND 260 MYR PERIODICITIES?

Rampino and Stothers (1984a,b) have recently analyzed geologic data going back typically 500 Myr to 2 Gyr and find two dominant, long-term periodicities, one at 33 Myr which they attribute to the half period of the solar system's oscillation through the galactic plane where it would encounter massive molecular clouds. For the second periodicity at 260 Myr, they have no explanation but they point out that it is similar to the period of revolution for the Sun about the center of the Galaxy.

The results presented in the previous sections immediately rule out the possibility of the 33 Myr period having anything to do with the passage of the Sun near or inside molecular clouds. In Sec. III we found that the mean encounter time for the Sun with typical GMCs was 1 Gyr and thus a 33 Myr period could not show up in record going back just one or two collision times. In addition, Thaddeus and Chanan (1985) have pointed out that the amplitude of the solar motion relative to the molecular cloud layer is not sufficiently large to produce a strong modulation in the local number density of molecular clouds during its cyclic motion in Z. Based upon the present solar Z motion, they estimate that the Sun will reach a maximum height of 75 pc which is almost exactly equal to the half width of the molecular cloud layer as given in Sec. III. Thus the modulation in the encounter rate would be 50% at most.

A similar argument can be made with respect to the longer period of 260 Myr. In order for such a period to arise due to the Sun's passage through molecular clouds situated in the spiral arms, these arms must be regularly spaced with just two arms and the molecular clouds must be spaced closely within those arms so that nearly every time the Sun passes through the arm, it encounters a molecular cloud, and when it is outside the arm it does not encounter clouds.

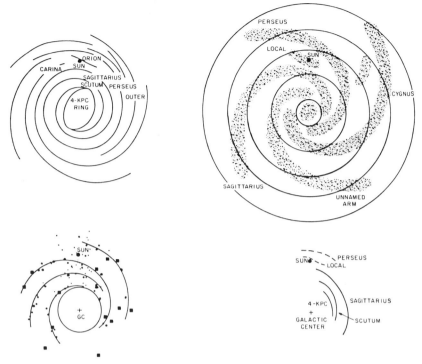

Fig. 6. Comparison of four schematic models proposed for the spiral arms in our Galaxy. *Upper left:* model proposed by Simonson (1976) for 21 cm data; *upper right:* model of Blitz (1982) based on 21 cm data primarily in the outer Galaxy; *lower left:* model of Georgelin and Georgelin (1976) based on optical and H II regions; and *lower right:* that of Cohen et al. (1980) to model the CO.

In Fig. 6 we show four separate models for the spiral structure in our Galaxy as derived from 21 cm, radio H II region, and molecular-line data. The generally poor agreement between the sketches strongly suggests that there is no simple two-arm pattern in our Galaxy. In the local region of the Galaxy three-arm segments have long been known from studies of optical and radio H II regions. These are referred to as the inner or the local arm, and the Perseus or outer arm. The local arm is sometimes classified as a galactic spur, yet it contains the massive molecular clouds of Orion and Taurus and its total mass is entirely comparable with that of the Perseus arm segment. The spacing of these arms is such that the Sun would encounter them on a time scale of 50 to 100 Myr, not on a 250 Myr period.

Aside from the problem that the arm structure is much more complex than a simple two-arm spiral, molecular-line data from the inner Galaxy show that possibly a large fraction of the molecular clouds may in fact exist in the inter-arm regions. A recent study of CO clouds in the inner Galaxy by Sanders et al. (1985a) clearly shows a population of hot clouds closely correlated with the positions of H II regions and by inference, the best locations of the spiral arms. However, a second population of colder clouds is seen to spread rather uniformly through the galactic disk. The extent to which the similar bimodal distribution pertains at the solar circle is not yet determined; nevertheless, if there did exist a large percentage of the clouds outside the arms, then no strong modulation at the arm passage would be seen.

In addition to all the above problems in accounting for a 250 Myr periodic effect, the spiral arms or more precisely, the molecular clouds within the arms, cannot be viewed as a nearly opaque picket fence through which the Sun must pass. The typical spacing between clouds, even within the arms is 5 to 10 times their typical diameters. Even in the most ordered picture of spiral structure (see, e.g., Cohen et al. 1980,1985; Dame 1983) where the molecular clouds are lined up like beads on a string in the Sagittarius arm, the mean spacing between the largest clouds typically exceeds 1 kpc as compared with their diameters of $\sim$ 100 pc. The probability of passing through any cloud during a spiral arm passage is therefore $\sim$ 10%.

In summary, then, there are three problems with attributing the 260 Myr period to encounters of the solar system with molecular clouds in a spiral pattern: (1) the clouds are not highly confined to the arms; (2) even in the arms the clouds are not sufficiently packed to insure that the Sun will pass through at least one cloud on every arm passage; and (3) the spacing between the armlike structures is in fact more like 100 Myr than 260 Myr.

Acknowledgments. It is a pleasure to acknowledge a helpful discussion and reading of this manuscript by S. Tremaine. Partial support for this research is provided by a grant from the National Science Foundation (NS), and through the Jet Propulsion Laboratory, California Institute of Technology, under contract with the National Aeronautics and Space Administration (DS).

INTERSTELLAR CLOUDS NEAR THE SUN

PRISCILLA C. FRISCH and DONALD G. YORK
The University of Chicago

The region of space near the Sun is relatively free of interstellar material. Stars within the nearest 75 to 100 pc of the Sun seldom show foreground neutral hydrogen column densities in excess of 5×10^{19} cm^{-2}. The motion of the solar system is carrying it away from an extended low-density region relatively free of neutral interstellar material, and towards a region of the sky interlaced with interstellar clouds. Several independent lines of data indicate that the nearest interstellar material is low density ($n \lesssim 0.1$ cm^{-3}), warm ($T \sim 10^4$ K) and moving both with respect to the Sun and the local standard of rest. This nearest gas flows through the solar system and resonantly scatters solar H I (Lyman α) and He I (λ584) radiation in weak emission features. This weak emission is alternatively termed backscattered emission and the local interstellar wind. Optical interstellar absorption lines in stars within 100 pc of the Sun show evidence of an analogous flow, although the flow vector differs somewhat from the backscatter vector. The flow may be driven by a relic superbubble (supernova remnants and/or stellar winds) centered on the Scorpius-Centaurus association. Either a separate supernova remnant, or the same relic superbubble, has expanded to the general vicinity of the solar system where its effects are felt as soft X-ray emission and shock-front activity. From observations of more distant interstellar material, we surmise that this low-density flow probably contains higher-density cloud cores which, if they encounter the solar system, may perturb the chemical equilibrium of the terrestrial oxidizing atmosphere, quench the ionosphere and confine the solar wind to within 1 AU of the Sun.

I. INTRODUCTION

The solar system is presently located in a region of the Galaxy relatively free of interstellar gas and dust. Distant giant molecular clouds have spatial densities on the order of $\gtrsim 1000$ cm^{-3} and typical diffuse clouds have densities ~ 10 to 100 cm^{-3}. In the very local interstellar gas, however, average densities $\lesssim 0.1$ cm^{-3} are encountered. As an example of the densities involved, a cubic centimeter of sea-level atmosphere spread out over the nearest 50 pc (that is, a column 1 cm$^2 \times 50$ pc) would have the same spatial density as nearby interstellar gas. The low-density region around the Sun apparently arises because most interstellar material (ISM) is found near regions of star formation which does not occur in the nearest 100 pc. The absence of nearby star formation, in turn, may be an artifact of local spiral structure, or possibly related to the kinematic properties of Gould's belt stars (see, e.g., Frogel and Stothers 1977). The journey of the solar system through a low-density environment may change, however, since we are heading for a region of space containing a larger average density of matter than we have experienced over the past several million years. An encounter with a relatively dense interstellar cloud may possibly alter the solar system environment, and influence the terrestrial atmosphere and climate. If planetary climates prove to be fragile, encounters with dense interstellar clouds may significantly perturb the equilibrium of our terrestrial atmosphere and generally be a factor in the evolution of planetary atmospheres. (Should intelligent life prove to be found only on planet Earth, rather than assuming that our uniqueness arises from a fatal flaw in intelligence, perhaps we can blame our solitude on capricious galactic weather patterns.)

This chapter is organized as follows: in Sec. II, the distribution and characteristics of interstellar dust and gas near the Sun are described. Interstellar wind flowing through the solar system is discussed in Sec. III, and evidence for a flow of more distant interstellar material past the solar system is presented in Sec. IV. In Sec. V, possible effects of an encounter between the solar system and a cloud core embedded in this low-density flow are discussed. Section VI contains a summary of this chapter.

II. INTERSTELLAR MATERIAL WITHIN THE NEAREST 100 pc

A. Nearby Dust and Gas

The overall distribution of interstellar material in the solar neighborhood ($d \lesssim 500$ pc) appears to be dominated by material associated with the Scorpius-Centaurus, II Perseus, I Orion, I Lac, II Mon and IC2602 associations of massive early-type stars of ages < 15 Myr. These associations form an apparent band or ring on the sky known as Gould's belt (Stothers and Frogel 1974; Frogel and Stothers 1977; Goulet 1984). The interior of this ring contains most of the interstellar material within 100 pc of the Sun, and contains relatively little interstellar gas or dust. The existence of this region of low average density, or hole, in interstellar material around the Sun is widely accepted. The hole is seen in both the dust and gas distributions and extends to several hundred pc from the Sun in the third quadrant ($\ell = 180°$ to $\ell = 270°$)

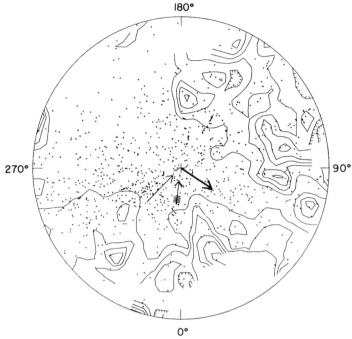

Fig. 1. Color excess, $E(B - V)$, distribution in the galactic plane from Lucke (1978). The radius of the diagram is 500 pc. The Sun is at the center of the figure and the galactic center, corresponding to $\ell = 0°$, at the bottom. The contour interval is $E(B - V) = 0.1$ mag. Hatch marks represent depressions in the contours, while dots are the positions of the OB stars used in the calculation. Three vectors are shown in this figure: the plain arrow is the LISW velocity in the LSR; the feathered arrow is the wind vector derived from interstellar line data (see text) with respect to the LSR and the bold-faced arrow is the vector of solar motion in the LSR.

(FitzGerald 1970; Stothers and Frogel 1974; Lucke 1978; Bohlin et al. 1978; Bruhweiler 1982; Perry and Johnston 1982; Perry et al. 1983; Frisch and York 1983: Paresce 1984).

The hole in nearby interstellar dust is illustrated by Fig. 1 where the color excesses $E(B - V)$ (or amount of reddening of starlight by interstellar dust grains) for O and B stars within 500 pc of the Sun are plotted. In this figure, contours of $E(B - V)$ (units of magnitude), as derived from UBV observations of ~ 2000 stars are shown. The minimum contour $E(B - V) = 0.1$ mag corresponds to $A_V = 0.3$ mag (using the standard ratio $A_V/E(B - V) = 3$). Typical errors on $E(B - V) \sim 0.1$ mag are about 30%. The minimum contour $E(B - V) = 0.1$ mag corresponds to $N(H) = 5 \times 10^{20}$ cm^{-2} using the ratio $N(H)/E(B - V) = 5 \times 10^{21}$ cm^{-2}/mag from Bohlin et al. (1978). According to Fig. 1, the nearest interstellar cloud complexes are in the galactic longitude intervals 300° to 30°, corresponding to the Scorpius-Centaurus region, and 150° to 200°, corresponding to the Perseus-Pleiades-Taurus region. The solar

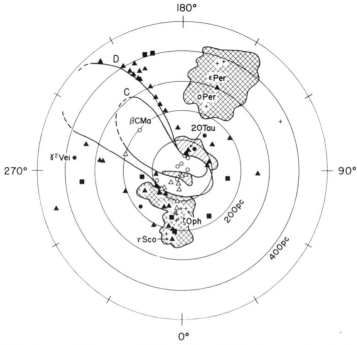

Fig. 2. Nearby neutral hydrogen distribution as viewed looking down on the galactic plane. Contours C and D correspond to $N(H) = 0.5$ and 5×10^{19} cm^{-2}, respectively. The details of this plot are discussed in Frisch and York (1983).

velocity vector with respect to nearby stars of 20 km s^{-1} towards $\ell = 57°$, $b = +20°$ (Allen 1973) is shown plotted as the bold-faced arrow on Fig. 1. It is immediately evident that the motion of the Sun is carrying the solar system away from a region of space with low ambient interstellar density and towards a region with much higher average density of interstellar material.

Observations of neutral and molecular hydrogen gas show the same overall distribution of nearby interstellar gas as seen for dust. Observations of neutral interstellar gas made by the *Copernicus* and *IUE* satellites are summarized in Fig. 2. Selection effects influence the data displayed in this figure because these satellite data are derived almost entirely from observations of bright $V < 5$ mag stars (*Copernicus* was unable to observe faint stars because of guidance limitations), biasing these data against regions of high density. Also, the sensitivity limits between Fig. 2 and Fig. 1 data are different because the ultraviolet absorption lines (upon which Fig. 2 is based) are more sensitive to low column density material than the Fig. 1 $E(B - V)$ measurements. However, despite the selection effects and difference in sensitivity to low-density material, the dust and gas distributions shown in the two figures are roughly similar, with the Sun located in a low-density region of space where $n_H \lesssim 0.1$ cm^{-3}.

Ultraviolet observations of gas in front of the nearest stars, $d < 20$ pc, indicate that the ambient low-density neutral gas is warm ($T \sim 10^4$ K) and may be partially ionized, although inhomogeneities may be present (see, e.g., Landsman et al. 1984; Bruhweiler 1984; York and Frisch 1984; Landsman et al. 1986). This low-density warm gas is formed where diffuse cloud gas is subject to ultraviolet and X-ray ionizing photons (McKee and Ostriker 1977). Larger clouds embedded in supernova remnants will have cool cores ($n > 50$ cm^{-2}, $T \lesssim 100$ K; see Sec. IV.B) with warm neutral and ionized interface gas. Another place where the warm interface gas will be found is the interior of superbubble H I shells that are filled with hot, low-density X-ray emitting gas ($n \sim 5 \times 10^{-3}$ cm^{-3}, $T \sim 10^6$ K; see Secs. II.B and II.C). (Another possible type of interface gas that produces O VI absorption lines is low-density hot gas [$n \sim 10^{-3}$, $T \gtrsim 10^5$ K; Jenkins 1984].)

There is evidence that in the general direction of the galactic center more nearby interstellar material, including dust, is present than in the opposite direction. Observations of the polarization of starlight by interstellar dust grains for $d \lesssim 35$ pc stars show a large patch of dust covering 30° to 60° of the sky, centered at $\ell \sim 0°$ and $b \sim -20°$, with the nearest dust associated with this patch within 5 pc (Tinbergen 1982); this patch is the small hatched area in Fig. 2. The degree of polarization caused by this dust, $P \sim 0.03\%$, corresponds to a visual extinction of $A_V \sim 0.01$ mag, or $N(H) \sim 1.7 \times 10^{19}$ cm^{-2}, in rough agreement with the amount of H I at this distance in Fig. 2. Light from many stars behind this nearby dust patch is unpolarized, so the dust cloud is probably clumpy; in addition, the magnetic field thought to align the dust grains may change in direction (see Sec. IV.A below). Since some of the nearby dust may be associated with ionized gas rather than H I, the gas-to-dust ratio may not be constant (see, e.g., Bohlin et al. 1978).

The gas associated with this dust patch should be detectable as one component of the interstellar gas seen in front of the 100 pc distant star λ Sco which is behind the patch. York (1983) reports a warm H I cloud with $v_{LSR} = -32$ km s^{-1} in front of λ Sco. This velocity is near the velocity expected for nearby interstellar material in this direction (see Sec. IV.A). This component has $N(H\ I) = 1.7 \times 10^{19}$ cm^{-2} and $T \sim 10^4$ K, in rough agreement with the gas column density through the dust patch and the gas temperature in front of nearby stars. The density $n_H > 2$ cm^{-3} determined in this -32 km s^{-1} component suggests that the gas cloud associated with the patch is < 2.7 pc thick and clumpy.

The motion of the Sun with respect to this low-density extended region produces the interesting result that the Sun has probably been in a region of low average interstellar density for several million years. The solar system motion shown in Fig. 1 indicates that the solar system is leaving a region of low average interstellar dust and gas density and entering a region of space with a higher average density of interstellar material. The interstellar material distribution of Figs. 1 and 2 is probably representative of only the last few million years both because of the physical evolution of interstellar clouds and because the overall cloud distribution is strongly affected by rapid stellar evolution in nearby OB star associations. During the last few million years spent in this low-density region of space, the Sun will have traveled ~ 50 to 100 pc since the solar velocity of 20 km s^{-1} corresponds to 20 pc Myr^{-1}. On time scales longer than a few million years, this picture has little predictability.

B. Scorpius-Centaurus or Loop I Bubble

The positions of the stars used to make Figs. 1 and 2 are shown projected onto the galactic plane and convey no sense of the three-dimensional structure of nearby interstellar material. The geometry of this material indicates that it is strongly dominated by the evolution of stars in nearby OB associations. In the fourth quadrant ($\ell = 270°$ to $\ell = 0°$), most nearby interstellar gas ($d \lesssim 100$ pc) appears to be associated with an incomplete shell-like or bubble feature called Loop I observed most easily at high galactic latitudes, $b > 20°$. This bubble structure appears as enhanced 820 Mhz radio continuum and H I 21-cm emission, and causes the polarization of starlight by aligned dust grains (Mathewson and Ford 1970; Berkhuijsen 1972; Weaver 1979). Estimates of the distance of this shell suggest that it is centered about 130 pc from the Sun near the position $\ell = 329°$, $b = +17°.5$, and has a radius 90 to 115 pc (Berkhuijsen 1972). The North Polar Spur region of intense radio continuum emission is a segment of Loop I. The interior of the shell is a source of excess soft X-ray emission in the 0.4 to 1.0 keV range arising from low-density hot gas ($n \sim 10^{-3}$, $T \sim 10^6$ K; McCammon et al. 1983). Loop I is concentric with the 170 pc distant Sco-Cen association (Weaver 1979). Loop I defines a large cavity in the interstellar material formed by intense evolutionary activity in young clusters in the Sco-Cen association. The effect of Loop I is seen as a hole in the interstellar gas distribution in the fourth quadrant in Fig. 1. This feature appears to dominate the distribution of nearby interstellar material (see, e.g., York and Frisch 1984 and references therein). Similar giant bubbles are found in the Eridanus-Orion region (Cowie et al. 1979; Nousek et al. 1982; Reynolds and Ogden 1979) and Cygnus region (Cash et al. 1980), and may be found through X-ray emission or high-velocity gas components (Cowie et al. 1981b).

The Loop I bubble is 0.1 to 2 Myr old, with the age dependent on the scenario assumed to cause the superbubble. One set of postulates assumes at least two events, with the first event either a supernova explosion (Iwan 1980) or intense stellar wind and supernova activity in the young clusters in the Sco-Cen association (Weaver 1979; Bruhweiler et al. 1980; Kafatos et al. 1980). The first event(s) occurred ~ 2 to 4 Myr ago, while the last supernova explosion which reheated the remnant interior to current X-ray emitting temperatures occurred ~ 0.2 Myr ago (Iwan 1980). A second view is that Loop I expanded asymmetrically with the most rapid expansion occurring into the vicinity of the Sun where ambient gas density is low (Frisch 1981). Other models consider the present-day X-ray emission and the shell radius. Based on a rough radius of 90 pc for the Loop I supernova remnant and the soft X-ray emission, Cowie et al. (1981a) model Loop I (or the North Polar Spur) as a middle-aged remnant about 0.2 Myr old, with interior clouds still evaporating.

C. Local Supernova Remnant

There may be, in addition to the Loop I superbubble, a separate supernova remnant immediately around the Sun. Very soft X-ray emission (< 0.3 keV) is also observed throughout the sky. Since an optical depth of 1 is achieved by $N(H) = 5 \times 10^{19}$ cm^{-2} at 0.2 keV, most of this emission must arise relatively close to the Sun (Sanders et al. 1977), although soft X-ray brightening at the galactic poles suggests some of the emission is nonlocal (McCammon 1984). This ambient soft X-ray emission is alternatively explained as arising from $n \sim 5 \times 10^{-3}$ cm^{-3}, $T \sim 10^6$ K gas caused by a separate $\sim 5 \times 10^{50}$ erg supernova explosion about 0.1 Myr ago centered near the Sun (Sanders et al. 1977; Cox and Anderson 1982; Edgar and Cox 1984) or the residual emission from the cooling of the old superbubble formed by a 10^{52} erg explosion(s) ~ 4 Myr ago (Innes and Hartquist 1984). This existence of a separate remnant around the Sun appears to be supported by observations of excess γ-ray emission from radioactive ^{26}Al in the general direction of the galactic center (Mahoney et al. 1982). This emission may arise from a past supernova event at a distance of 14 pc from the Sun (Clayton 1984; see chapter by Clayton et al.).

III. LOCAL INTERSTELLAR WIND

Low-density interstellar gas is found within the solar system. In the 1960s, terrestrial geocoronal Lyman α emission observations revealed an additional weak asymmetrical component of apparent extraterrestrial origin (Morton and Purcell 1962). Subsequent observations found this asymmetry to be due to an all-sky Lyman α glow caused by solar photons scattering resonantly from low-density neutral gas flowing through the solar system, with the glow showing a broad intensity maximum in the upwind direction (the direction from which the gas is flowing) and a minimum in the opposite downwind direction. The flow has a bulk motion with respect to both the solar system and the local standard of rest (LSR) suggesting that it is an interstellar cloud flowing through the solar system. The interstellar origin of the flow was confirmed by the discovery of a weak sky-glow from the He I λ584 line, which showed an intensity maximum in the downwind direction, such as predicted for the interaction between interstellar He I and the solar system. This flow of gas through the solar system has been dubbed the local interstellar wind

TABLE I
Local Interstellar Wind[a]

	H I	He I
n(H I or He I)$_\infty$ [atoms cm^{-3}]	0.02–0.14	0.008–0.023
⇒n(H)$_\infty$ [atoms cm^{-3}][b]	0.02–0.15	0.12–0.34
T [K]	11 ± 4 × 10^3	15 ± 9 × 10^3
Heliocentric[c]:		
$\quad |V_{HC}|$ [km s^{-1}]	19–27	20–30
$\quad$ downwind α, δ [deg]	—	72, +18 (±5)
$\quad$ downwind ℓ, b [deg]	—	183, −17
$\quad$ upwind α, δ [deg]	263, −20 (±5)	—
$\quad$ upwind ℓ, b [deg]	6, +7	—
LSR[d]:		
$\quad |V_{LSR}|$ [km s^{-1}]	17–22	17–23
$\quad$ downwind α, δ [deg]	—	∼ 32, +63
$\quad$ downwind ℓ, b [deg]	—	∼ 134, 1
$\quad$ upwind α, δ [deg]	∼ 240, −71	—
$\quad$ upwind ℓ, b [deg]	∼ 317, −14	—
Upwind cavity boundary[e] [AU from Sun]	2–6	∼ 0.5
Intensity maximum		
$\quad$ upwind [Ra][f]	∼ 600–800	—
$\quad$ downwind [Ra]	∼ 200–400	12
H I or He I lifetime at 1 AU [s]	2.8 × 10^6	7–16 × 10^6
$\langle\mu\rangle$[g]	∼ 0.7–0.9	∼ 0

[a]H I values are derived from Lyman α observations; He I values are derived from λ584 observations. Values given for infinity mean outside of the influence of the solar system and are taken from the following references: Ajello et al. (1979); Adams and Frisch (1977); Paresce et al. (1974); Meier (1977); Clarke et al. (1984); Wu and Judge (1980); Holzer (1977); Bertaux et al. (1972); Thomas (1971); Meier (1977); Weller and Meier (1981); Dalaudier et al. (1983); Freeman et al. (1980).

[b]Values for the total hydrogen spatial density were derived assuming a cosmic helium abundance [He/H] = 0.068 (Zombeck 1982).

[c]With respect to the Sun, or the apparent velocity; α, δ are equatorial coordinates. This position lies near α Cen.

[d]With respect to the LSR, or the true velocity; α, δ are equatorial coordinates. $V_{LSR} = V_{HC} + V_\odot$ where V_{LSR} and $V_\odot$ are the wind and Sun vectors in the LSR and V_{HC} is the heliocentric wind vector.

[e]The distance at which n(H I)/n(H) equals e^{-1} times the value at infinity. This distance depends on the variable solar particle and photon flux. Note that the region of maximum emission occurs within this boundary because of the r^{-2} dependence of the solar radiation field.

[f]1 Rayleigh (Ra) = $10^6/4\pi$ photons cm^{-2} s^{-1} ster^{-1}.

[g]The exact value of $\langle\mu\rangle$, the ratio of radiation pressure to gravitational attraction, depends on the solar cycle.

(LISW). The properties of the LISW have been discussed and reviewed, for example, by Tinsley (1971), Fahr (1974), Meier (1977,1981), Blum et al. (1975), Holzer (1977), Bertaux (1984), Ajello and Thomas (1985) and Lallement et al. (1984).

A range of LISM parameters derived from LISW observations of both H I and He I are summarized in Table I. From these values, we pick best values of $T \sim 12{,}000$ K, $n \sim 0.1$ cm^{-3} and $V_{HC} \sim -25$ km s^{-1} from the heliocentric (HC) upwind direction of $\ell = 3°$, $b = +17°$, corresponding to an upwind velocity vector in the LSR of $V_{LSR} \sim -20$ km s^{-1} from $\ell = 314°$, $b = -1°$. Generally, the models of the H I and He I glows differ because neither the atomic trajectories nor the loss mechanisms for these two neutral atoms within the solar system are the same. Consequently, H I and He I models are discussed separately below. Generally, the LISM parameters are derived from observations by fitting the backscatter data with models of the trajectories and ionization of H I and He I in the solar system. The free parameters in these models include the density, temperature and velocity of the ambient interstellar gas before interacting with the solar system (sometimes denoted as the values at infinity). In Table I, values for the upwind velocity vector are summarized in terms of two coordinate systems. One system represents the upwind direction which we observe (corresponding to the heliocentric velocities, but also known as the *apparent* upwind direction) and the other is the upwind direction in the LSR (also known as the *true* upwind direction).

There may be disagreement between the LISM parameters derived from H I versus He I data. For instance, the average values of n_∞(H I) and n_∞(He I) in Table I indicate a ratio for the nearby interstellar medium, n(H I)$_\infty/n$(He I)$_\infty$ ~ 5, that is below the cosmic ratio of 15 (Zombeck 1982). This difference, if real, may indicate that the inflowing interstellar hydrogen gas is partially ionized before entering the solar system. However, more recently Shemansky et al. (1984) report that Pioneer 10 and Voyager 2 observations of H I and He I in the outer solar system yield densities consistent with each other, which would then indicate that the models of He I and H I backscattering in the inner solar system are incomplete.

Since both the models and observations of the H I and He I backscattering differ significantly, these LISW components are discussed separately below. Also discussed are possible extrasolar system contributions to these emissions.

A. Neutral Hydrogen

The backscattered H I Lyman α is observed as an all-sky emission brighter in the upwind than downwind direction. Ranges for n, T and V derived from H I observations are presented in Table I. The temperature, density and velocity of the LISW are parameters which are generally derived simultaneously from model fits to broadband all-sky surveys (Ajello et al. 1979; Paresce et al. 1974; Meier 1977; Wu and Judge 1980; Holzer 1977; Bertaux

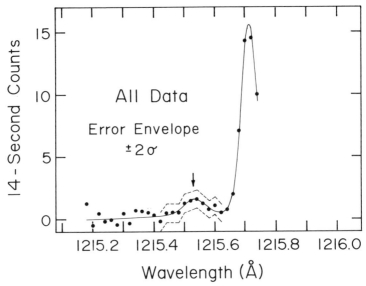

Fig. 3. *Copernicus* observations of Lyman α backscattered radiation from the local interstellar wind (LISW) by Adams and Frisch (1977). These data show the geocoronal emission plus the short wavelength LISW emission. The strong geocoronal line is from the geocorona outside of the 750 km altitude of the satellite.

1984; Thomas 1971; Weller and Meier 1981; Dalaudier et al. 1983), although some high-resolution observations have been made (Adams and Frisch 1977; Clarke et al. 1984). Most of these H I observations were made from Earth orbits and during the data analysis it is necessary to simultaneously extract the weak LISW feature and strong geocoronal emission features. Two sets of high-resolution observations have been made, however, using the spectra on board the *Copernicus* and *IUE* satellites. The *Copernicus* observations, made in a position close to the upwind direction, are shown in Fig. 3. These data illustrate the relative weakness of the Lyman α backscattered line with respect to the terrestrial geocorona at the 750 km altitude of the *Copernicus* spacecraft. All other determinations of the LISW velocity are more indirect and involve extrapolating the velocity as one of three variables from models of the all-sky Lyman α or He I λ584 emission. The *Copernicus* and *IUE* sets of high-resolution data were acquired during solar-cycle minimum conditions. To within the errors, they agree and yield a heliocentric (HC) wind velocity of $V_{HC} \sim -25$ km s^{-1} when projected to the heliocentric upwind direction.

The extraction of the parameters of nearby interstellar material requires accurate modeling of the backscattered emission, which in turn requires an understanding of the history of the interstellar gas as it enters the outer solar system. As the interstellar H I enters the solar system, it crosses the poorly understood heliopause region. This region, the boundary between the solar wind and ambient interstellar material, is a series of one or more shock fronts that are probably supersonic (see, e.g., Ripken and Fahr 1983). As the neutral hydrogen crosses the heliopause region, up to 20% of the H I atoms may be ionized by charge exchange (Keller et al. 1981). Within the heliopause, slow H I atoms are destroyed by charge exchange with fast solar wind protons, and, less importantly, by ionization by solar ultraviolet photons. The H I trajectories within the solar system are a function of a parameter μ, the ratio of the solar radiation pressure to the gravitational attraction. Generally $\mu \sim 1$ with values ranging from ~ 0.5 during periods of low solar activity to ~ 1.5 during periods of high solar activity. The upwind H I ionization cavity boundary is 4 to 6 AU from the Sun during periods of high solar activity and ~ 1 to 2 AU during periods of low solar activity; the volume emissivity, or source function, becomes a maximum inside of the ionization cavity boundary at 2 to 3 AU during solar maximum and $\lesssim 1$ AU during solar minimum (Blum and Fahr 1970; Thomas 1971; Thomas and Krassa 1974; Fahr 1974). Hence, the backscattered emission we observe originates quite close to the Sun. In the upwind Ophiuchus direction, the emission has a maximum intensity of ~ 600 to 800 Ra (1 Rayleigh = $10^6/4\pi$ photons cm^{-2} s^{-1} ster^{-1}). In the opposite downwind direction a minimum emission strength of ~ 170 Ra is found. The upwind Lyman α maximum region is closer to the Sun than the downwind minimum region since the maximum shows a parallax $\sim 70°$ over a seven month period, while the minimum varies by only 15° (Thomas and Krassa 1971). The shape of the H I Lyman α emission profile is distorted from the incident Doppler distribution of the particles by multiple scattering of the photons, velocity-dependent ionization effects, and by some secondary recombination of fast protons (see, e.g., Keller et al. 1981).

The H I Lyman α sky-glow models must incorporate a number of secondary corrections to accurately model the emission. As each of these corrections is incorporated into the models, the fit between model and observations improves. Examples of the corrections that improve the fit between models and observations of the LISW are anisotropies in the solar Lyman α radiation field and solar wind proton flux (Ajello et al. 1979), ionization of LISW neutral atoms in the heliopause region (Ripken and Fahr 1983) and multiple scattering of solar photons (Keller et al. 1981). Observations of an optically thin H I line would simplify the modeling of the backscattered radiation. The Lyman β line is a candidate because it has a smaller optical depth than Lyman α (a factor of 5) and will not be multiply scattered (Keller et al. 1981). High-resolution observations of the $\lambda 1025$ Lyman β emission should yield a more accurate model of the local interstellar wind (Shemansky et al. 1984).

B. Neutral Helium

Neutral helium, the second most abundant species in interstellar gas produces a weak $\lambda 584$ backscattered emission with different characteristics than the Lyman α line. Table I summarizes results derived from He I $\lambda 584$ observations. He I in the solar system is essentially collisionless so most He I is lost to photoionization within the Earth's orbit. The continuum emission from the Sun at λ584 is much weaker than at λ1216 so μ for He I is negligibly small. The hyperbolic He I orbits cross the downwind axis to create a gravitationally focused enhanced density region, or cone, downwind of the Sun. The λ584 emission pattern shows a narrow peak $\sim$ 9 Ra brightness downwind, with a broad $\sim$ 1 Ra minimum upwind. In the cone, the density is enhanced roughly by a factor of five. The downwind radius of the He I ionization cavity is roughly 0.5 AU with respect to the Sun, and relatively independent of the solar cycle so that the Earth passes through the tail of the He I cone density enhancement.

The size and location of the He I downwind density cusp is sensitive to the temperature and velocity vector of the LISW. The position of the well-defined downwind He I maximum is a more accurate diagnostic for the wind vector direction than is the broad upwind Lyman α maximum because the shape of the emissivity distribution is sensitive to the gas temperature and the cloud-Sun velocity vector as well as the ionization rate. Higher gas temperatures will broaden the λ584 maximum while larger velocities move it downwind.

C. Contamination by Extrasolar System Sources

Contamination of the backscattered emission by extrasolar sources can generally be dismissed because, for an ambient density of $n_{H\,I} \sim 0.1$ cm^{-3}, pathlengths of >1.5 pc are optically thick at Lyman α and λ584. However, some galactic continuum sources may contribute particularly in the direction of ionized material. Ajello et al. (1979) observed galactic emission of $\sim$ 20 R from the direction $\alpha = 200°$, $\delta = -62°$ with a FWHM $\sim$ 20°. This continuum source was confirmed by simultaneous observations in the adjacent 1048 Å channel. This peak is symmetric about the galactic equator and corresponds to a portion of the Milky Way at $\ell \sim 305°$, $b \sim -1°$ midway on the sky between α Cru and β Cen. This region is also a bright source of diffuse galactic ultraviolet light according to Henry (1977). Ajello et al. point out that this secondary ultraviolet source may also be related to the south ecliptic pole since this look direction corresponded to the closest observation of the spacecraft to the direction of the south ecliptic pole. However, Shemansky et al. (1984) also report Pioneer 10 measurements of a galactic contribution to the Lyman α emission.

D. Cause of Local Interstellar Gas Flow

When the apparent upwind direction given above is converted to a true upwind direction by the vector subtraction of the solar motion with respect to the LSR, the upwind direction in the LSR is found to be $\ell \sim 314°$, $b \sim -1°$ using He I values. This direction is $\sim$ 25° from the center of the Loop I H I and radio continuum bubble discussed above. One possible cause of the LISW is that the near side of the Loop I bubble is driving material past the solar system (Frisch 1981; Crutcher 1982), and the LISW may be associated with the shell. The nearby Tinbergen dust patch (Sec. II.A) may also be associated with this shell. This argument is supported by observations of interstellar clouds near the Sun. We now review these observations of interstellar absorption-line components in nearby stars ($d < 100$ pc) and find that they support the idea that the backscattered radiation arises from our local sampling of a flow of material past the solar system.

IV. FLOW OF MATERIAL TOWARDS THE SOLAR SYSTEM

A. Flow Vector

There is a tendency for nearby ($d \lesssim 100$ pc) interstellar clouds to show interstellar line features at velocities indicative of a flow of material past the Sun. In addition, some of the more distant of these stars will also have components at ~ 0 km s^{-1} LSR. The coincidence or near coincidence of interstellar absorption-line velocities with the local flow velocity has been discussed in several papers (see, e.g., Adams and Frisch 1977; Frisch 1981; Crutcher 1982). The interstellar absorption lines appear to show a somewhat different wind vector than the LISW. Crutcher first made a least squares fit to Ti II absorption features found in seven nearby ($d \lesssim 100$ pc) stars and found they could be described by a common velocity vector. Using a larger set of 21 nearby ($d \lesssim 100$ pc) stars, we find that the heliocentric vector $V_{HC} = -27$ km s^{-1} from the direction $\ell = 34,°$ $b = +15°$, corresponding to the LSR vector $V_{LSR} = -12$ km s^{-1}, $\ell = 354°$, $b = +3°$ gives a better fit to the extended set of data (Frisch and York, unpublished). In Fig. 4, the observed heliocentric velocities of interstellar components towards 21 stars within 100 pc of the Sun are plotted against the best-fit flow vector to the interstellar line data. These observed velocities are taken from optical Ti II, Na I and Ca II absorption-line data from Stokes (1978), Hobbs (1969,1974,1978) and Frisch and York (1984, and unpublished). The best-fit flow vector was derived from a least squares fitting procedure using only one velocity component from each interstellar line spectrum. When more than one component is found toward a given star, only one component was used in the fit (the other component is shown plotted as an "x" in Fig. 4). Based on these velocity data shown in this figure, it appears that there is a net flow past the solar system of neutral interstellar gas from a direction which corresponds to the interior of the Sco-Cen bubble. This flow is not perfectly uniform, however, as can be seen by the scatter in the data shown in the figure and additional interstellar components found even in nearby stars.

This flow encompasses the solar system where it is locally manifested as the LISW. These optical lines, however, are not necessarily formed in the warm low-density LISW material. Some of the Na I absorption lines are too

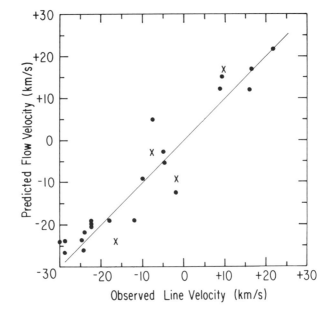

Fig. 4. Velocities of optical interstellar Na^0, Ca$^+$ and Ti$^+$ absorption lines plotted against the predicted flow velocity for nearby ($d <$ 100 pc) stars. The predicted flow velocity is based on the interstellar line flow vector projected to the star sight line (see text). Interstellar lines used to compute the predicted flow velocity are plotted as dots. The predicted flow vector in heliocentric coordinates is $V_{IS} = -27$ km s^{-1} from $\ell = 34°$, $b = +15°$. Where more than one interstellar velocity component is found in a star, only one component was chosen for the fitting procedure. The extra component, which was not used in the fit, is plotted as an "x".

narrow (e.g., towards δ Cyg where FWHM ~ 0.73 km s^{-1}; Blades et al. 1980) to be formed in 10^4 K gas and may arise in cooler clouds embedded in a warmer substrate. A plausible scenario is that there is a general motion of nearby interstellar material past the solar system driven by the expansion of the relic supernova remnant(s) which caused the Sco-Cen bubble. Both the warm partially ionized material around the solar system and the more distant cooler and mostly neutral clouds share this motion. More distant H II regions found within the nearest 100 pc also tend to show a net negative velocity from the Loop I direction (York and Frisch 1984). The difference between the observed upwind directions determined from interstellar lines versus backscatter data may indicate local variations in the flow.

B. Possible Cloud Core

The low-density optically thin material currently flowing through the solar system is too tenuous to have any known effect on the solar wind or terrestrial atmosphere. However, numerous studies have discussed possible effects from an encounter between the solar system and an $n \gtrsim 1000$ cm^{-3} cloud.

Clouds with density clumps $n \sim 1000$ cm^{-3} are common in the interstellar medium. As noted earlier, the Tinbergen polarizing cloud is likely to be clumpy. Such a density clump, or cloud core, may be embedded in the ambient interstellar gas, and the local interstellar wind may be the halo of such a core. A core-halo structure for interstellar clouds is supported observationally by detections of both low-density warm and high-density cool cloud material, and theoretically by the McKee and Ostriker (1977) model for three-phase interstellar material regulated by supernova explosions.

The properties of a nearby cloud core of density $n \sim 1000$ cm^{-3} can be constrained by the fact that material of this type must have a small physical extent, or it would have been detected. The closest candidates for the core material are the Magnani et al. (1985) small CO clouds. These clouds have average sizes of 1.7 pc, average masses of 40 M$_\odot$, and typical extinctions in the range $A_V = 0.4$ to 4.0 mag. They have average volume densities of 35 to 500 cm^{-3}. However, they are probably clumpy with density variations of at least a factor of two over the average volume density since they have structure on scales smaller than the 2′.3 beamwidth. Thus, this type of cloud is a good candidate for a cloud core that could be embedded in the local interstellar wind. Magnani et al. estimate that the CO clouds which they observe are $\sim$ 100 pc distant, although other clouds may be closer. Structures this small have a low probability of being detected in interstellar absorption-line studies which traditionally target bright OB stars as background sources.

In the absence of magnetic fields and in an evolving supernova remnant interior, a $n = 500$ to 1000 cm^{-3} clump would be evaporating and not be in pressure equilibrium with low-density warm gas. Magnetic fields are present, but of unknown strength, in the upwind direction (Tinbergen 1982), and may be expected to contribute to the pressure equilibrium.

V. POSSIBLE EFFECTS OF A CLOUD CORE ON THE SOLAR SYSTEM

The effects of a $n \sim 1000$ cm^{-3} cloud core on the solar system fall into three main categories, determining the distribution of the solar wind, accretion of heavy elements onto the solar surface, and the altered chemistry of the upper atmosphere of the Earth. These effects have long been discussed in the literature by pioneering astronomers who bravely tried to hold the heavens responsible for terrestrial climatic catastrophes (see, e.g., Shapley 1921,1949; Hoyle and Lyttleton 1939). We draw from these previous studies to discuss the possible effects of an encounter between the solar system and a dense core embedded in the low-density interstellar wind. Other possible ways that interstellar clouds may influence the solar system would be if comets are found to be physically associated with interstellar clouds. This possibility is not discussed here.

A. Solar Wind

The solar wind partially shields the inner solar system against both interstellar clouds and low-energy cosmic rays. The magnetosphere of the Earth is also in equilibrium with the solar wind. If either the density of the interstellar cloud or the relative Sun-cloud velocity are large enough, however, the ram pressure of the interstellar cloud collapses the solar wind to within 1 AU of the Sun, leaving the Earth unprotected from the low-energy galactic cosmic-ray flux (Begelman and Rees 1976). Using typical 1 AU values for the solar wind flux of 5 particles cm^{-3} traveling at 400 km s^{-1}, the ram pressure from a 25 km s^{-1} interstellar wind will confine the solar wind to within 1 AU for neutral cloud densities of $\gtrsim$ 1000 cm^{-3}. The solar wind excludes low-energy charged cosmic rays from the inner solar system. For instance, cosmic-ray protons lose about 600 MeV energy traversing to 1 AU during average solar cycle conditions (Urch and Gleeson 1973). In the absence of a solar wind at 1 AU, cosmic rays with energies $\lesssim$ 600 MeV will be much more abundant near Earth. The terrestrial magnetosphere, with properties partly determined by the solar wind flux, will vary with possible consequences for lower atmosphere temperatures (see references in Begelman and Rees 1976).

B. Heavy Element Accretion

Shapley (1921, 1949), Hoyle and Lyttleton (1939), and later Talbot and Newman (1977) discuss the accretion of heavy elements onto the surface of the Sun. Butler et al. (1978) discussed the enrichment of planetary atmospheres by the accretion of interstellar material.

In the original Hoyle and Lyttleton scenario, the release of gravitational energy from interstellar material falling into the Sun raises the radiant energy of the Sun, causing increased precipitation on Earth resulting in the cooling of the terrestrial surface, and therefore ice ages. McCrea (1975) rediscussed Hoyle and Lyttleton's suggestion, and found that a cloud of density $n \sim 10^3$ cm^{-3} and relative Sun-cloud velocity of 25 km s^{-1}, would not enhance the solar radiation field. However, for higher densities or lower velocities (e.g., $V < 10$ km s^{-1} and $n > 10^4$ cm^{-3}) the solar luminosity would change by $\gtrsim 1\%$, and affect the terrestrial climate. McCrea suggests past encounters with such clouds when the Sun traversed a spiral arm in our Galaxy.

In the context of the accretion of interstellar material onto stellar surfaces, Talbot and Newman (1977) considered the alteration of solar surface heavy-element abundances from repeated passages of the Sun through interstellar clouds. They conclude that there is a nonnegligible probability that the Sun has accreted $\gtrsim 10^{-2}$ M$_\odot$ of interstellar material, based on $\gtrsim 100$ encounters with clouds during the lifetime of the Sun. This accreted material should be relatively metal-rich, masking the relative solar abundances at the birth of the solar system and possibly reducing the solar flux through increased line blanketing.

C. Effects on the Upper Atmosphere of the Earth

Atomic hydrogen evaporates from the exosphere at a rate of $\sim 2 \times 10^8$ $cm^{-2}\,s^{-1}$ (Lewis and Prinn 1984). If a $n \sim 1000\,cm^{-3}$ cloud core embedded in the local interstellar wind surrounded the Earth, this H I escape would be overwhelmed by an inflow from the cloud of $\sim 2.5 \times 10^9\,cm^{-2}\,s^{-1}$. This additional hydrogen concentration which would probably be a mixture of molecular and atomic hydrogen, although only a fraction of the oxidizing atmosphere, alters the chemistry of water in the thermosphere. McKay and Thomas (1978) predict that the chief results of this altered chemistry are to destroy the ionospheric F-region through enhanced electron recombination, lower the ozone concentration above 50 km, and therefore lower the temperature and average height of the mesosphere by allowing deeper penetration of ultraviolet radiation. A further effect may be the creation of a permanent worldwide layer of mesospheric ice clouds (i.e., noctilucent clouds). Naturally occurring variations by factors of $\lesssim 2$ in the H_2O content of the mesosphere are known to enhance noctilucent cloud formation, which may cause the cooling of the mesosphere by >15 K and result in a global surface cooling (McKay 1985). The equilibrium within the dry lower stratosphere and across the tropopause are poorly understood. However, one possible result of the enhanced concentration of ice particles in the upper mesosphere would be a slow downward flow of ice particles through the troposphere, nucleating water condensation in the clouds being formed in the cooling troposphere and causing global precipitation.

VI. SUMMARY

The solar system is presently in a region of the Galaxy relatively free of interstellar gas and dust. Stars within the nearest 75 to 100 pc of the Sun seldom show foreground neutral hydrogen column densities in excess of $5 \times 10^{19}\,cm^{-2}$ (Frisch and York 1983). The motion of the Sun is carrying the solar system away from an extended low-density region relatively free of interstellar material, and towards a region of space where neutral interstellar clouds frequently occur. Shapley (1921) first noted a possible relation between interstellar material and the terrestrial climate. He observed that the solar motion carried the solar system away from the Orion region of space (formerly believed to be near the Sun) where dense interstellar nebulosity is found which, in the past, may have affected the Earth's climate. He commented "from the present direction and amount of [Orion stars and solar] motions, we compute that a few million years ago our sun was in the vicinity of the Orion nebula; at its present speed the Sun would require nearly a million years to pass through that particular nebulous region. . . . An 80 percent change [in the solar radiation due to irregular interstellar nebulosity], unless counteracted by concurrent changes in the terrestrial atmosphere, would completely desicate or congeal the surface of the Earth" (Shapley 1921). We now know that the solar system is approaching a region of higher average density than in the recent past, so effects of this sort may occur in the future.

Several independent lines of data show that the interstellar material now around the solar system is low density, $n \lesssim 0.1$ cm^{-3} and warm, $T \sim 10^4$ K. It moves both with respect to the Sun and the local standard of rest showing a general flow pattern. The flow manifests itself in the solar system as low-density wind scattering solar Lyman α and He λ584 photons. The local interstellar wind appears to flow from an upwind direction (in the LSR) near $\ell \sim 314°$, $b \sim -1°$ in comparison to the upwind direction $\ell \sim 354°$, $b \sim +3°$ inferred from interstellar lines. This flow appears to be symptomatic of a net flow of optically thin material past the solar system from this general direction of the sky. The flow appears to be driven by the relic superbubble (formed by successive supernova events or the combined stellar winds from the Sco-Cen association) called Loop I or the Sco-Cen Bubble. This superbubble originated around the Scorpius-Centaurus association $\gtrsim 0.2$ Myr ago. Observations of more distant interstellar material show that this low-density flow may contain high-density cloud cores. Upon encountering the solar system, these cloud cores will suppress the solar wind, lead to the accretion of interstellar material onto the solar surface, and perturb the equilibrium of the terrestrial oxidizing atmosphere enough to alter the climate on Earth. The near side of the Loop I superbubble shell may coincide with the $d \lesssim 5$ pc Tinbergen dust patch, indicating that the solar system may encounter dense interstellar material within 0.2 Myr.

The properties of the interstellar medium near the Sun are consistent with conditions needed to trigger global effects on terrestrial weather. The key elements present are a general flow of material at ~ 25 km s^{-1} relative to the Sun and the existence of small dense condensations ($n_H \gtrsim 10^3$ cm^{-3}) in the flow. While the flow has been detected, and condensations in the flow are known, data are insufficient to predict accurately the frequency of Sun-cloud encounters, either in the past or the future. We do know that during the past 1 Myr the Earth has traveled through a much more rarefied medium than it is heading into now. It is likely that the solar system is heading for condensations within 100 pc and that encounters may occur in the next 0.1 to 1 Myr. Our current knowledge of the distribution of nearby interstellar material do not allow statements to be made in more detail.

Acknowledgments. This research has been supported by a grant from the National Science Foundation and several grants from the National Aeronautics and Space Administration to the University of Chicago.

PART III
Other Galactic Features

THE CHEMICAL COMPOSITION OF COMETS AND POSSIBLE CONTRIBUTION TO PLANET COMPOSITION AND EVOLUTION

J. MAYO GREENBERG
University of Leiden

The theory of cold aggregation of interstellar dust into comets is shown to support the ideas that comets provided not only a substantial part of the Earth's water but also of its complex organic surface material. Comparison of the density of interstellar dust with meteor densities leads to the conclusion that comets are very fluffy bodies whose solid matter occupies no more than 2/5 of the total volume and possibly substantially less. This implies an extremely low comet albedo of ~ 0.05 in the visual even for slightly absorbing individual components.

I. INTRODUCTION

It is well recognized that many comets as well as asteroids have impacted the Earth (Wetherill 1975). The Moon's surface bears obvious evidence for the latter, and the continuous appearance of new comets from the Oort cloud implies a certain probability for comet collisions with the Earth. There has been a good deal of discussion on the subject of the periodic occurrence of showers of comets. Whether or not this theory is ultimately accepted, it has been suggested that in the early stages of the Earth's evolution large numbers of comets added material to the Earth's atmosphere and oceans (Chang 1979). The principal aim in this chapter is to make some estimates both as to the chemical distribution as well as to the amount of such material which an individual comet brought to the Earth. This involves first assuming a chemical and morphological model for comets, and second answering the question of what happens to such bodies when they strike a planet at ~ 50 km s^{-1}. Sections II, III, IV and V provide the basis for the comet model based on aggregated interstellar dust. In Sec. VI, I consider the impact phenomenon and in Sec. VII some speculations on the relevance of comet deposition on the Earth both in its early and later stages of evolution.

II. THE BIRTHPLACE OF COMETS

The basic ingredients of all objects in a star or planetary system are derived from the interstellar medium. How direct the connection between them is depends on the degree by which the interstellar components have been modified or metamorphosed. This dependence is largely a function of the temperature prevailing before, during or after the birth process of the object relative to the volatility of the various interstellar dust components.

Of all the objects of the solar system, the comets are presumed to be the most primitive; i.e., they are believed to represent most closely the presolar-system environment (Whipple 1979). The first question is how cold it was where comets were born. If the comets were born in the primitive solar accretion disk at distances of the order of the outer planets, say between Uranus and Neptune, and we assume that the temperature distributions were as predicted by Cameron (1978b), then the maximum would be about ~ 60 K which is too low to evaporate H_2O ice. However, if turbulent transport plays an important role (Morfill and Völk 1984), then some of the constituents which were evaporated closer to the center including even such refractories as silicates, but certainly H_2O, would have percolated outwards and recondensed, but *not* in their original configuration. In such a case, *only* presolar material may not necessarily exist even where temperatures never exceeded ~ 50 K.

On the other hand, if comets were born at a distance of 10^4 AU, as suggested by Biermann and Michel (1978), then the comet ingredients would be unlikely to have been at temperatures > 16 K (Greenberg 1983). If their birthplace was even farther out, in a gravitationally connected (to the presolar nebula) cloud fragment with its own accretion disk (Biermann 1981; Donn 1976), or, as suggested by Clube and Napier (Clube 1983b; also see their chapter), in dense molecular clouds, the temperatures to be considered are $< \sim 15$ K. In the next section I summarize what is known about the chemical constituents of interstellar dust and consider them in relation to various temperature regimes.

III. INTERSTELLAR DUST COMPONENTS

I restrict myself here to the model of interstellar grains developed by Greenberg and coworkers because it appears well suited to explain the evolutionary characteristics of grains in collapsing and protostellar clouds (Greenberg 1982a). Based on the widely observed properties of the wavelength dependence of the extinction of starlight from the far infrared to the ultraviolet, we conclude that there are at least three populations of particles (Greenberg and Chlewicki 1983):

1. Large particles of sizes $\simeq 0.15$ μm (mean radius) which are responsible for the more or less linear dependence of extinction in the visual wavelengths and which contain the major bulk of the solid matter in space;
2. Small carbon particles (probably molecular and amorphous) responsible for the 2160 Å hump, as well as possibly part of the infrared extinction, whose sizes are $\leqq 0.01$ μm;
3. Small silicate particles of sizes $\leqq 0.01$ μm, responsible for the far ultraviolet extinction.

The structure of the large particles consists of a silicate core of ~ 0.05 μm radius with an inner mantle of complex organic molecules and, in molecular clouds, an outer mantle containing predominantly H_2O (Greenberg and Van de Bult 1984; Van de Bult et al. 1985). Comparison of observed infrared absorption spectra with spectra in the laboratory gives clear evidence that the inner mantle is an (as yet) unspecified complex organic mixture or polymer (Greenberg 1982b; Agarwal et al. 1985), and that the outer mantle contains not only H_2O, but also CO (Lacy et al. 1984), a molecule containing the cyanogen (C≡N) group (Lacy et al. 1984), and two sulfur-containing molecules, OCS and H_2S (Geballe et al. 1985). Evidence also exists for the presence of H_2CO (Van der Zwet et al. 1985).

The basic processes that govern the evolution of the core-mantle grains in molecular clouds are:

(a) accretion of the condensibles which include O, C, N and S;
(b) ultraviolet photoprocessing of the accreted ices at the prevailing grain temperatures $T = 10$ to 15 K; and
(c) grain explosions which replenish molecules in the gas phase.

Considerations of the gas-phase ion molecule reactions plus grain surface interactions and explosions lead to the predominance of H_2O in the outer grain mantles (Tielens and Hagen 1982; d'Hendecourt et al. 1985). It is as a result of the photoprocessing of the H_2O rich mantles that there occurs a buildup of complex organic material which is tough enough and refractory enough for a substantial fraction to survive the destructive processes occurring during the grain's existence in the diffuse clouds. Figure 1 shows a schematic drawing of the structure of core-mantle grains in dense molecular clouds.

Along the way to producing complex organic molecules, the effect of ultraviolet photoprocessing is to modify the molecular composition of the grain mantles. In the laboratory it is seen (see Fig. 2), for example, that CO is converted into CO_2 (Greenberg and d'Hendecourt 1985). Another product of the photoprocessing is the diatomic sulfur molecule S_2 as demonstrated in the

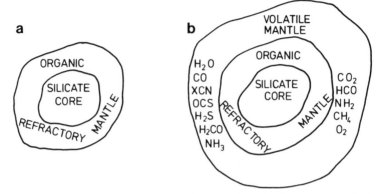

Fig. 1. Structure and chemical composition of core-mantle grains in precometary clouds: (a) diffuse cloud grain; (b) precometary, grain with the list of molecules on the left *observed* and that on the right *inferred*.

TABLE I
Molecules Directly Observed
in Interstellar Grains
and/or Strongly Inferred
from Laboratory Spectra and Theories
of Grain Mantle Evolution

Molecule	Comment[a]		
H_2O	O		M
NH_3	O	I	M
H_2S	O		M
CO	O		M
H_2CO	O	I	M
$X-N\equiv C$	O		M
OCS	O	I	M
CO_2		I	M
CH_4		I	M
S_2		I	M
Complex organic	O	I	M
"Silicate"	O		C,B
"Carbon"	(O)		B

[a] O ≡ observed; M ≡ mantle; B ≡ small bare; I ≡ inferred; C ≡ core; parentheses indicate uncertain observations.

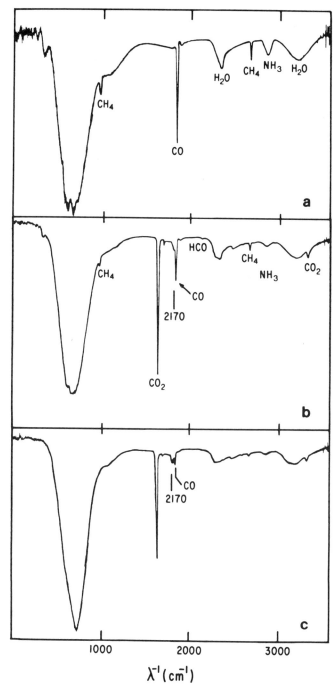

Fig. 2. Spectra showing the production of CO_2 along with other molecules by ultraviolet irradiation of a dirty ice ($H_2O:CO:NH_3:CH_4$ = 6:2:1:1) grain mixture: (a) unirradiated; (b) photolyzed (irradiation); (c) photolyzed and warmed up to 95 K. The feature at 2170 cm^{-1} is due to a cyanogen containing molecule.

TABLE II
Mass and Volume Fractional Distributions of the Principal Chemical Constituents of Precometary Dust

Component	Mass Fraction	Volume Fraction
Silicates	0.14 + (0.06)[a]	0.06 + (0.03)[a]
Carbon	(0.06)[a]	(0.03)[a]
Organic residue	0.19	0.21
H_2O	0.20	0.28
CO	0.03	0.04
CO_2	0.04	0.05
Other molecules and radicals (H_2CO, NH_3, CH_4, XCN, XS, H_2S, HCO ...)	0.27	0.31

[a] Values in parentheses correspond to very small particles ($a \lesssim 0.01$ μm).

laboratory from irradiation of low temperature ($T \simeq 10$ K) dirty ices containing H_2S (Grim et al. 1986).

Table I gives a list of molecules which may be shown from observations and laboratory spectra to be definitely present in interstellar grains. The relative proportions of the various chemical constituents are estimated in Table II for precometary grains in the latest stage of cloud contraction.

IV. VOLATILITY OF MAJOR DUST CONSTITUENTS

Some of the more volatile constituents of grains are CO, CO_2, NH_3, H_2O, CH_4. Expressing their vapor pressures in the form

$$\log_{10} P = -\frac{A}{T} + B \log_{10} T + C \quad (1)$$

(where P is in mm Hg), we can derive the values of A, B and C shown in Table III from published tables (Honig and Hook 1960). Note that A, B and C are not unique; they depend on which P, T pairs are chosen from the tables, but the predicted values of P derived for a given T are well within 1% of the tabulated values. The abundant species Ne is not included in the table because it requires too low a temperature to be expected to survive. At $T = 16$ K, its pressure is already ~ 1 torr. One can show that the time required to evaporate Δa of a pure (molecule M) grain of radius a is (Greenberg 1983)

$$\tau = \frac{4sa}{P}(8kT/\pi m_H M)^{1/2}\left(\frac{\Delta a}{a}\right)$$

$$= \frac{1.26 \times 10^{-12}}{P}\left(\frac{s}{1}\right)\left(\frac{\Delta a}{10^{-6} \text{ cm}}\right)\left(\frac{T}{15}\right)^{1/2}\left(\frac{20}{M}\right)^{1/2} \text{ yr} \quad (2)$$

TABLE III
Vapor Pressure Parameters and Log Pressures at Several Temperatures for Some Possible Primordial Comet Volatile Components

Molecule	A	B	C	$\log P_{15}$ (torr)	$\log P_{20}$ (torr)	$\log P_{50}$ (torr)
CH_4	467	+1.12	4.91	−24.9	−17.1	—
N_2	378	+0.73	6.97	−17.4	−11	—
O_2	481	+0.20	8.80	−23	−15	—
CO	404	+3.59	2.01	−20.7	−13.5	—
CO_2	1364	−0.02	9.96	—	—	−17
H_2S	1326	−1.68	13.17	—	—	−16
NH_3	1627	+0.08	9.80	—	—	−22.6
H_2O	2455	+2.44	3.72	—	—	−41.2

where s = specific density, P = vapor pressure (torr), M = molecular weight, a = particle radius, m_H = hydrogen mass, k = Boltzman's constant. If one uses as a (conservative) criterion that < 0.1 μm of a grain is evaporated in 10^5 yr, we arrive at the maximum temperatures for pure grains by noting where $P > 10^{-17}$ torr, as listed in Table III. Below ~ 50 K we would expect all volatiles, with the exception of CO, CH_4, N_2 and O_2, to survive intact, although even these volatiles could survive as trapped species in H_2O at higher temperatures.

The most direct evidence we have for the formation of comets at a very low temperature is the observation of the diatomic molecule S_2 in comet IRAS-Araki-Alcock by A'Hearn et al. (1983; see also A'Hearn and Feldman 1985). It appears that the production of this molecule cannot occur by gas phase reactions because the abundance of S/H ~ 10^{-5} is so low. However, laboratory simulation of irradiation of dirty ices containing H_2S (Grim et al. 1986) show that S_2 is created by photoprocessing of interstellar grains at temperatures $T_d \simeq$ 10 to 15 K and that warmup beyond about 30 K in an argon matrix eliminates it (Hopkins and Brown 1975). This will be further studied in a dirty ice mixture. If we accept this and other (Greenberg 1983) evidence that comet aggregation occurred at temperatures well below 30 K, it may only be useful to consider higher-temperature regimes if the origin of comets or cometlike bodies occurred in more than one kind of environment.

However, if we do consider temperatures as high as but no higher than ~ 500 K to be relevant for comet aggregation, one can assume that all refractory components like carbon and silicates are totally unaffected, even though recondensation effects of H_2O on silicates could modify them to produce hydrated silicates. With temperatures as high as $T \simeq 500$ one could possibly evaporate some of the organic refractory molecules because this is nearly where our laboratory-created organics display a finite vapor pressure. However, in view of the fact that the interstellar organic refractories are certainly less volatile than the laboratory-created samples because of their substantially greater ultraviolet processing, it appears reasonable, as a first approximation, to assume that at $T \simeq 500$ *only* the true volatiles may be evaporated and subsequently recondensed. We cannot exclude the possibility that there exist such primitive bodies that are similar to but not identical with the majority of comets.

V. MORPHOLOGICAL STRUCTURE

Not only the chemical composition but also the morphological structure of cometesimal material depends on the maximum temperature at formation. Thus, the presence of S_2 in a comet not only requires the aggregation of previously irradiated interstellar grains but also coagulation at low temperature. Grain-grain collisions at $V > 40$ m s^{-1} lead to runaway reactions among the free radicals in the grain matrix and resulting temperatures > 70 K (Greenberg 1979; d'Hendecourt et al. 1982). Therefore we must assume that collision speeds < 40 m s^{-1} prevail in cometesimal formation. Since collision ($V < 40$ km s^{-1}) induced temperature spikes up to 25 K last < 10 s, we expect that even the very volatile grain constituents like CO are preserved. Surface molecules bound with energies of only 500 K (~ 0.06 eV) and surface vibration frequencies $\sim 10^{-10}$ s would also not be substantially evaporated (d'Hendecourt et al. 1985).

The core-mantle grains are very nonspherical as implied by the degree of polarization of starlight. Considered to have elongated rather than flat shapes, their length/thickness ratio is at least 2 or 3 to 1 (Greenberg 1968). Elongated large individual grains are certainly not likely to clump into compact structures but rather to give rise to a highly porous configuration as shown in Fig. 3. A piece of a cometesimal also contains (shown in Fig. 3 as black dots) the very abundant small carbon and silicate grains ($\sim$ hundreds to thousands small for 1 large) imbedded in the mantles of the large grains. In Table II the relative proportion of volatile and nonvolatile components by volume are indicated.

It has for some time been assumed that comets are low-density fragile objects (Donn 1963). One of the facts leading to this conclusion is the tidal disruption of Sun-grazing comets (Öpik 1966). Another is the occurrence of fragmentation among comets at heliocentric distances up to 9 AU (Sekanina 1982). The morphological structure of interplanetary particles (Brownlee 1978) suggests an initial porosity of comet debris. Wallis and Macpherson (1981) claim that Whipple's (1977) values for the nongravitational force can be generally reconciled with H_2O outgassing only if the mean comet density is low, under 0.5 to 0.7 g cm^{-3}. Making use of the aggregated interstellar dust model defined in Fig. 3 and Table II, it now becomes possible to derive di-

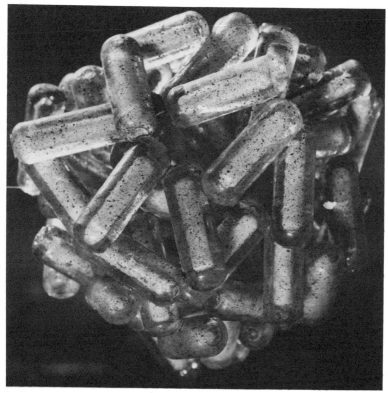

Fig. 3. Clump of interstellar grains. Individual core-mantle grains are represented in the mean as ~ 0.4 μm thick with elongation $e = 3.5:1$. Porosity is $p = 0.4$. Diameter of ensemble is ~ 4 μm. Black specks are representative of very finely divided carbon and silicate particles which accrete on the grain's outer mantles in the last stages of preaggregation. See also Plate 5 in the color section.

rectly quantitative estimates of the porosity of comets. We do this by comparing the density of a compact body made of the chemical constituents itemized in Table II with the density of such comet fragments as interplanetary dust and meteorites. The mean density of the mixture given in Table II, if fully packed ($c \equiv$ compact), is

$$\overline{\rho_C^c} = 1.36 \text{ g cm}^{-3}. \quad (3)$$

After a piece of comet breaks off, its volatile components (at 1 AU) are rapidly evaporated leaving only the skeleton structure of core-organic mantle grains behind (if we assume that the very fine particles are carried away with the volatiles). The density of the skeleton structure is $\rho_1^{sk} \simeq 0.46$ g cm^{-3}. Including the very small particles gives $\rho_2^{sk} \simeq 0.57$ cm^{-3}. We shall use the mean value $\bar{\rho}^{sk} = (\rho_1^{sk} + \rho_2^{sk})/2 = 0.5$ g cm^{-3} for the skeleton structure with empty spaces where the volatiles had been, as representative of comet dust from an initially compact comet. The dust experiment on the Helios spaceprobe (Grün et al. 1980; Grün 1981) observed a low-density component of dust grains orbiting the Sun with high eccentricities. Such dust grains are presumed to be relatively recently released from a comet and their densities can be stated as $\rho_d \lesssim 1$ g cm^{-3} according to Fechtig (1982), which is consistent with the "bird's nest" model (Greenberg and Gustafson 1981; Greenberg 1983; Mukai and Fechtig 1983), but is not quantitative enough to define the degree of porosity. On the other hand, the evidence for densities of $\bar{\rho}_m = 0.2$ g cm^{-3} for 85% of the meteors associated with short period comets (Verniani 1973) and even lower densities in the range $\rho_m \simeq 0.01$ g cm^{-3} for the Draconids, associated with Comet Giacobini-Zinner, imply a very substantial volume of empty space in the original particles. The fluffy nature of the cometary nonvolatile distribution is firmly established although Whipple (1982) assumes that some comets have $\rho > 1.3$ g cm^{-3} which I would consider too compact. Since it is not unreasonable to assume that the outer surface of short-period comets contains more, rather than less, refractory (higher-density) materials than do pristine (or long-period) comets, it would appear to be conservative to estimate the initial packing factor p ($\equiv$ ratio of volume of solid matter to total volume) of comet material by comparing its skeletal density $\bar{\rho}^{sk} = 0.5$ g cm^{-3} with the meteor density $\rho_m < 0.2$. This leads to a packing factor of $p < 0.2/0.5 = 0.4$ and an upper limit on the mean pristine comet density $\bar{\rho}_C < 0.4 \times 1.36 = 0.54$ g cm^{-3}. If some meteors really have densities of 0.01 g cm^{-3}, one would conclude that some comets may be predominantly empty space with packing fractions as low as $p = 0.01/0.5 = 0.02$ and densities of $\bar{\rho}_C \simeq 0.03$ g cm^{-3}. This is a very suspicious result. Nevertheless, there is a strong possibility that comets are as fluffy as freshly fallen snow whose packing is in the range of ≤ 0.1 (Seligman 1936). We note that even high reflectivity of the basic interstellar grain material leads to very low albedos when so loosely assembled. A rough estimate of the effective (Bond) albedo of the aggregate of interstellar dust envisioned here is $R_B \sim \alpha(1-g)/2$ where, α and g are the mean albedo and asymmetry factor for individual grains. Using $\alpha = 0.6$ and $g = 0.8$, we get $R_B \simeq 0.05$. It is difficult to reconcile high porosity of comets with the theory that they may have resulted from breakup of larger bodies other than very large comets.

VI. COMET IMPACTS

What happens to a comet when it impacts a planet is difficult to answer with great confidence, particularly if we are trying to determine what fraction of it is evaporated or pyrolyzed by the intense shock-generated heat. The comet structure postulated in the previous section possesses features at various scales (see Daniels and Hughes [1981] for similar types of structures) which must play a role on the effects of the shock of impact at speeds $\simeq 50$ km s^{-1}.

We have shown that the packing of a comet nucleus must be in the range 0.02 to 0.4 ($\bar{\rho}_C$ = 0.03 to 0.54) so that the assumption of a mean comet density $\bar{\rho}_C$ = 0.1 made by Lin (1966) for considering comet impacts was not unreasonable. At this and lower densities, the impact with the atmosphere will already be expected to disintegrate a considerable outer fraction of the comet before it impacts the planet.

It has been shown (Zel'dovich and Raizer 1967) that the strong heating by shock compression of porous bodies can lead to sharp anomalies on the Hugoniot curve. Whereas, for compact materials, the pressure rises with decreasing volume (compression), for increasing porosity (decreasing packing), the pressure rises more and more steeply and finally the anomalous state is reached in which the volume increases with increasing pressure. This leads to the generation of a rarefaction shock rather than a compression shock. Another factor which must play an important role in the effect of the shock is that on the smallest (interstellar grain) scale, the composition of the grains is chemically heterogeneous and the sound speed in the more volatile components is substantially less than in the refractory components. The shock dissipates more energy in the low sound velocity components, which therefore provide an energy sink, so that they are quickly evaporated with less heat going into the refractories. In a manner of speaking, each grain may act as a free surface unloading material in the direction of the initial motion of the shock. In general, boundaries between the hierarchy of sizes of cometesimals (Hughes 1983) treated in this way would provide fragile dividing layers in the comet nucleus structure leading to splitting of the comet into many fragments following a similar hierarchy. Thus, rather than the shock leading to complete evaporation of the comet, this distribution of the total energy allows a great deal of the energy to be dissipated by unloading at the enormously increased surface area. Conversion of the kinetic energy of the comet head by the reflected shock would be completed in a time scale $\tau_c = 2 R_c/V_c \simeq 0.4$ s (for R_c = 10 km) when the rear of the comet head reaches the reflected shock wave. Beyond this time, the comet is entirely fragmented. The energy being taken up by comet fragments could be radiated and ablated away so that, if this scenario is correct, we could believe that a substantial fraction of the comet material preserves its chemical and particulate integrity, rather than being totally evaporated.

VII. COMET CONTRIBUTION TO THE EARTH

As early as 1961 Oró suggested that comets could have supplied part or all of the initial inventory of organic matter for chemical evolution. Chang (1979) stated that, if the hydrogen/carbon ratio in comets is within an order of magnitude of values between 90 and 4, then comets must have provided a major fraction of the volatiles of the present atmosphere and oceans and that bound in the biosphere and the crust. Following Greenberg (1983), it may be shown that the interstellar dust model predicts H/C for comet volatiles alone to be (H/C) = 9.1/(0.28 × 3.7) ≃ 9, while including refractories leads to (H/C) = 3. Both of these values are well within the bounds given by Chang. Vanysék (1983) has shown that by assuming an enhancement of D in comets as due only to molecules XH other than H_2O, then in the comet model derived in Table II, a ratio of $XD/XH \simeq 10^{-3}$ leads to an upper limit of D/H $\simeq 2 \times 10^{-4}$ in comets; this is similar to the ratio D/H for the water in the Earth's ocean and an order of magnitude higher than in the interstellar medium. Even though, in the current epoch, the number of comet impacts is too small to contribute much to the Earth's volatiles (see, e.g., Zimbelman 1984, and references therein), in the past, whether or not periodic comet showers have occurred, there must have been an early period of large numbers of comet impacts.

Let us now consider the contribution of one comet. Using a mean density $\rho_C = 0.1$ and $\bar{R}_C = 10$ km gives a total mass 5×10^{17} g. Thus, one large comet supplies (see Table II) $\sim 0.3 \times 5 \times 10^{17} = 1.5 \times 10^{17}$ g H_2O. The current ocean mass is $\sim 1.3 \times 10^{24}$ g (International Phys. Tables) which is equivalent to about 10^7 10-km comets. It is interesting to note that the quantity of complex organic matter accompanying this amount of water is $\sim 0.21 \times 1.2 \times 5 \times 10^{17} \times 10^7 \simeq 0.5 \times 10^{24}$ g which is about 10^6 times the current biomass of the Earth. If it is indeed true that comets provided the early oceans and therefore also the initial inventory of organic matter for chemical evolution (Oró 1961), then permitting even a small fraction f of the organic refractories, in the form of amino acids, to survive intact, they would have given an enormous impetus to prebiotic evolution. Note that if f is only 10^{-5}, this estimate leads to the presence of a 10^{-4} molar solution of amino acids on the primitive Earth.

A corollary of the possibility that comets were responsible for life's origins is, of course, the possibility that comets were responsible for extinction and concomittant evolutionary jumps. The dust injected into the Earth's atmosphere by one comet can provide a very substantial extinction of sunlight if the fragments are small enough. An upper limit to the extinction is obtained by reducing the comet to its original interstellar dust nonvolatile constituents. The number density of 0.1 μm grains in a ($\rho = 0.1$) porous comet is $\sim (0.1) R_c^3/a_d^3 = 10^{32}$. Distributed evenly within the Earth's atmosphere, this leads to a column density of $10^{32}/4\pi R_e^2 \simeq 2 \times 10^{13}$ cm^{-2} and an optical depth in the ultraviolet $\simeq 10^4$. Since much smaller optical depths may be sufficient to trigger an ice age, the effect of one comet would be catastrophic, even allowing only a small surviving fraction of comet debris with, or without, considering the additional debris from the impacted Earth.

Without periodic comet showers, the mean frequency of Earth collisions is $\simeq 1 \times 10^{-8} \times$ number of new comets per year, which is of the order of unity (see Weissman 1982a). This is about one-half the value resulting from taking the ratio of the projected area of the Earth to the area of its orbit. To obtain this value, I have multiplied the probability derived by Zimbelman (1984) per passage (1.33×10^{-9}) with the factor of 8 corresponding to the mean number of Earth orbit passages for a new comet (Everhart 1967a). During a comet shower as large as 2×10^9 as postulated by Davis et al. (1984), the number of comet collisions could be as large as 20. This number is based on the presence of $N_o = 10^{13}$ comets in the Oort cloud which is a rather high estimate compared with Weissman's (1982a) value of $N_o = 1$ to 2×10^{12} for the outer Oort cloud, although comparable with his estimate of the number of comets in an inner Oort cloud between 20 and 30 thousand AU. The mean frequency of shower comets based on 20 per 26 Myr is $\simeq 80 \times 10^{-8}$ yr^{-1} which is ~ 100 times as large as the general average frequency of comet impacts. This frequency of shower comets of 80×10^{-8} extended over the full 4.6 Gyr age of the solar system leads to a total H_2O input of (at most) 4×10^{-4} of the current ocean mass so that the presumption of an exceedingly high initial comet flux is required if the comets contributed substantially to the Earth's water. On the other hand, the same shower frequency could have easily provided, over an initial period of $\sim 5 \times 10^8$ yr, an inventory of complex organic molecules comparable to or greater than the current biomass.

VIII. CONCLUSION

Evidence for very low-temperature comet aggregation appears more and more to support the theory that interstellar dust grains in pristine form are the basic building blocks of comets. The fact that the mean density of such material even after removal of its volatiles is higher than most meteors leads to the conclusion that comets are probably as fluffy as freshly fallen snow. The implied comet densities between $\rho \simeq 0.5$ g cm^{-3} and $\rho \simeq 0.03$ g cm^{-3}, have strong implications for the albedo, the nongravitational forces and the physics of comet impacts. The assumption that comets have the same chemical composition as fully accreted interstellar dust confirms the idea that comets contributed very substantially both to the Earth's water and also to the early complex molecular composition of its surface. Simple disintegration of a comet into its refractory components upon collision with the Earth is expected to have provided enough small particles (even allowing only 10^{-4} survival) in the Earth's atmosphere to have readily generated catastrophic attenuation of solar radiation.

Acknowledgments. I wish to thank V. Icke for helpful advice on shock problems and J. Klinger for some ideas on snow. The aggregated dust model shown in Fig. 3 was constructed by F. Robbers and others in the Huygens Laboratory machine shop. The photograph was taken by L. Zuyderduin. This work was supported in part by a grant from the National Aeronautics and Space Administration.

TEMPORAL VARIATIONS OF COSMIC RAYS OVER A VARIETY OF TIME SCALES

J. R. JOKIPII
University of Arizona

and

K. MARTI
University of California at San Diego

The intensity of energetic charged particles (cosmic rays) in the inner solar system is observed to vary with time over a variety of time scales. The Sun is the cause of many of these variations, but some of the observed longer-term ones may in part reflect changes in the local interstellar medium. Galactic cosmic rays dominate the average intensity of energetic particles above about 200 MeV. They are modulated by the Sun and have their lowest intensity during high solar activity. It is clear that changes in the intensity of cosmic rays in the local interstellar medium could be masked by solar modulation. In addition, observed changes in the intensity in the inner solar system could actually reflect changes in the Sun. Cosmic-ray intensity has been studied using satellites for short-term variations and terrestrial and extraterrestrial materials for the longer-term variations. Atmospheric ^{14}C activities suggest modulation cycles of 200 yr and possibly $\approx 10^4$ yr. Studies of chondritic meteorites indicate that the average galactic flux in the inner solar system was constant for the past 10^7 yr. Iron meteorites have been used as monitors over the 10^8 to 10^9 yr time period. The radionuclide ^{40}K (1.3×10^9 yr) indicates that the flux was smaller in the distant past. The current view of the origin of cosmic rays is that they are produced in supernova shock waves. Hence, depending on the frequency of supernova explosions in our vicinity and the parameters of the interstellar medium, the intensity of cosmic rays should vary in the vicinity of the Sun. These variations will be modified by the solar modulation, but no flux variability is observed on a 10^5 yr time scale, as would be implied by a local recent supernova event.

Cosmic rays consist chiefly of energetic protons with energies above roughly 100 KeV energy, which originate either at the Sun (solar cosmic rays) or in interstellar space (galactic cosmic rays). Galactic cosmic rays have a typical energy of ~ 1 GeV and are present continuously, but fluctuate on a variety of time scales. Solar cosmic rays are produced sporadically in solar flares, and are intense mainly at considerably lower energies than the galactic particles. Their spectrum is also a much more rapidly decreasing function of energy. After a solar flare produces solar cosmic rays, they are present in the solar system only for a relatively short time and decay away on a time scale of day (low energies) to hours (GeV particles).

The average intensity of these two types of cosmic rays, as a function of energy, is illustrated in Fig. 1, where the solar particles are a solar-cycle average. It should be noted, in addition, that the galactic cosmic rays are isotropic to a very high degree, and bathe the Earth nearly equally from all directions.

The intensity of galactic cosmic rays in the inner solar system is observed to vary over a wide variety of time scales. The time variations of galactic particles are due to variations in the solar wind and its entrained magnetic field, which are accessible to direct measurement. There exists a generally accepted physical model which can account quantitatively for these modulations.

I. THE OBSERVED VARIATIONS AT VARIOUS TIME SCALES

In this section the cosmic-ray variations observed in the inner solar system, using a variety of techniques are discussed. Their relation to the basic theory is discussed in Sec. II.

A. Direct Measurements

Direct measurements of the time variations of cosmic rays are available for the time period including the last two sunspot cycles.

The most important known periodic variation is the variation of galactic cosmic rays in anti-phase with the 11 yr sunspot cycle. The variation of the galactic cosmic-ray intensity over the past two sunspot cycles is illustrated in Fig. 2; there is no doubt about the variation and its association with the sunspot index. The interpretation in terms of transport theory will be discussed in detail in Sec. II.

Figure 3 shows a power spectrum of the variation in the intensity of galactic cosmic rays observed at Earth with the Climax neutron monitor. The main conclusion one can draw from this plot is that there is a continuous spectrum of fluctuations. If one were to obtain such a curve covering all tem-

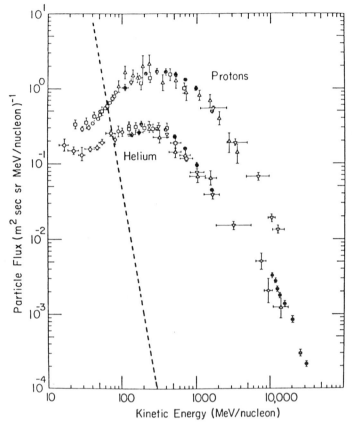

Fig. 1. Typical spectrum of cosmic-ray protons and helium nuclei obtained during sunspot minimum (from Meyer 1969). The low-energy end of the spectrum will decrease somewhat as sunspot maximum is approached. Superposed (dashed line) is the approximate intensity of solar cosmic rays averaged over a sunspot cycle (J. King, personal communication).

poral frequencies, it would be similar—consisting of a smooth background of variations, with stronger or weaker peaks at certain frequencies (not shown in Fig. 3), corresponding, for example, to the 11 yr sunspot cycle. The background variations illustrated in the figure are real and can be shown to be the result of the continuous bubbling and variation of the solar wind with its entrained magnetic field. The peaks in the spectrum are the result of periodic or nearly periodic variations in the Sun which affect the solar wind, and hence change the modulation of the cosmic rays.

Probably the most important generally nonperiodic variation of galactic cosmic rays are the "Forbush decreases." These are produced by outward-propagating shock waves in the solar wind, which sweep the cosmic rays

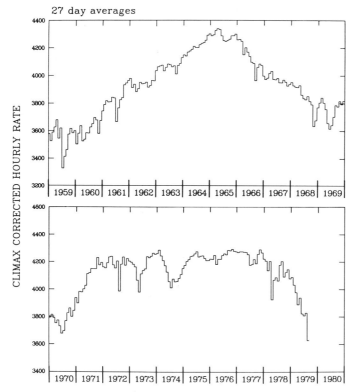

Fig. 2. Illustration of the variation of the 1 GeV cosmic-ray counting rate as given by the Climax neutron monitor over the last two solar sunspot cycles (from J. A. Simpson, personal communication).

ahead of them, causing a temporary decrease in intensity. The intensity then slowly fills in behind the shock. The shock waves may be the result of solar flares, or may be associated with quasi-steady fast wind streams in which case the decreases may show a quasi-periodic behavior with the rotation of the Sun. The interpretation of these Forbush decreases in terms of the general diffusive transport theory appears to be successful.

B. Utilization of Nuclear Reactions

Galactic cosmic-ray particles have sufficient energy ($\sim$ GeV) to induce nuclear reactions in solid solar system matter. On the other hand, as discussed above, the solar cosmic-ray flux at these energies have much lower intensity and solar cosmic-ray effects can only be seen in surface layers of extraterrestrial matter. The cosmic ray record in solar system matter has been reviewed by Reedy et al. (1983). The records were studied in terrestrial samples (e.g., the ^{14}C record) and in extraterrestrial samples (lunar rocks and mete-

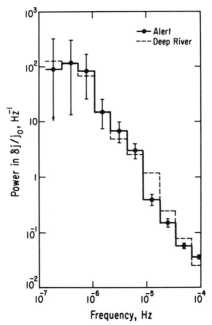

Fig. 3. Temporal power spectrum of the relative variation of the cosmic-ray intensity observed at the Alert and Deep River neutron monitors (reported in Owens 1973).

orites). The record regarding the cosmic-ray flux can only be obtained if the exposure geometry is known and has remained constant; limitations are set, for example, by erosional processes. Information on the longer time scales of 10^8 to 10^9 yr can be obtained from iron meteorites, since the times of exposure to cosmic rays range up to 2×10^9 yr. The energy of an interacting particle and the chemical composition of the target determine the cascade of nuclear reactions taking place and the distribution of reaction products. An important question which will be addressed later (Sec. II) is how the galactic cosmic-rays flux is modulated inside the heliosphere and how one can tell apart interstellar flux variations from solar modulation effects.

1. Variations on the 10^1 to 10^4 yr Time Scale. Correlations are known among solar activity indices and ^{14}C activity. ^{14}C increases during solar minima and decreases during solar maxima, reflecting solar modulation effects on galactic cosmic rays. On a 10^3 to 10^4 yr time scale, the dominating effect appears to be a 10^4 yr period, at least a half-cycle. The variation is not exactly sinusoidal, but wiggles of a higher frequency are superimposed (Fig. 4). The largest "Suess wiggle" amplitude is observed at about 200 yr (Neftel et al. 1981; Sonett 1985). The activities of the radionuclides ^{22}Na (2.6 yr), ^{46}Sc (83 d) and ^{54}Mn (312 d) were measured in a number of meteorites by Evans et al.

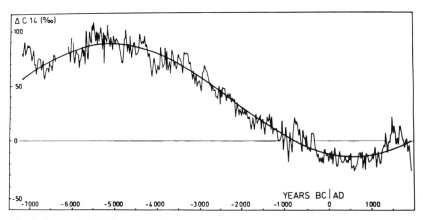

Fig. 4. Atmospheric ^{14}C concentrations during the last 9000 yr. The data are obtained from dated Bristlecone pine and floating European oak series by the La Jolla Radiocarbon Laboratory (Suess 1980). The curve is a best-fit sine function. (Figure courtesy of H. E. Suess.)

(1982). The observed activities, when corrected for target chemistry and for shielding differences, indicate variations of the spallation production rates by a factor of 2. These variations are correlated with the sunspot cycle from 1967 to 1978. Therefore, the production rates of cosmic-ray-produced nuclides also vary with the solar 11 yr cycle.

2. Variations on the 10^5 to 10^7 yr Time Scale. The record of the galactic cosmic-ray flux on a million year time scale can be inferred from induced nuclear reactions in extraterrestrial matter of known exposure geometry, such as lunar rocks or meteorites. Nuclear reactions produce a variety of radioactive and stable nuclei that can be measured and related to the incident cosmic-ray flux. The radionuclides ^{81}Kr (2.1×10^5 yr half-life), ^{36}Cl (3.0×10^5 yr), ^{26}Al (7.2×10^5 yr), ^{10}Be (1.6×10^6 yr) and ^{53}Mn (3.7×10^6 yr) represent a good set of monitors for cosmic-ray flux variations on this time scale. Among the chondritic meteorites which were studied extensively, the production rates of the above radionuclides can vary because of differences in size and shielding conditions. Appropriate shielding corrections need to be considered in the evaluation of cosmic-ray fluxes. Systematic ^{21}Ne production-rate calibrations were carried out by several groups (Nishiizumi et al. 1980; Muller et al. 1981; Moniot et al. 1983) and the cosmic-ray fluxes inferred from the nuclides ^{22}Na, ^{81}Kr, ^{10}Be and ^{53}Mn are in excellent agreement with each other and reflect a constant ($\pm$ 10 to 15%) galactic flux over the 10^5 to 10^7 yr time scale, which also agrees with average present-day flux. The ^{26}Al activities disagree with this result, but it is most difficult to understand this anomaly in terms of a flux variation, since ^{26}Al has an intermediate half-life. Among possible explanations are biases in data selection and preirradiation effects before the last cosmic-ray exposure.

3. Variations on the 10^7 to 10^9 yr Time Scale. There are few radioisotopes with appropriate half-lives that can be used for this time scale and only ^{129}I (1.6 × 10^7 yr) and ^{40}K (1.3 × 10^9 yr) have been studied so far (Nishiizumi et al. 1983,1985; Voshage 1962,1978).

Chondritic meteorites cannot be used to study variations in the cosmic-ray flux on longer time scales, because their exposure ages are typically less than a few tens of million years. Fortunately, there are numerous recovered iron meteorites which were exposed in space as small bodies up to two billion years and which are well suited for this purpose. There exists an excellent data base due to the work of Voshage and coworkers who have obtained isotopic compositions of potassium in some 80 meteorites, including the radioisotope ^{40}K, together with information on shielding of the samples. The measurement of all three isotopes of potassium permits the detection of the cosmic-ray-produced component which is superimposed onto potassium initially present in the meteorite. As already inferred by Voshage (1962) and confirmed by recent systematic studies, the measured ^{40}K/^{41}K isotopic data show that the rates of cosmic-ray-induced nuclear reactions have changed by 50%. Results obtained by Marti et al. (1984) and Lavielle et al. (1985) suggest that this is

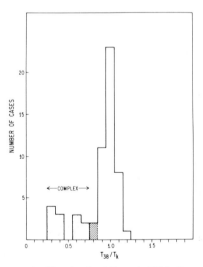

Fig. 5. T_{38}/T_k systematics: complex histories. Lavielle et al. (1985) discussed a method that allows recognition of meteorites with complex exposure geometries, which are unsuitable cosmic-ray detectors. They used a cross calibration of ^{38}Ar-based exposure ages (T_{38}) with those based on the ^{40}K/^{41}K method (T_k). The histogram shows that a good match is obtained (T_{38}/T_k = 1.00 + 0.15). The inferred average production rates P_{38} of ^{38}Ar is only about two-thirds of the production rates during the last 10^7 yr.

due to an increase in the cosmic-ray flux less than about 200 Myr ago. For the time period of 0.2 to 1.0 Gyr ago, these authors observed essentially constant ^{38}Ar production rates and agreement between ^{38}Ar ages and ^{40}K–^{41}K ages (Fig. 5). They argue that space erosion or changes in the meteorite geometry, due to collisions, cannot be the cause, except in "complex" cases (Fig. 5).

There is a limitation regarding the applicability of the potassium method. The age determination is based on the measured deviations of the growth curve of ^{40}K from that expected for a stable isotope. Since this deviation becomes very small for young ages, it is not feasible at this time to determine, with high precision, ages of less than about 200 million years, a time interval of obvious interest. A new method, which involves the radioisotope ^{129}I (1.6 × 10^7 yr) and stable ^{129}Xe, is now developed to make up for this deficiency (Nishiizumi et al. 1983,1985; Marti 1985). Monitors for the 10^7 to 10^8 yr interval are required to further constrain the past cosmic-ray flux. The ^{129}I half-life is quite appropriate and shielding corrections are not required.

II. THE PHYSICAL MECHANISM OF COSMIC-RAY VARIATIONS

The observed variations in the extraterrestrial cosmic-ray intensity discussed in Sec. I have been ascribed variously to changes in the effects of the Sun and solar wind in modulating galactic cosmic rays impinging on the outer boundary of the heliosphere, or to actual variations in the local interstellar intensity. In this section we outline the current theory of solar modulation, and demonstrate that *all* of the observed changes could easily be the result of solar variations on long and short time scales.

A sophisticated, quantitative physical model of cosmic-ray transport has been developed over the past two decades which now appears capable of relating most of the shorter-term variations discussed in Sec. I to physical processes in the solar wind and its imbedded magnetic field. The basic idea, illustrated schematically in Fig. 6, is based on the fact that the motion of cosmic rays is dominated by the interplanetary magnetic field, and, furthermore, that this field is highly irregular and turbulent. Hence, the direction of motion of the particles is quickly randomized by the random fluctuations in the magnetic field, and the resulting net motion is described in terms of a diffusion or random walk. The convection of the irregular magnetic field with the solar wind leads to two further effects: convection of the cosmic rays with the wind, and an energy loss due to the expansion of the radially flowing wind. Finally, the particles drift in the large-scale magnetic field. These basic effects can be combined into the fundamental equation of cosmic-ray transport (Parker 1965; Axford 1965; see, e.g., Jokipii [1971], Völk [1975], and Fisk [1979] for reviews of the theory)

$$\frac{\partial n}{\partial t} = \frac{\partial}{\partial x_i}\left[K_{ij}\frac{\partial n}{\partial x_j} - (V_i + V_{d,i})n\right] + \text{div}(\mathbf{V})\frac{\partial n}{\partial \ln(p)} \quad (1)$$

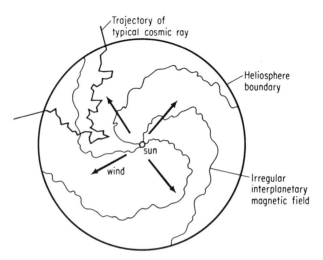

Fig. 6. Schematic illustration of the motion of a typical cosmic-ray particle in the solar wind.

where $n(r,T,t)$ is the density of charged particles having energy T at time t and position r. $\mathbf{V}$ is the convection velocity, K_{ij} the diffusion tensor and $\mathbf{V}_d$ is the guiding center drift motion in the ambient magnetic field. Equation (1) may be shown to be a good approximation if the scattering mean free path is small compared with the other length scales (Jokipii 1971). The basic solutions to Eq. (1) then describe cosmic rays random walking in the irregular interplanetary magnetic field, and hence spreading out, while they are simultaneously being systematically convected with the wind and losing energy due to the expansion of the wind as it flows outward.

The decrease of the ambient cosmic-ray intensity during a Forbush decrease can be understood reasonably well in terms of Eq. (1). In this picture, the passage of an interplanetary shock wave sweeps cosmic rays ahead of it, in a snow-plow effect caused by the increased convection speed, magnetic field and turbulence behind the shock. Gradually, after the shock has passed, the cosmic rays diffuse back into the depleted region, until the intensity has essentially reached its previous intensity. The details of the effect depend substantially on the parameters. Nonetheless, it appears that the Forbush decrease can be well explained in terms of the basic theory. As the number of shock waves and associated Forbush decreases rises toward solar maximum, they can contribute to the overall 11 yr sunspot cycle modulation.

In many ways, the most sophisticated solutions to the transport equation have been to the 11 yr modulation problem. The general approach has been to find the steady-state solution to Eq. (1) with an assumed unmodulated galactic cosmic-ray spectrum imposed as a boundary condition at some outer radius D, and with a plausible variation of solar-wind parameters and diffusion coefficients inside the boundary (see Fig. 6).

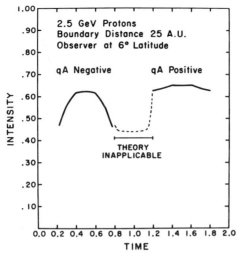

Fig. 7. The result of a numerical solution of the cosmic-ray transport equations showing the calculated intensity of galactic cosmic rays at Earth during a complete solar magnetic cycle, or two successive sunspot cycles. The values of qA refer to the change in sign of the solar and interplanetary magnetic field from one sunspot cycle to the next. Compare the general shape of the curves with the observed variation in Fig. 2. For a detailed discussion of this model, see Kota and Jokipii (1983).

Present knowledge and available computer power make necessary idealizations. A series of comprehensive solutions have been obtained over the past few years at the University of Arizona (see, e.g., Kota and Jokipii 1983). These calculations are the only ones to include *all* of the known effects in a plausible steady-state calculation. Previous models were incomplete in that they neglected the guiding-center drift effects entirely, and furthermore were at most two-dimensional. A typical result of the model calculations, showing a model 22 yr solar magnetic cycle is shown in Fig. 7. This model should be compared with the observed variation shown in Fig. 2. In both figures qA (or A) positive corresponds to the period from the 1970 solar maximum until the last maximum, during which the northern solar polar magnetic field was directed outward. qA negative corresponds to the previous 11 yr period. Although there are still areas of disagreement, this and other similar agreements suggest that the model may be an approximate representation of the modulation phenomenon. An unexpected result of these computations has been the discovery that, for plausible values of the various parameters, the guiding-center drift effect due to the large-scale structure of the interplanetary magnetic field, is the dominant effect in determining the motion of the particles. Hence the structure of the large-scale magnetic field, which determines the drifts, is the most important parameter. Indeed, in constructing the model illustrated in Fig. 7, only the magnetic field structure was varied, with the wind velocity, diffusion coefficient, etc., held constant.

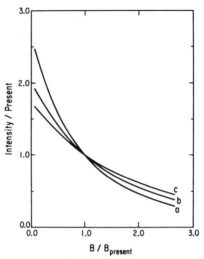

Fig. 8. Illustration of the expected cosmic-ray intensity variation relative to present at 3 different energies (a = 0.5 GeV, b = 1 GeV, c = 2 GeV), if the magnitude of the interplanetary magnetic field were varied from its present value. Clearly, a substantial change in intensity can result from a relatively modest change in interplanetary parameters.

Much of the detail in the above picture depends on our attempts to extrapolate observations carried out in the vicinity of the solar equatorial plane to high heliographic latitude. Nonetheless, we believe that the general theory is correct and that it can account for a wide range of observed phenomena.

A consequence of the probable importance of drifts is that the general structure and magnitude of the interplanetary magnetic field may be perhaps the most important parameter determining the modulation of galactic cosmic rays. Figure 8 shows the results of an approximate calculation of the effects of varying the magnitude of the interplanetary magnetic field on the intensity of cosmic rays of various energies at the distance of Earth. It is clear that varying the magnetic field by a factor of 2 can account for the long time scale variations discussed in Sec. I. In addition, it should be stated that similar variations in other parameters such as the solar-wind velocity may cause similar effects. For this reason, the basic conclusion—that variation of the cosmic-ray intensity by a factor of 2 can be produced by plausible solar variations—is rather insensitive to the precise model used.

We conclude that present understanding of cosmic-ray modulation, although uncertain in some areas, nonetheless indicates that quite reasonable variations in interplanetary parameters can account for all the observed fluctuations in the cosmic rays, without invoking any changes in the galactic flux or spectrum. This is not to say that such changes have not occurred.

The bath of interstellar cosmic rays in which the solar system exists can be expected to vary as the result of two basic causes: events and motion through spatial structure. It is easy to show that variations by observable amounts could have occurred in the past. For example, if cosmic rays are more intense in spiral arms than between them, then the motion of the solar system around the galactic center would produce variations on an approximately 250 Myr time scale. This may indeed be the cause of the slow variation seen in meteorite observations.

It is now widely held that the bulk of galactic cosmic rays are produced at shock fronts produced by supernova explosions in our Galaxy. This picture has been made even more attractive in recent years by the introduction of the concept of diffusive shock acceleration, which has the virtues of being highly efficient and widespread throughout the galactic disk, and which produces a spectrum close to that observed. As pointed out by Axford (1981), the frequency of supernovae (approximately one every 30 to 50 yr in our Galaxy), and the variation of the shock-wave parameters with distance result in a prediction that the intensity of cosmic rays at a typical spot in the galactic disk should increase by more than a factor of 2 for 35% of the time every 10^5 yr. Such variations appear to be ruled out by the observational evidence cited in Sec. I, and this would appear to be a problem for the theory. Solar effects might fortuitously mask such fluctuations, but this seems unlikely. In addition, there have been idealizations made in the acceleration models which may invalidate the conclusions. Nonetheless, the contrast between the predicted variations and those allowed by observations is disturbing.

III. A LOCAL RECENT SUPERNOVA?

Clayton et al. (see their chapter) discuss the possibility that the local interstellar medium (100 pc from the Sun) may have been reheated by a supernova event some tens of parsecs from the Sun about 10^5 yr ago. This model is based on the recently observed ^{26}Al spectral line data from HEAO-3 (Mahoney et al. 1982), which is interpreted as being too intense to be produced by the average concentration of ^{26}Al in the interstellar medium if supernova explosions are the origin of ^{26}Al (which is, of course, not certain).

Location of the Sun within a young supernova remnant would provide a most suitable mechanism for cosmic-ray shock-wave acceleration. We have discussed the evidence for a recent increase in cosmic-ray flux compared to 10^9 yr average flux. We also have discussed that radionuclides in meteorites indicate no flux variability on a 10^5 to 10^7 yr time scale. One of the monitors is ^{81}Kr (2.1×10^5 yr half-life) which would be expected to record a local supernova effect only 10^5 yr ago. If such an event had taken place more than 10^7 yr ago, this would satisfy the evidence for an observed flux increase, but then it could not explain the 1.809 MeV gamma-ray flux, since ^{26}Al will have decayed. The HEAO-3 (high-energy astronomy observatory) cosmic-ray elemental abundance data (Binns et al. 1982) determine upper limits for the actinide elements (including radioisotopes of $T_{1/2}$ 10^5 yr half-life). The data are consistent with solar system abundances and do not indicate the presence of freshly synthesized r-process nuclides.

IV. CONCLUSIONS

This review shows that there is evidence for variations of cosmic rays on a variety of time scales, ranging from days to 10^8 yr. Our present understanding of the effects of the Sun and solar wind on cosmic rays is sufficient to demonstrate that *all* of the time variations observed could very well be of solar origin. However, our knowledge is also insufficient to say that variations due to other causes have not occurred.

Acknowledgments. This work was supported, in part, by grants from the National Aeronautics and Space Administration (JRJ) and (KM) and by a National Science Foundation grant (JRJ). We thank J. A. Simpson and J. King for providing unpublished data.

A LOCAL RECENT SUPERNOVA: EVIDENCE FROM X RAYS, ^{26}AL RADIOACTIVITY AND COSMIC RAYS

DONALD D. CLAYTON, DONALD P. COX, and F. CURTIS MICHEL
Rice University

Several lines of astrophysical investigation suggest that the solar system is immersed in a low-density cloudlet (n ~ 0.1 cm^{-3}, T ~ 10^4 K) extending roughly one pc from the Sun, that this cloudlet is immersed in a much lower-density medium of high temperature (n ~ 4 × 10^{-3} cm^{-3}, T ~ 10^6 K) extending roughly 100 pc from the Sun, and that the latter medium was reheated by a supernova event some tens of parsecs from the Sun about 10^5 yr ago. If this picture is correct, it could cast doubt on the interpretation of both the ^{26}Al gamma radiation and the locally observed cosmic rays as samples of the average galactic distributions. Alternatively, if the observed cosmic rays require an uncontaminated nonlocal source, then either there was no recent local supernova or such an event in the local environment has had remarkably little effect on local cosmic rays, given that the collection of such events is regarded to be their general source. In this chapter we present possible ways in which cosmic rays could have been contaminated by a local recent supernova and speculate on ways in which this contamination may be affecting our interpretation.

The solar neighborhood is astrophysically very interesting. An up-to-date description of its many significant problems may be found in the collection of observations reported in Kondo et al. (1984). It is fortunate for the history of astronomy that we reside in a hot low-density medium of great optical clarity out to about 100 pc. In the last decade or so we have come to interpret such hot low-density phases in terms of supernova remnants that overlap before radiative cooling becomes too important (Cox and Smith 1974; McKee and Ostriker 1977). The detection of hot ($T = 10^5$ to 10^6 K) superbubbles in the interstellar medium (Heiles 1979; Cash et al. 1980), containing the energetic ejecta of perhaps 100 massive stars, suggests that such bubbles follow clumped bursts of star formation by several million years, the evolution time of the massive stars. Elmegreen and Lada (1977) described a scenario of sequential star formation capable of producing such an outburst. It seems reasonable that the Sun now moves through such a general region of the Galaxy.

Soft X-ray Background

One especially interesting aspect of this hot gas is that it emits soft X rays, a radiation to which the cooler neutral medium is quite opaque. Observations (Burstein et al. 1976; McCammon et al. 1983) show the local level of emission to be somewhat higher than expected in the general interstellar diluted phase. In an attempt to explain this high level, Cox and Anderson (1982) concluded that the Sun is located within a specific recent supernova remnant, which injected about 5×10^{50} erg into the local medium only about 10^5 yr ago. By fitting the emission intensity at the shock, which now surrounds the solar system at least 100 pc from the event, they found that the preshock density was about $n_0 \simeq 4 \times 10^{-3}$ cm^{-3}, already diluted because of the local superbubble. This explanation not only seems capable of solving the immediate problem of the X-ray background, but imagining this last local event as the finale of the general local outburst that has created a superbubble, changes the *a priori* statistics, which would otherwise have rendered improbable a recent nearby supernova. Innes and Hartquist (1984) have in fact argued that the superbubble which formed about 4×10^6 yr ago by correlated explosions is itself the source of the X-ray background. McCray and Kafatos (1986) have discussed similar considerations. The view presented by Cox and Anderson (1982) is that the last supernova creates most of the soft X rays today.

^{26}Al Gamma Rays

Clayton (1984) was recently led to suggest on quite other grounds almost exactly the same nearby supernova about 10^5 yr ago. Figure 1 shows the ^{26}Al spectral line data from HEAO-3. What Clayton (1984) argued is that this 1.809 MeV gamma-ray line flux produced by the decay of 10^6 yr ^{26}Al (Mahoney et al. 1982, 1984; Share et al. 1985) is too intense to be produced by the average concentration of ^{26}Al in the interstellar medium if supernova explosions are the origin of the ^{26}Al. Clayton suggested that either the ^{26}Al nucleosynthesis results from some other common specific source, probably novae, or its source is instead a local feature. Although supernovae cannot, owing to their rarity, maintain an adequate Galaxy-wide concentration, as a single object they do nonetheless eject the largest mass of ^{26}Al, about 10^{-5} M$_\odot$ of new ^{26}Al per massive star. Setting the resulting gamma-ray flux (Clayton 1982)

A LOCAL RECENT SUPERNOVA   131

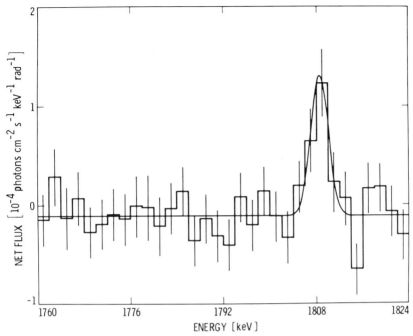

Fig. 1. HEAO-3 energy spectrum near 1809 keV of diffuse galactic gamma radiation (Mahoney et al. 1984). Background has been subtracted, yielding negative statistical fluctuations. Each channel has a width of 2 keV. The solid curve is the best fit to background plus a Gaussian line at 1809 keV owing to the decay of ^{26}Al. This is the first detection of interstellar radioactivity. The question we raise here is whether or not this radioactivity is in a nearby supernova remnant.

$$F_\gamma(^{26}\text{Al}) \simeq \frac{0.12 \text{ cm}^{-2} \text{ s}^{-1}}{D^2 \text{ (pc)}} \qquad (1)$$

equal to the flux 4.8×10^{-4} cm^{-2} s^{-1} measured by both HEAO-3 and Solar Maximum Mission spacecraft places the ^{26}Al about $D = 15$ pc away. This close event must be young enough that the ^{26}Al still survives and, more stringently, that the neon burning shell within the evolved core (Woosley and Weaver 1980) remains primarily on the galactocentric side of the Sun. Observational arguments that the radiation is not isotropic, force this constraint. Assuming no relative motion between Sun and center of mass of the ejecta and also that the core is expanding at 10^2 km s^{-1}, then requires the remnant age to be $< 1.4 \times 10^5$ yr. This places the event as being recent, perhaps 10^5 yr ago, in close agreement with the interpretation of the soft X-ray background. To ascertain the correctness of this interpretation will require an angular distribution of F_γ. Leising and Clayton (1985) have described the expected distribution for Galaxy-wide nucleosynthesis as a part of the plan to make this measurement a key objective for the Oriented Scintillation Spectrometer Experiment (OSSE) spectrometer (Kurfess et al. 1984) to be launched on the Gamma Ray Observatory. If the source is local, as described here, rather than galactic, as described by Leising and Clayton (1985), its distinguishing feature will surely be that of extending to high galactic latitudes, probably even centering above or below the galactic plane, rather than being concentrated in the plane. The same can be said for a recent reinterpretation in a preliminary preprint by Morfill and Hartquist (1985), who advocate a similar ^{26}Al concentration within the entire 100 pc superbubble as a result of ^{26}Al ejection from the 100 events that created the superbubble. The angular extent of that distribution would be nearly isotropic if we reside within it, in contrast to that of a single expanded supernova core. We could suppose that the ^{26}Al gamma rays come instead from the nearby Loop I superbubble, but that is not specific enough for the soft X-ray background.

Cosmic-ray Acceleration

It will be clear that any of these interpretations directly influence our understanding of the interaction of the Galaxy and the solar system. In the present work we relate these issues to possible implications for the cosmic-ray record. Clayton et al. (1985) have described the reasonable expectation that several aspects of the cosmic rays may be measurably influenced by the location of the solar system within a 10^5 yr old supernova remnant, and we now turn to these possibilities.

Location of the Sun within a young supernova remnant would place it within one of the most promising sources for cosmic-ray acceleration, supernova remnants propagating in a very low-density background medium. The corresponding shock-wave acceleration (Axford et al. 1976; Bell 1978; Blandford and Ostriker 1978; Michel 1981; Drury 1983) is now thought to energize much of the cosmic-ray spectrum. Such acceleration occurs both outwardly and inwardly from the shock surfaces and would, therefore, vigorously bombard the solar system with newly accelerated cosmic rays. We discuss this possibility as a natural consequence of the attempt to develop a self-consistent picture of the solar environment. We cannot attempt to review here the full field of cosmic-ray physics, however. (The biannual *Proceedings of International Conferences on Cosmic Rays* constitute the best source to the general cosmic ray literature.) The specific question of their time dependence is reviewed in this volume (see chapter by Jokipii and Marti) because of the special relevance of that subject to the question of local sources that may or may not have affected the solar system.

The deep interior of a cosmic-ray accelerating blast wave potentially contains four separate cosmic-ray populations of interest; these components are listed below and are shown schematically in Fig. 2

1. True galactic cosmic rays (CR) which have interacted with the shock front (further acceleration) and then diffused inward against the remnant outflow;

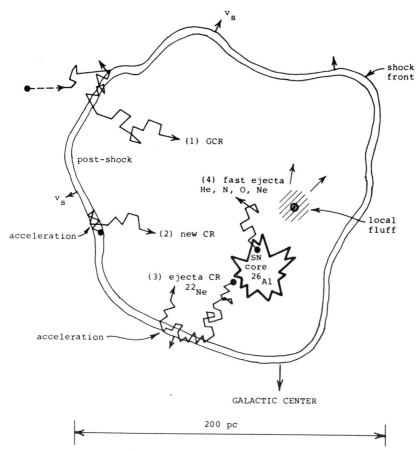

Fig. 2. A cartoon of the bubble created by the local supernova 10^5 yr ago. Soft X rays and cosmic-ray acceleration occur at the shock front, whereas the radioactive ^{26}Al is still confined to the galactocentric side by the slower expansion of the core of thermonuclear debris. The relevant cosmic rays fall into four separate populations (components), as explained in the text.

2. Freshly accelerated cosmic rays arising from the ambient gas at the shock, and which have also diffused into the interior;
3. Independently accelerated high-energy particles from the central source which not only fill the local bubble but may also be reaccelerated at the shock boundary and diffuse back in;
4. Independently accelerated low-energy particles (10-100 MeV) from a central source which are slowly diffusing outward.

We see the following possible roles for these components as portions of the observed cosmic rays: (1) is an older (high-grammage) component containing the abundant spallation secondaries; (2) is a young (low-grammage) component representative of the ambient interstellar medium; (3) is a component having the composition of those parts of the supernova that preferentially provide particles accelerated to high energies, presumably from the outer layers, and perhaps provide the source of the ^{22}Ne excess in cosmic rays; and component (4) is an outflow of low-energy cosmic rays, remembering the bulk composition of their type-II source, He, O, N and Ne.

Reinterpretations of the Solar Environment

In Table I we list several features of the solar environment along with the reinterpretations we are suggesting based on the solar system's location within

TABLE I
Reinterpretations via Local Supernova Bubble

Phenomenon	Average Interstellar Medium	Local Supernova (SN)
Soft X ray	enough visible hot gas	within post-shock gas
^{26}Al gamma ray	galactic novae distributed on a line on the sky	core of nearby SN ejecta making circular area on sky
Low-density outside local fluff	hot region of interstellar matter	located in post-shock local bubble
$E < E_D$ cosmic ray (CR)	galactic (GCR)	GCR plus locally accelerated cosmic rays
Grammage path	monotonic distribution of GCR path lengths	higher grammage GCR diluted by lower grammage local CR
^{22}Ne excess	feature of GCR reflecting source compositions	bubble feature caused by local nucleosynthesis and acceleration in SN envelope
^{10}Be	10^7 mean GCR age within exponential distribution	older GCR having little ^{10}Be plus locally produced ^{10}Be
Anomalous component	neutral drift into heliosphere plus local acceleration	high bubble flux of nucleosynthesis products He, N, O, Ne by local acceleration to modest energy
Ti-Cr/Fe decrease with E	deficiency of path lengths < 1 g cm^{-2}	Ti-Cr excess at lower E from acceleration of local Type II SN ejecta (Fe deficient)

a supernova remnant. These reinterpretations are speculative, but we discuss them as advocates in what follows.

Because the cosmic rays generated are in approximate equipartition with the thermal gas inside the shock boundary, and both are decompressed in the interior to a total pressure about one third the post-shock value, the expected cosmic-ray pressure in the interior of even the most active bubbles is expected to be similar to the ambient thermal pressure of the interstellar medium. This equivalence is confirmed for the local bubble where the average thermal pressure (found from matching the soft X-ray background) is indeed of order 10^{-12} dyn cm^{-2}, suggesting a cosmic-ray energy density of that same magnitude if the bubble is actively involved in acceleration. Thus the fact that the observed cosmic rays have that same energy density is consistent. From gamma-ray observations (Stecker et al. 1975; Bloemen et al. 1984) and the ancient cosmic-ray record in solar system matter (Reedy et al. 1983; chapter by Jokipii and Marti), it seems clear that the cosmic-ray energy density in the Galaxy as a whole and the average cosmic-ray flux in the past are very similar to our local values. In the reinterpretations we are considering, that is no less expected than if we were not surrounded by an accelerator. If galactic cosmic rays and the magnetic field are in approximate equipartition with the ambient pressure, they will be similar in energy density to the interiors of strongly accelerating remnants in general and of the local bubble in particular.

If, as we argue below, the local acceleration has only doubled (approximately) the local cosmic-ray flux, the gamma-ray emissivity per H atom should be somewhat less in the outer Galaxy than expected from the local cosmic-ray flux measurements; and a hint of this is indeed observed in the papers cited below. Wolfendale (1985) argued on the basis of high-energy gamma-ray observations that molecular clouds are less massive by a factor 2 to 3 than commonly inferred from their CO temperatures. We note here that this discrepancy could also be resolved if one instead assumes that the cosmic-ray flux is a factor 2 to 3 smaller in the clouds than the value measured locally. This resolution strengthens the plausibility of taking the local cosmic-ray flux to be higher than average by a modest factor. In a related vein, Bhat et al. (1985) have recently argued that a similar cosmic-ray enhancement exists in the Loop I supernova remnant, further strengthening the plausibility of moderate enhancements in supernova remnants generally. Furthermore, the cosmic-ray record in the iron meteorites should show a somewhat greater flux during the past 10^{5-6} yr than previously; and a hint of this possibility is also observed. However, Jokipii and Marti (see their chapter) point out in their review that 2×10^5 yr ^{81}Kr has been observed in concentrations that are consistent with a constant flux for the past 10^6 yr. The situation is difficult, however, because a neighboring supernova 10^5 yr ago would not have had time to build ^{81}Kr up to its saturation value, so that it perhaps records the preevent flux. The question is then how accurately one is able to compare that flux to the values observed by other indicators today.

Discussion

The supernova rate of 1 per 30 yr in the Galaxy implies one supernova within 100 pc of a typical location per 10^6 yr. If essentially all supernovae evolve to become active acceleration regions of radius ~ 100 pc at age $\sim 10^5$ yr, then the chance of being somewhere inside an accelerator of that size and age is of order 0.1; therefore the postulated configuration is not a highly improbable one. Only when we demand the 10^5 yr event to be much nearer, say 20 pc, does the *a priori* probability drop below one percent. But it is just here that the presence of the local superbubble alters the *a priori* statistics. If 100 supernovae have occurred over a 10^6 pc^3 volume during a few-million-year period several million years ago, the odds against one recently from a slightly less massive (15–20 $M_\odot$) star nearby are much improved. Ours might thus have been a trailing supernova whose birthplace $\sim 10^7$ yr ago is not now evident.

The ^{10}Be-measured cosmic-ray trapping lifetime of 8×10^6 yr (Wiedenbeck and Greiner 1980) is usually taken to indicate that the spallation grammage in local cosmic rays belongs to a component of that mean age, presumably the true galactic cosmic rays. In the current context we must suppose that the true galactic cosmic-ray grammage is somewhat higher and has been diluted in our sample by freshly accelerated particles, assuming that the newly accelerated cosmic rays at the bubble boundary have traversed essentially zero matter. We suppose that the degree of this dilution is probably moderate, perhaps a factor of 2, after considering the general possibilities. The possibility should not be overlooked, moreover, that the ^{10}Be $\simeq 0.16$ ^{9}Be in the moderate-energy measurements actually arises from component (3), the local supernova cosmic rays propagating at the source through about 1 g cm^{-2} during their preacceleration, and that the true galactic cosmic rays are much older and contain very much less ^{10}Be. The question is how much ^{10}Be can be produced from flarelike or shock activity in the supernova envelope itself and preaccelerated along with the rest of component (3). However, such an origin of ^{10}Be may even be required if we accept the recent argument (Garcia-Munoz et al. 1984) that path lengths < 1 g cm^{-2} are absent in the cosmic rays near 100 MeV, because 100 MeV cosmic rays with path lengths > 1 g cm^{-2} must all be older than 10^7 yr if the entire path is accumulated in the present solar environment having $n_H < 0.1$ cm^{-3}. The problem of recent propagation needs fresh study. (See Streitmatter et al. 1985 for an alternative approach to grammage accumulation in superbubble walls.)

The mode in which a particle interacts with the blast wave depends on the ratio of its diffusion coefficient to the quantity $R_s v_s$, where R_s is the shock radius and v_s its expansion velocity. With a diffusion mean free path λ, diffusion coefficient $\lambda c/3$, $R_s \sim 100$ pc, and $v_s \simeq 300$ km s^{-1}, the crossover is roughly $\lambda \simeq 0.3$ pc, the gyroradius of a 3×10^{14} eV particle in a field of one microgauss. Fresh cosmic-ray acceleration can proceed only to energies whose mean free paths are comparable to the above value. Particles with diffusion coefficients in excess of $R_s v_s$ do not have time for significant acceleration (Lagage and Cesarsky 1983). Similarly, only ambient cosmic rays with diffusion coefficients lower than $R_s v_s$ interact strongly with the blast wave. Longer mean-free-path particles simply pervade the system at their ambient densities, unaffected by the 10^5 yr event.

The net result is that the highest cosmic-ray energy in the fresh sample is comparable to the lowest energy of ambient cosmic rays that can easily penetrate the remnant. We shall suppose this crossover energy E_D is greatly in excess of 1 GeV, although certainly $< 10^{15}$ eV. As shown by Lagage and Cesarsky (1983), the exact value is rather uncertain, but we may justifiably think of its value as $E_D \simeq 10^3$–10^5 GeV.

It should not be implied that deep within the remnant the cosmic-ray spectrum would contain entirely fresh particles for $E \ll E_D$. In this energy regime, both fresh and ambient particles interact strongly with the shock and diffuse slowly and together into the interior. On the whole, we anticipate that the energy spectra in the external environment, the immediate post-shock environment, and the deep interior can be *qualitatively* represented as shown in Fig. 3. (The details depend on the behavior of the diffusion coefficients deep in the remnant interior, the length of time that acceleration has been active, the current shock strength, etc.)

It is interesting to consider the impact of grammage dilution on estimates of the cosmic-ray power requirement. Supposing that the local bubble as now observed has diluted the galactic cosmic-ray grammage by a factor D (the true galactic cosmic-ray grammage thus being $\sim 5D$ g cm^{-2}), we immediately note that $1 \lesssim D \lesssim 20$ from the nuclear cross-section limit on total grammage. The usual assumption, that $D = 1$ and that the observed cosmic-ray energy density is representative of the trapping volume as a whole, coupled with the leaky-box model indicate that the cosmic-ray power required to overcome leakage P_{CR} is roughly 10% of the available supernova power P_{SN}. Considering, however, that the true galactic cosmic-ray density may be ϵ times that observed locally, and that the grammage may have been diluted, the present picture requires an average power input

$$P_{CR} \sim 0.1 \frac{\epsilon}{D} P_{SN}. \tag{2}$$

Somewhat paradoxically, the better our bubble has been at accelerating fresh cosmic rays (the larger is D), the less cosmic-ray power is required from the ensemble of supernova occurrences. This situation makes possible a rough consistency estimate of the dilution. In particular, the *excess* cosmic-ray energy within the local bubble (considering the higher density near the shock) is perhaps of order $0.3 \, E_0$ where E_0 is the explosion energy. Assuming that a fraction β of all supernovae evolve as ours has and that the cosmic rays suffer

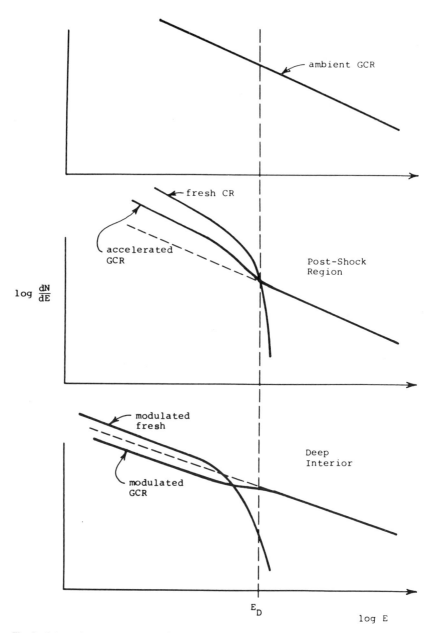

Fig. 3. Schematic energy spectrum of cosmic radiation in the general interstellar medium (upper panel), the post-shock region of a supernova remnant (middle panel), and deep within the supernova remnant after modulation against the continuing outflow of ejecta (lower panel). The energy E_D (probably 10^3 to 10^5 GeV) is the maximum acceleration possible in this single event, so that only cosmic radiation with $E < E_D$ has contributions from this local source.

expansion losses which reduce their energy by a factor f in leaving the bubble to join the ambient component, the total cosmic-ray power would be

$$P_{CR} \sim 0.3 f\beta \, P_{SN}. \qquad (3)$$

Furthermore, the maximum likely expansion loss equals the present bubble radius divided by the cosmic-ray scale height. From the grammage and ^{10}Be trapping time, that scale height is thought to be of order 1 kpc/D. Thus $f \gtrsim D/10$ making

$$\frac{P_{CR}}{P_{SN}} \sim 0.1 \frac{\epsilon}{D} \sim 0.3 f\beta \gtrsim \frac{0.3 \, \beta D}{10}. \qquad (4)$$

Consistency thus requires $D^2 \lesssim 3\epsilon/\beta$. Since ϵ appears not to be very different from unity (from gamma rays and history), D cannot be large unless β is small, that is, unless few supernovae interact as vigorously with the cosmic rays as our bubble would be inferred to have done (by its large D). In short, either the grammage dilution in the local bubble is small ($D \lesssim 2$) or few remnants behave as ours. The latter possibility seems reasonable only until it is realized that our chances of being inside such a bubble are reduced to order 0.1 β. The closed-box limit would require $D^2 \sim 300$, $\beta \sim 10^{-2}$, giving us only one chance in 10^3 of seeing what we do. Thus, the extreme possibility that the vast majority of the cosmic rays we observe are freshly accelerated by our local bubble seems to be excluded both by the observation that the ambient density cosmic rays are sufficient to contribute significantly in the bubble and, independently, by general energetics of the cosmic-ray system. We point out, in passing, that the presence of an equipartition density of galactic cosmic rays might significantly inhibit the shock-wave injection of fresh cosmic rays. With no precise constraints to guide us, we think of $D \approx 2$, although the sketch in Fig. 3 was made, for conceptual clarity, to correspond more nearly to $D \approx 3$.

The possibility that much of the cosmic-ray flux has a recent local origin has been considered many times. Forman and Schaeffer (1979) took supernovae to be point sources of promptly accelerated cosmic rays which then diffuse from their points of origin; they concluded that only supernovae nearer than 100 pc could leave a temporal imprint on the cosmic-ray record. Now that supernova shocks are thought to be the main acceleration mechanism, any temporal record is even more obscured, and, in fact, may not be expected on most models. Streitmatter et al. (1985) have considered trapping and confinement in the local superbubble. In their description, cosmic-ray acceleration is partly local but not to the extent that it violates the temporal record. We have been led to a different but very explicit idea, however. A single supernova whose shock front passed the Earth almost 10^5 yr ago and which is accelerating cosmic rays could have telltale features in energy spectra, in composition, and in time history.

Turning to the possibility that the observed cosmic rays may be significantly enriched by a population arising from supernova preacceleration (component 3), we note first that the mass of ambient material enclosed by the shock is roughly 500 $M_\odot$ and, in addition, that the shock front, where the main acceleration takes place, is well outside the envelope of the ejecta. We do not thus propose that unaccelerated ejecta material are now significantly contaminating the cosmic-ray source at the shock front. On the other hand, there have been suggestions for supernova acceleration of cosmic rays at earlier epochs, by explosion directly (Colgate and Johnson 1960) or in a Rayleigh-Taylor unstable interaction between the ejecta and the surrounding medium at the time of comparable swept-up mass – the Cas A phase (Scott and Chevalier 1975). Much of the source preacceleration may occur for reasons similar to solar high-energy particles. We are not aware that such early acceleration has been shown not to take place; it seems out of favor for maintaining galactic cosmic rays primarily because it is thought that the particles would suffer far too much expansion loss before joining the galactic cosmic rays. However, we now see that the shock front can reaccelerate component (3) high-energy particles as they continuously arrive after outward diffusion.

A very significant observation impacted by this idea is the ^{22}Ne richness of locally observed cosmic rays. Mewaldt (1983) reviews data showing that the ^{22}Ne/^{20}Ne ratio is about four times solar. The upper panel of Fig. 4 shows this ^{22}Ne excess clearly in the energy range 28 to 169 MeV nucleon^{-1}. This enrichment is far greater than that characterizing the post-solar evolution of the interstellar medium as a whole and thus constitutes a cosmic-ray anomaly. As we shall see, this degree of enhancement is easily available from early cosmic-ray acceleration involving supernovae ejecta. The cosmic-ray particle density is roughly 1 eV cm^{-3}/10^9 eV (particle)$^{-1}$ or 10^{-9} proton cm^{-3}. The relative abundance of ^{22}Ne at solar composition is 1×10^{-5}. Thus a volume of radius 100 pc in the bubble interior would contain 10^{-8} $M_\odot$ of ^{22}Ne in cosmic rays. Woosley and Weaver (1981) have considered the ^{22}Ne production in massive presupernovae and have applied it to the possibility of supernovae as the general cosmic-ray source. Taking, for example, 20 $M_\odot$ for the preexplosion star, there would have existed 0.03 $M_\odot$ of ^{22}Ne in the He shell, between 5 and 7 $M_\odot$ in radial coordinate. The larger ^{20}Ne mass lies deeper, within the carbon-exhausted core, so that, along with the ^{26}Al in our hypothesis, it lags behind in the more slowly expanding dense core.

To be specific for purposes of numerical estimation, assume that this He shell is ejected at 3000 km s^{-1}, carrying 10^{50} erg/$M_\odot$ of kinetic energy, of which half (equipartition) is converted to type (3) cosmic rays. At 1 GeV per amu, the accelerated mass could then be 3×10^{-5} $M_\odot$ per solar mass ejected. Applying this fraction to the 0.03 $M_\odot$ of ^{22}Ne within the helium shell yields 10^{-6} $M_\odot$ of ^{22}Ne entering the cosmic-ray pool at that time, exceeding by 10^2

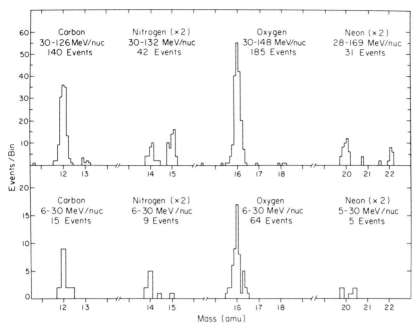

Fig. 4. Mass spectra of C, N, O and Ne cosmic rays measured by Mewaldt et al. (1984) in two separate energy intervals. The top panel shows the ordinary cosmic rays as detected between about 30 and 140 MeV per nucleon. Excess ^{22}Ne is clearly evident, as is excess ^{15}N resulting from spallation of oxygen; i.e., this is a high-grammage component with an excess nucleosynthetic anomaly at ^{22}Ne. The lower panel between 5 and 30 MeV shows the counts dominated by the anomalous cosmic-ray component. It is a low (zero?)-grammage component having little carbon. Its origin is unknown.

the ^{22}Ne cosmic-ray mass expected within the local bubble. Thus the ^{22}Ne excess in observed cosmic rays poses no insurmountable problem and, in fact, should be expected within the model context. And because ^{22}Ne has the largest overabundance factor in the He shell, it should have the largest cosmic-ray enhancement factor; however, the smaller observed excess in heavy Mg isotopes may be accounted for by the same model. Application of the same efficiency argument for ^{10}Be is not so straightforward because it requires first an estimate of the amount of ^{10}Be created during the explosion and shortly after by nonthermal processes. Estimation of the ^{10}Be yield thus requires an independent study. This scenario for ^{22}Ne enrichment may be improved by considering presupernova loss of the H envelope which can, by prior double-shell convective mixings, carry away much ^{22}Ne and leave the He shell as the fastest part of the later supernova ejecta. In a variant of this idea, Cassé and Paul (1982) use ^{22}Ne-rich winds from Wolf-Rayet stars as the first step in ^{22}Ne acceleration. The overtaking of these Wolf-Rayet ejecta by the eventual supernova shock could be the event that fills the bubble with supernova-associated cosmic rays. If this local enhancement of ^{22}Ne is correct, the ^{22}Ne excess should disappear at energies in excess of $E_D \simeq 10^3$–10^5 GeV (unless most galactic cosmic-ray sources have the same properties, so that the true galactic cosmic rays are ^{22}Ne-rich at all energies). Finally, it is also possible that the general abundance of ^{22}Ne has been enriched within the local bubble by the numerous earlier supernovae which excavated it, leading directly to ^{22}Ne enrichment of component (2).

Anomalous Component

A second identifiable component of the cosmic radiation that may find new interpretation in this model is the so-called anomalous component. The anomalous component is characterized by a peak in the low-energy (5-20 MeV nucleon^{-1}) flux containing a large overabundance of He, O, N and Ne, but not of C (see review by Gloeckler 1979). Figure 5 shows the energy spectra for H, He, C, N and O from Mewaldt et al. (1984). The broad peak near 200 MeV is a compromise between the extra-solar system power-law flux and the difficulty of penetrating the solar wind below 100 MeV. Below 30 MeV, however, He, N and O fluxes show an anomalous increase that is the defining signature of the anomalous component. H and C are absent or very deficient in this component, which also has a peculiar isotopic composition shown in the lower panel of Fig. 4. The absence of ^{13}C, ^{15}N and ^{22}Ne excesses shows the anomalous component to be a low-grammage one. The He is also almost isotopically pure ^{4}He (not shown). The most frequently heard interpretation is that only neutral atoms having ionization potentials greater than that of hydrogen can penetrate from the interstellar medium into the heliosphere, where they are then ionized and accelerated (Fisk et al. 1974). But others (McDonald et al. 1977; Webber and Cummings 1983) have suggested that this component may reflect an anomalous low-energy component external to the heliosphere. It is especially interesting for the present discussion that the anomalous component appears to be a low-grammage component. We are moved to speculate that the anomalous component within the heliosphere reflects a high flux of *low-energy* (~ 100 MeV) He, O, N and Ne within the local supernova bubble; in fact, it is just these elements that are greatly enriched (relative to H and C) in the bulk ejecta from massive stars (Woosley and Weaver 1981). H and C are underabundant in massive-star ejecta. If a portion of these bulk ejecta are accelerated only to the moderate cosmic-ray energies that translate, after heliospheric modulation, to 5 to 20 MeV amu^{-1}, a new interpretation of the anomalous component may be possible. A detailed understanding may involve the coupled effects of the local source and the high ionization potentials of the injected elements, perhaps the injection of fast neutral atoms into the heliosphere, or perhaps a smaller Z/A ratio for He, O, N and Ne ions than for other elements, leading to a lesser modulation for them. The new thrust here is the possibility of a nearby source in our local bubble for just the requisite elements, injected in large numbers with a unique energy spectrum.

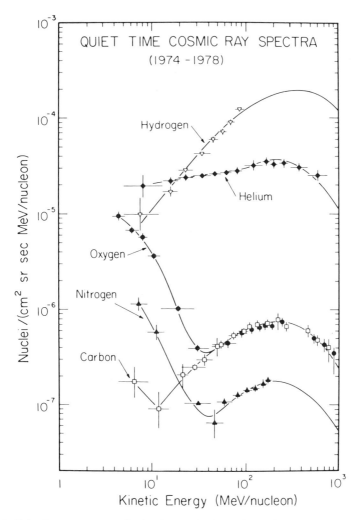

Fig. 5. Quiet-time energy spectra for the elements H, He, C, N and O measured by Mewaldt et al. (1984) at 1 AU between 1974 and 1978. Although the H and C fluxes drop off as the energy is lowered, a result of the modulation by the solar wind, the He, N and O show a new excess near 10 MeV, called the anomalous component. It is isotopically distinct as well, as shown in Fig. 4. The anomalous helium is essentially pure ^{4}He, unlike the helium in ordinary cosmic rays.

We might also entertain the idea that the anomalous component is the local interstellar gas accelerated from rest to modest energies by the shock front. The isotopic data is consistent with this suggestion. The deficiency of carbon might be the result of its being trapped in grains while the shock passes, but the deficiency of hydrogen has no obvious cause in this picture, unless perhaps its greater difficulty (at fixed speed) in penetrating the solar wind. We therefore display this interpretation for more expert consideration.

Conclusions

The picture that we have suggested here seems damaged by the recent argument (Garcia-Munoz et al. 1984) that for the observed cosmic rays near 1 GeV there is an absence of path lengths $\gtrsim 0.5$ g cm^{-2} in the usual exponential distribution of path lengths, because we have assumed that the component (2) of accelerated ambient circumstellar gas is a zero-grammage component. This argument is based on the observation that the ratio of sub-Fe elements (Ti-Cr) to Fe increases more rapidly with decreasing energy near 1 GeV than would be expected in a conventional propagation model. While acknowledging this difficulty, we speculate on two resolutions worthy of more study: (1) the sub-Fe enhancement could perhaps be a nucleosynthesis result of oxygen and silicon burning in the ejecta, accelerating a small fraction of these newly synthesized elements to join component (3), thereby becoming an abundance anomaly similar to ^{22}Ne while not synthesizing the solar complement of Fe; and (2) the newly accelerated particles may actually have a grammage within their local source of order 1 g cm^{-2} for reasons not yet clearly seen (as tentatively suggested above for "new ^{10}Be").

One may wonder about the past and present effects on the solar system of such a nearby supernova event. We are presently shielded dynamically from the fluid flow (but not from cosmic rays) by the local fluff, a small cloud of density about 0.1 to 0.3 cm^{-3} surrounding the solar system and of about 1 pc radial extent around the Sun (Weller and Meier 1981; Kondo et al. 1984). Other small diluted clouds also exist nearby. The local fluff could possibly have achieved its present velocity component through interaction with the remnant, but whether it could have done so without total disruption is unclear. What is clear is that such an event offers the possibility of a number of exciting revisions to our understanding of locally measured astrophysical quantities. The event itself would have been quite dramatic, potentially 0.1 to 1% as bright as the Sun, a possibly stressful situation for nocturnal creatures and the emerging human race alike.

Acknowledgments. This work was supported in part by grants from the National Aeronautics and Space Administration, from the Naval Research Laboratory for work on the OSSE spectrometer for the Gamma Ray Observatory, and from the National Science Foundation.

PART IV
The Oort Cloud

DYNAMICAL INFLUENCE OF GALACTIC TIDES AND MOLECULAR CLOUDS ON THE OORT CLOUD OF COMETS

MICHAEL V. TORBETT
University of Kentucky

The influence of the Galaxy, quite negligible for the planets, can be significant for the loosely bound collection of comets known as the Oort cloud. The tides of the Galaxy impose a maximum size of order 0.5 pc for the Oort cloud as well as effecting rapid and significant changes in cometary angular momenta. In particular, the vertical tides seem to be at least as efficient in perturbing comets to enter the inner solar system as are random passing stars. The tidal limit to the Oort cloud, which is roughly the same as the limit imposed by passing stars, is, however, factors of several larger than the observed extent of the cometary cloud. The smaller size of the actual Oort cloud can be attributed to the action of gravitational perturbations produced by encounters of the solar system with interstellar molecular clouds.

I. INTRODUCTION

The Oort cloud of comets was proposed in 1950 by Oort to explain the equilibrium influx of very long-period or "new" comets apparently entering the solar system for the first time. These comets, comprising a vast reservoir ranging up to one-third the distance to the nearest star, were assumed to be delivered to the inner solar system by the gravitational perturbations of passing stars (see the chapter by Weissman for details). A detailed estimate of the efficiency of stars in delivering these new comets, with semimajor axes ranging from 20,000 to 50,000 AU, led Oort to conclude that the population of the cloud of comets must be $\sim 10^{11}$. More complete statistics on new comets with fainter limiting magnitudes (Marsden et al. 1978; Everhart 1967) together with Monte Carlo simulations (see Weissman's chapter for references) indicate that this number may be more like 10^{12}.

The origin of the Oort cloud is still uncertain; yet, theories for the origin of comets can be divided into two broad categories: (1) capture of interstellar cometesimals; and (2) production as by-products of the processes leading to solar and planetary formation. The hypothesis of the capture of interstellar comets during passages through molecular clouds has been championed recently by Clube and Napier (1984a; see also their chapter) who claim that the distributed comet density required for this rather inefficient process is not inconsistent with the liberal upper limit of $\sim 10^{-4} AU^{-3}$ derived from the lack of observations of clearly hyperbolic comets (Sekanina 1976). The consensus of opinion, however, is that this process is very improbable and by Occam's razor, this sentiment will likely remain so unless some future isotopic analysis clearly indicates extrasolar origin. The alternative hypothesis of comet origin, being a naturally attendant consequence of solar system formation, can be further subdivided into two principal scenarios: (a) *in situ* formation and (b) scattered Uranus- and Neptune-zone planetesimals. In the former, comets are seen as forming either in much the same manner as did the planetesimals that gave rise to the planets, although in dense, satellite fragments of the protosolar disk (Cameron 1973), or in the outskirts of the solar nebula beyond the planetary system (Cameron 1962; Hills 1981,1982). In the latter, a fraction of the icy planetesimals in the outer solar system that eventually accreted to form Uranus and Neptune were scattered by close encounters with the growing planetary cores to large distances from the Sun (Oort 1950; Kuiper 1951; Safronov 1969; Shoemaker and Wolfe 1984a).

The distribution of the orbital elements of Oort cloud bodies is as unclear as is their origin. The distribution of observed semimajor axes of long-period comets are summarized in the chapter by Weissman with the conclusion that the Oort cloud extends from roughly 20,000 to 50,000 AU or approximately 1000 times the size of the planetary system. As was pointed out by Oort (1950), Bailey (1977) and Hills (1981), the inner limit is probably a selection effect associated with the relative stability to stellar perturbations of orbits within 20,000 AU. Thus, there could exist an unobserved and possibly dense inner Oort cloud (see Weissman's chapter). In fact, numerical simulations of planetesimal scattering produce Oort clouds with number density increasing dramatically inward to several hundred AU (Shoemaker 1984a).

The distribution of orbital eccentricities is highly uncertain. If comets formed *in situ*, then modest eccentricities may be the rule. If, however, comets formed in the planetary system and were subsequently ejected to the Oort cloud, then large eccentricities may predominate. As we shall see, though, information on the original distribution is likely to be largely erased over the age of the solar system as the Oort cloud is randomized by the perturbations due to passing stars and other perturbations. Hence, the details of the Oort cloud are far from being resolved; yet, there are some general features to be noted that are of interest to the present discussion:

1. The outer boundary of the Oort cloud seems to be in the region 40,000 to 50,000 AU;
2. Due to the randomization by stellar and other perturbations mentioned above, the distributions of orbital elements should be reasonably smooth and consistent with statistical equilibrium;
3. When selection effects are taken into account, the distribution of angular orbital elements are essentially random despite some reports of asymmetries (Fernández and Jockers 1983; see also the chapter by Delsemme). Caution is urged (See Sec. II) in interpreting distributions of angular elements.

Thus, to first order, one can assume an essentially spherically symmetric cloud with the semimajor axis distribution a monotonically decreasing function of semimajor axis, the distribution of eccentricities e given by statistical equilibrium or $f(e) = 2e$ (Heggie 1975), and a random distribution of inclinations and perihelia directions. The subject of this chapter is the stability of such a comet cloud to galactic tidal effects and encounters with giant molecular clouds (GMC). In Sec. II the dynamical limits imposed by the galactic tides are examined and compared to observed limits. It will be seen that neither the tides nor stellar perturbations can limit the Oort cloud to its observed dimensions. In Sec. III it is argued that the additional perturbations provided by gravitational encounters with interstellar molecular clouds can tidally strip the Oort cloud to its present size. Various conclusions are discussed in Section IV.

II. DYNAMICAL INFLUENCE OF GALACTIC TIDES

The tidal effects of the Galaxy as a whole, negligible for planetary orbits, may have important consequences for the Oort cloud. In addition to providing an outer boundary of stability for cometary orbits, the tides of the Galaxy—both radial and vertical—can effect significant variations on the orbital elements.

Boundary of Stability

Classical analysis (see, e.g., Brouwer and Clemence 1961) of the restricted three-body problem (massless third body) indicates that comet orbits are formally stable only within the gravitational sphere of influence defined by the zero-velocity surface. This surface is defined as a level curve of the Jacobi integral

$$\Omega = \tfrac{1}{2}v^2 + C \qquad (1)$$

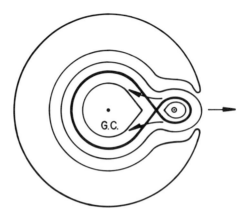

Fig. 1. Zero-velocity curves for comets orbiting the Sun moving about a point-mass galaxy situated at the point labeled G.C. Comets inside a given curve will remain inside (see text). The heavy line indicates the maximum volume around the Sun within which comets remain bound to the Sun. As comets become less bound, the curves open up permitting comets to be lost to the Galaxy as a whole as indicated by the arrows.

where v is particle velocity, C is the Jacobi constant, and Ω is an effective potential consisting of the gravitational potential plus a rotational energy term, and hence a function of distance from the secondary. For given values of the Jacobi constant C, the zero-velocity surface constitutes a boundary across which the particle may not cross because this would require imaginary velocities. The constant C plays the role of an energy and thus there exists a critical value corresponding to a maximal surface for which the motion is bound to the Sun as shown by the heavy line on Fig. 1. For values of C less than critical, the zero-velocity curves open up and allow comets to be lost to the Galaxy as a whole, as shown by the light lines and arrows in Fig. 1. One can thus obtain a limiting tidal radius in this manner (see, e.g., King 1962) given by

$$r_t = \left(\frac{m}{3M}\right)^{1/3} R \qquad (2)$$

where m/M is the mass ratio of secondary to primary and R is the separation. Applying this to the situation at hand by idealizing the Galaxy as a point mass of mass $M = 1.3 \times 10^{11} M_\odot$ at a distance of 8.2 kpc yields $r_t = 1.1$ pc $\sim 220{,}000$ AU. Note that this applies to a point mass galaxy.

A similar three-dimensional analysis, taking into account the distributed nature of the Galaxy, yields an ellipsoidal limiting volume. By writing the potential as

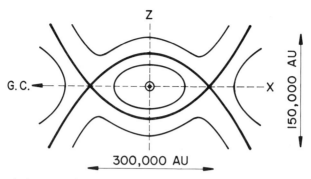

Fig. 2. Zero-velocity curves for a more realistic galactic potential. X denotes radial distance away from the galactic center labeled G.C. and Z denotes distance from the Sun perpendicular to the plane of the Galaxy. The limiting boundary is indicated by heavy lines (refer to explanations given for Fig. 1). (Figure after Antonov and Latyshev 1972.)

$$V = \frac{-GM_\odot}{r} - 1/2(\alpha x^2 + \gamma z^2) \qquad (3)$$

where r is comet-Sun distance and where $\alpha = 4A(A - B)$ and $\gamma = C^2$, with A and B being the Oort constants describing galactic rotation, Antonov and Latyshev (1972) find a zero-velocity ellipsoid as shown in Fig. 2 with dimensions $x_m = 293{,}000$ AU, $y_m = 196{,}000$ AU, and $z_m = 152{,}000$ AU; x is in the radial direction and z is perpendicular to the galactic plane. An interesting extension of the zero-velocity technique by Innanen (1980), by focusing on *zero relative acceleration,* leads to different limiting radii for prograde and retrograde orbits. Due to the difference in the direction of the Coriolis force, retrograde orbits are more stable than prograde and form a nested sequence with the tidal radius defined in Eq. (2) and given by

$$r_{pro} = r_t/1.44$$
$$r_{retro} = 1.44 r_t. \qquad (4)$$

Thus, one should expect an excess of retrograde bodies over prograde near the limits of stability.

These general stability criteria, however, are necessary but not sufficient conditions and thus serve only to suggest the boundary of stability. The standard analytical methods of celestial mechanics are valid when the perturbations are not too large. The orbits of bodies near the boundary of stability, however, are only vaguely reminiscent of Keplerian orbits and thus to explain properly the outer boundary one must resort to numerical simulations in order to model actual orbits.

The first such attempt in a cometary context was a two-dimensional numerical integration of the equations of motion conducted by Chebotarev

(1965). In an inertial frame fixed to the Sun, the effect of the Galaxy on a cometary orbit was assumed to be approximated by artifically constraining a point mass of $1.3 \times 10^{11} M_\odot$ to orbit the Sun. Aside from not being self-consistent, this analysis is flawed due to the neglect of the Coriolis and other inertial forces introduced by the transformation to the Sun-centered reference frame. Understandably, then, the results erroneously indicate stability for prograde orbits for semimajor axes $< \sim 230{,}000$ AU and $\sim 100{,}000$ AU for retrograde orbits. In a subsequent reanalysis (Chebotarev 1966), this conclusion is retracted with a somewhat vague statement of stability for semimajor axes $a \lesssim 80{,}000$ AU.

The effective boundary of stability for cometary orbits has recently been determined by numerical integration of the equations of motion over the age of the solar system (Smoluchowski and Torbett 1984). As detailed there, the integrations were done in an inertial Cartesian reference frame whose origin coincides with the galactic center. The Sun and a comet were given approximate initial conditions for their relative motion and then the solar system was launched on a circular orbit about the Galaxy. This self-consistent procedure automatically incorporates tidal forces as well as Coriolis and other inertial forces. The equations of motion for the comet are given by

$$\ddot{x} = \frac{-GM_G x}{r^3} - \frac{GM_\odot (x - x_\odot)}{r_c^3}$$

$$\ddot{y} = \frac{-GM_G y}{r^3} - \frac{GM_\odot (y - y_\odot)}{r_c^3} \qquad (5)$$

$$\ddot{z} = \frac{-GM_G z}{r^3} - \frac{GM_\odot (z - z_\odot)}{r_c^3} + f(z)$$

where x, y, z and $x_\odot, y_\odot, z_\odot$ are the Cartesian coordinates of the comet and the Sun, respectively, $M_G = 1.3 \times 10^{11} M_\odot$ is the mass of the Galaxy idealized as a point mass, $r = (x_\odot^2 + y_\odot^2 + z_\odot^2)^{1/2} = 8.2$ kpc is the Sun-galactic center distance and $r_c = [(x - x_\odot)^2 + (y - y_\odot)^2 + (z - z_\odot)^2]^{1/2}$ is the Sun-comet distance. The term $f(z)$, accounting for the vertical forces of the galactic disk, was derived from recent analysis (Bahcall 1984a) of the local galactic mass density and is given per unit mass by

$$f(z) = -2.565 \times 10^{-11} z + 1.66 \times 10^{-16} z^3 \qquad (6)$$

where z is in parsecs.

The boundary of stability was determined by isolating the maximum semimajor axis remaining bound to the Sun over the age of the solar system. As already mentioned, orbits near the boundary are poorly described by Kep-

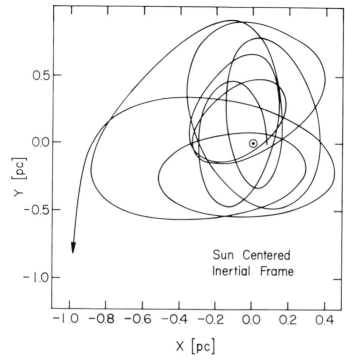

Fig. 3. Representative trajectory of an unstable comet in an inertial reference frame illustrating the highly non-Keplerian nature of the orbits. The inclination with respect to the galactic plane was 40° and the initial semimajor axis was just outside the boundary shown in Fig. 4.

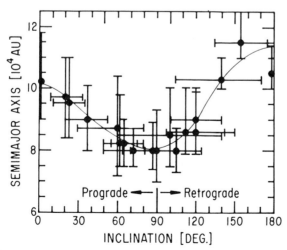

Fig. 4. Maximum stable semimajor axis as a function of galactic inclination. Error bars denote range of variables encountered during calculation while points represent average values.

lerian ellipses. Shown in Fig. 3 is a representative unstable orbit in the Sun-centered inertial frame with galactic inclination of 40° and semimajor axis just outside the boundary. In this nonrotating reference frame, a Keplerian ellipse would retrace itself. The resulting stability boundary from the study by Smoluchowski and Torbett (1984) is shown in Fig. 4, where semimajor axis is plotted against inclination with respect to the galactic midplane. The error bars denote the ranges of eccentricity and semimajor axis encountered during the calculation while the points indicate average values. The solid line is a suggested fit to the data. The stability boundary is rather sharp in that for semimajor axes roughly 10% smaller than the limits shown by the curve in Fig. 4, one recovers Keplerian ellipses. Several features can be discerned from Fig. 4. A slight retrograde excess is observed consistent with greater stability for retrograde motion due to the asymmetry in the direction of the Coriolis force. Furthermore, a reduced stability at inclination $\sim 90°$ is evident. This last result is, on the surface, surprising because inclined orbits on average sense less of the disruptive radial tide than do orbits lying in the plane; yet there are at least two explanations.

First, writing the potential of Antonov and Latyshev (1972) in terms of galactic coordinates, r,b,ℓ Bailey (1977) finds

$$V = \frac{-GM_\odot}{r} - 1/2 r^2 (\alpha \cos^2 b \cos^2 \ell + \gamma \sin^2 b). \tag{7}$$

A comet in this potential will escape the influence of the Sun provided that it passes a critical semimajor axis a_c at which

$$\frac{\partial V}{\partial r}\bigg|_{b,\ell} = 0. \tag{8}$$

This condition reduces to

$$\alpha \cos^2 b \cos^2 \ell + \gamma \sin^2 b = \frac{GM_\odot}{a_c^3}. \tag{9}$$

The critical a_c is thus dependent on the direction of the major axis (ℓ, b) and has a minimum at $b = \pm 90°$. The value of a_c becomes large along lines given by

$$\tan^2 b = \frac{-\alpha}{\gamma} \cos^2 \ell \sim \frac{1}{3} \cos^2 \ell \tag{10}$$

which are confined to galactic latitudes $b \leq 30°$. Hence, one expects the boundary of stability to be at larger semimajor axes when the major axes are lying near or in the galactic plane than at high galactic latitudes.

As for the second explanation, the vertical force due to the disk given by Eq. (6), which leads to vertical oscillations about the galactic midplane, departs slightly from a linear force law. The vertical tidal stress (which is actually a compression) on a comet is thus a function of z, being a maximum at the galactic plane. Hence, in its excursions about the midplane, the solar system is subject to varying tidal stresses similar to the gravitational compression shock experienced by globular clusters during passage through the disk (Ostriker et al. 1972). Furthermore, the semiperiod of the vertical motion of the solar system of $\sim$ 30 Myr roughly corresponds to the orbital period for marginally bound objects. This presents the possibility for resonant accumulation of these perturbations producing decreased stability for inclined orbits. This effect of resonance between orbital and vertical oscillation periods was, in fact, observed in the calculations from which Fig. 4 was constructed. Orbits with semimajor axes considerably larger than the limit of stability were observed to remain bound for longer times than objects just outside the stability limit.

Galactic Influence on Orbital Elements

The effect of the Galaxy on the orbital elements of long-period comets can be pronounced, especially for the angular elements. Numerical integrations supported by analytical analysis (Byl 1983) for the restricted three-body problem show interesting long-period effects for highly eccentric orbits at 25,000 and 50,000 AU. The variation of the semimajor axes was observed to be meager while the direction of the major axis and the magnitude of the perihelion distance q were highly variable. This is a reflection of the well-known fact that it much easier to change the angular momentum than it is to change the energy. The most interesting aspect, however, was that orbits with initial perihelia $<$ 3 AU had their perihelia raised within several orbits to distances lying outside the solar system. For a wide variety of initial conditions, the perihelion distance underwent periodic variation with the longest periods approaching 10^9 yr. Thus, within a time scale of just a few orbits, these long-period comets were perturbed out of the solar system not to return for $\sim 10^9$ yr. In a recent numerical study of 150 long-period orbits, Harrington (1985a) has extended this work by including the vertical forces of the disk which dominate over the radial forces. The results indicate that orbits with small perihelia have their perihelia raised out of the solar system in a *single orbit*. The exceptions to this rule are low-inclination orbits with their major axes aligned with the direction towards the galactic center. This tends to explain the pronounced clustering of orbits when integrated backwards (Ovenden and Byl 1978) as a novel form of selection effect. Harrington concludes, therefore, that the observed distribution of perihelia is not inconsistent with an Oort cloud that is uniformly distributed in directions over the sphere. Matese (1985) in an independent study of the effect of the vertical forces of the galactic disk likewise concludes that such perturbations can impart a galactic lati-

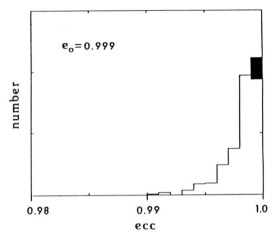

Fig. 5. Effect of vertical tide of the Galaxy on eccentricities of Oort cloud bodies. Initial eccentricity was $e_o = 0.999$ and initial semimajor axis was $a_o = 30{,}000$ AU. The solid part of histogram denotes the fraction of objects perturbed to perihelia < 5 AU.

tude dependence to comets injected from a spherically symmetric cloud. The lesson in this is that care must be taken to account for all significant perturbations when interpreting the distribution of orbital elements, especially angular elements.

Given the large effect of the galactic tides on the angular momenta of comets found by Harrington (1985a), then conversely, the possibility exists for the tides to be capable of injecting comets with perihelia outside the solar

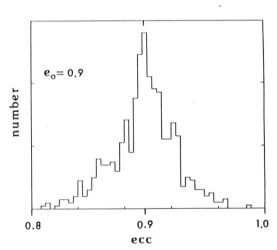

Fig. 6. Same as Fig. 5 except $e_o = 0.9$. Note the marked redistribution in eccentricity induced by the galactic tides.

GALACTIC INFLUENCE ON THE OORT CLOUD 157

system directly into the solar system. This process can, therefore, compete with passing stars in delivering "new" comets. In Oort's (1950) model and subsequent studies (see Weissman chapter for details) concerning the equilibrium influx of long-period comets, it has customarily been assumed that injection of new comets is solely due to perturbations by random passing stars. However, as emphasized by Tremaine and Muller (see their respective chapters), the action of the galactic tides can significantly influence the angular momenta of Oort cloud comets. In subsequent work, this effect has been quantitatively studied (Morris and Muller 1985; Torbett 1986; Heisler and Tremaine 1985) with the conclusion that the galactic tides, and in particular the vertical tides, seem to be at least as efficient at delivering new comets to the inner solar system as are passing stars. The efficient redistribution of eccentricity induced by the action of galactic tides can be seen in Figs. 5 and 6 (Torbett 1986). There, the Oort cloud is modeled by a 600 particle ensemble uniformly distributed over the sphere. The equations of motion were integrated for two orbits subject to the vertical forces given by Eq. (6) for comets at semimajor axes of 30,000 AU and a single initial eccentricity as indicated. Note the marked redistribution of eccentricities that occurs within just two orbits which can very effectively randomize the Oort cloud. The shaded portion of Fig. 5 denotes the fraction of comets injected to perihelia < 5 AU. Shown in Fig. 7 is the injection efficiency $\epsilon(e)$ as function of initial eccentricity. Numerical integration of these curves, interpolated for a mean semimajor axis of the Oort cloud $\bar{a} = 25,000$ AU, yields an injection efficiency given by

$$\eta = \int_0^{e_{max}} \epsilon(e) f(e) de \simeq 3.3 \times 10^{-11} \text{ yr}^{-1} \text{ comet}^{-1} \qquad (11)$$

where $f(e) = 2e$ is the distribution of eccentricities in statistical equilibrium and $e_{max} = 0.9998$ corresponding to $q = 5$ AU. In order to account for the observed influx of long-period comets, this efficiency requires that the Oort cloud need consist of only $\sim 6 \times 10^{11}$ comets (Torbett 1986). That this is roughly a factor of two smaller than population estimates based on stellar perturbations alone, allows one to conclude that the action of galactic tides seem to be roughly twice as efficient in delivering new comets as are passing stars. This conclusion was independently arrived at by Heisler and Tremaine (1986) who obtained a factor of 1.4 for the increase in injection compared to stars.

The action of the tides are thus strong enough to surmount, in a single leap, the planetary barrier against diffusion of cometary perihelia into the inner solar system (see the chapter by Weissman). This efficient redistribution of angular momenta also means that most new comets making a first pass near the Sun are prevented from ever making second passes. Hence, the equi-

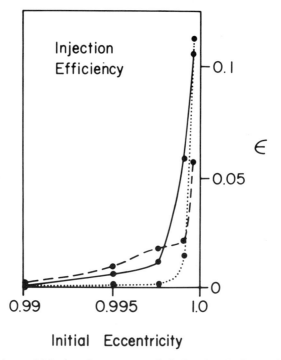

Fig. 7. Efficiency of injection of comets to perihelia less than 5 AU as a function of initial eccentricity for various semimajor axes. Dotted: 20,000 AU; solid: 30,000 AU; and dashed: 40,000 AU. Dotted: 20,000 AU; solid: 30,000 AU; and dashed: 40,000 AU.

librium distribution of comets as a function of perihelion distance will look rather different than that derived by Weissman (see his chapter, Fig. 4) where only stellar and planetary perturbations are considered. In short, then, the influence of the tides of the Galaxy, not considered until recently, can produce effects which require reassessment of Oort cloud dynamics.

Oort Cloud Size and Stability Limits

As noted in Sec. I, the maximum observed semimajor axes for Oort cloud comets is roughly 40,000 to 50,000 AU. However, as we have seen, the tidal limit imposed by the Galaxy is, at $\sim 10^5$ AU, significantly in excess of that value. The dynamical limit due to perturbations by passing stars also turns out to be insufficient to explain the size of the Oort cloud. Monte-Carlo studies of the orbital evolution of comets over the age of the solar system subject to gravitational influences of stars indicate that bodies with semimajor axes $\lesssim 10^5$ AU are stable (Weissman 1980). Thus, both galactic tides and passing stars seem to be unable to limit the Oort cloud to its observed size. Accordingly, we next turn our attention to the dynamical influence of the best candidate for the missing perturber—giant molecular clouds.

GALACTIC INFLUENCE ON THE OORT CLOUD 159

III. GRAVITATIONAL ENCOUNTERS WITH GIANT MOLECULAR CLOUDS

The missing perturber of the Oort cloud was probably identified by Biermann (1978) when he drew attention to the fact that the average density of the molecular gas in the Galaxy is roughly comparable to the effective density of the stars that can significantly perturb Oort cloud comets. For a passing star to provide sufficient relative impulse to change the energy of a comet enough to unbind it, a rather close approach of the star to either the Sun or the comet is required. Biermann showed that this requires restricting impact parameters $\lesssim 1/12$ pc with respect to the comet or the Sun. The fraction of the mass of passing stars that approach that close to the Sun or comet was noted to be approximately equal to the density of interstellar molecular matter. The mass contained in molecular clouds, unknown at the time of Oort's original analysis, was suggested to be equally as important as stars in delivering new comets to the inner solar system. Furthermore, the statistical increase in velocity due to gravitational stirring by giant molecular cloud encounters was suggested as sufficient to unbind Oort cloud comets with semimajor axes in excess of the observed limit.

Giant Molecular Cloud Properties

In order to evaluate adequately this premise, it is necessary to consider the molecular cloud environment through which the Sun moves. Recent carbon monoxide surveys of the Galaxy put the mean surface density of molecular matter in the region of the Sun at $< \sigma(H_2) > \sim 2$ to $5.2\ M_\odot\ pc^{-2}$ (Sanders et al. 1984; Dame and Thaddeus 1985; see also Blitz [1985] and the chapter by Sanders and Scoville). This translates to a mean volume density of $\bar{\rho} \sim 1$ to $3 \times 10^7\ M_\odot\ kpc^{-3}$. The size distribution of clouds for moderate-to-large masses is given roughly by $n(m) \sim km^{-1.5}$ (Thaddeus and Dame 1983; Drapatz and Zinnecker 1984). This distribution can readily be integrated to give the number of clouds per cubic kiloparsec larger than given mass m

$$N_m = \frac{\bar{\rho}}{m_2}\left[\left(\frac{m_2}{m}\right)^{1/2} - 1\right] \qquad (12)$$

where m_2 is the maximum mass expected for clouds. The number N_m is sensitive to the upper mass cutoff and is shown in Table I for several values of m and m_2.

The number of encounters with clouds more massive than a given mass is given, then, by the product of the number density N_m times the volume swept out by a circle of radius equal to a given impact parameter b. Taking the velocity of the solar system with respect to the local standard of rest as $v \sim 20$ km s^{-1}, then the volume swept out by a cylinder of radius b over the age of the solar system T is

$$V = \pi b^2 \mathrm{v} T \sim 0.26 \text{ kpc}^3 \left(\frac{b}{30 \text{ pc}}\right)^2. \qquad (13)$$

Hence, the number of encounters expected over the history of the solar system with clouds of mass $> 10^5$ $M_\odot$ is given by

$$N = VN_m \sim 2 - 15\left(\frac{b}{30 \text{ pc}}\right)^2 \qquad (14)$$

where the mass cutoff is assumed to be $m_2 = 10^6$ $M_\odot$. The intervals between cloud encounters will then be

$$\Delta t = \frac{4.5 \times 10^9 \text{ yr}}{N} \sim 300 \text{ to } 2500 \text{ Myr}. \qquad (15)$$

In addition to the objection by Thaddeus (see his chapter), this interval is large enough to cast doubt on the model for periodic extinctions wherein galactic plane crossings modulate comet influx via enhanced cloud encounters (Rampino and Stothers 1984a; Schwartz and James 1984). The solar system will simply not encounter giant molecular clouds frequently enough for that model to work. The size of this interval can also imply that the Sun's galactic motion may have been roughly the same for the time span covered by the geologic record.

The velocity of the solar system relative to the molecular cloud system is also an important quantity in determining the dynamical effect of collisions with clouds. The ratio of collision time scale to cometary orbital period will turn out to be an important parameter describing an encounter. A recent determination of the velocity dispersion of massive clouds finds $\Delta v \sim 6$ to 8 km s^{-1} with the motion of the Sun relative to nearby clouds being $v_{rel} \sim 4$ km

TABLE I
Number of Clouds (N_m) with Mass Greater than a Given Mass m kpc^{-3} for Various Values of m and m_2 (Maximum Mass Expected for GMC's)

m \ m_2	5×10^5	10^6	5×10^6
5×10^4	130	104	54
10^5	74	65	36
5×10^5	—	25	18

GALACTIC INFLUENCE ON THE OORT CLOUD 161

s^{-1} (Stark 1984). If there is a systematic motion of the clouds with respect to the stars, as presumed in the ballistic particle model (Bash 1979), then a systematic motion of $\sim$ 10 to 20 km s^{-1} is expected between clouds and stars. The largest velocity component will be that due to stellar velocity dispersion itself. The rms velocity of GO dwarfs is v_{rms} = 37 km s^{-1} (Mihalas and Binney 1981). This is to be compared to the Sun's present motion as determined by Hut and Tremaine (1985) to be $v_{\odot rms}$ = 18 km s^{-1}. The disparity in these values suggests that the Sun's motion in velocity space has evolved with perhaps a recent (i.e., within the last 300 to 2500 Myr interval discussed above) scattering event to low velocities. Taken together, these results indicate that the relative velocity between the solar system and giant molecular clouds might be expected to have been anywhere in the range $v_r \sim$ 1 to 30 km s^{-1} with the extremes rather rare and velocities $v_r \sim$ 10 to 20 km s^{-1} most frequent.

Collisions with Giant Molecular Clouds and Tidal Stripping of the Oort Cloud

Close collisions of the solar system with giant molecular clouds as indicated previously, have the potential of seriously influencing and possibly disrupting the Oort cloud of comets. While superficially appearing similar to the gravitational encounters between galaxies with the resulting tidal stripping, the collision of a molecular cloud with the solar system more closely resembles the interaction of binary stellar systems with passing stars. This is due to the fact that galaxies are true n-body problems and hence subject to collective effects inherent to such systems while the comet-cloud system can be considered to be a superposition of individual 3-body problems because comet-comet interactions are negligible. The dynamics of binary-single star scattering have been extensively studied in the past decade (see, e.g., Heggie 1975; Hut and Bahcall 1983; Hut 1983a,b, and references therein). These analyses apply well to perturbations due to passing stars; however, close inspection shows that they have limited applicability to the present problem due to the large mass of the perturber and the long duration of the encounter.

In the case of the gravitational effect of passing stars on a long-period comet, the length of time the star is in range where its influence is important is short compared to an orbital period. An encounter time scale can be defined as the Oort cloud size (justified below) divided by a representative stellar velocity of $\sim$ 20 km s^{-1} yielding $\tau \lesssim 5 \times 10^4$ yr whereas Oort cloud orbital times scales are typically 3 to 10 Myr. Accordingly, the assumption normally made for passing stars is that the gravitational effect can be considered to be an impulse delivered instantaneously at one point in the orbit of the comet in what is known as the impulse approximation. The disturbing influence relative to the Sun on the comet is given by the difference between the individual velocity increases experienced by the Sun and the comet. The change in velocity experienced by the Sun is given in the impulse approximation by $\Delta v_\odot$

TABLE II
Velocity increment, $\Delta\Delta v$ in m s^{-1}, suffered by a comet in the impulse approximation due to an encounter with a 10^5 M$_\odot$ interstellar cloud at $b = 30$ pc impact parameter for various values of encounter velocity[a] and comet semimajor axis

a (AU)	$\Delta\Delta v$ (m s^{-1}) v_r (km s^{-1})				V_{esc} (m s^{-1})
	1	10	20	30	
30,000	256	26	13	7	149
50,000	420	42	21	10	98
70,000	590	59	30	15	64

[a]Compare the $\Delta\Delta v$ value with the columns at the right which gives the escape velocity for given semimajor axes.

$= 2GM/bv_r$ (Oort 1950) where M is the mass of the perturber passing by at a distance b with velocity v_r and G is the gravitational constant. The velocity change suffered by the comet is $\Delta v_c = 2GM/dv_r$, where d is the comet-perturber distance. In the impulse approximate the net velocity change of an optimally situated comet relative to the Sun is then

$$\Delta\Delta v = \Delta v_c - \Delta v_\odot = \frac{2GMa}{b^2 v_r} \quad (16)$$

where a is the comet semimajor axis. In order to induce values of $\Delta\Delta v$ on the order of orbital velocities requires a star to pass within a distance ~ 0.3 pc to either the comet or the Sun and thus validates the assumption of interaction distances being roughly equal to Oort cloud size. Listed in Table II are values of $\Delta\Delta v$ imparted to comets in the impulse approximation by a collision of a 10^5M$_\odot$ cloud at an impact parameter of 30 pc for various semimajor axes and encounter velocities. Escape velocities are also listed for comparison.

For giant molecular clouds (GMC), the impulse approximation is not such a good assumption since in that case the encounter time scales are on the order of orbital periods. To effect an influence comparable with that of a passing star in the impulse approximation (Eq. 16), a molecular cloud must encounter the solar system only within a distance of

$$b \sim 130 \left(\frac{M}{10^5 M_\odot}\right)^{1/2} \left(\frac{V_r}{10 \text{ km s}^{-1}}\right)^{-1/2} \text{pc}. \quad (17)$$

The time scale for such an encounter is roughly $\tau \sim 2b/v_r \sim 26$ Myr. Thus, the impulse approximation is clearly inappropriate for such an encounter.

The applicability of the impulse approximation will depend on the ratio of encounter time scale τ_{enc} to orbital period τ_{orb}. For $\tau_{enc} \ll \tau_{orb}$, the impulse approximation is a good description of the motion. For $\tau_{enc} \gg \tau_{orb}$, however, the comet will make many orbits during the encounter and adiabatic invariance will insure that the dynamical influences will be minimal. If $\chi = \omega_c/\omega_{GMC} \sim \tau_{enc}/\tau_{orb}$ is the ratio of the angular velocity of the comet ω_c to the angular velocity of the interstellar cloud ω_{GMC} as seen from the Sun, then the energy change is smaller than that expected on the basis of the impulse approximation by a factor $\chi^{3/2}e^{-\chi}$ (Hut and Tremaine 1985; Yabushita 1972). Thus, for $\chi \gtrsim 1$ the impulse approximation seriously overestimates tidal stripping. However, for $\tau_{enc} \sim \tau_{orb}$ the possibility exists for a resonant enhancement of the perturbation. An early numerical study of the stripping of stars by interstellar clouds from a stellar cluster with a quadratic gravitational potential found that a weak, broad resonance exists for $\chi \sim 1$ (Spitzer 1958). Similarly, one also expects a resonant enhancement of stripping efficiency for $\chi \sim 1$ in the present context of a $1/r$ potential. The resonance is expected to be relatively weak and broad because in a hyperbolic encounter there is a range of frequencies with the angular velocity at closest passage weighted most heavily. Self-consistent modeling at the limits of stability of this dual tidal interaction from GMCs and the Galaxy at the limits of stability is being planned.

Single Encounters

Under the assumption of the impulse approximation, the fate of a comet in an encounter with an interstellar cloud can be determined by balancing the increase in energy relative to the Sun with its gravitational binding energy. The energy increase suffered by a comet is given by

$$\Delta E = \mathbf{v}_o \cdot \Delta \Delta \mathbf{v} + \tfrac{1}{2}(\Delta \Delta \mathbf{v})^2 \qquad (18)$$

(Bailey 1983). The first term is a random walk or thermalizing effect whose sign and magnitude depends on the orbital orientation of the comet but whose net effect averages to zero. The second term is a systematic energy increase leading to an outward drift. Focusing on the second term, in order for the comet to become unbound requires that its velocity increment per unit mass, satisfies

$$\tfrac{1}{2}(\Delta \Delta v)^2 \sim \Delta \Phi \qquad (19)$$

where $\Delta \Phi$ is the required change in gravitational potential. Recall that the comet need not be removed to infinity to become unbound, but must only to go out to $\sim 10^5$ AU at which point the comet is lost to the Galaxy as a whole. For the average comet, this effect makes a $\sim 25\%$ difference in velocity from that required to go to infinity. Solving for a_c, the critical semimajor axis beyond which the comet becomes unbound, one obtains

$$a_c = \left(\frac{M_\odot v_r^2}{2GM_c^2}\right)^{1/3} b^{4/3} \quad (20)$$

or

$$a_c \sim 160{,}000 \left(\frac{v_r}{10\ \text{km s}^{-1}}\right)^{2/3} \left(\frac{b}{30\ \text{pc}}\right)^{4/3} \left(\frac{M_c}{10^5 M_\odot}\right)^{-2/3}\ \text{AU}. \quad (21)$$

Careful inspection shows that a_c can be $\sim 35{,}000$ AU or roughly the boundary of the Oort cloud if $v_r = 1$ km s^{-1} or $M_c = 10^6 M_\odot$. Since the number of

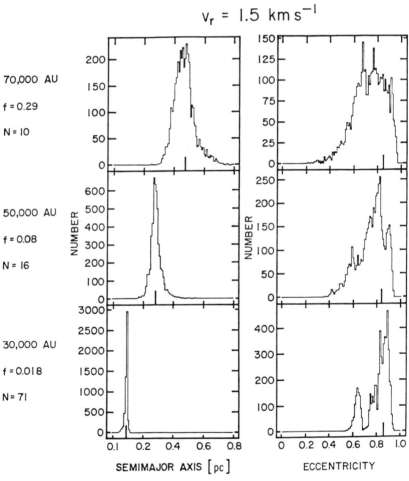

Fig. 8. Resultant distributions in semimajor axes and eccentricity following an encounter with a $10^5 M_\odot$ giant molecular cloud at impact parameter $b = 30$ pc. Initial values for semimajor axes and eccentricity are indicated by heavy tick marks. The encounter velocity was $v_r = 1.5$ km s^{-1} at infinity.

$v_r = 11$ km s^{-1}

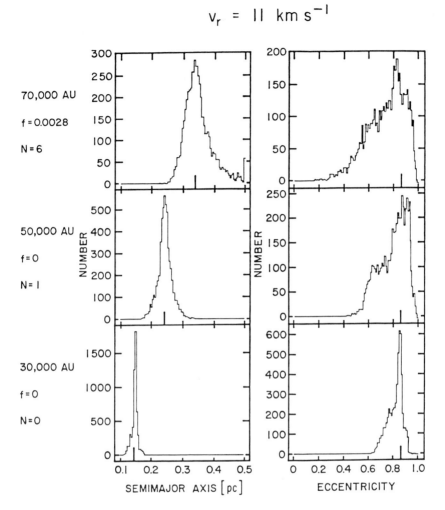

Fig. 9. Same as Fig. 8 except that $v_r = 11$ km s^{-1}.

encounters expected is low, a 1 km s^{-1} collision is highly unlikely; yet, an encounter with a 10^6 M$_\odot$ interstellar cloud is somewhat more probable. Thus, although possible, it is unlikely that a single encounter with a giant molecular cloud will have caused serious damage to the Oort cloud. Such an encounter can, however, sufficiently reshuffle the Oort cloud to induce a comet shower.

In order to model an encounter between the solar system and an interstellar cloud, the numerical code detailed earlier (Smoluchowski and Torbett 1984) was modified to include a fourth point mass to serve as interstellar cloud. Representative collisions were adopted with the parameters $M_c = 10^5$ M$_\odot$, $b = 30$ pc, $v_r = 1.5$ and 11 km s^{-1}. The response of the Oort cloud to the collision was modeled by constructing a spherical cloud of ~ 5700 comets distributed uniformly over the sky. The integrations were started at a time 4 to 6 orbits, depending on semimajor axis, prior to closest approach and continued for 6 to 8 orbits following the encounter. The semimajor axis, eccentricity and perihelion distance were monitored and the average values for the integrations calculated. The results are plotted in Figs. 8 and 9. Note the resultant distributions in semimajor axis and eccentricity for single initial values of a_o = 70,000, 50,000 and 30,000 AU and e_o = 0.85 indicated by heavy tick marks on the horizontal axes. The other quantities listed in these Figs. are f, the fraction of the Oort cloud that is tidally stripped and N, the number of comets perturbed to have perihelion < 5 AU and hence injected as new comets. The breakdown in the impulse approximation can be seen in the smaller disruption efficiency for v_r = 1.5 km s^{-1} than expected in the impulse approximation (see Table II). From these simulations it can be seen that the dynamical influence of a molecular cloud encounter is to impart a random walk to the semimajor axes of Oort cloud comets with an outward drift towards the stability limit as mentioned earlier. The number of comets removed from the Oort cloud ($a \sim$ 20,000 to 50,000) appears to be rather small even with the small relative velocities adopted. Thus, single encounters are not expected to do much damage to the Oort cloud.

Multiple Encounters

Molecular clouds are an order of magnitude more disruptive than stars as was pointed out by Spitzer (1958) for stellar clusters with quadratic gravitational potentials. As mentioned earlier, Biermann (1978) was the first to estimate the net dynamical effect of multiple encounters with interstellar clouds in a cometary context. Summing the squares of the $\Delta\Delta v$'s in the impulse approximation with the interstellar medium properties known at that time, the energy increases due to traversals through clouds were claimed to be roughly comparable to the binding energy of comets at semimajor axes $a \sim$ 0.25 pc or roughly the observed outer limit. Thus, the present outer boundary might be understood as due to "pruning" by molecular clouds.

The dynamical consequences of encounters, strictly penetrating encounters, with giant molecular clouds has been examined with numerical integrations of the equations of motion (Napier and Staniucha 1982). The Oort cloud was modeled by adopting a grid of randomly oriented orbits uniformly distributed in semimajor axis and eccentricity and the interstellar clouds were assumed to be collections of point-mass cloudlets of mass $M = 2 \times 10^4$ M$_\odot$. Considering 28 encounters with mean relative velocity v_r = 12.5 km s^{-1} and dispersion $\Delta v_r \sim$ 8 km s^{-1} at impact parameters $b \sim$ 5 pc, they concluded that the Oort cloud would be removed down to a limiting size of $\sim$ 8000 AU. They further concluded that if such is the case, then clearly these results contradict a primordial origin hypothesis for comets and is taken to imply an interstellar capture hypothesis (see also, Clube and Napier 1984a). However, this conclusion is not compelling because, if indeed the Oort cloud is removed, it could be continually resupplied from the dense, inner cloud discussed in Sec. I (see also Sec. IV below).

A subsequent analytical analysis (Bailey 1983a, 1986) of the dynamics of Oort cloud comets subject to both passing stars and molecular clouds concludes that the mean energy transfer rate to comets is dominated by the contribution from molecular cloud encounters. In the limiting case for a single close encounter, Bailey writes the net relative energy gain per unit mass of a comet under the assumption of impulse approximation and point-mass perturbers as

$$\Delta E = 1/2 (\Delta \Delta v)^2 = 2 \left(\frac{GM}{v_r} \right)^2 \frac{r^2 [1 - (\mathbf{r} \cdot \mathbf{v}_r)^2]}{b^2 d^2} \quad (22)$$

where r is the Sun-comet distance at the time of impulse and the other quantities are as defined earlier. Considering only the systematic energy increase, the mean energy transferred per encounter as a function of semimajor axis is given by

$$\Delta E(a) \sim \frac{4 G^2 M^2 n \langle r^2 \rangle}{3 v_r^2 b^2 d^2} \quad (23)$$

where n is number density of interstellar clouds and where $\langle r^2 \rangle$ is the average Sun-comet distance or roughly $(1 + \frac{3}{2} e^2) a^2$. Integrating over impact parameters yields the mean energy transfer rate

$$\dot{E}(a) \sim \frac{8 \pi^2}{3} \beta_1 (G \bar{\rho})^2 t_0 \left(1 + \frac{3}{2} e^2 \right) a^2 \quad (24)$$

where $\bar{\rho}$ is mean mass density of molecular clouds, β_1 is a factor of order unity arising from the integration over b, and t_o is the time over which the energy transfer rate is being considered. This last factor derives from the fact that, for longer time intervals, a smaller impact parameter can be expected which leads to a correspondingly higher $\dot{E}$. Integrating this mean energy transfer rate from an initial semimajor axis to infinity, Bailey finds a half-life $t_{1/2}$ for a given semimajor axis

$$t_{1/2}(a) \simeq 0.05 \left(\frac{M_\odot}{G} \right)^{1/2} \frac{a^{-3/2}}{\bar{\rho}}. \quad (25)$$

Assuming a mean value of $\bar{\rho} = 0.02 \, M_\odot \, pc^{-3}$, this can be solved for the critical semimajor axis a_c for which the half-life is equal to the age of the solar system and thus a limit beyond which comets are expected to have been removed. The evaluation gives

$$a_c \sim 10,000 \text{ AU} \quad (26)$$

and suggests that the survival of a primordial Oort cloud may be in doubt. Its present existence would require continual resupply from the dense, inner cloud.

The two preceding studies have been criticized on grounds that to assume that giant molecular clouds or their substructure can be considered as point masses grossly overestimates their dynamical effect. The point-mass approximation seriously overestimates the number of close, i.e., most destructive encounters. Clearly, the hierarchical structure of interstellar clouds (Scalo 1985) will weaken somewhat the conclusions concerning disruption based on the point-mass approximation. A recent study attempting to account for this and other corrections concludes that the Oort cloud has probably not been seriously disrupted (Hut and Tremaine 1985). Employing analytic formulae for the half-life for disruption of binary stars by passing objects, they write for the half-life

$$t_{1/2} = 1/8 \times 10^7 \text{ yr} \left(\frac{\rho}{M_\odot \text{ pc}^{-3}}\right)^{-1} \left(\frac{a}{25{,}000 \text{ AU}}\right)^{-3/2}. \quad (27)$$

Note that a straightforward evaluation of this expression for $a = 8400$ AU gives a half-life equal to the age of the solar system, close to the result found by applying the analysis of Bailey (1986). In order to account for a variety of effects likely to influence disruption, the authors write the corrected half-life $t^*_{1/2}$ in terms of the above value as

$$t^*_{1/2} = \frac{t_{1/2}}{f_z f_e f_t f_s f_p f_g} \quad (28)$$

where the terms in the denominator are efficiency factors representing various modifications to the half-life. The factors f_z and f_e account for variations in disruption rate due to excursions vertically and radially, respectively, to regions of different gas density. The factor f_t describes the departure of the disruption rate from that predicted on the basis of the impulse approximation. The finite structure of the clouds is taken into account by the factor f_s, while possible enhanced gas density in the past and thus increased disruption is described by f_p. Finally, the increase in encounters and hence, disruption at given impact parameters due to gravitational focusing is accounted for by f_g. The values adopted are $f_z = 0.5$, $f_e = 0.7$, $f_t = f_g = 1$, $f_p = 1.5$ and $f_s = 0.5$ yielding $t^*_{1/2} = 2.8 \times 10^9$ yr for a half-life and leading the authors to conclude that giant molecular clouds have had a substantial but not a devastating effect on the Oort cloud. From a comparison with the effect of passing stars, Hut and Tremaine conclude that giant molecular clouds are only about as effective as stars in pruning the Oort cloud.

The principal difference in these treatments can be attributed to the discrepant values used for f_s, the factor describing the effect of not considering the clouds or their substructure to be point masses, the gravitational focusing factor f_g and to a lesser degree the value adopted to account for the radial excursions f_e. Bailey (1986) adopts $f_s = 1$ corresponding to treating the clouds as point masses, $f_g = 0.5$ and $f_e = 1$. Hut and Tremaine (1985) take

$$f_s = \min\left(\frac{\Sigma_{av} C}{\Sigma_{crit}}, 1\right) \qquad (29)$$

where Σ_{av} is the mean surface density of a cloud, Σ_{crit} is the surface density corresponding to a maximum column density $N = 10^{23}$ cm^{-2} and C is factor describing the clumpiness. Comparing observed column densities to the maximum value and considering clumping factors $C \sim 2$ to 3 leads to their adopted estimate of $f_s = 0.5$, with the assertion that the correct value may lie in the range $0.01 < f_s < 1$. The uncertainty in the value for this factor, linked to the uncertainty in the internal structure of the clouds, is critical for determining the survival of the Oort cloud. As noted by Hut and Tremaine, if the clouds are rather clumpy, then the half-life can be considerably shortened. By assuming uniform internal density for the clouds, the half-life is significantly overestimated. If, however, one assumes subcondensations of finite size possessing density contrasts of several or more with respect to the rest of the cloud, then the dynamical effect of these cloudlets can become problematical for primordial Oort clouds. More effort must be devoted both to an observational and a theoretical level in order to begin resolving this issue.

Work has begun in an attempt to answer the question of just how important is the clumpiness of clouds on the process of disruption. Employing a hybrid Monte-Carlo scheme with distant encounters treated by a Fokker-Planck analysis and close encounters handled individually by the impulse approximation, the time evolution of wide binary stellar systems has been computed (Weinberg et al. 1986). This study includes perturbations from stars as well as molecular clouds. Realistic distributions for masses and densities for both perturbers are adopted. The galactic tidal field was included by assuming a maximum stable semimajor axis of $a_{tidal} = 1$ pc, which, as we saw in Sec. II, is likely a factor of 2 too large and thus values obtained for the half-life may be overestimated. The internal structure of the clouds are modeled by uniform distribution of point masses each of mass 2×10^4 M$_\odot$ leading to underestimation of the half-life which tends to oppose the effect of an excessive tidal radius. The calculations involved tracking an ensemble of objects with a variety of orbital elements where: (1) both stellar and cloud perturbations are included; and (2) where stellar perturbations are considered alone. Comparison of the two simulations yield markedly different values for the half-life leading the authors to assert that cloud encounters do indeed significantly decrease the half-life. They further conclude that the likelihood of finding binaries with separations greater than 0.05 pc ($\sim 10,000$ AU) is small if their half-life is to exceed half the age of the Galaxy. For case (1) above, the half-life is given by

$$t_{1/2} = 1.7 \times 10^8 \text{ yr} \left(\frac{a_o}{\text{pc}}\right)^{-1.4}. \tag{30}$$

This value is not very sensitive to the density of molecular matter $\bar{\rho}$ and scales as $(\bar{\rho})^{-1/2}$.

In these simulations it was possible to obtain information on the temporal evolution of the probability, $P(a,a_o,t)$, of finding an object with initial semimajor axis a_o at a semimajor axis a after a time t. From this a probability of survival $P(t)$ can be constructed at time t for a given a_o. Shown in Fig. 10 is $P(t)$ for a variety of semimajor axes. Note that P becomes small for all semimajor axes corresponding to the observed Oort cloud and suggests that molecular cloud encounters may indeed disrupt the Oort cloud. In future papers the authors intend to relax the assumption of point mass for the cloudlets to see what effect that assumption has upon disruption.

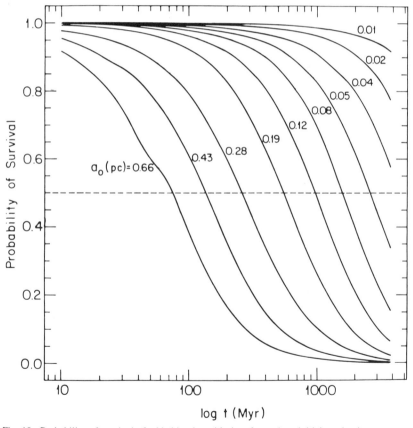

Fig. 10. Probability of survival of wide binaries with time for various initial semimajor axes a_o. Perturbations from both stars and molecular clouds are included. Note behavior for representative Oort cloud object $a_o = 0.12$ pc and the putative solar companion, $a_o = 0.43$. (Figure taken from Weinberg et al. 1985.)

IV. DISCUSSIONS AND CONCLUSIONS

As we have seen, the dynamical influence of the Galaxy can have important consequences for the evolution of a loosely bound cloud of comets around the Sun. There exists an interesting observational constraint that can be brought to bear on this issue: the existence of wide binaries.

Wide Binaries and the Oort Cloud

Wide binary stars with small binding energies are effective sensors of their dynamical environment and can thus serve as good indicators of the strength of perturbations likely to have influenced the Oort cloud. Surveys of binary stellar systems have shown that binaries exist with separations up to ~ 0.1 pc $\sim 20,000$ AU (Bahcall and Soneira 1981; Latham et al. 1984). Hut and Tremaine (1985) claim that these binaries are almost certainly primordial because, based on analytical treatment (Retterer and King 1982) of this problem, scattering of binaries from smaller semimajor axes would produce a flat distribution in binding energy. However, this need not be the case for two reasons. First, the sample number of the observational surveys, 8 and 19, are too low to place much confidence in derived energy distributions. Second, and most importantly, the solutions of the Fokker-Planck equation showing flat distributions in the logarithm of binding energy (see Fig. 4 in Retterer and King 1982) were obtained for the case where the outer limiting radius was essentially infinite since disruption required the binding energy to become zero. In the case of the binaries within the Galaxy, as we saw in Sec. II, the galactic tides impose an outer boundary at binding energies only factors of several smaller than the binding energies of Oort cloud comets themselves. This outer boundary serves as a "one-way membrane" through which diffusive loss would significantly alter the energy distribution function. Specifically, the outer boundary would be roughly in the middle of the flat portions of the distribution functions in the above mentioned figure. This would imply that scattering from smaller semimajor axes need not necessarily produce flat distribution functions and hence, *wide binaries need not be primordial.* Thus, the potential disruption of the Oort cloud by molecular clouds and resupply from the inner Oort cloud is not inconsistent with the existence of wide binaries. In fact, the presence of a cutoff in binary semimajor axes close to the limits of the Oort cloud, i.e., well within the galactic tidal limit or the limit imposed by stars, is significant in that it suggests a common cause. Accordingly, it is reasonable to conclude that the outer boundary of the Oort cloud has been generated by the cumulative tidal stripping by interstellar clouds.

Solar Companion

It has been proposed, in order to account for the putative periodic extinctions, that there exists a low-mass stellar companion to the Sun (Whitmire and Jackson 1984; Davis et al. 1984). In order to match the possible periodicity of extinction events in the geologic record, the companion star would have to have a semimajor axis in the range 80,000 to 90,000 AU. The stability of such an orbit has been called into question. Self-consistent integration of the equations of motion over the age of the solar system incorporating galactic tidal influences alone indicate that orbits with only a limited range of inclinations with respect to the galactic plane are stable (Torbett and Smoluchowski 1984). For galactic inclinations $> \sim 30°$, the orbits do not survive and for inclination of $\sim 90°$ the lifetimes are ~ 500 Myr. Similar calculations that purport to show stability (Hut 1984a) can be criticized on grounds that the tidal limit imposed by the Galaxy may not have been properly taken into account. Inspection of Fig. 4 of Hut (1984) shows that the companion is considered stable for semimajor axes well in excess of 1 pc. This is in disagreement with the results of the study by Smoluchowski and Torbett (1984) wherein particles at that distance are observed to fail to complete a single orbit. The difference in the two simulations may stem from the differences in the adopted coordinate systems, i.e., rotating and inertial, respectively.

At any rate, Hut (see his chapter), Bailey (1986) and others point out that a solar companion is not likely to have started at its presumed present semimajor axis of $a \sim 0.4$ pc but will, over the history of the solar system, have been gradually unbound from smaller semimajor axes as discussed above. This possibility has been criticized by various authors (see, e.g., Bailey 1986) on the grounds that repeated passage of the companion through the inner Oort cloud might effectvely remove most of the comets or at least violate upper bounds for cratering on the outer solar system satellites. Whether or not these objections merit attention remains unclear and further study of the evolution of a comet cloud in a potential double-star system would seem necessary.

In conclusion, contrary to the prevailing view a few years ago, the gravitational influences of the Galaxy and giant molecular clouds are now seen as playing a major role in the evolution of the Oort cloud of comets. The tidal fields of the Galaxy, in particular the vertical tide, can significantly affect the angular momentum of long-period comets enough to assist stars in randomizing the Oort cloud. Moreover, these perturbations can be as efficient as stars in delivering new comets to the inner solar system. The gravitational perturbations due to encounters with interstellar molecular clouds may dominate those due to stars and serve to tidally strip the Oort cloud to its observed size.

Acknowledgments. The author gratefully acknowledges helpful discussions with M. Bailey, S. Tremaine, M. Weinberg, D. Gilden, R. Smoluchowski, P. Hut, and E. M. Shoemaker. Appreciation is graciously extended to W. Taylor at Murray State University for his assistance. The author wishes to thank R. Smoluchowski and the Department of Physics and Astronomy of the University of Texas at Austin for computing funds for some of the lengthy numerical integrations.

COMETARY EVIDENCE FOR A SOLAR COMPANION?

A. H. DELSEMME
The University of Toledo

> A set of 126 cometary orbits that have been the least influenced by the planets has been selected; their orbital angular momenta show a large anisotropy in a plane almost perpendicular to the ecliptic. This anisotropy would dissipate by orbital diffusion in 10 to 20 Myr, and therefore it cannot come from primordial or galactic effects; it must be due to a recent impulsive event in the Oort cloud. Gravitational perturbations due to fast-moving stars or molecular clouds (20 to 30 km s^{-1}) are demonstrated not to produce generally any anisotropy; a massive body is needed that is slow enough to be likely bound to the solar system (200 to 300 m s^{-1}). The strip of the sky centered on its presumed orbit reveals large anomalies in the ratio of retrograde to prograde comets; extending across half of the sky, these anomalies suggest the position of the perihelion of an eccentric orbit. Identification with the Nemesis of the geologists is proposed. Other possibilities are mentioned, although none stands out as more likely than the existence of a slow massive body (not necessarily Nemesis) piercing the Oort cloud < 20 Myr ago.

Nemesis is the name proposed for a hypothetical stellar companion of the Sun. Its periodic passage through the Oort cloud (the huge reservoir of comets surrounding the solar system) would trigger each time a cometary shower in sufficient amount to produce one or a few direct hits on the Earth, every 26 to 33 Myr. The reason for the Nemesis hypothesis (Davis et al. 1984; Whitmire and Jackson 1984) is the apparent periodicity, or at least quasi-periodicity, of the mass extinction of species on Earth, which has been reported by Raup and Sepkoski (1984), and which has attracted much attention and curiosity. We will not take a position here in the discussion about the actual periodicity, which seems to be at best somewhat irregular and of uncertain period, but will turn to the proposed astronomical mechanisms. Only one of these, the Nemesis hypothesis, seems to have survived reasonably well the general discussion among astronomers (chapters by Hut and by Tremaine; see also Hut 1984a), although the stability of its orbit is severely restricted by various conditions (Torbett and Smoluchowski 1984).

The discussions among astronomers have remained theoretical (chapter by Tremaine; Hut 1984a and his chapter; Hills 1984a and his chapter; Clube and Napier 1984c and their chapter). So far, no existing astronomical observations have been used that could directly or indirectly establish the existence of hypothetical Nemesis. However, the assumed passage of slow-moving Nemesis through the Oort cloud some 5 to 15 Myr ago, could have left a signature still visible now. It would have transferred enough forward angular momentum to the perturbed comets to produce some anisotropy in their direction of revolution. After all, the same mechanism has worked, on an entirely different scale, to flatten the system of short-period comets onto the ecliptic, and make it turn in the same direction as the perturbing planets. In order to detect statistically the existence of the impulse of hypothetical Nemesis, we must use young comets, that is, comets that have been the least influenced by the planets (Delsemme 1985a).

I. EXISTENCE OF ANISOTROPY

Among the list of original orbits of comets given by Marsden and Roemer (1982), 126 very young comets were selected, using as a criterion that their present binding energy to the solar system should be < 850 (in 10^{-6} AU^{-1}). The final results have been checked to be rather insensitive to the choice of this somewhat arbitrary limit. The anisotropy discussed later remains very large for the smaller set of 89 purely new comets ($a^{-1} < 90 \times 10^{-6}$ AU^{-1}); but when only 89 comets are considered, the statistical significance is diminished. Standing in contrast, a much larger number of those comets with periods $<30,000$ yr diminishes the anisotropy, presumably because a larger fraction has been more perturbed by the planets.

Since new comets come from an average nominal aphelion distance (Marsden et al. 1978) of 43,500 AU, the orbital diffusion due to nearby stars is not large enough at that distance to bring perihelia in one single period, from beyond the planetary system (> 40 AU) down to 1 or 2 AU where we usually discover them. Our definition of a young comet remains therefore very close but not quite identical to that of a new comet in Oort's (1950) sense; we call them young to avoid confusion, although new comets have also usually diffused through the outer solar system for several revolutions before being discovered.

The orbital angular momenta of 126 young comets have been computed from their velocity at perihelion (in m s^{-1}) multiplied by their perihelion distance (in AU). These momenta have been projected onto three orthogonal planes: plane Oxy of the ecliptic, axis Ox to the vernal point, axis Oy to ecliptic latitude 90° and axis Oz to the north pole of the ecliptic.

The projections q_{xy} and v_{xy} on the plane of the ecliptic, of the perihelion distance q and of the velocity v at perihelion, are given by

$$q_{xy} = q\sqrt{1 - \sin^2\omega \sin^2 i} \qquad (1)$$

$$v_{xy} = v_q\sqrt{1 - \cos^2\omega \sin^2 i} \qquad (2)$$

where ω and i are the argument of perihelion and the inclination of the orbit. The reader can easily deduce the momentum of the projection by multiplying Eq. (1) by Eq. (2), as well as the corresponding equations for planes Oyz and Oxz. Except for a (original) which is found in Marsden and Roemer's (1982) table, the parameters q, ω and i have been taken from the osculating orbits of Marsden's (1983) catalog. Our 126 comets (see chapter appendix) have not been influenced enough by the solar system for their osculating parameters to be very different from the original parameters (except barely for a). The velocity at perihelion is computed by using q and a (original) in the energy equation.

The prograde rotation is defined in each plane, as usual, by the counterclockwise direction, as seen from the positive side of the axis perpendicular to the plane. To avoid confusion, when needed, the term "direct" instead of "prograde" will occasionally be used, only in the plane of the ecliptic.

The prograde and retrograde momenta p_P and p_R have been added separately and compared. The results are summarized in Table I. The orbital angular momenta have been reduced, per unit mass of each comet, in m s^{-1} of their transverse velocities at the nominal distance of the mean aphelion,

TABLE I
Number of Comets and Angular Orbital Momenta, per Unit Mass, of the Set of 126 Young Comets[a]

Plane	Number of Comets			Angular Orbital Momenta		
	R	P	R/P	p_R	p_P	p_R/p_P
Oxy	61	65	0.94	−60.72	+61.97	0.98
Oxz	62	64	0.97	−52.94	+52.52	1.01
Oyz	70	56	1.25	−69.26	+45.01	1.54

[a] Units: m s^{-1} at nominal distance of 43,500 AU.
Resultant angular momentum vector: $(1.25\ z - 0.42\ y - 24.25\ x)$.
Pole of rotation (ecliptic coordinates): $\ell = 181°$, $b = +3°$.
The 126 comets used here are the 126 first entries in Marsden and Roemer's (1982) list of 220 original orbits. Orbital parameters are from Marsden's (1983) *Catalog of Cometary Orbits*.

43,500 AU. This is a convenient way to compare directly with transverse velocities at the distance of the Oort cloud.

In plane Oxy as well as in plane Oyz, the ratios R/P of the numbers of retrograde R to prograde P comets, as well as the ratios p_R/p_P of their retrograde to prograde momenta are not significantly different from unity. However, a surprisingly large excess of 25% more retrograde comets ($R/P = 1.25$) appears in plane Oyz perpendicular to the ecliptic. This excess becomes even more spectacular when the ratio of the momenta ($p_R/p_P = 1.54$) is considered. The resultant vector obtained from the six sums of orbital momenta is also given in the footnote at the bottom of Table I. It points to a pole of rotation whose ecliptic coordinates are $\ell = 181°$, $b = +3°$.

Due to small-number statistics, deviations are possible. The rms deviation $\sigma = \sqrt{Npq}$, with $N = 126$ comets, is $\sigma = 5.6$; therefore the observed deviation from the mean is 1.25σ, implying that the odds are 8 to 2 that the observed distribution is not random. However, the skewness to the momenta is much stronger; to assess its deviation, we still use the binomial distribution (because of the discrete number of comets), but we attribute to each comet a weight in proportion to its momentum. We now reach 1.65σ, or a confidence level of 90%. It will appear later that this confidence level is again considerably enlarged by the specific concentration of the anisotropy in the equatorial zone of the rotation axis.

The accuracy of the results has been assessed by computing the three rms deviations of the momentum distribution in the three planes. The x component of the resultant momentum vector (rotation in the Oyz plane) happens to be very large; therefore the uncertainty in the direction of this vector comes mainly from the uncertainties attached to the y and z components. The rms distribution in the planes xy and xz yields an error ellipse with $\sigma_z = 2°.5$ and $\sigma_y = 2°.2$ (in arc degrees).

II. ORIGIN OF ANISOTROPY

What is the origin of this anisotropy? Three possibilities have been examined:

(1) If a large primordial rotation had existed in the Oort cloud, it would still be superimposed on the diffusion of the transverse velocities (due to passing stars) which grows with the square root of time. The existence of a sink of new comets (by escape and by decay) within the planetary system has produced a well in the velocity distribution, for those transverse velocities that are small enough (-6.3 m s^{-1} < v < $+6.3$ m s^{-1}) to have perihelia within the planetary system. The ongoing diffusion from the two walls of the well is responsible for the present flux of young comets; the larger number of retrograde comets could be explained from the height difference in the two walls of the well, which grows with the asymmetry of the distribution due to the primordial velocity. The observed ratio $R/P = 1.25$ in Oyx would require a

retrograde primordial velocity much larger than the escape velocity from the Oort cloud; the model is therefore completely unrealistic. It does not explain either why the observed revolution is almost perpendicular to the plane of the ecliptic, or why the momentum of *each* retrograde comet is on the average 23% larger than that of *each* prograde comet.

(2) Could tidal or epicyclic effects in the Galaxy have induced a differential depletion in the Oort cloud? Tidal effects distort orbits and change periods, but their symmetry excludes a preferential direction of revolution. Epicyclic orbits turning in a direction opposite to the galactic rotation are the stablest, implying a potential depletion of all the other orbits. However, orbital diffusion due to passing stars and molecular clouds works against such a preferential depletion. The time scales of the two processes must therefore be compared. The time scale for orbital diffusion induced by stars can be assessed from the total impulse Δv due to perturbations (Delsemme 1985a) as

$$\Delta v = 1.8\ T^{1/2}\ \mathrm{m\ s}^{-1} \tag{3}$$

with T in Myr. The transverse velocity change in one specified direction (for instance, that of the galactic plane) is $\Delta v/\sqrt{3}$, that is, close to 1 m s^{-1} per Myr. The time scale T_0 that completely reverses the direction of the orbit of one of the young comets is therefore

$$T_0 = 4q \tag{4}$$

with q in AU and T_0 in 10^6 yr. For perihelia q already within the solar system and observed from Earth, the time scale for orbital diffusion varies therefore between 1 and 20 Myr. The time scale T_1 for preferential depletion by galactic rotation is much larger and thus cannot maintain any anisotropy in the orbits ready to become observable from Earth. An example of a $T_1 > 100$ Myr can be deduced from recent studies (Torbett and Smoluchowski 1984) of the differential depletion of Nemesis-like orbits. The orbits of comets in the Oort cloud usually have even shorter periods, which make them more strongly bound to the Sun; Torbett and Smoluchowski's study recognize that they would be much more stable against galactic effects.

(3) The passage of a massive body could in principle produce an impulse on many comets in the same general direction as its motion. If the passage is recent (3 to 20 Myr) this impulse has not yet been totally dissipated by orbital diffusion. However, a fast-moving body does not produce such a drag along its direction of motion; not only can its impulse be considered to be instantaneous, but it also acts in a direction perpendicular to the body's trajectory and therefore does not produce any change in the R/P ratio of the perturbed comets. A demonstration of this fact is given in Sec. III. Hence, stars or molecular clouds grazing or piercing the Oort cloud cannot explain the observations.

It could be argued that the slowness of a body should be estimated by its relative *angular* velocity; a larger distance might balance a larger linear velocity (Bailey 1983). This is certainly true for the total impulse, but it is not true for anisotropies which depend on a differential impulse, as demonstrated by the theory given below. As the distance grows, the differential impulse becomes vanishingly small.

III. THEORY OF THE ANISOTROPY DUE TO THE PASSAGE OF A MASSIVE BODY

The theory is based on a generalization of the impulse formula. Already used by Oort (1950), the impulse formula is easily derived by integration of the gravitational acceleration due to mass M during its passage along a rectilinear trajectory. It is rigorous only if the comet's displacement can be neglected during the passage of mass M:

$$\Delta v = \frac{2GM}{Dv_*} \tag{5}$$

where G is the gravitation constant and v_* the star's velocity. Acceleration components along the direction of the star's passage cancel each other by symmetry, hence the impulse Δv is directed along D (D is the perpendicular drawn from the comet to the star's trajectory; it is usually called the impact parameter). For simplicity, we assume $D \ll r$ (which is true most of the time) so that we can neglect the Sun's acceleration; see Oort (1950) for the case $r < D$.

To introduce a differentiation between R and P comets, the velocity v of the comet must be used. Equation (3) remains rigorous (even for large v's) if we use it in the inertial frame moving with the comet, where it becomes

$$\Delta v = \frac{2GM}{D|\mathbf{v}_* - \mathbf{v}|} \tag{6}$$

where $|\mathbf{v}_* - \mathbf{v}|$ means the absolute value of the vector $\mathbf{v}_* - \mathbf{v}$. For simplicity, we will discuss the planar case; the reader may check that any v component perpendicular to v_* somewhat diminishes Δv but, for symmetry reasons, does not change the anisotropy ratio, because the latter will be derived from a projection onto the *transverse* velocity plane. In the inertial system moving with the Sun, the impulse Δv has not changed its direction (because it is a velocity *difference*, it is not affected by translation), but $\mathbf{v}_*$ is not generally seen in the same direction as $\mathbf{v}_* - \mathbf{v}$. Therefore $\mathbf{v}_*$ is no longer at right angles to Δv; this is what makes the anisotropy appear.

In order to become eventually observable from the Earth, comets must reach the loss cone, whose cross section in the transverse velocity plane has a

COMETARY EVIDENCE FOR A SOLAR COMPANION

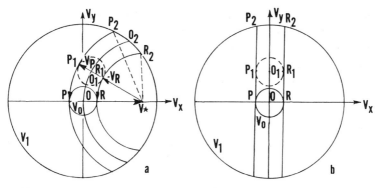

Fig. 1. Transverse cross section in velocity space. The ring between circles with radii v_1 (escape velocity) and v_0 (sink to planetary system) represents all possible transverse velocities in the Oort cloud. The retrograde velocities have been arbitrarily defined by $v_x > 0$. In diagram (a), a perturbing body is assumed bound to the solar system. Its retrograde transverse velocity is $v_* < v_1$. For a given impact parameter D, only the dashed circle P_1R_1 will be perturbed into a circle of radius v_0. However, the locus for comets perturbed into the solar system, at all distances D, is the ring between circles PP_1P_2 and RR_1R_2. To reach P, Δv is larger than to reach R, by the ratio of the two radii v_*P/v_*R. From Eq. (9), D is in proportion to the square of this ratio. The volume of perturbed comets is itself in proportion to D^2, hence the reciprocal of the ratio of retrograde to prograde comets is in proportion to the fourth power of the same ratio. In diagram (b), the perturbing body is a passing star. Its velocity is represented by a point several yards to the right, on v_x. The locus for comets perturbed into the solar system is now limited by two straight lines. The R/P ratio is indistinguishable from unity.

radius v_0 (Fig. 1). A comet with transverse velocity **v** must receive an impulse $\Delta \mathbf{v}$ given by

$$\Delta \mathbf{v} + \mathbf{v} = \varepsilon \qquad (7)$$

with ε being any small transverse velocity with $\varepsilon < v_0$. Therefore, for an arbitrary large **v**, it must be almost (anti) parallel to $\Delta \mathbf{v}$, hence almost perpendicular to $(\mathbf{v}_* - \mathbf{v})$. For clarity, let us assume $\varepsilon = 0$, although the following reasoning can be extended in all generality for any ε, with identical anisotropy results to the first order because of the symmetry of $\pm \varepsilon$ around the origin. In Fig. 1, we assume that the circle of radius v_1 represents the locus of the escape velocities from the Oort cloud at a given heliocentric distance. The Oort cloud comets are all within the ring between the two circles of radii v_0 and v_1. Beyond v_1, comets are lost on hyperbolic orbits. Within v_0, they reach the planetary system and decay quickly. Let us assume that the positive v_x direction is retrograde. The velocity of a retrograde perturbing body, bound to the solar system, can be represented by a point $v_* < v_1$. Let us call R and P the prograde and retrograde velocity limits within which comets enter the planetary system.

As seen in the inertial frame of the comet, the apparent velocity of the perturbing body $\mathbf{v}_* - \mathbf{v}$ can be constructed vectorially by drawing vector $-\mathbf{v}$

(alternately $-\mathbf{v}_P$ or $-\mathbf{v}_R$) from the extremity of vector $\mathbf{v}_*$. Now, the radius $\mathbf{v}_*R_1$ of the circle whose center is $\mathbf{v}_*$ remains parallel to $\mathbf{v}_* = \mathbf{v}_R$, and therefore the direction of the elementary impulse dv, which is known to be perpendicular to $\mathbf{v}_* - \mathbf{v}_R$ by the use of Eq. (4), is also perpendicular to radius $\mathbf{v}_*R_1$ and hence tangent to the circle centered on $\mathbf{v}_*$. During the velocity change due to a large impulse Δv, the changing direction of $\mathbf{v}_* - \mathbf{v}$ is matched by the changing direction of each elementary impulse dv, which remains tangent to the circle everywhere. Large impulses Δv can therefore be represented by vectors with a curvature matching that of the circle at P_1. The same reasoning applies to point R_1, but the circle is accordingly smaller.

The locus of all those comets that will eventually reach the loss cone is therefore the area between the two circles PP_1P_2 and RR_1R_2, both centered on v_*. Those on the left of circle OO_1O_2 will become prograde, those on the right retrograde.

The results are quite general. For instance, if the perturbing body is bound to the solar system, its orbit is not a straight line but the results are not affected to the first order because its projection onto the plane of the sky remains a great circle. Only the position of v_* will be secularly changing along v_x, giving different results at different points of its trajectory. We could also repeat the previous reasoning for angles slightly smaller or slightly larger than the right angle (in order to introduce ε); but the figures remain geometrically similar with the same similitude ratio. Another limitation of the theory is that the frame bound to the comet is not perfectly inertial, but the comet's changing velocity remains fortunately extremely slow in respect to the inertial frame of the Sun.

The ratio R/P of observed comets is essentially derived from the ratio of the segments RR_1 and PP_1. The geometry is such that $v_R/v_P = (\mathbf{v}_* - \mathbf{v}_0)/(\mathbf{v}_* + \mathbf{v}_0)$; in order to compute an *average* ratio R/P, we note that the circle of radius $v_0/\sqrt{2}$ has half the area of the circle of radius v_0, therefore the average v_R/v_P is

$$\frac{V_R}{vP} = \frac{v_* - v_0\sqrt{2}}{v_* + v_0\sqrt{2}} \tag{8}$$

From Eq. (6) we have

$$\frac{D_R}{D_P} \simeq \left(\frac{v_P}{v_R}\right)^2 \tag{9}$$

This is true because, in order to reach the loss cone, $\Delta v \simeq -v$ (the corrections $\pm \varepsilon$ cancel each other to the first order). The volume v perturbed down to a Δv threshold is in proportion to the square of D, hence

$$\frac{v_R}{v_P} = \frac{D_R}{D_P^2} \simeq \left(\frac{v_P}{v_R}\right)^4. \tag{10}$$

With Oort, we assume that the number density of comets per AU^3 follows a Maxwellian velocity distribution with its rms velocity close to the cutoff v_1. It can therefore be approximated by $\propto v^2$, neglecting the exponential term which is always close to unity; hence the number density of R and P comets is constant in velocity space, and $n_P = n_R$ at all points if there is no primitive rotation; therefore

$$\frac{R}{P} = \frac{n_R v_R}{n_P v_P} \simeq \left(\frac{v_P}{v_R}\right)^4 = \left(\frac{v_* + v_0/\sqrt{2}}{v_* - v_0/\sqrt{2}}\right)^4. \tag{11}$$

Taking into account the exponential term of the Maxwellian distribution would bring an exponent larger than 4 and smaller than 6 in Eq. (11) and would therefore enhance the R/P ratio. Since v_0 varies with q^{-1} and v_1 with $q^{-1/2}$, the ratio R/P depends also on the perihelion distance of the perturbing body. If we use $v_* = 0.87\, v_1$ (a bound orbit which is reasonably elongated), one has from Eq. (11) the values given in Table II.

These ratios are valid for most affected comets, because they were perturbed from large distances; however, they are lower limits because of the breakdown of the impulse theory for intermediate distances, and also because close passages have been neglected. Close passages transfer to comets a larger amount of momentum in the direction of the perturbing body; as shown by Tisserand's approximation for the 3-body problem, the whole ring of the velocity space is transported to the right, producing a large depletion of prograde orbits; since the number of comets involved in close passages is probably small, we have not tried to assess this contribution. On the other hand, the observed anisotropy $R/P - 1 = 0.25$ has probably been diluted by a factor of two, if we assume that only half of the 126 comets have indeed been perturbed by a slow-moving body, the other half being due to the standard mechanism of stellar perturbations.

TABLE II
Dependence of the Ratio of Retrograde to Prograde Comets on Perihelion Distance q

q ($\times 10^4$ AU)	R/P
1	≥ 1.53
2	≥ 1.35
3	≥ 1.28
4	≥ 1.24
5	≥ 1.21

All that can be said is that the observed anisotropy is consistent with a perihelion distance of at least 10,000 AU, that could possibly be as large as 30,000 to 40,000 AU. The present simple theory has no other pretense than to show the existence of the effect and its consistency with observational data.

In contrast, if the same theory is used for a fast-moving star, or molecular cloud, their velocity is of the order of 20 to 30 km s^{-1}, thus $v_* \geq 100\ v_1$ and is represented by a point much outside of Fig. 1b. Therefore the two circles $RR_1 v_*$ and $PP_1 v_*$ have an immense radius because their common center v_* has receded so far away. They can be represented by two parallel lines drawn in the vertical direction from R and from P (Fig. 1b). If we assume as an illustration that the perturbed comets are at 30,000 AU, then the ratio of similitude for the mean radii is 1.0003, and $R/P = 1.0012$. Such an anisotropy would clearly be totally undetectable for a fast-moving star as well as for a molecular cloud, whatever its size or its distance.

It is concluded that, in order to produce the observed anisotropy, the massive perturbing body must be slow enough to be probably bound to the solar system; the mean tangential impulse transmitted to the system of perturbed comets will, of course, be exactly in the direction of the body's motion. Its orbit should therefore be retrograde and in the plane of the great circle whose pole is $\ell = 181° \pm 4°.5$ and $b = +3° \pm 5°$ (ecliptic coordinates). This means that it moves in a direction opposite to that of the Galaxy, and at an angle of 28° from the galactic plane.

The previous theory suggests that the original directions of the aphelia of the perturbed orbits should be concentrated in a strip of the sky centered on the great circle whose pole is $\ell = 181°$, $b = +3°$. After an average of a few orbital periods (needed to become visible from the Earth), a fraction of these orbits may have been perturbed enough by the giant planets to scatter their aphelia in direction as well as in distance. If the original event is recent (a few million years), a sizable fraction of aphelia must still be concentrated in the original strip.

IV. MODEL OF THE SKY DISTRIBUTION OF PERTURBED COMETS

In order to understand the observations better, we need a model predicting the *relative* number of comets perturbed at different places in the sky, by the passage of a massive body.

The R/P ratio is of no immediate interest because it has already been established. Then Eq. (5), although less accurate for absolute values, is much easier to use and gives the same results for the *relative* number of comets perturbed along a trajectory.

Knowing that no fast-moving body could induce the observed anisotropies, we will approximate the passage to perihelion of the perturbing body by a parabola. In particular, its velocity v_* varies with distance r as

$$v_* = v_0 r^{-1/2} \tag{12}$$

and the angle α of the parabola with the normal to the radius vector is

$$\sin \alpha = \frac{\sqrt{2}}{2}\sqrt{1 - \cos \theta} \qquad (13)$$

θ being the angle from perihelion to present position.

Let us consider any distance D perpendicular to the trajectory (but not to the radius vector) for which the transverse component Δv_t of the impulse Δv reaches a given threshold.

At perihelion (Fig. 2) the trajectory is perpendicular to the radius vector to the Sun, and the perturbed volume is given by two cylinders tangent to each other and to the trajectory. Their cross section is two circles because $\Delta v_t = \Delta v \cos\phi$ (ϕ being the angle between distance D at point A, and distance D_0 at a right angle to the radius vector to the Sun). For all points, angle A is hence a right angle subtending D_0 as a diameter, and the locus of A is the circle of diameter D.

When far from perihelion, the projection of Δv on the transverse plane (perpendicular to the radius vector to the Sun) is

$$\Delta v_t = \Delta v \sqrt{1 - \sin^2\phi \, \cos^2\alpha}. \qquad (14)$$

Therefore, for a constant Δv_t and from values in Table II and Eq. (13) (Fig. 2)

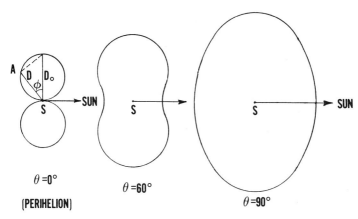

Fig. 2. Three cross sections of the cylinder of perturbed comets, centered on the perturbing body's trajectory, for three different locations: one at its perihelion, the other ones at 60° and 90° off perihelion. Point S is the trace of the trajectory; the Sun's direction is also shown. The curves represent the maximum distance D (from S) where the transverse impulse Δv_t due to the perturbing body's passage is down to the minimum threshold needed for the comet eventually to reach visibility from the Earth. The scaling factor D_0 at perihelion is the perturbed distance at right angles to both the radius vector and the trajectory. The number of comets perturbed is in proportion to the area of the curves, but it also depends upon the number density law of comets in the Oort cloud.

$$D = D_0 \, r^{1/2}(1 - \sin^2\phi \cos^2\alpha)^{1/2}. \tag{15}$$

Here D_0 is the perturbed distance, for perihelion, at right angles to both the radius vector and the trajectory. The area A of the cross section is given by integrating Eq. (14) over ϕ. This yields

$$A = \frac{\pi D_0^2}{2\sqrt{2}} \frac{3 - \cos\theta}{\sqrt{1 + \cos\theta}}. \tag{16}$$

The number density n of comets per AU^3 in the Oort cloud varies with heliocentric distance r. For distances $< 10,000$ AU, the density law is totally unknown; whereas from the distribution of cometary aphelia, there is at least a crude indication that (Delsemme 1977, p. 461, Fig. 4), with $n = n_0 r^{-k}$, the exponent k is close to 1 between 10,000 and 30,000 AU, close to 2 from 30,000 to 60,000 AU, and grows beyond limits from 60,000 to 100,000 AU. Since this is the only information available, it seems reasonable to keep k variable in the number density law, before multiplying it by Eq. (16)

$$N = 2^{1/2-k} N_0 \frac{(3 - \cos\theta)}{(1 + \cos\theta)^{k+1/2}}. \tag{17}$$

N is the number of perturbed comets, per AU along the trajectory. Since the results given in Table II suggest that the perihelion of the perturbing body is probably between 10,000 and 30,000 to 40,000 AU, the exponent k is likely to vary from 1 near perihelion, to larger values with increasing θ. Figure 3 shows four curves for $k = 1.0, 1.2, 1.5$ and 2, illustrating the relative number of perturbed comets $N(\theta)/N_0$, as a function of angle θ to perihelion. Following the previous remark about the variation of k, the dotted line suggests the likely variation of the actual curve, with $k = 1$ from perihelion to $\theta = \pm 30°$ or $40°$, followed by a sharp drop to an abrupt limit near $\theta = \pm 90°$. Only a value of $k > 2$ could diminish sizably the length of the perturbed strip.

The comets perturbed by a slowly moving massive body should therefore meet these two criteria:

1. Those comets should display an almost uniform concentration of aphelia in a strip of the sky centered on the massive body's perihelion; the length of the strip would be about 180°; the width of the strip depends on the mass of the body; order of magnitude estimates show that 30° would be consistent with 10 to 90 times Jupiter's mass (see Sec. IX). Fluctuations due to small-number statistics may make it difficult to identify the shallow minimum centered on perihelion and displayed in Fig. 3 for small values of k.
2. Those comets in the same 180° strip should display a retrograde to prograde ratio R/P much larger (or much smaller) than one. The direction of the massive body is retrograde if $R/P > 1$, prograde if $R/P < 1$.

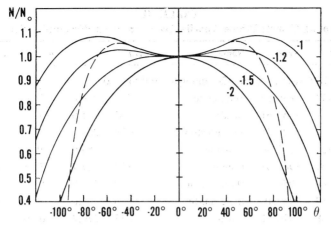

Fig. 3. Relative number of perturbed comets N/N_0 along the trajectory of the perturbing body, as a function of the angle θ, before or after perihelion. The shallow minimum at perihelion comes from the smaller size and odd shape at perihelion of the cross section of the perturbed cylinder surrounding the trajctory. The different curves correspond to different exponents $k = 1.0$; 1.2; 1.5 and 2.0 of the density law $n = n_0 r^{-k}$ of comets in the Oort cloud. Data imply that k grows from 1.0 near 30,000 AU to very large values beyond 60,000 AU; the dotted line is merely suggestive of the probable behavior of the actual curve.

The second half of the great circle which lies outside of the 180° strip should show essentially similar properties as the rest of the sky.

V. COMPARISON WITH THE OBSERVATIONS

Table III tabulates the orbital properties of those 38 young comets whose aphelia are in a strip 30° wide, 15° on each side of the assumed orbit. These comets indeed show a much stronger anisotropy: 25 are retrograde and only 13 prograde; $R/P = 1.92$. The ratio of the two sums of retrograde and prograde momenta is $p_R/p_P = 2.06$. Since the orientation of the strip was imposed by the presumed orbit, it cannot be argued that it was selected by chance or design among the numerous possible subsets of the set of 126 comets. Only its width is somewhat arbitrary, but a width modification of ±20% does not intrinsically change the results. For instance, a width of 24° still contains 31 comets, 21 retrograde and 10 prograde.

However, the major anomalies are not distributed uniformly all along the 360° × 30° strip. They are concentrated in one half of it (see Table IV). This half strip of 180° × 30° goes roughly from $\ell = 270°$, $b = -15°$ through the south pole of the ecliptic to $\ell = 90°$, $b = +15°$. Since it is mainly south of the ecliptic, it will be called for short the southern strip; the other half will be called the northern strip.

TABLE III
Young Comets Whose Aphelia Are < 15° from Presumed Orbit[a]

N	N∘	Year	Name	1/a	ℓ	b	V_{zx}
Southern 180° strip (b from −20° through +90° to +20°)							
1	114	1880.50	Schaeberle	+534	279.0	−28.6	−1.154
2	64	1955.61	Baade	+42	271.9	−34.7	−1.879
3	39	1925.75	Van Biesbroeck	+24	269.2	−46.7	−0.556
4	126	1892.36	Denning	+846	253.9	−50.7	−1.327
5	51	1903.65	Borelli	+33	287.7	−52.5	−0.524
6	60	1956.07	Haro-Chavira	+39	267.9	−55.9	+1.877
7	83	1959.19	Burnham-Sl.	+76	254.6	−59.5	−0.755
8	58	1976.51	Lovas	+37	279.2	−61.0	−2.276
9	123	1975.68	Kobayashi-B.-M.	+821	278.2	−61.6	−0.580
10	110	1863.09	Bruhns	+521	314.4	−73.8	+0.768
11	13	1911.77	Beljaws	−74	250.4	−70.6	+0.532
12	28	1863.99	Baeker	+14	215.2	−76.4	+1.066
13	69	1886.96	Barnard-H.	+46	245.9	−77.8	+0.525
14	84	1941.67	van Gent	+78	32.9	−83.5	−0.881
15	122	1949.82	Bappu-Bok-New.	+735	40.7	−74.2	−1.078
16	15	1932.73	Newman	−56	94.1	−66.8	−1.139
17	53	1948.37	Pajdu.-Mrkos	+34	60.1	−66.7	−1.310
18	87	1890.42	Brooks	+89	88.3	−53.5	−0.884
19	102	1970.80	Abe	+283	100.1	−52.8	+0.375
20	86	1954.23	Abell	+82	78.2	−52.8	−1.337
21	24	1952.54	Peltier	+2	107.3	−45.2	−0.169
22	106	1910.71	Melcalf	+474	77.6	−41.7	−1.224
23	62	1959.94	Humason	+40	95.2	−36.2	−1.164
24	90	1914.42	Kritzin	+126	90.1	−22.7	−0.360
25	44	1941.58	Neujmin	+27	95.4	−13.2	−1.776
26	25	1897.11	Perrine	+5	93.6	−4.3	+0.567
27	16	1975.97	Bradfield	−56	90.0	+0.8	−0.427
28	40	1948.13	Bester	+24	98.3	+6.2	−0.542
29	66	1976.84	Harland	+45	91.1	+8.3	+0.788
Northern 180° strip (b from +20° through −90° to −20°)							
30	108	1949.33	Wirtanen	+498	82.3	+35.7	+1.266
31	10	1957.27	Arend-Roland	−98	67.0	+42.5	−0.383
32	77	1898.70	Codd.-Pauly	+68	99.4	+48.8	+1.205
33	56	1975.64	Lovas	+36	88.2	+49.9	+0.388
34	96	1847.87	Mitchell	+212	81.8	+70.7	−0.132

(*continued*)

TABLE III (Continued)

N	N°	Year	Name	1/a	ℓ	b	V_{zx}
35	57	1954.66	Kresák	+36	81.2	+74.7	+0.808
36	7	1953.07	Mrkos	−125	319.6	+72.3	−0.337
37	1	1976.01	Sato	−734	278.0	+35.4	−0.892
38	14	1898.72	Chase	−71	280.8	−1.8	+0.570

[a]N: running number; N°: number growing with 1/a; Year: epoch of perihelion passage; Name: name of comet shortened to ≦ 13 letters; 1/a: bonding energy in 10^{-6} AU^{-1} units; ℓ and b: ecliptic coordinates of aphelion; V_{zx}: O_{zx} component in m s^{-1} of comet's velocity when at nominal distance of 43,500 AU.

The southern strip contains 29 aphelia, 21 prograde and 8 retrograde; the northern strip contains 9 aphelia, 4 prograde and 5 retrograde. The total number of aphelia is 3.22 times as large in the southern strip as in the northern strip, to be compared with the 1.38 south–north ratio of the aphelia of the 126 young comets. This latter ratio represents rather well the observational bias introduced by the lack of observers in the southern hemisphere. Correction of this bias leaves an anomalous concentration of aphelia by a factor of at least 2.3 in the southern strip. Second, the southern strip is centered on a point more than 30° from the south celestial pole. These two data seem to imply that the concentration is real.

The anisotropy in the aphelion numbers is absent from the rest of the sky and that in the angular momenta is much smaller. The R/P anisotropy detected among the 126 comets distributed in the whole sky comes therefore mostly from the southern strip.

A concentration of aphelia around the antapex has been sometimes mentioned for long-period comets (see Biermann et al. 1983; Lüst 1984). Five of

TABLE IV
Statistical Data on Cometary Aphelia < 15° from Presumed Orbit Plane

	Number of Comets				Momenta in O_{yz} Plane		
	N[a]	R[b]	P[b]	R/P	p_R[c]	p_P[c]	p_R/p_P
360° strip	38	25	13	1.9	−23.09	+10.74	2.15
Southern 180° strip	29	21	8	2.6	−21.34	+6.50	3.28
Northern 180° strip	9	4	5	0.8	−1.75	+4.24	0.41
Sky minus southern strip	97	49	48	1.02	−47.92	+38.51	1.24

[a]N: number of cometary aphelia < 15° from presumed orbit.
[b]R and P: number of retrograde R or prograde P orbits.
[c]p_R and p_P: sums of orbital angular momenta for retrograde R or prograde P orbits.

our aphelia are indeed in a circle of 30° diameter surrounding the antapex; this apparently represents a factor of 2.1 concentration in respect to the whole sky. However, three of these five comets belong to the alignment identified by Biermann et al. as coming from the recent passage of a nearby star; discounting them, the antapex concentration vanishes completely. The influence of Biermann et al.'s (1983) star trek will be considered in Sec. VI below. For now, it is concluded that the antapex concentration is low enough not to introduce any significant bias.

To summarize, after correcting for known biases, the southern strip contains:
  (a) a number of aphelia 2.3 times larger than expected;
  (b) an anomalously large number ratio R/P of 2.6; and
  (c) an even larger momentum ratio p_R/p_P of 3.28.

The rest of the sky (excluding the southern strip) contains:
  (a) a much smaller density of aphelia;
  (b) a quasi-isotropic number ratio R/P of 1.02; and
  (c) a lower momentum ratio p_R/p_P of 1.24.

The southern strip is aligned along an arc of the great circle across half of the sky, defined by the pole of rotation (Table I) of the ensemble of the 126 comets. Its position along the great circle is rather well defined:
  (a) by the conspicuous gaps at its two ends (each of about 27°);
  (b) by the distance of 160° separating the first and the last aphelion present at its two extremities; and
  (c) by its center, that can be defined in different ways. For instance, the mid-distance between the first and last aphelion is near $b = -72°$, but this is based on two aphelia only; the position of the median number of comets, projected onto the presumed orbit, is near $b = -75°$, whereas the mean latitude of the 28 comets is $b = -76°$.

Because the mean latitude takes the distribution of *all* aphelia into better consideration, we accept:

$$\ell = 104° \pm 5° \quad b = -76° \pm 5° \tag{18}$$

as the probable position of the perihelion of the perturbing body.

VI. POISSON PROBABILITY TEST FOR THE DISTRIBUTION OF COMETARY APHELIA

The distinction between various observational biases and the observed clustering and anisotropy in the southern strip has been discussed semi-quantitatively in the previous section. The unreliability of small-number statistics makes it imperative to supplement the previous considerations by a Poisson probability test, to establish whether the density of the aphelia numbers can be explained by statistical fluctuations. For this purpose, the sky is divided into

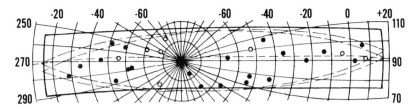

Fig. 4. The southern 30° × 180° strip of the sky which contains the largest concentration of anisotropies. Black circles are the aphelia of those 21 young comets that are retrograde; white circles are the aphelia of those 8 young comets that are prograde; the ratio $R/P = 2.6$. These 29 comets (also found in Table IV) are close to the assumed trajectory of the perturbing body (the central dashed line) found independently as the equator of the mean rotation of the system of 126 young comets. They also show a density concentration 2.3 times larger than expected, after correction of the well-documented north-south bias. Galactic coordinates are used. In order to use the Poisson distribution test, the major anisotropy must be enclosed in one of twelve sectors of equal areas (180° × 30°). The dashed sectors show two alternate solutions, one with 27 aphelia, the other with 26. Even if 26 aphelia only are used, they deviate from a random distribution by 4.78 σ (see text).

twelve 180° × 30° sectors (and also into twenty-four 90° × 30° half-sectors).

To avoid the objection that the limits of the sectors have been made *a posteriori*, we will not select the first sector limits on the basis of its large density of aphelia numbers, but only on the basis of the largest observed concentration in R/P anisotropy. If it is objected that the density of aphelia numbers is not statistically independent from the R/P anisotropy, we will also have made our point, since this fact is predicted by the model and not expected from random distributions. For this reason, the two ends of the Fig. 4 strip are selected for the sector poles, since this strip contains the largest anisotropy. Fig. 4 shows that there is a 4° to 5° leeway to maximize anisotropy. The two possible sectors contain 26 or 27 aphelia. To remain on the safe side, we adopt the sector with the smallest aphelion density. Then the sector poles are $\ell = 96°$, $b = +15°$ and $\ell = 276°$, $b = -15°$. Sector 1 is centered on the meridian of maximum anisotropy (the one that goes through the sector poles and through the south pole of the ecliptic). Sectors are numbered from 1 to 12 in the prograde direction, as seen from axis Oy. For convenience, the 12 sectors are also divided into 24 half-sectors by the great circle perpendicular to the two selected poles.

The resultant distribution in aphelia numbers is given in Table V. The last two lines of Table V give the number of sectors containing, respectively, 1, 2, 3... *n* cometary aphelia.

The number *n* of comets perturbed into visibility in each sector is of course an extremely small fraction of the total number of comets *N* perturbed in the Oort cloud. Although their individual probability *p* of coming into visibility is also extremely small, the product $\lambda \equiv Np$ is finite because several

TABLE V
Poisson Probability Test

Distribution of the 126 young comets into 12 sectors (respectively, 24 half-sectors) of equal area covering the whole sky

Sector Name	1	2	3	3	5	6	7	8	9	10	11	12
Southern half (S)	14	12	6	3	9	3	1	4	4	3	3	2
Northern half (N)	12	5	4	5	6	2	5	2	0	3	7	11
Totals per sector	26	17	10	8	15	5	6	6	4	6	10	13

Number of half-sectors containing a given number of aphelia

Number of aphelia per half sector	0	1	2	3	4	5	6	7	8	9	10	11	12	13	14
Number of half-sectors with this number of aphelia	1	1	3	5	3	3	2	1	0	1	0	1	2	0	1

Number of sectors containing a given number of aphelia

Number of aphelia per sector	4	5	6	8	10	13	15	17	26
Number of sectors with this number of aphelia	1	1	3	1	2	1	1	1	1

comets have been observed in each sector. This is a typical case where the Poisson approximation is excellent. Poisson's equation is

$$P_n(\lambda) = \lambda^n e^{-\lambda}/n! \quad (19)$$

where $P_n(\lambda)$ is the probability of finding n comets in one sector (or in one half-sector) for a given value of λ.

Considering first the 24 half-sectors, the expected probabilities P_n have been plotted as a function of n in Fig. 5a, and compared with square boxes representing the observed numbers of half-sectors that contain each 1, 2, 3,... n comets. The observed numbers are so far away from the Poisson curve for randomness that a Chi Square test is utterly useless. Four half-sectors contain the largest anomaly; they are (Table V): 1S (with 14 comets), 1N (with 12), 2S (with 12) and 12N (with 11). Sector 2N contains a large fraction of the concentration that has been well documented around the solar antapex. Sector 12N contains a large fraction of the comet alignment reported and studied

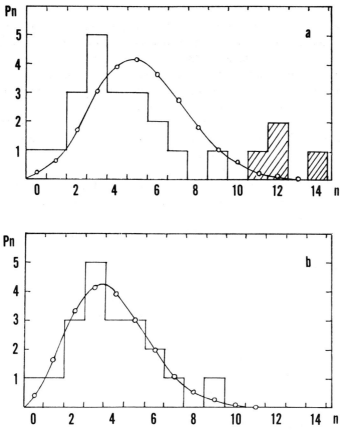

Fig. 5. (a) Poisson probability test. The sky has been divided into 24 half-sectors (90° × 30°). The rectangular boxes represent the number of sectors, each containing n aphelia (the number of aphelia in one half-sector varies from zero to fourteen). The small circles (joined by a continuous curve just to help the eye) represent the probability distribution of the 24 half-sectors to contain 1, 2, 3 . . . n aphelia, *if* the distribution were random. It appears clear that the distribution is not random. The four half-sectors that are completely out of step are those (shadowed with slanted lines) that contain 11, 12, 12 and 14 aphelia, respectively. (b) The four half-sectors that contain the nonrandom distribution of aphelia have been removed. The Poisson probability test, applied this time to the 20 remaining half-sectors, now shows a good fit, demonstrating that the remaining aphelia are distributed at random in the rest of the sky. See text for interpretation of the four half-sectors containing too many aphelia.

conclusively by Biermann et al. (1983) as coming from the recent perturbation of a single star. The two last half sectors 1N and 1S are the two parts of sector 1 which precisely contains the 180° strip with the major anomalies.

If these four half sectors are removed from the distribution, the anomalies disappear from the Poisson distribution (Fig. 5b), and the rest of the sky appears to be reasonably random. In particular, the well-known geographic N-S bias is well discounted, probably because the sectors cut across the two N and S hemispheres at a large angle and also because this bias does not introduce local anomalies of the size of one half-sector.

For our purpose, it is now reasonable to concentrate on the two half-sectors that contain the largest residual anomalies. They are, as if by accident, aligned along 180° of a great circle, since they are the two half-sectors belonging to sector 1.

For this reason, we can reduce the number of boxes to the 12 whole sectors, and consider the Poisson probability of having sector 1 containing (Table V) $n = 14 + 12 = 26$ comets. This is by far the largest number in a sector (the next one contains 17 comets). For 12 sectors, $\lambda = 126/12 = 10.5$ and the rms of the distribution is $\sigma = \sqrt{\lambda} = 3.24$. Here, $P_{26} = 2.43 \times 10^{-5}$ per sector, and $12P_{26} = 2.92 \times 10^{-4}$. Among the 12 sectors, the mere existence of one sector containing 26 aphelia has therefore a probability 3427 times higher than that which would be expected by chance alone. The deviation from the mean is $26 - 10.5 = 15.5$ or, in sigma units: $15.5/3.24 = 4.78 \sigma$.

Of course, because of the connectivity of sectors 12, 1 and 2, the influence of extraneous causes coming from sectors 12 and 2 must now be taken into account.

In particular, Biermann et al.'s alignment of aphelia is well documented, and it goes across the width of sector 1 with three well-aligned aphelia, namely those of comets 1952 VI Peltier, 1954 V Abell and 1890 Brooks. Second, the well-documented concentration near the antapex implies that some of the comets close to the antapex may come from an extraneous cause. But the three comets removed from Biermann's alignment are already most of the antapex concentration. If we remove the two next comets that are the closest to the antapex, namely comets 1970 XV Abe and 1910 IV Metcalf, we produce in sector 1 a large gap of some 30° diameter around the antapex, where there are no more cometary aphelia left. We have clearly overcorrected for the presence of extraneous comets from the antapex concentration; hence, we can reasonably assume that only three comets come from extraneous causes. Excluding them from the statistics, we are left with only 23 comets in sector 1; the deviation still is $23 - 10.5 = 12.5$, or in sigma units $12.5/3.24 = 3.86 \sigma$. This is a confidence level of 99.96% that the concentration of aphelia along the great circle of sector 1 is neither due to chance alone, nor to extraneous causes like Biermann et al.'s alignment or the antapex concentration. Another way to express this result is to emphasize that the concentration of aphelia is not statistically independent, but directly linked to the largest R/P anisotropy of the 180° × 30° strip, a fact that is predicted by our interpretation.

VII. PERIOD AND EPOCH OF LAST PASSAGE

Let's consider now the various uncertainties that cannot be resolved by the data.

Period

Any elongated orbit with a perihelion that pierces the Oort cloud beyond 10^4 AU, but whose aphelion goes at least beyond the outer fringe of the Oort cloud near 50,000 AU, must have a period longer (and probably much longer) than 5 Myr. Orbital stability studies like those of Torbett and Smoluchowski (1984) show that stable orbits do not exist for a period > 50 Myr (and probably much less). The alternate possibility is an unbound orbit, but the velocity distribution of nearby stars and molecular clusters yields a vanishingly small probability of having a velocity at infinity of a few meters per second only. Such a small velocity would, however, be required to produce a perihelion velocity of 200 or 300 m s^{-1}, needed to explain the observed anisotropy.

Epoch of Last Passage

Comets perturbed in the Oort cloud take a highly variable time to come down to the outer planetary system. Indeed, the transit time could be as short as 1 Myr only if the perturbed comet was already on the leg coming back from its aphelion; but it could easily be 3 Myr if the comet was perturbed on the leg going away to its aphelion. The subsequent 2 to 10 passages bringing comets into Earth's visibility do not add much: typically from 0.1 to 1 Myr (Everhart 1967a) spreading somewhat more the duration of the shower. For an individual event (like the one claimed to explain the Cretaceous-Tertiary layer) there is therefore an added uncertainty of at least 2 Myr on the date of the perihelion passage. For the whole cometary shower, we must expect a total spread of 4 Myr at least, 1 to 5 Myr after perihelion passage, a fraction of this duration resulting from the time needed for the perturbing body to go through the Oort cloud.

Because of the random walk of the orbit parameters, the tail of such a shower may subside but remain observable until the anisotropy subsides, that is with a time constant of 10 Myr (Eq. 3). From our data, a smaller shower could have been induced by a perihelion passage 2 to 4 Myr ago. But we could still see the remnants of much larger fireworks dating back from a passage 10 or 15 Myr ago. All that can be said at this stage is that nothing so far is inconsistent with a Nemesis-type object.

VIII. THE NEMESIS HYPOTHESIS

The previous discussion is suggestive of a Nemesis-type object. In spite of the recent dispute concerning Nemesis' existence, its long-term stability or even its necessity (see the chapter by Tremaine), it is clear that we must now consider this hypothesis in earnest.

In principle, an identification with the Nemesis of the geologists brings information about the missing parameters, namely its period and the epoch of its last passage. Unfortunately, the 26 to 28 million-year period originally proposed by Raup and Sepkoski (1984) is now contested; the last of the mass extinctions is imprecise, and cratering evidence rather suggests 28 to 32–33 Myr, depending on whether to trust the 11 impact craters selected by Alvarez and Muller (1984) or the 20 craters used by Shoemaker and Wolfe (see their chapter). Without trying to solve the dilemma at this stage, we will use first a 29 million-year period, which represents the mean between the two extremes. Counting from the best documented mass extinctions, that of the Cretaceous-Tertiary layer, and taking a 2 Myr transit time for the comet shower, Nemesis should have passed its perihelion 9 Myr ago.

The last missing parameter to establish where Nemesis is now in the sky, is its perihelion distance. The anisotropy theory of Sec. III suggests a range rather between 10,000 and 30,000 AU, but is inconclusive because of the theory breakdown at intermediate and short distances; whereas the aphelion distances of those new comets that represent the central half of the southern strip, although very scattered, presumably by orbital diffusion, are rather suggestive of 30,000 AU. For instance, the 6 new comets' aphelia closest to the presumed perihelion have a median distance of 31,000 AU, whereas the median of the central strip ($\pm 40°$) is near 43,000 AU, consistently larger, as expected. The perturbation is maximum near the comet's aphelion because any radial velocity diminishes the duration of its interaction; however, comets may be perturbed elsewhere, as mentioned earlier.

A period of about 29 Myr implies a semimajor axis of 94,400 AU. Accepting the assumption of $q = 30,000$ AU then implies an eccentricity of 0.68, putting the present position of Nemesis 20° before its aphelion. Within the previous assumptions, the location of Nemesis in the sky would therefore be

$$\ell = 278° \pm 5° \quad b = +56° \pm 5°. \tag{20}$$

The safest known date seems to be the perihelion passage preceding the Cretaceous-Tertiary date by 1 to 3 Myr (Sec. VII). With the large uncertainty on the periods, the present position could be 8° after its recent aphelion, for the shortest period that has been proposed, and as far away as 40° before its future aphelion, for the longest period. Taking into account all these uncertainties, the strip of the sky that should be explored is centered on the arc traced between these two extreme positions (in ecliptic coordinates):

$$\ell = 276° \pm 5° \quad b = +84° \pm 5° \tag{21}$$

$$\ell = 275° \pm 5° \quad b = +36° \pm 5°. \tag{22}$$

IX. THE MASS OF NEMESIS

The observed anisotropy in the aphelia identifies a certain number of comets as coming from the tail of the Nemesis shower. This allows two possible verifications:

1. Is the detected shower consistent with mass extinctions on the Earth? (This requires a reasonably high probability of collision with the Earth.)
2. Is the mass of the perturbing body consistent with a Nemesis-type object?

From the orbits observed now, it is possible to deduce how many comets were aimed at the inner solar system during the Nemesis shower. Since the retrograde excess $p_R - p_P$ is 1.63 times as high for the whole sky (including the strip) as for the strip alone, we assume that 39% of the aphelia have diffused away from the strip. This implies that half of all young comets, or 64 comets per century are from the shower. If we distribute the poorly defined parabolic orbits in the proper proportion, then 141 comets have been observed per century belonging to the shower. Because after a few periods they must be equally distributed along all possible orbits, with an average first period of 3.2 Myr, there are still now 4.4 million comets in the tail of the shower. With 50% loss to hyperbolic orbits per passage, and about 4 passages in the outer planetary system before coming into Earth visibility, the retrograde excess now observed corresponds to an event of 70 million comets some 5 to 10 Myr ago. The few comets whose perihelia are larger than 1 AU are probably more than compensated by the incompleteness of the observations, so we conclude that the comets of the shower that reached the Earth's orbit are of the order of 10^8. If we then take into account the captures by Jupiter and Saturn producing multiple passages on short-period orbits, our number is consistent with several collisions with the Earth during a 4 Myr period, 5 to 10 Myr ago. The contribution of the passing stars is therefore diminished to one-half of what was believed before and the rate of new comets must be considerably modulated by the period of Nemesis.

The shower of comets reaching the solar system represents a fraction of the total number of perturbed comets that is given, within the assumption of the Maxwellian distribution discussed in Sec. III, by the ratio in velocity space of the two areas of the circles with radii R_0 and R_2. Because the Maxwellian distribution falls off sharply beyond its rms velocity, it is reasonable to use for the cut-off R_2 a value between the rms velocity (110 m s^{-1}) and the circular velocity (174 m s^{-1} at 30,000 AU). Using $R_2 = 140$ m s^{-1} and $R_0 = 8.9$ m s^{-1} at 30,000 AU, then $(R_2/R_0)^2 = 250$. The total number of perturbed comets is therefore 2.5×10^{10} per passage. It is important to note that although all of these comets have their velocity changed somewhat, almost none become hyperbolic.

With $q = 30,000$ AU, the length of the elliptical orbit piercing the Oort cloud (assuming a radius of 60,000 AU) is 150,000 AU. We prefer to use the parabolic approximation of Sec. IV, and we limit the integration at a distance of 60,000 AU (covering therefore from $\theta = -90°$ to $\theta = +90°$, in general agreement with the observations); then the integration path is 140,000 AU.

The number N_0 of comets perturbed at perihelion, per AU of orbital path, can be deduced from Eq. (3) by using 2.5×10^{10} for the total number of comets in the volume perturbed along the trajectory. Assuming $k = 1.5$, an average value for the density law between 30,000 and 60,000 AU, we find $N_0 = 1.8 \times 10^5$. Using a density of 2×10^{-4} comets AU^{-3} in the Oort cloud at 30,000 AU, the cross section area of the perturbed cylinder is $A = 9 \times 10^8$ AU^2.

Finally, from Eq. (3) and if we use $\Delta v_{min} = 8.9$ m s^{-1} for the maximum distance D_0 of the perturbed comets (at 30,000 AU) that will reach the solar system, it follows that the mass of Nemesis

$$M = \frac{D_0 v_* \Delta v_{min}}{2G} = 28 \; m_{2\!\!\!/} . \qquad (23)$$

Of course the cumulative uncertainties of this assessment are very large, probably more than a factor of 3; therefore it is only suggestive of a mass in the general range of 10 to 90 times the mass of Jupiter, $m_{2\!\!\!/}$. Unfortunately, this range goes across the limit between black dwarfs, through brown dwarfs to genuine red dwarfs, making the prediction of the brightness of Nemesis even more uncertain. Our nominal result suggests to search for a very faint object detectable in infrared (near 10 to 12 µm) for large telescopes only, but we must be ready for surprises.

X. DISCUSSION

Existence and Cause of Anisotropy

The existence of the anisotropy is established for the whole sky at the level of 1.25σ for numbers of comets, at the level of 1.65σ for their momenta; this is not enough to carry conviction. But the local anomalies of anisotropy and of number of aphelia along a 180° arc of the great circle, come out of the noise of statistical fluctuations at the level of 3.86σ (confidence level of 99.96%) as established by a Poisson distribution test.

However, anisotropies are not easy to produce and they do not survive for a long time; individual stars or molecular clouds move too fast to induce anisotropies of any significance. Collectively, their random motions are a powerful stirring mechanism dissipating anisotropies quickly, by diffusion of the parameters of cometary orbits. Galactic rotation introduces an asymmetry that could conceivably deplete orbits in a preferential direction; but this possible cause has a time scale of at least one order of magnitude larger than the stirring mechanism. Besides, it is difficult to imagine how galactic effects would produce such a large concentration in both anisotropy and aphelia numbers, along half a great circle asymmetrically located at 28° from the galactic plane.

The only possible cause that has finally been identified is the tangential impulse of a slow-moving body. This cause implies that the anisotropy was concentrated, at least initially, along the trace on the sky of the body's trajectory. A search along this great circle has revealed that it was indeed still the case. Concentrated in a 180° sector 30° wide centered on the great circle, the deviation from isotropy as well as the concentration of aphelia is maximum. An important argument is that this was predicted by the model. If another explanation were found, it would have also to explain this local anomaly, in concentration and in isotropy, along a narrow strip of the sky, perfectly aligned along a great circle.

Another surprise is that the anisotropy is not distributed equally along a 360° strip (as a perturbing body would do if moving on a circular orbit), but only along half of it, the other half being close to isotropy. This is suggestive of a perihelion passage on an elongated orbit, piercing the Oort cloud only across half of the sky. A model was developed for a parabolic orbit (of course, it also represents well the perihelion of any elongated ellipse). The model shows the observational feature of a sharp cut-off after a 160° to 180° length. In this southern strip, the anisotropy is very high (a ratio of three). In contrast, the northern strip is depleted of comets, and is as close to isotropy as it could be, with 4 R and 5 P comets. Certainly, it could be argued that this apparent isotropy is due to a fluctuation; because of the small numbers involved, the deviation would only be 1.06σ from a 3 to 1 R/P distribution; hence its isotropy has no hard significance. If aphelia numbers were neglected, it could be argued that a quasi-circular orbit would explain the observations almost as well; however, we cannot neglect the fact that the total number of cometary aphelia is 2.2 to 2.3 times larger than expected in the southern sky (Sec. V), even after correction for the well-understood north–south bias. This has a natural explanation in the elongated ellipse hypothesis; whereas this anisotropy cannot be explained by Whitmire and Matese's (1985; see also the chapter by Matese and Whitmire) Planet X.

Identification with Nemesis

A body, whose mass could be in the range of 10 to 90 times the mass of Jupiter, has passed through the Oort cloud some 2 to 15 Myr ago, at a perihelion distance of 30,000 ± 10,000 AU; it was moving so slowly that it is bound to the Sun. This scenario seems the only reasonable interpretation so far of the observed anisotropy in the orbits of young comets, and it is highly suggestive of the Nemesis required by geologists. In particular we note that:

1. Section III has established that the anisotropy becomes substantial only when the perturbing object's velocity is very close to the comets' ve-

locities; it diminishes to negligible values when the object has a velocity at infinity which is not close to zero. Although it does not require a bound orbit, the chances of a stellar passage meeting the previous condition is vanishingly small.
2. The epoch of the perturbation is ill defined (2 to 15 Myr ago), but consistent with the also imprecise (3 to 13 Myr ago) last extinction of species on the Earth.
3. The period of the perturbing body is not known, although it should be between 5 and 50 Myr. This is consistent with all possible periods proposed for Nemesis.
4. As soon as we accept a period and a date for the last passage, the intensity of the cometary shower and the mass of Nemesis can be deduced from the observational data. The previous computations were based on the mean period of 29 Myr. The shortest period pushes the last extinctions back to 13 Myr ago and Nemesis' passage 2 Myr earlier (comet transit time). Because of steady losses to hyperbolic orbits (50% per passage), the intensity of the main shower was about 4 times as large, making the chances of several direct hits on the Earth much greater. This enlarges the mass of Nemesis by a factor of 2, to (nominally) 56 Jupiter masses. Vice versa, if we use the longest period of 32 Myr, the last extinction took place 1 Myr ago and Nemesis' passage only 3 Myr ago. The intensity of the shower is reduced 4 times, making a direct hit with the Earth a more improbable event. It also diminishes the mass of Nemesis by a factor of 2, nominally to 14 Jupiter masses. These are only crude assessments and round figures to illustrate the general trend.

Orbital Stability of Nemesis

A major argument against the existence of Nemesis is the absence of orbital stability. A recent capture seems to be excluded; the mass of the Oort cloud is too tenuous to involve a substantial braking action on a passing star, and any other type of capture would involve an unlikely third body. Therefore, if Nemesis exists, it must have existed for the duration of the solar system. However, analytical estimates (Heggie 1975; Retterer and King 1982; Bailey 1983) as well as numerical experiments (Hills 1984a) confirm the survival in 4600 Myr of only 3 to 5% of the orbits.

If this result is not encouraging, it is nevertheless not a fatal blow, in particular because here the orbit turns in the opposite direction to the galactic rotation, at an angle of 28° from its plane; hence it meets the criteria for the stablest orbits (Torbett and Smoluchowski 1984). It is clear that as a group, a very large fraction of the stablest orbits, that were first closely bound and have evolved recently to $a > 50,000$ AU, is able to survive for the duration of the solar system. Besides, if the Sun was born in a stellar association, nothing guarantees that Nemesis *alone* was born on a bound orbit. If several bodies like Nemesis existed earlier, it is only natural that at least the stablest orbit has survived. We conclude that the question of the orbital stability of Nemesis has weathered recent criticism well enough to consider here the Nemesis hypothesis in earnest.

Star Tracks through the Oort Cloud

The well-known clustering on the sky (Lüst 1984) of very long comet aphelia that remains after taking observational bias into account has been convincingly explained recently by Biermann et al. (1983) as the result of perturbations due to the recent passage of a very small number of stars that have pierced the Oort cloud, 1 to 4 Myr ago. The strongest identification is that of a cluster of 17 comets, which contains an excess of some 13 comets, properly aligned along an arc of great circle covering about 90 to 100°. We have verified that the R/P ratio of this clustering does not show any significant deviation from isotropy; this fact, as well as the short length of the arc, is the signature of a fast-moving star. In contrast, the 29 comets we identified in the southern 180° strip (Fig. 4) have an excess of some 19 comets; they are also a star track, but with a difference: the anomalous strip length of at least 160° is the signature of the track's curvature around perihelion; the large R/P ratio is the signature of a slow-moving body; the two features confirm each other and are suggestive of a Nemesis-type body.

Biermann et al.'s clustered aphelia are within ±3° of a great circle; ours are ±15°. This fact, as well as the fraction of the R/P anisotropy that has diffused outside of the strip, is suggestive of the last remnants (that have diffused least) of a stronger shower, most members of which have been scattered and lost. For those few comets that still are in the strip, none of the orbital parameters has changed greatly. In particular, their periods have remained close to that of new comets. The present diffusion of their aphelia suggests a longer lifetime than that of Biermann et al.'s clustering. This remains consistent with a time scale of 10 Myr for orbital diffusion.

Cratering Versus Mass Extinctions

The puzzling discrepancy between the periodicity found from mass extinctions and that from cratering rates has been mentioned in Sec. VIII. The large difference between the 26 Myr period and the 33 Myr period must have an explanation. First, it must be realized that the very small number of crater ages used in the cratering statistics makes any Fourier analysis of the data rather futile. Second, most of the craters used so far seem to come rather from random asteroid impacts, if we trust their mineralogy. This is understandable because not only have comets a more chondritic nature, but also an inner fragility and a much lower density that implies that most of them would explode in the atmosphere. For this reason, they are unlikely to leave a crater signature as asteroids do.

The minimum size for a cometary nucleus to reach the ground is not known, but it is likely to be in the range of 10 to 20 km, whereas most asteroids down to the 1 km size would reach the ground and make craters without difficulty. Of course, the Cretaceous-Tertiary layer is likely to be linked with an exceptionally large cometary impact that may have produced a crater somewhere. However, other mass extinctions are not necessarily linked to the cratering ability of the impacting body, but rather to a global distribution of dust. Since small comets are likely to explode at the top of the troposphere, their dust is more easily dragged away globally by horizontal stratospheric winds. The size distribution of comets is rather well known in the 1 to 10 km range (Vsekhsvyatskii 1964; Delsemme 1985b); for each 10 to 20 km size comet, there are about one hundred 1 km size comets and many more smaller objects. It is therefore reasonable to think that most (but not all) global extinctions have been produced each by some 10 to 20 upper atmospheric events within a spread of 3 to 5 Myr (duration of the comet shower), whereas small asteroids would produce craters spaced in time at random, inducing extinctions localized to 500 to 5000 km spots.

This discussion suggests that, if a periodicity is accepted, more weight should be given to the mass extinction data. For instance in the search for Nemesis, first priority should be given to the 26 Myr value given by Raup and Sepkoski (1984), and therefore to position given in Eq. (21), whose error bars should probably be extended to $7°$ or $8°$ to take into account unknown systematic errors.

XI. CONCLUSION

The observed anomalies in R/P anisotropy and in aphelion concentrations come out of the noise of statistical fluctuations: Section VI establishes that their significance level is 3.86 σ (99.96%). Section II excludes explanations not related to the recent passage of a massive body. Section III establishes that the perturbing body must be moving so slowly that it is likely to be bound to the Sun. The model of Sec. IV fits in with the observations only if the body is on an elongated orbit whose perihelion pierces the Oort cloud between 10,000 and 40,000 AU. The position of the perihelion is identified in the sky, probably within $5°$.

Since recent discussions (see chapters by Shoemaker and Wolfe and by Tremaine) have cast some doubt on the necessity of Nemesis and even on the existence of a periodicity in the mass extinctions, the possibility that the identification with the "death star" is wrong remains open. In this case, all that can be said is that no alternative has been found to a body with a mass 10 to 90 times that of Jupiter and that could probably be on an orbit with a period between, say, 5 and 50 Myr. Since its present position would then be unpredictable, the search should be extended to the whole great circle ($\pm 5°$) which is the trace of its trajectory on the sky.

However, the identification with the Nemesis of the mass extinctions remains so suggestive, that a more localized search for a brown dwarf with no proper motion and a parallax between 1.3 and 2.0 arcsec is certainly worth attempting. Nemesis' present position, given at the end of Sec. VIII, takes care of the uncertain period of the geologists by extending the search to a 50° strip along its presumed orbit; however, the position given in Eq. (21) is recommended on a first priority basis.

Acknowledgments. B. Soonthornthum has computed the projections of the momenta for the 126 comets. Discussions and/or correspondence with T. Gehrels, M. Bailey, M. V. Torbett, P. Hut, R. Smoluchowski, Z. Sekanina, P. Weissmann are gratefully acknowledged. This research was undertaken under grants from the National Science Foundation and the National Aeronautics and Space Administration.

Note added in proof: Using the 126 comets as well as a larger set of 152 comets, I have now analyzed in collaboration with M. Patmiou the aphelia distribution in several new ways in order to establish whether any significant bias or nonrandom effect had been previously neglected. Only one significant new effect was found. It occurs when perihelia are classified in zones of constant b (galactic latitude). The two galactic poles, as well as the zone close to the galactic plane, are rather depleted in aphelia, whereas there are more aphelia than expected in the two zones, $+30°<b<+60°$ and $-60°<b<-30°$. If the vertical tide of the Galaxy on the Oort cloud brings an important contribution of new comets into visibility, then their aphelia density should be proportional to $\sin 2b$ (chapter by Torbett); this fits in better with the observed data than the random distribution. Our data can therefore be interpreted as bringing the first observational detection of the vertical galactic tides on the Oort cloud. Tides diminish the comets' transit time from the Oort cloud into visibility; apart from this fact, none of the previous conclusions is changed. In particular, the large concentration of aphelia and the large anisotropy in the strip at a 28° angle with the plane of the Galaxy, shown in Fig. 4, has not received any other explanation than the slow-moving perturbing body proposed here.

APPENDIX

List of the 126 "young" comets, classified in order of their original binding energy. The figure after the comet's name is $1/a$ in 10^{-6} AU^{-1}. The table comes straight from Marsden and Roemer (1982).

1976	I	Sato	−734	1976 XIII	Harlan	+45
1955	V	Honda	−727	1932 VI	Geddes	+45
1959	III	Bester-H.	−446	1912 II	Gale	+45
1895	IV	Perrine	−172	1886 IX	Barnard-Hartwig	+46
1971	V	Toba	−142	1889 I	Barnard	+48
1960	II	Burnham	−135	1954 VIII	Vozarova	+49
1953	II	Mrkos	−125	1967 II	Rudnicki	+49
1940	III	Okabayasi-Honda	−124	1972 VIII	Heck-Sause	+49
1899	I	Swift	−109	1975 II	Schuster	+51
1957	III	Arend-Roland	−98	1900 I	Giacobini	+57

(*continued*)

APPENDIX (Continued)

Year		Name	Value	Year		Name	Value
1968	VI	Honda	−82	1937	IV	Whipple	+62
1904	II	Giacobini	−75	1898	VII	Coddington-Pauly	+68
1911	IV	Beljawsky	−74	1972	IX	Sandage	+69
1898	VII	Chase	−71	1959	IX	Mrkos	+69
1932	VII	Newman	−56	1954	X	Abell	+70
1975	XI	Bradfield	−56	1973	IX	Gibson	+71
1942	VII	Oterma	−34	1915	II	Mellish	+75
1892	VI	Brooks	−27	1959	I	Burnham-Slaughter	+76
1849	II	Goujon	−25	1941	VII	van Gent	+78
1886	I	Fabry	−18	1966	V	Kilston	+78
1946	I	Timmers	−13	1954	V	Abell	+82
1947	I	Bester	−1	1890	II	Brooks	+89
1941	I	Cunningham	+1	1980b		Bowell	+120
1952	VI	Peltier	+2	1937	V	Finsler	+124
1897	I	Perrine	+5	1914	II	Kritzinger	+126
1974	XII	van den Bergh	+11	1910	I	Great Comet	+135
1853	III	Klinkerfues	+12	1976	XII	Lovas	+142
1863	VI	Baeker	+14	1882	I	Wells	+144
1942	IV	Whipple-B.-K.	+16	1908	III	Morehouse	+174
1917	III	Wolf	+17	1847	II	Colla	+180
1957	VI	Wirtanen	+17	1847	VI	Mitchell	+212
1921	II	Reid	+18	1904	I	Brooks	+227
1944	IV	van Gent	+18	1947	VI	Wirtanen	+234
1936	I	Van Biesbroeck	+19	1977	XIV	Kohler	+234
1973	XII	Kohoutek	+20	1958	III	Burnham	+256
1919	V	Metcalf	+20	1950	I	Johnson	+263
1922	II	Baade	+21	1970	XV	Abe	+283
1975	V	Bradfield	+23	1977	X	Tsuchinshan	+314
1925	VII	Van Biesbroeck	+24	1973	II	Kojima	+320
1948	I	Bester	+24	1886	II	Barnard	+332
1962	III	Seki-Lines	+25	1910	IV	Metcalf	+474
1907	I	Giacobini	+25	1972	XII	Araya	+476
1903	II	Giacobini	+26	1949	I	Wirtanen	+498
1914	III	Neujmin	+27	1969	IX	Tago-Sato-Kosaka	+507
1902	III	Perrine	+27	1863	I	Bruhns	+521
1948	II	Mrkos	+28	1930	IV	Beyer	+524
1905	IV	Kopff	+28	1948	IV	Honda-Bernasconi	+525
1978	XXI	Meier	+29	1973	X	Sandage	+531
1914	V	Delavan	+29	1880	II	Schaeberle	+534
1977	IX	West	+33	1970	III	Kohoutek	+555
1903	IV	Borrelly	+33	1959	IV	Alcock	+593

(*continued*)

APPENDIX (Continued)

1979	VII	Bradfield	+33	1900	II	Borrelly-Brooks	+610
1948	V	Pajdusakova-Mrkos	+34	1927	IV	Stearns	+623
1947	VII	Wirtanen	+34	1905	VI	Brooks	+630
1925	VI	Shajn-Comas Sola	+35	1966	II	Barbon	+643
1975	VII	Lovas	+36	1974	III	Bradfield	+690
1954	XII	Kresák-Peltier	+36	1949	IV	Bappu-Bok-Newkirk	+735
1976	IX	Lovas	+37	1975	IX	Kobayashi-B.-M.	+821
1951	I	Minkowski	+37	1844	II	Mauvais	+824
1956	I	Haro-Chavira	+39	1968	I	Ikeya-Seki	+842
1925	I	Orkisz	+40	1892	II	Denning	+846
1959	X	Humason	+40				
1979	VI	Torres	+42				
1955	VI	Baade	+42				
1946	VI	Jones	+44				

THE OORT CLOUD AND THE GALAXY: DYNAMICAL INTERACTIONS

PAUL R. WEISSMAN
Jet Propulsion Laboratory

In contrast to most of the known solar system which is deep within the Sun's gravitational potential well, the Oort cometary cloud extends out to the limits of the Sun's sphere of influence. The evolution of cometary orbits in the cloud is affected by perturbations from the galactic nucleus and disk, from random passing stars, and from occasional encounters with GMCs. Over the history of the solar system these perturbations have randomized the orbits in the cloud, leaving little evidence of where the comets might have formed. Dynamical studies give a population estimate for the cloud of 2×10^{12} comets with a total mass of 7 to 8 $M_\oplus$. There is a growing consensus on the existence of a massive inner Oort cloud with a population up to 10^2 times that of the dynamically active outer cloud. This inner cloud can help to solve several important cometary problems such as the source of the short-period comets, or how the outer Oort cloud is replenished after a close disruptive encounter with a GMC. IRAS observations of dust shells around Vega and some 40 other main sequence stars in the solar neighborhood may be the result of comets forming in inner Oort clouds around each of these stars. Also, IRAS has detected clouds of cool material in the outer solar system which may be part of the Sun's own inner Oort cloud. Speculations about the existence of a small unseen solar companion star or a tenth planet causing periodic comet showers from the inner Oort cloud are not supported by dynamical studies or analyses of the terrestrial and lunar cratering record.

I. INTRODUCTION: THE OORT HYPOTHESIS

On the galactic scale of things, the solar system is a rather small place. At the Sun's distance from the galactic center, roughly 8.5 kpc, the mean spacing between stars is about 1 pc while the orbital radius of the most distant

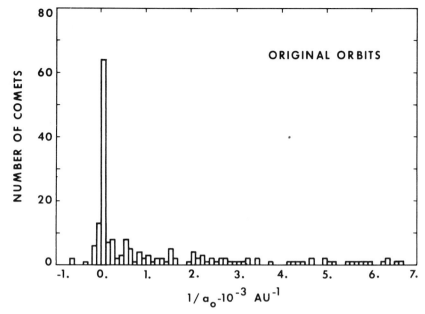

Fig. 1. The distribution of original inverse semimajor axes $1/a_o$ for 190 long-period comets as found by Marsden et al. (1978) and Everhart and Marsden (1983). The large spike at near-zero energy represents the dynamically new comets from the Oort cloud. The low continuous distribution is composed of returning comets which have already been perturbed in orbital energy by the planets.

known planet, Pluto, is less than 1.9×10^{-4} pc, or 39.4 AU. However, one part of the solar system does extend to interstellar distances, the cloud of comets surrounding it, first described by Oort (1950).

Oort was attempting to explain the observed distribution of orbital energy, expressed as the inverse of the original semimajor axis, $1/a_o$, for the long-period comets, as shown in Fig. 1. It had been observed that long-period comets passing through the planetary region had osculating, or instantaneous, orbits ranging from eccentric ellipses, to parabolas, to modest hyperbolas. However, when the orbits were integrated backwards to a time prior to their entering the planetary region and being perturbed by the major planets, and then converted to a barycentric rather than a heliocentric coordinate system, most of the hyperbolic orbits and some of the elliptical ones became very eccentric ellipses with semimajor axes between 10^4 AU and infinity. These orbits are represented by the spike of comets at near-zero, but still gravitationally bound energies in Fig. 1.

Van Woerkom (1948) had shown that planetary perturbations, mostly by Jupiter, would cause the long-period comets to diffuse in $1/a$ to form a flat, uniform distribution of orbital energies. This is represented by the low, continuous distribution of comets in Fig. 1. Oort recognized that the spike of comets at near-zero energy could not be the result of random perturbations but must represent the source of the long-period comets. He suggested that a massive cloud of comets surrounded the solar system and that comets were slowly being fed into the planetary region by perturbations from random passing stars, and then diffused in $1/a$ by Jupiter perturbations. The typical change in the energy of a cometary orbit on one perihelion passage due to Jupiter is $\pm 685 \times 10^{-6}$ AU^{-1}, almost seven times the width of the spike of new comets in Fig. 1.

Comets perturbed to the left of the spike in Fig. 1 escape the solar system on hyperbolic orbits and do not return. The few, apparently hyperbolic orbits in this figure are believed to be the result of errors in the orbit determination, and the failure to account for the effect of nongravitational (jetting) forces on the nuclei of comets which make the orbits appear more eccentric than they actually are.

Oort calculated the change in cometary orbits due to stellar perturbations using the impulse approximation

$$\Delta V = 2 \, GM_*/DV_* \tag{1}$$

where ΔV is the velocity impulse imparted to the comet, G is the gravitational constant, M_* is the mass of the passing star, D is the closest approach distance between the star and comet, and V_* is the relative velocity of the star. For a typical 1 M$_\odot$ star passing at a distance of 1 pc at a velocity of 20 km s^{-1}, ΔV = 43 cm s^{-1}. The passing star also imparts an impulse on the Sun, and the change in the comet's orbit results from the net difference between the two impulses.

The number of stars passing within a given distance of the solar system is given by

$$N = \pi R^2 \, \rho V_* T \tag{2}$$

where R is the maximum encounter distance, ρ is the density of stars in the local solar neighborhood, and T is time. For $R = 1$ pc, $V_* = 20$ km s^{-1}, and ρ = 0.08 stars pc^{-3} (Allen 1973), $N = 5.1$ passes per Myr. Over the history of the solar system about 2.3×10^4 stars would be expected to pass within 1 pc of the Sun. Oort estimated that the sum of the random impulses on the comets over the solar system's history would be equal to an average impulse of 136 m s^{-1}, equal to the circular orbit velocity at 4.8×10^4 AU from the Sun, or the escape velocity at twice that distance.

Oort showed that the perihelia of comets in the cloud would diffuse into the planetary region due to the random stellar perturbations, where the comets would then be removed from the cloud by planetary perturbations. He calculated that the radius of the cloud had to be between 5×10^4 and 1.5×10^5 AU

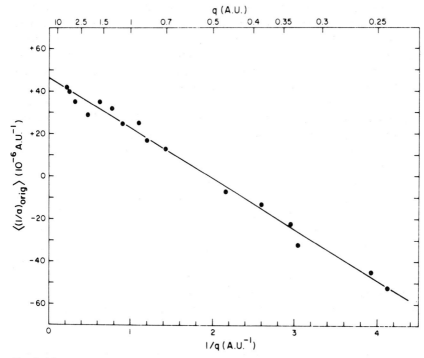

Fig. 2. Mean $1/a_o$ for groups of dynamically new comets versus inverse perihelion distance as found by Marsden et al. (1978). Extrapolating the data to large perihelion distances shows the mean aphelion distance of new comets from the Oort cloud is about 4.3×10^4 AU.

for stellar perturbations to be great enough to keep a continuous flux of new comets flowing into the planetary region from the cloud. Based on an estimate of one dynamically new comet per year observed with perihelion < 1.5 AU, Oort estimated the population of the cloud to be 1.8×10^{11} comets, with a total mass $M = 0.01$ to 0.10 Earth masses ($M_\oplus$). Oort also showed that the random stellar perturbations would result in the perihelion distribution of new comets in the planetary region being uniform with increasing perihelion distance.

Marsden et al. (1978) were able to compensate for the effect of nongravitational forces on cometary orbits and find the average semimajor axis of observed new comets from the Oort cloud. They grouped the new comets by perihelion distance and showed that the mean value of $1/a_o$ increased with increasing perihelion. This result is shown in Fig. 2. Extrapolating the fit to the data to large perihelion distances gives a mean $1/a_o$ of 46×10^{-6} AU^{-1}, corresponding to a mean semimajor axis of 2.2×10^4 AU, or an aphelion distance of 4.3×10^4 AU. Note that this figure is somewhat less than the radius originally calculated by Oort, suggesting the existence of other perturbers to the comet cloud not included in Oort's calculations.

Oort also constructed a simple analytical model of the evolution of long-period comets from the cloud under the influence of planetary perturbations. He showed that by assuming a disruption probability per orbit of about 0.014, he could explain the slow decay of the continuous distribution in Fig. 1 as comets diffused towards larger binding energies. One problem Oort could not resolve was the relative height of the spike of dynamically new comets and the low continuous distribution. The observed number of older, returning comets was too few as compared with the observed number of dynamically new comets. Oort suggested that "the new or almost new comets which come for the first few times near the Sun have a greater capacity for developing gaseous envelopes, and that a large number of these would not be rediscovered at subsequent passages when they would be much less brilliant."

In the 35 years since it was first proposed, the Oort hypothesis has achieved wide acceptance. Kendall (1961), Whipple (1962), Weissman (1978,1979), and Fernández (1981) developed a variety of analytical and numerical models which confirmed the major features of the Oort hypothesis, in particular, its ability to recreate the observed $1/a_o$ distribution. A computer generated $1/a_o$ distribution based on a Monte Carlo simulation model of the physical and dynamical evolution of hypothetical long-period comets entering the planetary region from the Oort cloud (Weissman 1979) is shown in Fig. 3.

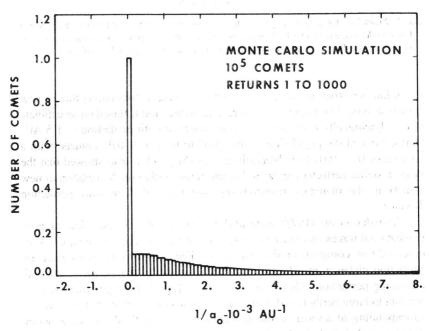

Fig. 3. Computer-generated $1/a_o$ distribution for long-period comets originating from the Oort cloud, using Weissman's (1979) Monte Carlo simulation model.

The problem of the discrepancy in the observed height ratios of the spike and the low continuous distribution in the observed and modeled cases has tended to diminish as better statistics have become available for the observed orbits, and the dynamical models have been improved. Additionally, Oort was correct about the ability of new comets to produce brighter comae, leading to higher discovery probabilities. This is evidenced by the fact that only dynamically new comets are discovered with perihelia beyond about 3 AU, the distance within which water ice sublimation is significant.

Weissman (1978,1979) showed that about 65% of the dynamically new comets which enter the observable region, $q < 4$ AU, from the Oort cloud are ejected on hyperbolic orbits due to Jupiter perturbations. The other major loss mechanism is random disruption which accounts for about 27% of the comets, 10% of them on their first passage through the planetary region. Other loss states include: development of nonvolatile crusts which cut off sublimation, 7.1%; capture to short-period orbits, 0.04%; and perturbation to a Sun-impacting orbit, 0.02%. Weissman showed that the average long-period comet made approximately five returns to the observable region and had an average comet's lifetime of 6×10^5 yr between its first passage and its last.

II. DYNAMICAL MODELING OF THE OORT CLOUD

In recent years the subject of cometary dynamical studies has turned from the evolution of long-period comets under the influence of the major planets, to the evolution of comets in the Oort cloud and the manner in which they are injected into the planetary region. The primary tool in these studies has been Monte Carlo simulation modeling.

The first to apply the Monte Carlo technique to the Oort cloud problem was Schreur (1979) who simulated the passage of 100 stars through a hypothetical Oort cloud of 10^5 comets in a random, spherical distribution, and then extrapolated results to the total history of the solar system. The perturbations by the stars on the Sun and comets were calculated by the impulse approximation. Schreur showed that the comets initially beyond about 10^5 AU from the Sun would be rapidly stripped from the Oort cloud, and that all of the orbital elements of the remaining comets would be randomized by the stellar perturbations with the exception of the inclinations (which would nevertheless be considerably changed). She also demonstrated that encounters with slow moving massive stars would result in the largest perturbations on the cloud, and that the effects of stellar encounters at distances $\lesssim 2$ pc would dominate over the more numerous perturbations from more distant passing stars.

An important result found by Schreur was that most of the comets whose perihelia were perturbed into the planetary region came from highly eccentric orbits with perihelia already close to the outer planets. Thus, it is the slow random walk of perihelia which brings comets into the observable region, rather than large changes caused by close (but rare) stellar encounters.

Weissman (1982a, 1985b) developed a Monte Carlo simulation model to follow the dynamical history of large numbers of hypothetical comets, using a statistical model of the effect of stellar perturbations. He modeled the stellar perturbation on each comet as a randomly directed velocity impulse at the aphelion of each orbit. The magnitude of the velocity impulse was chosen randomly from a Maxwellian velocity distribution whose rms value was a function of the comet's orbital period. Orbit elements were then recomputed and tested for one of three possible dynamic end-states: (1) hyperbolic ejection; (2) diffusion of aphelia to distances beyond the Sun's sphere of influence (taken to be 2×10^5 AU); or (3) perturbation into the planetary region where Jupiter and Saturn perturbations would likely eject the comets on hyperbolic orbits or capture them to shorter-period orbits immune from stellar perturbations. Comets which survived repeated orbits were marked as survivors (a fourth possible end-state) when the total evolution time exceeded the age of the solar system.

The comets were started in orbits typical of the three major hypotheses for cometary origin:

1. Formation as icy planetesimals in the Uranus-Neptune zone with subsequent ejection to the Oort cloud;
2. Formation in an extended nebula accretion disk extending several hundred AU from the Sun;
3. Formation *in situ* in distant subfragments of the protosolar nebula.

Typical runs involved 2×10^4 hypothetical comets or more.

Results for four possible end-states are shown in Table I. For comets begun with perihelia in the Uranus-Neptune zone, stellar perturbations cause the perihelia of the comets to diffuse rapidly into Jupiter- and Saturn-crossing orbits, resulting in their ejection from the solar system. For the other comet origin hypotheses, as the initial perihelia of the comets increases, the fraction lost back into the planetary region decreases and the other end-states grow. In particular, stellar perturbations have sufficient time to diffuse slowly the ap-

TABLE I
Oort Cloud End-States

End-State	Initial Perihelion		
	20 AU	200 AU	10,000 AU
Ejected	0.0	0.0	0.0
Planetary loss	0.773	0.477	0.076
Diffusion of aphelia	0.068	0.068	0.214
Survivor	0.159	0.455	0.710

helia of the orbits to distances beyond 2×10^5 AU. Also, the fraction of survivors increases.

The number of comets directly ejected by stellar perturbations is near zero in each case. The use of the Maxwellian velocity distribution for stellar perturbations tended to underestimate the fraction of comets receiving large impulses from very close (but rare) stellar encounters, on the order of 500 AU or less. As discussed below in Sec. III on stellar perturbations, the actual fraction of comets directly ejected is expected to be about 9%.

Note that for a cometary origin in the Uranus-Neptune zone, only about 1/6 of the original cloud population has survived over the history of the solar system. Most of the lost comets have been ejected to interstellar space, forming a slowly expanding cloud of unbound comets in the local galactic region. Even for comets formed far from the Sun, 1/3 to 1/2 have been lost to interstellar space.

By comparing the rate of perturbation of comets into the planetary region in the dynamical model with the observed rate, it is possible to derive an estimate for the current population of the Oort cloud. Everhart (1967a) showed, after correction for observational selection effects, that about 16 comets brighter than absolute magnitude 11.0 pass within 1 AU of the Sun per year, and that the perihelion distribution is either uniform or slowly increasing with increasing perihelion distance.

Population and mass estimates for the Oort cloud are shown in Table II. The three different formation hypotheses each yield very similar current populations, about 2×10^{12} comets. This indicates that each of the hypothetical initial Oort clouds have been randomized and appear essentially the same after 4.5 Gyr of stellar perturbations. Assuming a mean mass for cometary nuclei of about 2×10^{16} g (Weissman 1985a) the total mass of the present Oort cloud is estimated to be 7 to 8 $M_\oplus$. Applying the depletion factors in Table I, the original mass of the Oort cloud was between 10 and 45 $M_\oplus$.

The Monte Carlo simulation model can also be used to find the distributions of orbital elements in the Oort cloud. The perihelion distribution for comets passing within 200 AU of the Sun (Weissman 1985b) is shown in Fig.

TABLE II
Oort Cloud Population and Mass

	Initial Perihelion		
	20 AU	200 AU	10,000 AU
Current population (10^{12})	1.8	2.1	1.9
Original population (10^{12})	11.7	6.5	3.7
Current mass ($M_\oplus$)	7.1	8.2	7.4
Original mass ($M_\oplus$)	45.5	18.4	10.5

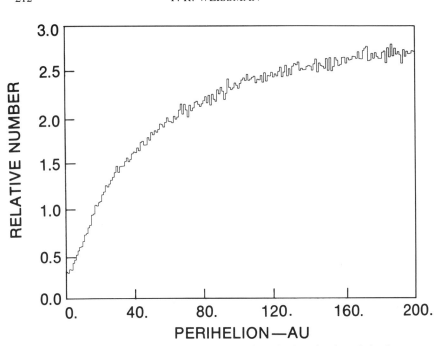

Fig. 4. Perihelion distribution for new comets from the Oort cloud passing through the planetary region over the history of the solar system, as found by Weissman (1985b). The number of comets increases rapidly with increasing perihelion distance.

4. Weissman found that the outer planets, in particular Jupiter and Saturn, acted as a barrier to the diffusion of cometary perihelia into the inner solar system. Comets must effectively "leap frog" across the orbits of Jupiter and Saturn to become visible and be seen by terrestrial observers. Only a fraction of the Oort cloud comets manage to do this. Thus, the inner solar system is undersupplied in new comets from the Oort cloud. This result was also demonstrated by Fernández (1982).

Weissman found that the average long-period comet from the Oort cloud made at least four passages through the planetary system during its lifetime, though virtually all those passes were confined to the Uranus-Neptune zone. Uranus and Neptune are not massive enough to remove comets from the Oort cloud (except in the rare case of very close encounters), but only to cause a modest diffusion of the aphelion distances.

Some appreciation for the degree to which cometary orbits are randomized by the combined effects of stellar and planetary perturbations is afforded by the data shown in Fig. 5. The figure is a scatter diagram of the energy ($1/a$) and eccentricity values for 7928 survivor comets out of an original sample of 5×10^4 hypothetical comets. Starting from an initial value at the top center of the diagram (Uranus-Neptune zone formation), the comets

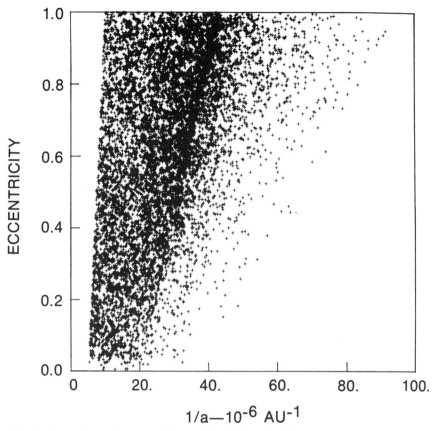

Fig. 5. Scatter diagram in energy and eccentricity for the orbits of 7928 hypothetical comets in the current Oort cloud, as found by Weissman (1985b). The comet orbits have been randomized and have diffused widely away from their initial values of $1/a_o$ and eccentricity of 50×10^{-6} AU^{-1} and 0.9990, respectively. The sharp cut-off at the left is caused by the aphelion limit of 2×10^5 AU in the simulation model.

have effectively diffused in energy and angular momentum to fill the available phase space.

Fernández (1982) developed a similar Monte Carlo simulation model but with the stellar perturbations modeled as actual passages of stars through a hypothetical cloud of comets. This technique is more costly in computer time so Fernández was forced to run smaller samples than Weissman had, on the order of 10^3 to 10^4 comets. Fernández's findings were generally similar to Weissman's (1985b). However, Fernández found a greater fraction of comets were lost by ejection and diffusion of aphelia to large solar distances, as a result of the more accurate modeling of the stellar perturbations. In addition, the fraction of the cloud surviving after 4.5 Gyr decreased, to only about 10% for a cometary origin in the Uranus-Neptune zone.

Estimates of the cloud population by Fernández were also different from Weissman's. Fernández estimated a current Oort cloud population of 6×10^{10} comets for a Uranus-Neptune origin, increasing to 3×10^{12} comets for an origin in distant nebula subfragments far from the Sun. Two points account for these differences. First, Fernández assumed a smaller observed flux of long-period comets than Weissman, based only on the well-established statistics for long-period comets brighter than absolute magnitude 7.0.

Secondly, Fernandez assumed that comets formed in distant nebula subfragments would be in initially circular orbits, rather than the eccentric orbits ($e_o = 0.60$) used by Weissman. Such orbits take longer to diffuse down to perihelia near the planetary region. As a result, Fernández's modeled Oort clouds for the different origin hypotheses have not been fully randomized, and require different populations to account for the observed cometary flux. It is unlikely, however, that nebula subfragments would be in circular orbits, because if the initial nebula had that much angular momentum, it would probably never collapse.

A fourth Monte Carlo simulation model for the Oort cloud is that by Remy and Mignard (1985) who built a model similar to Weissman's but with the perturbation by several thousand random stars calculated exactly as the comets circulated in their orbits, using the impulse approximation. Unfortunately, this method is even more consuming of computer time and typical runs were for samples of only 200 comets at a time.

Remy and Mignard showed that, by not requiring the perturbation to occur only at aphelion on each orbit, the randomizing effect of the perturbations was increased. As with Fernández, they found that the more realistic modeling of the stellar perturbations led to a smaller fraction of the cloud surviving, and a larger fraction being ejected or diffused beyond the Sun's sphere of influence. The fraction of ejected comets was typically 8 to 10%, those diffused to interstellar distances 43 to 80%, and those lost through planetary ejection and/or capture 4 to 35%. Note that even for an origin in the outer-planets zone, Remy and Mignard find only half as many of the comets are lost by reentering the planetary region. The larger effects of the stellar perturbations cause more comets to have their perihelia and aphelia rapidly pumped away from the Sun, lowering the probability of planetary ejection.

The estimate of the current Oort cloud population by Remy and Mignard is 1.4 to 2.3×10^{12} comets, close to Weissman's value, as would be expected considering the similarity of the approaches. The original cloud population is estimated to be an order of magnitude greater, reflecting the greater depletion factor found with their Monte Carlo simulation.

III. PERTURBATION OF COMETARY ORBITS

Cometary orbits are perturbed by the planets, by nongravitational forces, by random passing stars, by interstellar clouds, and by the galactic gravitational field. The first two of these tend to act primarily near perihelion, resulting in changes in the semimajor axis and period of cometary orbits, while the last three tend to act nearer to aphelion and cause changes in the perihelion distance and inclination of the orbits. The five different perturbation sources combine to keep the orbits ever changing, making impossible the exact prediction of the return of short-period comets or the origin of long-period comets.

Although planetary and nongravitational perturbations are important in understanding the evolution of cometary orbits once comets enter the planetary region, the Oort cloud's dynamical interaction with the Galaxy occurs chiefly through stellar, interstellar cloud, and galactic perturbations. The following discussion will be confined to a review of the work to date on those three perturbation sources.

Stellar Perturbations

Oort was not the first to consider the effect of perturbations by random passing stars on the orbits of the long-period comets. Öpik (1932) examined the problem and arrived at several results similar to Oort's, though Öpik overestimated the Sun's sphere of influence by about a factor of three. However, Öpik failed to link the dynamical processes that he described to the observed distribution of inverse semimajor axes for long-period comets, and thus did not predict the existence of a comet reservoir at large solar distances.

Subsequent to Oort, the problem of stellar perturbations was studied by Faintich (1971), Yabushita (1972), Rickman (1976), Weissman (1980a) and Bailey (1983a), each continuing to use the impulse approximation described in Sec. I. Using a variety of statistical and analytical approaches, these authors estimated the range of stellar perturbations, modeled as a total statistical velocity impulse on each comet orbit, to be about 110 to 170 m s^{-1}. They showed that the stellar perturbations would randomize the cometary orbits beyond about 10^4 AU from the Sun, leaving little or no trace of the original distribution of cometary orbits.

Oort (1950) and Weissman (1980a) suggested that the stellar perturbations would lead to comets being randomly scattered in velocity phase space near zero transverse velocity, and that this would result in a uniform distribution of cometary perihelia in the planetary region. However, as shown in Sec. II above, the effects of the outer planets acting as a barrier to cometary diffusion into the inner solar system, and the fact that the planetary region is a physical and dynamical sink for long-period comets, leads to a perihelion distribution that increases with increasing heliocentric distance, as shown in Fig. 4.

Weissman (1980a) also considered the effect of stars passing through the Oort cloud. He showed that each star would eject comets within a specific distance of its path, that distance being given by

$$D = (2GR/M_\odot)^{1/2} M_*/V_* \qquad (3)$$

where R is the heliocentric distance, $M_\odot$ and M_* are the mass of the Sun and star, respectively, and V_* is the velocity of the star. For 1 $M_\odot$ stars moving at a typical velocity of 20 km s^{-1}, $D = 2.1 R^{1/2}$, where R is in AU; at 5×10^4 AU from the Sun, $D = 470$ AU. From Eq. (2) Weissman estimated that 5.4×10^3 stars had passed within an assumed Oort cloud radius of 10^5 AU, ejecting 9% of the original Oort cloud population. This result was later confirmed by Remy and Mignard's (1985) Monte Carlo model which simulated passage of stars through the Oort cloud.

The effect of close stellar encounters, on the order of 2.5×10^4 AU or less, was investigated by Scholl et al. (1982). They integrated orbits for hypothetical close encounters between random stars and comets in the Oort cloud and showed that the impulse approximation was generally valid for encounter velocities > 20 km s^{-1}. They also found that close star-comet encounters would often result in very substantial changes in the comet's orbital elements, perhaps perturbing it close to the planetary region from a perihelion initially very far out in the Oort cloud.

A number of different Monte Carlo simulation models of the effect of multiple stellar perturbations on the comets in the Oort cloud have already been described in Sec. II above.

Interstellar Clouds

Although the perturbation of the Oort cloud by random passing stars has received a great deal of study since the Oort hypothesis was first proposed, perturbation by encounters with interstellar clouds has been relatively ignored until recently. Faintich (1971) considered perturbations by interstellar clouds, but only by the 10^2 to 10^4 $M_\odot$ H I clouds known at the time. He showed that a grazing encounter between the solar system and an interstellar cloud would be comparable to a close (on the order of 0.1 pc) stellar encounter. But Faintich pointed out that encounters with random stars would be far more frequent, and thus would still tend to dominate the comet's motion.

The giant clouds of molecular hydrogen in the Galaxy were essentially unknown prior to 1970, and a real understanding of their spatial and mass distributions has only been developed in the past decade. Biermann and Lüst (1978) suggested that encounters between the solar system and giant molecular clouds (GMCs) might lead to the ejection of large numbers of comets from the Oort cloud. This idea was extensively developed by Clube and Napier (1982b, 1984a) and Napier and Staniucha (1982) who believed that the Oort cloud has been repeatedly disrupted by GMC encounters, and then replenished by capture of interstellar comets from the same GMCs.

Clube and Napier calculated that the solar system had made on the order of 10 to 15 encounters with GMCs of mass 2×10^5 $M_\odot$ over the history of the system, and that each encounter was capable of stripping away the Oort cloud beyond about 10^4 AU from the Sun. At the same time they predicted that the solar system would capture on the order of 10^{11} new comets from the GMC, the comets having been formed in interstellar space by compressive shocks as the GMCs passed through the galactic spiral arms. Clube and Napier associated these repeated disruption and capture events with impact related biological catastrophes on Earth.

Weissman (1983a) noted a number of problems with Clube and Napier's work. They typically assumed low encounter velocities between the solar system and the GMCs, on the order of 5 or 10 km s^{-1}, considerably less than the Sun's random galactic motion of 16 km s^{-1}. Also, they calculated the effect of the total GMC acting as a point mass, and then the effect of substructure of the GMCs, and would then sum the results, thus counting each mass at least twice. Finally, Weissman noted that Valtonen and Innanen (1982) and Valtonen (1983) had shown that the probability of cometary capture at such high encounter velocities was extremely low, and would not supply sufficient comets (even if they existed at the high spatial densities required by Clube and Napier) to replenish the Oort cloud.

Bailey (1983a) reexamined Clube and Napier's (1982,1984) dynamical arguments. Bailey noted that the Sun's current random velocity of about 16 km s^{-1} with respect to the local standard of rest (LSR) at its galactic radius was significantly lower than the average of 60 km s^{-1} typical for similar G-type stars in the Galaxy. Assuming that the Sun had always traveled at a random velocity of 16 km s^{-1}, Bailey found that Clube and Napier had overestimated the magnitude of the GMC perturbations by a factor of two, and that the GMC perturbations would still be disruptive. On the other hand, Bailey argued, if the Sun had moved at a more typical velocity of 60 km s^{-1} in the past, then the GMC perturbations were reduced by a factor of ten, and were somewhat less than the perturbation from random passing stars. Bailey also showed that the Sun's vertical motion above and below the galactic plane carried it into regions where the local density of GMCs was lower, thus decreasing the encounter frequency. This point was expanded on by Hut and Tremaine (1985) who showed that additional factors, in particular the fact that the GMCs were not point masses but were actually quite extended, and the Sun's eccentric orbit about the galactic nucleus (the Sun is currently close to the periapsis of its galactic orbit), led to a substantial decrease in the expected perturbation from GMCs.

Hut and Tremaine (1985) estimated the half life of comets with initial semimajor axes of 2.5×10^4 AU against encounters with GMCs was 2.8 Gyr, or about 62% of the age of the solar system. They also showed that the half life against disruption by encounters with clouds of atomic hydrogen was more than an order of magnitude longer, the greater lifetime being due to the lower mass of the H I clouds. They noted that observations of wide binaries in the Galaxy with separations on the order of 0.1 pc (2×10^4 AU) or more confirmed the stability of bound systems of that dimension.

Thus, although GMCs remain a major perturber of the Oort cloud, their effect does not appear to be as catastrophic as predicted by Clube and Napier.

Comets perturbed to Oort cloud distances at the time of the solar system's formation would likely have been lost from the system by now; but more tightly bound comets would have been pumped up in energy to replace them. As we shall see in Sec. IV, the reservoir of comets available for replenishing the Oort cloud is probably quite large.

Galactic Perturbations

As with the effects of interstellar clouds, perturbation of comets in the Oort cloud by the galactic gravitational field received only modest attention until recently. An early study by Chebotarev (1965,1966) considered the limits on the solar system caused by perturbations from the galactic nucleus and set a dynamical limit of 2×10^5 AU (1 pc) on the aphelion distance of comets.

Chebotarev's work was expanded on by Antonov and Latyshev (1972) who calculated the 3-dimensional Hill surface for the solar system perturbed by both the galactic nucleus and the disk. They defined a triaxial ellipsoid with dimensions of the semiaxes of 3×10^5, 2×10^5 and 1.5×10^5 AU, with the long axis oriented towards the galactic center, and the short axis perpendicular to the plane of the Galaxy.

Innanen (1979) derived formulae for limiting direct and retrograde orbits of planetary satellites perturbed by the Sun, or analogously, comet orbits perturbed by the galactic nucleus:

$$r_d = R \, (9M/M_\odot)^{-1/3} \tag{4}$$

$$r_r = 2.08 \, r_d \tag{5}$$

where r_d and r_r are the limiting direct and retrograde radii for circular, coplanar comet orbits, R is the radius of the Sun's orbit about the galactic center, M is the mass of the galactic nucleus, and $M_\odot$ is the mass of the Sun. For a galactic nucleus mass of $1.3 \times 10^{11} \, M_\odot$ and an orbital radius of 10 kpc, $r_d = 2.0 \times 10^5$ AU, and $r_r = 4.1 \times 10^5$ AU. Innanen noted that the large stable radius for retrograde orbits only applied to orbits with inclinations $\gtrsim 150°$, i.e., retrograde orbits close to the galactic plane. Orbits oriented closer to perpendicular to the plane were far more unstable. Innanen also pointed out that orbits near these limiting radii would not resemble simple conics, and might be fairly chaotic. He also noted that observations of the orbits of satellites in the solar system and other dynamical systems in the Galaxy show that the orbits are typically a factor of two smaller than the calculated limiting radii. The same is likely true of the Oort cloud.

Smoluchowski and Torbett (1984) defined stability limits over the age of the solar system for orbits about the Sun when perturbed by both the galactic nucleus and the planar field of the galactic disk, based on integrations of a large sample of hypothetical comet orbits. They found that the effect of the disk field was greater than that of the galactic nucleus, and that it further reduced the stable radius for the Oort cloud. For direct orbits, they found a maximum stable semimajor axis of about 10^5 AU, decreasing to about 8×10^4 AU for inclinations between 60° and 120°, and then increasing again to 1.0 to 1.2×10^5 AU for retrograde comets.

Harrington (1985) showed that galactic perturbations, in particular those by the galactic disk, can result in substantial changes in the perihelion distance of cometary orbits. He demonstrated that comets in the Oort cloud might be perturbed into or out of the planetary region on a single orbit due to the planar galactic field. He also showed that the perturbation was near zero for orbits with their perihelia aligned along the radius vector to the galactic nucleus, possibly leading to apparent concentrations of observed orbit major axes in that direction. He suggested that this might have been incorrectly interpreted in the past by some observers as indicative of a departure of the Oort cloud from a random spherical distribution.

The problem was also studied by Heisler and Tremaine (1986) who found that the tidal force caused by the galactic disk was 15 times that from the galactic nucleus at the solar system's distance from the center of the Galaxy. They estimated that the galactic tide was more important than the perturbation by random passing stars in filling the loss cone of comets entering the planetary region.

IV. THE NEW OORT CLOUD

In the early 1980s, new dynamical studies of the Oort cloud, as described in Sec. II, have led to a considerable revision to the original Oort hypothesis. In our understanding, the cloud has grown in population and mass, while shrinking in total radius. Oort's previous view of the cloud slowly being stirred by a combination of planetary and stellar perturbations has been transformed into a more dynamic and violent place where stars passing through it and giant molecular clouds passing nearby may often result in major perturbations to a significant fraction of the cloud population.

Along with this improved view of the Oort cloud, a picture has begun to emerge of a massive inner cloud, interior to the active outer Oort cloud which produces the long-period comets that are observed. The inner Oort cloud is more difficult to detect dynamically because it is deeper in the gravitational potential well of the Sun and not so easily perturbed by the planets interior to it, or the stars and interstellar clouds around it. The inner Oort cloud may contain up to two orders of magnitude more comets than the dynamically active outer cloud, and may serve as a reservoir for replenishing the outer cloud as it is depleted by various loss processes. Also, the inner Oort cloud may be a more efficient source for the short-period comets than evolution of long-period comets from the outer cloud. Major arguments and evidence for the existence of a massive inner Oort cloud are listed in Table III and discussed below.

TABLE III
Inner Oort Cloud Arguments and Evidence

Icy planetesimals from extended solar nebula accretion disk	Cameron (1962, 1978a)
Perturbation of Neptune's orbit	Whipple (1964); Bailey (1983b)
Source of short-period comets	Fernandez (1980)
Source of comet showers	Hills (1981)
Centrally condensed distribution of observed orbits	Bailey (1983a)
Reservoir for replenishing outer cloud after GMC perturbation	Clube and Napier (1984a)
IRAS observations of cool gas clouds in outer solar system	Low et al. (1984)
Interpretation of IRAS observed dust shells around Vega and other stars	Weissman (1984); Harper et al. (1984)
Outer planets planetesimal swarm; impactors on outer planet satellites	Shoemaker and Wolfe (1984)

Cameron (1962, 1978a) proposed that the planetary system formed from an accretion disk of particulates, dust and ice grains, which settled to the equatorial plane of the rotating protosolar nebula. When material in the disk reached a critical density it would collapse due to Goldreich and Ward (1973) gravitational instabilities, forming planetesimals with rocky compositions in the inner planetary region, and ice-dust mixtures in the outer planets zone. A good review of these processes is given by Greenberg et al. (1984).

Cameron recognized that there was no reason why the nebula accretion disk should end at the orbit of the farthest known planet, about 35 AU from the Sun. Accretion of large, planetary-sized bodies would be slowed by the long orbital periods and low density of planetesimals far from the Sun, and would be halted by the dispersal of the solar nebula during the Sun's T Tauri phase; but the icy planetesimals at these large solar distances would already have formed by that time. Cameron argued that the disk could extend several hundred AU from the protosun, and the icy planetesimals would be the protocomets of the forming Oort cloud. He suggested that the dispersal of the solar nebula and additional mass loss during the T Tauri phase would enlarge the protocomet orbits as the central mass decreased.

Based on Cameron's work, Whipple (1964) suggested that a comet belt beyond Neptune might provide an explanation for the perturbations on Neptune's orbit, perturbations which cannot be explained by Pluto because of its low mass. Whipple estimated that the comet belt had a mass of 10 to 20 $M_\oplus$ if it was at 40 to 50 AU from the Sun and was inclined about $0°.8$ to the invariable plane of the solar system. Whipple also suggested that the belt might be a source of meteoritic material and that it may contribute to the zodiacal dust cloud.

Hamid et al. (1968) considered perturbations by a hypothetical comet belt on the orbits of periodic comets with large aphelion distances, including Halley's comet. They set upper limits of 0.5 $M_\oplus$ at 40 AU or 1.3 $M_\oplus$ at 50 AU. They recognized that their calculations were limited by the then unquantified nongravitational forces on these comets, in particular for Halley's comet with its very small perihelion distance.

Bailey (1983*b*) reiterated Whipple's suggestion of a comet belt being responsible for the perturbations on Neptune's orbit, and suggested that the belt or ring was the inner part of a massive, centrally condensed Oort cloud which he had suggested previously (Bailey 1983*a*) based on studies of stellar and GMC perturbations of the Oort cloud. Bailey (1983*c*) also suggested that if the comets in the inner cloud were close enough to the planetary region, they might have been detected as part of the sky background by the all-sky infrared survey of the Infrared Astronomical Satellite (IRAS). Low et al. (1984) stated that IRAS did, in fact, detect considerable structure in the infrared background at 100 μm wavelength which they interpreted as cold material in the outer solar system.

The existence of a comet belt beyond Neptune was used by Fernández (1980) to provide a more efficient source for the short-period comets than direct evolution from dynamically new orbits from the Oort cloud. Everhart (1972) proposed that low-inclination comets from the outer Oort cloud with their perihelia initially near Jupiter would evolve to short-period orbits in an average of 30 returns, before having their perihelia dumped into the terrestrial planets zone by a close encounter with one of the giant planets. The ability of this process to produce the observed number of short-period comets was questioned by Joss (1973) but defended by Delsemme (1973).

Fernández used a Monte Carlo simulation to show that one could maintain the observed steady-state number of short-period comets by assuming a ring of comets between 35 and 50 AU with a total mass of 1 $M_\oplus$. To maintain the required flux of new comets diffusing into the outer planetary region, Fernández required that some large protocomets in the cloud, on the order of the mass of Ceres (10^{24} g), had accreted, and served as scattering centers. Once comets became Neptune crossing, they would be handed down by planetary perturbations to the Uranus, Saturn and Jupiter zones, successively, over a period of several times 10^8 yr (Everhart 1977), eventually becoming visible as short-period comets. Fernández estimated that only about 6% of the initial Neptune crossers would evolve to observable orbits, but that this was roughly 300 times more efficient than assuming the comets originated from the outer Oort cloud, with initial semimajor axes of $\geq 2 \times 10^4$ AU.

The possible existence of a massive inner Oort cloud led Hills (1981) to suggest that random stars passing through the inner cloud would cause intense showers of comets to enter the planetary region, causing considerable cratering on the planets and possibly contributing to biological extinction events on the Earth. Hills argued that the apparent inner edge to the Oort cloud at about 1 to 2×10^4 AU was an observational artifact of the inability of distant random stars to severely perturb orbits closer to the Sun. He suggested that the Oort cloud could extend inwards of this boundary, particularly if comets had been formed at solar distances well beyond the planetary region, preventing them from easily diffusing into the planetary region and being removed by planetary perturbations. Hills estimated that the total number of comets in the inner Oort cloud was up to 10^2 times that in the outer cloud. He also found that the total number of shower comets entering the planetary region was an order of magnitude greater than the steady-state number of comets from the outer cloud, over the history of the solar system.

As discussed in Sec. III, Clube and Napier (1982b) suggested that occasional encounters with GMCs stripped away most of the Oort cloud comets with semimajor axes $> 10^4$ AU. They suggested that the Oort cloud was replenished by comets captured from GMCs. However, Clube and Napier (1984a) also showed that a massive inner Oort cloud like that described by Hills could alternatively serve as a replenishment source, the more tightly bound comets being pumped up to the outer Oort cloud by very severe GMC perturbations. Although, as noted in Sec. III, perturbations by GMCs are not likely to be as severe as calculated by Clube and Napier, the existence of an inner Oort cloud would still be a more efficient source for replenishing the cloud than interstellar capture.

Shoemaker and Wolfe (1984) used a Monte Carlo simulation to follow the dynamical evolution of a swarm of Uranus-Neptune planetesimals over the history of the solar system. By including stellar perturbations in their model for orbits with aphelia > 500 AU, they were able to get the perihelia of the planetesimals' orbits to detach from the Uranus-Neptune zone, greatly lengthening their dynamical lifetimes in those orbits. About 9% of the original population was found to survive over the age of the solar system, with 90% of them in orbits with semimajor axes between 500 and 2×10^4 AU; 85% of that group had semimajor axes $< 10^4$ AU. Shoemaker and Wolfe also found that the perihelion distribution of the orbits was still relatively peaked just beyond the orbit of Neptune. They estimated the mass of the inner cloud to be 100 to 200 $M_{\oplus}$.

The discovery by IRAS of dust shells around some main sequence stars in the solar neighborhood (Aumann et al. 1984; Aumann 1984) has been interpreted by Weissman (1984) and Harper et al. (1984) as material from comets in the inner Oort clouds around each of these stars. The photograph of one such dust shell around β Pictoris by Smith and Terrile (1984) has been interpreted as a nebular accretion disk around the star, seen edge-on. These observations will be discussed in more detail in Sec. VI.

The picture that has emerged for the solar system's inner Oort cloud is of some 10^{13} to 10^{14} comets, with an inner boundary just beyond the orbit of Neptune, and merging into the outer, dynamically active Oort cloud at about 2×10^4 AU from the Sun. The inner region of the cloud may be considerably flattened with orbits still relatively near the ecliptic plane, but the outer region has certainly been randomized by stellar and GMC perturbations, merging into the spherical outer Oort cloud.

Much of the evidence for the inner cloud is highly circumstantial, or inferred from theories of solar system formation and planetesimal scattering which we do not know to be completely correct. However, the observation of dust shells and possible accretion disks around other stars has given considerable credence to these ideas. Also, the fact that the existence of the inner Oort cloud can explain several dynamical problems with regards to comets is extremely alluring. This will clearly be an interesting area of study in the coming years.

V. VARIATIONS ON THE OORT HYPOTHESIS

Despite the wide acceptance of the Oort hypothesis, other hypotheses have occasionally been proposed to explain the origin of comets, the observed $1/a_o$ distribution, or some other aspect of cometary dynamics. In some cases the existence of the Oort cloud itself has been questioned, while in other instances the dynamics of the cloud has been modified by special perturbers to explain some observed phenomena. The major challenges to the Oort hypothesis are summarized in Table IV.

Prior to Whipple's (1950) icy conglomerate or dirty-snowball model for cometary nuclei, the leading hypothesis for the nature of cometary bodies was Lyttleton's (1948) sandbank model. The hypothesis stated that comets were small, gravitationally bound clouds of gas and dust in orbit about the Sun, which slowly lost a fraction of their mass to tidal, thermal, and other drag forces on each perihelion passage. The sandbank model has largely been abandoned because of the superiority of the icy conglomerate model in explaining such cometary phenomena as nongravitational accelerations (resulting from jetting of volatiles from the surfaces of rotating icy nuclei), random disruption, and comet lifetimes.

Coupled with the sandbank model was Lyttleton's idea that these small dust clouds formed as the Sun passed through interstellar clouds, gravitationally focusing material in the Sun's wake, 10 to 10^3 AU behind the Sun. According to Lyttleton, small variations in the initial velocity of the agglomerated clouds and random planetary perturbations served to create a spherical distribution of orbits over the celestial sphere. Some support for this idea was provided by the finding of a modest concentration of perihelion directions of cometary orbits near the solar apex, the direction of the Sun's relative motion in the Galaxy (Bogart and Noerdlinger 1982).

TABLE IV
Variations on the Oort Hypothesis

Comets accrete in solar wake as solar system passes through interstellar clouds	Lyttleton (1948)
Exploded planet in the asteroid belt as source of all comets, asteroids and meteors	Van Flandern (1978)
Disruption of Oort cloud by GMC encounters; capture of new comets from GMCs	Clube and Napier (1982b)
Unseen solar companion in distant eccentric orbit causes periodic comet showers	Whitmire and Jackson (1984); Davis et al. (1984)
Perturbation of Oort cloud during passage through galactic plane causes periodic comet showers	Rampino and Stothers (1984)
Tenth planet circulating in inner Oort cloud causes periodic comet showers	Whitmire and Matese (1984)

Lyttleton's gravitational focusing mechanism was unable to explain the observed $1/a_o$ distribution for the long-period comets, because it placed the formation site of the comets much closer to the planetary region. Also, since the comets must have formed relatively recently (within the last 10^7 yr) for them still to be circulating through the planetary region, there has not been sufficient time to randomize the perihelion directions of the comets to the degree that is observed. The small departure of the perihelia of comets from a completely random distribution have been ascribed by Biermann et al. (1983) to recent star passages through the Oort cloud.

Also, Weissman (1985a) has pointed out that the Sun's motion relative to the local group of stars may result in asymmetric perturbations on the Oort cloud such that more comets are brought into the planetary region with their perihelia aligned along the apex-antiapex line, or conversely, the perturbations may be most severe perpendicular to that line, and the Oort cloud population may be depleted in the perpendicular direction. The problem is greatly complicated by observational selection effects including a seasonal dependence on the number of comets discovered.

Van Flandern (1978) proposed that comets, asteroids, and meteorites all originated from a 70 $M_\oplus$ planet in the asteroid belt which exploded 5 to 15 × 10^6 yr ago. He explained the apparent Oort cloud comets as fragments ejected from the explosion on a very long-period ellipses making their first return to the planetary region at the present time. The problems with Van Flandern's idea are so massive that it has received little serious consideration. It does not explain the observed distribution of cometary orbits, either in energy or inclination. As noted by Harrington (1985) in Sec. III, galactic perturbations on very long-period orbits would be sufficiently severe that most would not even return to the planetary region. Van Flandern's hypothesis cannot be reconciled with the primitive ages or undifferentiated compositions of many meteorite types. It cannot explain the distribution of asteroid orbits or the distribution of compositional types within the asteroid belt. It predicts a recent heavy bombardment of planetary surfaces for which no evidence can be found. Lastly, it suggests no mechanism for causing the hypothetical planet to explode.

As noted in Sec. III, Biermann and Lüst (1978) suggested that the Oort cloud may be severely perturbed by close encounters with GMCs. This idea has been developed further by Clube and Napier (1982b, 1984a) and Napier and Staniucha (1982) who claim that each GMC encounter strips away the Oort cloud beyond about 10^4 AU from the Sun, and that the solar system then captures a new cloud of comets from the very same GMC. As also discussed in Sec. III, recalculation of the GMC perturbations by Bailey (1983a) and Hut and Tremaine (1985) have demonstrated that they are not likely as severe as claimed by Clube and Napier, and the probability of capturing a sufficiently large cloud of new comets is quite remote, as shown by Valtonen and Innanen (1982) and Valtonen (1983).

The latest variation on the accepted Oort cloud hypothesis has been the suggestion that periodic cometary showers from the cloud may result in biological extinction events on the Earth (see chapter by Shoemaker and Wolfe). Raup and Sepkoski (1984) demonstrated an apparent 26 Myr periodicity in the occurrence of major biological extinctions over the past 250 Myr, as deduced from the fossil record. Alvarez et al. (1980) had shown earlier that at least one of the major extinctions, the Cretaceous-Tertiary event 65 Myr ago, was associated with the impact of a large, on the order of 10 km diameter, asteroid or comet on the Earth. The discovery of microtektites and shocked quartz crystals in the fish clay at the Cretaceous-Tertiary boundary has helped to confirm the impact hypothesis. Evidence for major impacts at some other extinction boundaries has also been found, such as iridium layers or microtektites.

There are three proposed mechanisms for creating a periodicity in cometary showers from the Oort cloud. Whitmire and Jackson (1984) and Davis et al. (1984) independently suggested that the Sun has a distant, unseen companion in a 26 Myr orbit which periodically passes near the inner cloud and perturbs it severely enough to cause a comet shower, as first described by Hills (1981). On the other hand, Rampino and Stothers (1984a) suggested that the controlling factor was the Sun's epicyclic motion above and below the galactic plane, which has a half-period of 33 Myr; their shower-triggering mechanism was close encounters with GMCs located near the plane. (Innanen et al. [1978] had earlier noted an apparent coincidence between times of galactic plane crossings by the Sun and solar system, and the boundaries between geologic periods on the Earth.) Finally, Whitmire and Matese (1984) proposed that a tenth planet in an eccentric, highly inclined, precessing orbit at

100 to 150 AU from the Sun has cleared out a vacant zone in the inner Oort cloud, but that the precession of the planet's orbit periodically brings it close to the edges of the clear zone where it perturbs new showers of comets into the planetary region.

A nonimpact-related mechanism for the periodic extinctions was suggested by Schwartz and James (1984), associated with the Sun's epicyclic z-motion about the galactic plane. They argued that galactic cosmic rays or soft X rays might be more intense near the extremes of the Sun's z-motion where there is less intervening interstellar matter to absorb them, and that this might somehow bring about extinctions. However, this mechanism fails to explain the impact-related signatures seen at several of the extinction boundaries.

Each of the comet shower hypotheses above has a number of problems associated with it. Since the common factor in all three is the existence of periodic cometary showers as the extinction mechanism, we will begin by considering the expected consequences of such showers. The total number of comets perturbed into Earth-crossing orbits from comet shower varies from some 10^5 in Whitmire and Matese's (1984; also see chapter by Matese and Whitmire) Planet X hypothesis to 2×10^9 in Davis et al.'s Death Star idea. However, the total number of perihelion passages, equal to the number of Earth-crossing comets times the mean number of returns, must be on the order of at least several times 10^9 or more for each hypothesis in order for there to be a reasonable probability of one or more significant impacts on the Earth.

Weissman (1982b) calculated a mean impact probability per perihelion passage of 2.2×10^{-9} for random long-period comets, and 6.6×10^{-9} for short-period comets, based on the orbits of 20 observed Earth-crossing short-period comets. Although the short-period comets are three times more likely to strike the Earth (per perihelion passage), the long-period comets have a most-probable impact velocity of 56.6 km s^{-1}, as compared with 28.9 km s^{-1} for the short-period comets. Thus, long-period comet impacts are almost four times more energetic than short-period comets, for an equivalent nucleus mass.

The mean number of returns that a comet will make depends on how far from the Sun the shower originated. In the case of the unseen solar companion and galactic plane crossing hypotheses, the source of comets is the inner Oort cloud at 5 to 20×10^3 AU; for the Planet X hypothesis it is the inner Oort cloud but at only 100 to 150 AU from the Sun. Work in progress by Hut and Weissman (1985) has shown that the average shower comet with a semimajor axis of 3×10^3 AU would make about 12 returns, 10 returns for an initial semimajor axis of 10^4 AU. For the smaller orbits predicted by the Planet X hypothesis, the dynamical lifetimes would be much longer; Whitmire and Matese (1984) only set an upper limit of 3.6×10^4 returns which they argue is a plausible lifetime for the comets.

For a shower of some 10^9 comets (as stated in Davis et al. 1984) making an average of 12 returns each, there would be a total of 1.2×10^{10} perihelion passages. Compared with an observed steady-state flux (after correction for observational selection effects) of 16 Earth-crossing long-period comets per year brighter than absolute magnitude 11.0 as found by Everhart (1967a), one cometary shower would be equivalent to 7.5×10^8 yr of the steady-state flux. Assuming a constant background flux rate, 5.3 such showers would be equal to the total post-mare cratering on the Earth and Moon by long-period comets. If periodic comet showers had occurred every 2.6×10^7 yr over the entire 4.0×10^9 year post-mare period, then there would have been 153 cometary showers, equal to 28 times the total steady-state cratering in that period by long-period comets.

Grieve and Dence (1979) determined the crater production rate on the Earth based on counted craters on dated surfaces of 1.4×10^{-14} km^{-2} yr^{-1} for craters > 10 km in diameter, with a probable error in the rate of about 30%. The lunar production rate is approximately the same after scaling for differences in impact cross section and gravity field. Weissman (1982b) estimated that long-period comets accounted for approximately 5% of the total observed cratering rate on the Earth and Moon. However, Weissman (1985a) revised his estimates of comet masses up by a factor of 2.8, resulting in an approximate doubling in the effective cratering rate by long-period comets. In contrast, Shoemaker and Wolfe (1982) estimated that long-period comets might account for up to 100% of the terrestrial cratering, though more recently they have tended to revise that figure downwards.

The revised Weissman figure for the cratering rate by long-period comets gives a minimum cratering from periodic comet showers of 2.8 times the total observed cratering rate on the Earth and Moon. Adopting the Shoemaker and Wolfe estimate gives an even higher rate. Thus, periodic cometary showers resulting from an unseen solar companion or from galactic plane crossings predict a cratering rate considerably greater than the observed rate. The above calculation has ignored the number of shower comets which might be captured to short-period comet orbits, or even eventually evolved to Apollo asteroid-like objects, and thus must be considered a lower limit on the excess cratering flux expected from periodic showers.

In the case of the Planet X hypothesis the number of shower comets and average number of returns is specifically limited such that the periodic comet showers account for 50% or less of the total cratering on the Earth and Moon. As noted above, this requires that the mean number of returns for the shower comets must be 3.6×10^4 or less. That figure is rather large as compared with the sublimation lifetime of several times 10^3 returns for Earth-crossing long-period comets as found by Weissman (1980b), or 400 returns as estimated by Kresák (1981).

But if the lifetimes of the comets drop significantly, then they do not make enough returns to give a significant probability of several terrestrial impacts from the shower, in particular, at least one major impact. For example, if we assume average cometary lifetimes of 10^3 returns, then for 10^5

TABLE V
Periodic Terrestrial Craters[a]

Location	Diameter (km)	Age (Myr)	Projectile Type
Karla, USSR	10	7 ± 4	
Haughton, Canada	20	13 ± 11	
Ries, Germany	24	14.8 ± 0.7	Achondrite (?)
Mistastin, Labrador	28	38 ± 4	Iron or achondrite (?)
Wanapitei, Ontario	8.5	37 ± 2	Chondrite (C1, C2, LL)
Popigai, Siberia	100	39 ± 9	Iron
Lappajärvi, Finland	14	77 ± 4	C-chondrite
Steen River, Alberta	25	95 ± 7	
Boltysh, Ukraine	25	100 ± 5	
Logoisk, USSR	17	100 ± 20	
Mien Lake, Sweden	5	118 ± 2	Stony (?)
Gosses Bluff, Australia	22	130 ± 6	
Rochechouart, France	23	160 ± 5	Iron (IIA) or chondrite (?)
Obolon, Ukraine	15	185 ± 10	Iron
Puchezh-Katunki, USSR	80	183 ± 3	
Manicougan, Quebec	70	210 ± 4	

[a]Crater list: Alvarez and Muller (1984); Crater data: Grieve (1982) and Palme (1982).

comets in a shower we get a total of 10^8 returns and a probability of one Earth impact of 0.22 assuming the impact probability of 2.2×10^{-9} for long-period comets, or 0.66 assuming a short-period comet impact rate. The two figures probably represent lower and upper limits, respectively, for the probability of an impact during each comet shower.

Thus, it is far from likely that there will be *any* comet impact, much less a major impact, at every extincton cycle based on the Planet X hypothesis. By attempting to avoid the problem of making too many craters, Whitmire and Matese create the problem of probably not having enough major impacts to result in a biological extinction event.

A second problem with the periodic comet shower hypotheses involves the supposedly periodic cratering events on the Earth produced by these showers, as pointed out by Alvarez and Muller (1984). The 16 craters selected by Alvarez and Muller are listed in Table V along with their diameters and measured ages. Weissman (1985c) has pointed out that tentative identification of the compositions of the impacting bodies which caused eight of the craters have been made (Grieve 1982; Palme 1982) based on the siderophile signatures in the impact melts found in the craters. These identifications are also listed in Table V.

Of the eight identified impactors, six have compositions which are typical of highly differentiated meteorite types. Only two of the craters have impactors with likely primitive compositions. One of those two, at Lappajärvi, Finland, with an age of 77 Myr, falls almost exactly between two major extinction events 65 and 95 Myr ago, and is the only crater in the list identified by Alvarez and Muller as clearly not associated with the periodic cratering events.

Differentiated meteorites are believed to come from large asteroids which became sufficiently hot to melt and separate into nickel-iron cores and silicate mantles, and which were subsequently disrupted by collisions with other asteroids. Reflection spectroscopy studies of asteroids in the main belt and in Earth-crossing orbits have found many analogs for meteorite samples recovered on the Earth. However, differentiated meteorites are inconsistent with the icy conglomerate model for cometary nuclei; comets are believed to be pristine, virtually unprocessed material from the original solar nebula. Thus, the evidence from the crater impact melts is that most, if not all, were formed by asteroid and not comet impacts. No mechanism for creating asteroid impacts on the Earth with a period of 26 Myr has been suggested. In addition, although no specific crater has been identified with the K-T boundary, Palme (1982) has noted that the distribution of elements in the boundary "does not match the siderophile element pattern of unfractionated meteorites (chondrites)." An iron asteroid impactor has been suggested but is not a certainty.

Yet another argument against the existence of periodic comet showers in the past has been presented by Kyte (1984). He pointed out that background measurements of iridium levels in terrestrial sediments away from extinction boundaries was consistent with estimates from the slow accumulation of interplanetary dust particles by the Earth, and that those estimates were consistent with the estimated steady-state flux of long-period comets, the source of the dust, through the planetary region. However, the sudden influx of some 10^9 comets in a shower would lead to a tremendous increase in the amount of interplanetary dust and an increase in the accumulation of that dust in terrestrial sediments. Kyte predicted that this would give an iridium excess at the extinction boundaries 2 to 20 times more than that found at the Cretaceous-Tertiary boundary. Also, the width of the excess would be far greater than the narrow fish clay at the boundary, as the comet shower slowly decayed over several Myr.

Individually, specific dynamical problems can be pointed out with each of the comet shower hypotheses. In the case of the Death Star hypothesis, the orbital stability of the hypothetical unseen companion has been studied by Hills (1984*a*), Hut (1984*a*), Torbett and Smoluchowski (1984), Weissman (1985*b*) and Weinberg et al. (1986). With a 26 Myr period, the star would have a semimajor axis of 8.78×10^4 AU. For the median eccentricity of 0.7 expected by Davis et al. (1984), the star's perihelion would be about 2.6×10^4 AU and its aphelion would be 1.49×10^5 AU, roughly 3/4 pc. The dynamical studies all agreed that starting from this current orbit the Death Star would have a mean lifetime between 0.2 and 1.0 Gyr before it escaped from the solar system. The star is already so close to the edge of the Sun's sphere of influence that stellar and galactic perturbations cause its perihelion distance to random walk up and down on each orbit, and its orbital period to vary by 10% or more per orbit. The former fact results in a variation in the number of comets that might be perturbed in a comet shower and thus the severity of a likely biological extinction event on the Earth; the latter figure is consistent with an observed variation in the exact extinction and cratering periods of about 10%.

A possible problem with the variation in the star's orbital period is a tendency for the star to random walk away from the nominal 26 to 28 Myr period suggested by the extinction and cratering records. With a mean variation of 10% per orbit, after 10 orbits (the approximate number of cycles seen in the fossil and crater records) the star might be expected to have an orbital period ± 30% different from its initial value. The likelihood of it still being within 10% of its initial value is about 24%. Weissman (1985b) found that 23% of the Death Stars escaped within the first 10 orbits.

Hills (1984a) showed that the orbit of the hypothetical companion was likely to be even more eccentric than suggested by Davis et al., particularly if it was a low-mass, < 0.08 $M_\odot$, subluminous star. Hills estimated that the companion star's perihelion might be as small as 10^4 AU, and that there was a small probability (approximately 15%) that the star might actually pass through the planetary system if it had been in orbit around the Sun over the entire history of the solar system. Torbett and Smoluchowski (1984) showed that including the galactic field further decreased the expected stability of the star's orbit. Hut (1984a) showed that in at least one case the hypothetical companion star might survive for up to 3.2 Gyr, though the variations in the orbit during that time would be considerable.

If the companion star's orbit is unstable then it must have been captured relatively recently, or have been evolved outward from a more tightly bound initial orbit. As already mentioned in Sec. III, Valtonen and Innanen (1982) and Valtonen (1983) have shown that the probability of capture of random interstellar comets (or stars) by the Sun is very small, on the order of 10^{-13} for a random star passing at a typical velocity of 20 km s^{-1}. On the other hand, if the star had evolved from a more tightly bound initial orbit, it would have caused even more frequent comet showers in the past, further worsening the disagreement between the observed and predicted terrestrial and lunar cratering rates described above.

Yet another problem is the expected depletion of the Oort cloud population due to repeated passages of the star through it. If only 5% of the cloud is lost on each pass of the reputed Death Star (a conservative estimate based on some preliminary dynamical modeling experiments), then $< 5 \times 10^{-4}$ of the original cloud population would be left after 150 passages. Based on current estimates for the cloud population and mass (100 to 200 $M_{\oplus}$ for the inner and outer clouds), this implies an original Oort cloud mass of 0.67 to 1.35 $M_{\odot}$.

In the case of the galactic plane crossing hypothesis of Rampino and Stothers (1984a), Thaddeus and Chanan (1985) have shown that no strong periodic signal would result, certainly not one that would be detectable after only 9 or 10 cycles. This happens because the scale height of GMCs above and below the galactic plane, about 85 parsecs, is roughly equal to the magnitude of the solar system's epicyclic motion. Thus, the solar system is only slightly more likely to encounter GMCs near the galactic plane than at the extremes of its motion.

The Planet X hypothesis of Whitmire and Matese (1984; chapter by Matese and Whitmire) has not yet been studied as carefully as the others, but some likely problems are immediately evident. As the precession of the planet's orbit brought it into the populated inner Oort cloud zones, the planet would tend to scatter comets in the cometary disk, increasing the mean orbital inclinations, and decreasing the "sharpness" in time of the planet's interaction with the comets. Also, as the planet continued to eject these comets from its zone of possible interactions, it is not clear what mechanism would continue to feed new comets into the zone. Thus, the presence of the planet itself would likely tend to destroy the dynamical arrangement that allowed periodic showers in the first place, and eventually would remove all the local comet population available to supply the showers.

The numerous problems which have arisen with attempting to explain the periodic extinctions have led some, such as Shoemaker and Wolfe (see their chapter) to doubt the reality of the periodicity itself. They point out that the various geologic boundaries are not well dated, particularly between about 140 and 250 Myr ago where there are no dated tie points. Hoffman (1985) has reexamined the Raup and Sepkoski data and found that by using a different but equally well-accepted geologic time scale, the periodicity disappears. Possible biases introduced by the sampling methods used to define the extinctions in the fossil record, or the small-number statistics of the 16 craters used to define nine cratering cycles, cast doubt on the reality of the alleged periods. Stigler (1985) has questioned the correctness of the statistical techniques used by Rampino and Stothers to show a correlation between times of galactic plane crossings and major biological extinctions. Finally, the fact that periods between 18 and 36 Myr have been derived by different studies of the same data sets does not help to instill confidence in the statistical significance of the results.

If there is no evidence for periodic extinctions and cratering episodes, is there still some possibility that random comet showers as predicted by Hills (1981) can be detected in the fossil and crater records? The distribution of the measured ages of craters on the Earth is shown in Fig. 6, the top half of the figure for all dated craters, and the bottom half for those craters > 10 km

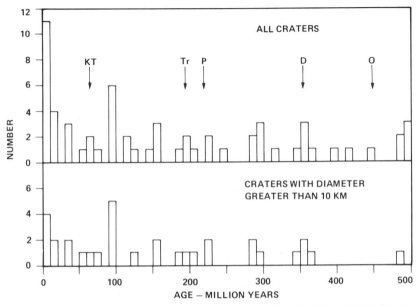

Fig. 6. Distribution of ages of known craters on the Earth as listed by Grieve (1982) for all known craters (top) and all craters > 10 km in diameter (bottom). The five major terrestrial extinctions: the Cretaceous-Tertiary (KT), late Triassic (Tr), late Permian (P), late Devonian (D), and late Ordovician (O), are indicated by the arrows. Coincidences between clusterings of craters and some extinctions may be evidence for random comet showers.

diameter. Also indicated by the vertical arrows in the figure are the five major extinctions in the past 5×10^8 yr.

There appear to be some clusterings of craters in time which may not be purely random. In particular, the five craters at about 3.6×10^8 yr ago, four of them larger than 10 km, may be associated with the late Devonian extinction which occurred at about that time. An iridium anomaly was recently detected associated with this geologic boundary (Playford et al. 1984). A second clustering of five smaller craters at about 4.9×10^8 yr ago may be associated with a series of lesser extinctions known as the Cambrian biomere events.

On the other hand, the late Ordovician extinction about 4.4×10^8 yr ago has no significant clustering of craters associated with it, and a cluster of five craters around 2.9×10^8 yr ago is not associated with any known extinctions, major or minor. It may be possible to explain some extinctions by random comet showers, but obviously not all. As noted earlier, the evidence seems to suggest more of a connection with single asteroid impacts rather than multiple comet ones. In fact, it is not at all clear that large impacts can serve as a universal panacea for explaining all biological extinctions. The fact that a major impact is definitely associated with the Cretaceous-Tertiary event does not immediately rule out other mechanisms for generating extinctions at other times.

As with Lyttleton's sandbank model and Van Flandern's exploding planet, it is expected that the Death Star and Planet X will soon be relegated to the graveyard of interesting but unproven hypotheses. Still, the various ideas proposed to explain the alleged periodic extinctions have been valuable in stimulating continued study of the dynamics of the Oort cloud and its interaction with the Galaxy. What has emerged is a clearer picture of the Oort cloud, its evolution, and its possible origins. In this regard, the various periodic extinction hypotheses have been quite useful.

VI. OORT CLOUDS AROUND OTHER STARS

Most of what we know about the solar system's Oort cloud is based on the very limited data set of approximately 600 long-period comet orbits, half of which we only have parabolic elements for. The rest of our knowledge is inferred from theories of solar system and cometary formation, and dynamical modeling of the interaction of the hypothetical comet cloud with random passing stars, interstellar clouds, and the galactic field.

If it is possible to generalize the current thinking on solar system formation to other main sequence stars in the Galaxy, then comet Oort clouds may have formed, or be in the process of forming, around those other stars. The first evidence of this has been provided by IRAS detection of cool dust shells around Vega (Aumann et al. 1984) and up to 10% of the other stars in the solar neighborhood (Aumann 1984). Weissman (1984) and Harper et al. (1984) have suggested that these dust shells may be material from comets in the inner Oort clouds around each star.

Data for the first four stars detected with such shells are given in Table VI. In each case, the temperature of the shell suggests that the dust does not extend all the way in to the star, but that a clear zone has been swept out close to the star, possibly by unseen planets formed in that region. The shells around α Lyra (Vega) and β Pictoris have been confirmed by other observers and, in each case, shown to extend farther from the stars than the original IRAS observations.

The temperatures of the shells are each low enough for volatile ices, particularly water ice, to condense out of protostellar nebulae. These icy grains, along with dust grains, would settle to the midplane of the nebula to form an accretion disk. When the disk approached a critical density, gravitational instabilities would cause it to break up and collapse into many individual icy planetesimals. These would be the protocomets of the forming Oort cloud.

In the case of the first two stars in the list, Vega and Fomalhaut (α PsA), the orbital periods at the inner radius of the dust shell are so long that it is unlikely that large planets have accreted in that zone in the stars' probable

TABLE VI
Circumstellar Dust Shells Detected by IRAS

Star	α Lyr	α PsA	β Pic	ε Eri
Stellar type	A0	A3	A5	K2
Star mass ($M_\odot$)	2.0	1.75	1.5	0.8
Luminosity ($L_\odot$)	58	13	6.5	0.37
Distance (pc)	8.1	7.0	16.6	3.3
Shell temperature (K)	85	55	100	45
Shell radius (AU)	85	94	20	23
Orbital period at shell (yr)	554	689	73	123

lifetimes, unless the surface density of the accretion disk material was very high. For the other two stars listed in Table VI, the orbital periods are short enough that planetary accretion and zone clearing may still be going on. Thus, what is being observed is likely analogous to the final accretion of Uranus and Neptune and the clearing of their respective zones in our own solar system.

The dust shell around β Pictoris has been photographed optically using a newly developed coronagraphic technique by Smith and Terrile (1984) and is shown in Fig. 7. The photograph is of what appears to be an edge-on accretion disk extending out 400 AU or more on either side of the occulted star. The measured thickness of the disk indicates orbital inclinations $\lesssim 5°$, suggesting that the disk is young and has not yet been perturbed much by random stars or other dynamical processes.

Assuming an asteroid-like particle size distribution, Smith and Terrile find that the disk has a mass of 200 $M_\oplus$, and a radial mass density varying as $r^{-3.1}$. Note that the estimated mass is in good agreement with Shoemaker and Wolfe's estimate for the mass of the inner Oort cloud around our own solar system. Based on the visual magnitude of β Pictoris which is seen through the surrounding dust disk, Smith and Terrile estimate that a clear zone around the star may exist, with the inner edge of the disk at about 30 AU, in rough agreement with the IRAS data.

If it is correct to interpret the β Pictoris photograph as a solar system in the late stages of outer zone clearing, then it indicates that Cameron's model of an extended accretion disk as the source of much of the cometary cloud is probably correct. Cometesimals ejected out of the Uranus-Neptune zone would contribute to the primordial inner and outer Oort clouds at all radii, while the bulk of the inner cloud would form in place from the extended accretion disk.

Further analysis of the IRAS data and additional photographs of other circumstellar disks is certain to add greatly to our understanding of the formation of planetary systems, and in particular, to the formation of cometary

Fig. 7. Photograph of the edge-on accretion disk around β Pictoris by Smith and Terrile (1984). The central star is blocked by an occulting disk with an effective radius of about 100 AU, and the remaining scattered light has been removed by digitally dividing the image by that of a similar star not known to have a circumstellar dust shell. The visible disk extends out about 400 AU on either side of the star, with a thickness of about 50 AU at 300 AU distance. Protocomets forming in the disk will likely populate the Oort cloud around this star.

clouds around main sequence stars. If enough such shells are observed, it may be possible to construct an observed sequence of the different stages in planetary formation and zone clearing. This is a very exciting prospect.

VII. DISCUSSION

Our understanding of the Oort cloud has evolved considerably from Oort's original view of a distant cloud of comets slowly being stirred by perturbations from random passing stars. The number of comets, mass, radial extent, number and variety of perturbers, and complexity of the dynamical interactions have each grown, while the absolute dimensions of the cloud have shrunk somewhat. This change has resulted, to some extent, due to the evolution of our understanding of the Galaxy itself. Giant molecular clouds which make up a major fraction of the interstellar medium were unknown when Oort wrote his description of the comet cloud in 1950. In addition, an increased understanding of the processes involved in the formation and subsequent dynamical history of the solar system has greatly clarified the important role that comets likely played in those events.

It now seems that the Oort cloud was populated from two different dynamical regimes within the early solar system. The outer shell of the original cloud was populated by icy planetesimals ejected from the Uranus-Neptune zone, while the inner cloud came from the extended accretion disk beyond Neptune which never accreted into planetary sized bodies (in this case we do not regard Pluto as a planet, but rather as among the largest of the accreted cometesimals). Perturbations by a variety of sources over the history of the solar system has led to the loss of that initial distant shell of comets, and the pumping up of cometary orbits from the inner cloud to fill the outer shell.

This leads to the interesting suggestion that the original Oort cloud was populated by comets less pristine, i.e., formed in a warmer region of the primordial solar nebula, than the comets which now fill the outer shell. If we could sample the comets in the original Oort shell we would find bodies representative of the conditions in the Uranus-Neptune zone. But by waiting 4.5 Gyr, we now find a population dominated by bodies from the more distant regions of the protosolar accretion disk. Thus, the Oort cloud comets now being sent back into the planetary system by stellar, GMC and galactic perturbations are more pristine than the original population which likely showered the planetary system during the late heavy bombardment.

There remain a number of major problems with regard to understanding the dynamical interaction between comets, the planetary system and the Galaxy. At this point we still do not know the nature or distribution of the missing mass in the Galaxy, and thus are unable to quantify its effect on the Oort cloud. There will likely be significant differences in the resulting dynamics if the missing mass is concentrated in low-mass brown dwarfs, higher-mass black holes, or some as yet undiscovered extended source in the interstellar medium. At present the missing mass is included in calculating the integrated effect of the galactic gravitational field, but it is also necessary to characterize the interaction with the Oort cloud of individual bodies (or clouds) of this missing material.

The comets themselves seem unlikely to account for a significant fraction of the missing mass. The present estimates for the mass of the total Oort cloud (inner and outer) are on the order of a few hundred $M_\oplus$, or about 10^{-3} $M_\odot$. Even if an order of magnitude more material was ejected in the process of forming the Oort cloud, and this fraction was repeated by every star formed, it still would amount to only about 1% of the star's mass per star, far too little to account for the missing mass.

On the other hand, if we consider the substantially shorter lifetimes of stars more luminous and more massive than the Sun, and the repeated generations of star formation that the Galaxy has likely undergone, then perhaps it is possible that cometary-sized bodies have been produced in such numbers to make up a sizable but difficult to detect fraction of the interstellar mass. This population of interstellar comets would only be detected when members randomly entered the observable region around the Sun. The failure to observe interstellar comets so far was used by Sekanina (1976) to set an upper limit on the mass density of these objects of 6×10^{-4} $M_\odot$ pc^{-3}, far too little to make any meaningful contribution to the missing mass in the Galaxy.

Until now, dynamical treatments of Oort cloud perturbers have tended to treat them individually and to regard them as competitors. However, the estimates of the effects of stars, GMCs and the galactic field have shown that they all are within an order of magnitude of one another in total effect. Clearly, a more integrated approach is necessary in the future, one that combines all three perturbers and looks for possible dynamical interactions between them.

Similarly, the inner and outer Oort clouds have tended to be treated as separate entities when in fact they are just different dynamical regimes of a single comet cloud which stretches from just beyond the orbit of Neptune to 10^5 AU or more. Future studies should look to obtaining a more integrated treatment of the entire Oort cloud and the evolution of cometary orbits within it.

Acknowledgments. This work was supported by the NASA Planetary Geology and Geophysics Program, and was performed at the Jet Propulsion Laboratory under contract with the National Aeronautics and Space Administration.

PART V
Perturbations of the Solar System

GEOLOGIC PERIODICITIES AND THE GALAXY

MICHAEL R. RAMPINO
New York University

and

RICHARD B. STOTHERS
Goddard Institute for Space Studies

New geologic and astronomical developments are reviewed that have recently led to the proposal of various galactic theories to explain the temporal pattern of impact cratering on Earth. Linear and harmonic time-series analyses have revealed that two dominant periodicities approximately equal to 33 ± 3 Myr and 260 ± 25 Myr underlie the geologic record of terrestrial impact cratering and global tectonic phenomena. These periodicities have been stable within ± 10% for at least the last 600 Myr (and probably the last 1800 Myr). Mass extinctions show a very similar, but less precisely determined, behavior. The most recent epoch of the 33 Myr cycle is found to be, in all cases, very close to the present time. Independent observations show that the inner solar system probably now lies immersed in the tail end of a comet shower. We argue that purely terrestrial mechanisms are unable to account for the cratering cycles as being preservational artifacts. On the contrary, cratering seems to influence tectonism, possibly through perturbations of mantle convection. The 33 Myr cratering cycle appears to be significant at the 0.1% level and is probably intrinsically quasi-periodic, although the underlying clock that causes it could be periodic.

Galactic models proposed to explain the two long-term periodicities are reviewed critically. The most likely explanation for the 33 Myr cycle involves the comparatively stable half period of vertical oscillation of the solar system about the galactic plane (which is astronomically determined to be 33 ± 3 Myr). This motion should modulate the (normally stochastic) encounters with intermediate-

sized and larger-sized interstellar clouds, which are concentrated preferentially toward the galactic plane. These encounters are expected to produce quasi-periodic gravitational perturbations of the Sun's extended family of comets, causing large numbers of comets to fall into the inner regions of the solar system, where some would strike the Earth. The 260 Myr cycle may be related to the rare encounters with galactic spiral arms during the revolution of the solar system around the center of the Galaxy. The probability of finding accidental agreement between the 33 Myr periodicities in tectonic phenomena, impact cratering, and galactovertical oscillation is $< 10^{-4}$.

The idea that the Earth's climatic and geologic systems somehow respond to changes in the galactic environment has a long history. Until recently, however, there was little reliable physical evidence that could be used to establish a clear connection. This situation has now changed markedly.

The discovery by Alvarez et al. (1980) of iridium enhancement at the Cretaceous-Tertiary boundary coincident with the mass extinctions at the boundary, and the subsequent discovery of sanidine spherules (Smit and Klaver 1981), shocked quartz grains (Bohor et al. 1984), and soot (Wolbach et al. 1985) at the boundary, have focused much attention on the possibility that biological evolution may be profoundly affected by the impact of large asteroids or comets on the Earth. Estimated ages of three large impact craters in the USSR (Grieve 1982) are also in approximate correspondence with the date of the Cretaceous-Tertiary boundary event. Urey (1973) among others had previously proposed that cometary collisions might mark some or all of the major boundaries of the geologic record. The visible remains of large impact craters have now provided one of the missing links in the full chain of deduction.

This chapter reviews recent geologic and astronomical developments that have led to our proposal of a galactic theory to explain the quasi-periodic time distribution of impact craters on Earth.

I. TIME SERIES: LINEAR ANALYSIS VERSUS HARMONIC ANALYSIS

Recent advances have depended heavily on the application of objective methods of statistical analysis to numerous geologic time series. In the studies reviewed here, all the time series consist simply of lists of dates when phenomena occurred. In most cases, the dates are too few in number for binning before analysis. The raw time series themselves (without preliminary smoothing) must be used for the analysis. Two types of time series may therefore be distinguished with regard to their intrinsic information content: the first type consists of the dates of N episodes (e.g., tectonic episodes) which have been determined by combining many radiometric and stratigraphic ages of individual events; the second type consists of the dates of N individual events (e.g., impact craters) each radiometrically dated, which may or may not belong to larger episodes. To yield the same amount of information about periodicity, N must be considerably larger in the latter case.

The method of analysis employed for most of the time series under discussion (Stothers 1979) seeks a best fit of the observed times t_i to an assumed linear periodic function of the form $t = t_o + nP$, where P is a trial period, t_o is a trial value for the most recent epoch, and n is an integer. A best fit occurs when the sum of the squared residuals $N\sigma^2$ is minimized. Groups of neighboring dates perform a function similar to binning, since they weight limited regions of the time domain. The present analog of a Fourier power spectrum can be computed for a sequence of trial periods P. In the resulting spectrum of the squares of the residuals indices $(\sigma_c - \sigma)^2/P^2$, where σ_c is a constant, a significant period may be identified by a high spectral peak. There should also appear high spectral peaks for the resonances, i.e., for the significant period's lowest harmonics, lowest integer multiples, and combinations thereof (caused by gaps and bunching in the data). Although more than one significant period can be detected in this way, it is often difficult to analyze a multiperiodic record unless the significant periods are widely separated.

The method, however, accommodates noise very well. Being nonparametric, it requires no specific information about the source or structure of the noise, or of the underlying periodicity. The true cycle may be, for example, intrinsically periodic but distorted by the presence of extraneous observational events or errors in the dates, or it may appear (or be) quasi-periodic like the sunspot cycle. Random errors, random extraneous events, and even missing events are compensated for by using a sufficiently large value of N. A nonperiodic episodicity among all the events would mean that the time intervals are random; hence the mean time interval $(t_N - t_1)/(N - 1)$ will never be, except by accident, a significant period.

The methods used by Raup and Sepkoski (1984) and Lütz (1985) are variants of the Stothers (1979) method. They depart from it in not solving for t_o at a trial period by formally minimizing the sum of the squared residuals. Raup and Sepkoski search for the arithmetical mean of the residuals that is closest to zero, while Lütz makes a circular transformation to obtain phases of the observed data whose dispersion is then minimized by performing two trigonometric summations, as in Fourier analysis.

Although Fourier analysis can also be used directly on a binary time series of the present kind, this classical method of analysis requires repeated evaluations of trigonometric functions, and so is computationally slow. Since it calculates only harmonics of the record length, it cannot pick the trial periods to be tested and is not accurate for locating long periods. The short-period cutoff at the Nyquist frequency is another limitation. The method also requires sampling at equal time intervals although modifications can be made. Generally, Fourier analysis is more suited to analyzing the kind of time series that has explicit amplitude information, which the linear method cannot handle.

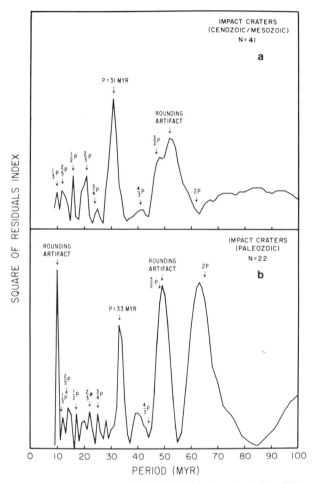

Fig. 1. Spectrum of the square of the residuals index for two time series of dated terrestrial impact craters. See text for identifications of the spectral peaks.

In actual time-series analyses, we have used samples as small as 7 episodes or as large as 283 individual events. (For Fourier analysis, we first binned the dates if the number of individual events was very large, and then we analyzed the histogram.) In all cases tested, the two methods of analysis yielded essentially identical normalized spectra at moderate to high frequencies. The closeness of the agreement is surprising because linear analysis fits a single, linear, discrete function for *time,* whereas harmonic analysis fits a multiple, sinusoidal, continuous function for *amplitude.*

Two typical examples of spectra produced by the linear method are shown in Fig. 1. These spectra refer to the analysis of terrestrial impact craters, as described in Sec. II below. One significant peak exists in each example and it is labeled with the period P. One or two other high peaks are artifacts caused by the divisibility by 10 or 50 of the published dates (given in round numbers) of some older craters. Note that these artificial peaks displace slightly the positions of the neighboring peaks that correspond to expected harmonics of the physically significant period. The anomalously large height of the peak for $2P$ in the Paleozoic spectrum arises solely from three exceptionally large clusterings of dates in the time series.

II. PERIODICITY OF LARGE-BODY IMPACTS

The most direct physical evidence of catastrophic extraterrestrial effects on the Earth consists of craters formed by the impact of planetesimals (asteroids or comets). Grieve (1982) has compiled data on the known terrestrial impact craters, including best estimates of their sizes and ages. We have analyzed Grieve's age data for possible periodicities by using the linear statistical method described above while excluding all ages given only as upper limits. Our adopted list is given in Table I. Initially, we analyzed all 41 suitable crater ages in the time interval 1 to 250 Myr (Cenozoic and Mesozoic times, excluding very recent events). This conservative analysis showed a significant mean periodicity of 31 ± 1 Myr (estimated mean error), with the most recent epoch of the mean cycle occurring at $t_o = 5 \pm 6$ Myr (Rampino and Stothers 1984a) (Fig. 1a). Analysis of all 65 suitable crater ages over the longer time interval 0 to 600 Myr (Phanerozoic time) yielded $P = 32 \pm 1$ Myr and $t_o = 5 \pm 4$ Myr. When we used only the 32 largest ($D \geq 10$ km) craters younger than 400 Myr, which have a greater survival probability and thus are more nearly uniformly sampled in time, the result of the analysis was identical. Finally, when we examined only the 22 craters (of all sizes) in the time interval 250 to 600 Myr (Paleozoic time), the dominant periodicity was 33 ± 1 Myr (Rampino and Stothers 1984b) (Fig. 1b). Despite the large estimated errors of the older craters' ages, our numerical tests consistently showed that the derived mean period is quite robust (and not a harmonic of 100 Myr). This is a consequence of N being sufficiently large and the period being sufficiently well defined.

In a study done two months after our initial investigation, Alvarez and Muller (1984) used Fourier analysis on a subset of 11 well-dated impact craters with $D \geq 10$ km in the time interval 5 to 250 Myr, and found a periodicity of 28.4 ± 1 Myr with $t_o = 13 \pm 2$ Myr. Essentially the same period and phase, $P = 28.5 \pm 1$ Myr and $t_o = 13 \pm 3$ Myr, are found from these data by using the linear method described above (Rampino and Stothers 1984b; Sepkoski and Raup 1986). However, these selected data are demonstrably biased by the small number of craters used and by the arbitrary exclusion of three large ($D \geq 10$ km) craters formed in the last 5 Myr. When these three craters (dated at 1.3, 3.5 and 4.5 Myr) are included in the analysis, the best-fitting period becomes 30 ± 1 Myr and the most recent epoch changes to 8 ± 3 Myr (Rampino and Stothers 1984b), in near agreement with our original result.

TABLE I
Ages (Myr Ago) of 65 Terrestrial Impact Craters[a]

0.0003	14.8	100[b]	200	360[b]
0.002	15[b]	100[b]	210[b]	360
0.05	37	100[b]	225[b]	365[b]
1.0	38[b]	100	230[b]	400
1.3[b]	39[b]	118	250	420
2.5	57[b]	120	290[b]	450
3.1	60[b]	130[b]	290[b]	485[b]
3.5[b]	65[b]	150	300[b]	490
4.5[b]	65	160[b]	300	495
5.0	70	160[b]	300	500
10.0[b]	77[b]	160	320	500
12.0	95[b]	183[b]	350[b]	550
14.8[b]	100[b]	200[b]	360[b]	600

[a]Table from Grieve (1982).
[b]$D \geq 10$ km.

A preliminary study of 25 well-dated Cenozoic and Mesozoic craters (including the three largest young craters) by Shoemaker and Wolfe (see their chapter) yielded an approximate period of 30 to 32 Myr. We find with these data and the linear method $P = 30 \pm 1$ Myr and $t_o = 5 \pm 5$ Myr.

The results of all these analyses indicate that impact cratering on Earth contains an underlying mean periodicity of 30 to 33 Myr, with the most recent epoch of the mean cycle being statistically indistinguishable from the present time. Although arbitrary omissions and groupings of crater ages can shift the derived period and phase slightly, the periodicity is basically robust and is almost certainly not a result of small-number statistics as suggested by Weissman (1985c).

These periodic (or quasi-periodic) large-body impacts are superposed on a background of nonperiodic impacts. It is important to try to estimate the relative strengths of the periodic and nonperiodic components, as well as the duration of the episodes of increased impact rate, because these are important constraints on any theories seeking to explain the complete record of impacts. The cratering statistics (which are supported by microtektite evidence) suggest that the episodes of increased impact rate last for several million years (perhaps as long as 10 Myr) and that the total number of episodic impactors during the times of increased flux may be of the same order of magnitude as the integrated flux of background impactors (Rampino and Stothers 1984a). The periodic showers are therefore of relatively low intensity. This is consistent with the fact that current estimates of the cratering rate on both Earth and

Moon agree roughly with the observed flux of Earth-crossing asteroids and comets (Shoemaker et al. 1979; Weissman 1985c,d). If, however, the inner solar system is now immersed in a comet shower (see, e.g., Yabushita 1979b; Clube and Napier 1982b; Biermann et al. 1983; see the chapter by Delsemme) and if most Earth-crossing asteroids are derived from comets, the background cratering rate could actually be negligible. The large uncertainty about the background rate arises partly from the fact that is is difficult to infer either from the size, type and geochemistry of the impact craters (Grieve 1982; Weissman 1985c) or from the statistics, orbits and physics of known asteroids and comets what might be the relative contributions of periodic-shower comets, background-shower comets, nonshower comets, interstellar comets, strayed main-belt asteroids, and comet-derived asteroids.

Improvement in our knowledge about the rate of impact cratering, free from problems of sampling, could ultimately come from a study of craters on the Moon. Baldwin's (1985) time series of dated lunar impact craters back to 3850 Myr contains a built-in regularity of 77 Myr for $t > 300$ Myr and one of 23 Myr for $t < 300$ Myr. These artificial periodicities are the indirect result of a paucity of known absolute dates for lunar craters (only seven absolute calibration points are available).

III. PRESERVATIONAL BIAS IN IMPACT CRATERS?

It has recently been suggested that the detected periodicity in impact cratering might be the consequence of a preservational artifact caused by purely terrestrial processes. For example, Jablonski (1984) proposed that at times of lowered sea level, a larger continental target area would be presented to the impactors and hence more impacts would be recorded. Dent (1973) had previously argued that impact craters on stable continental areas stood a better chance of preservation if they were buried quickly by marine sediments; this might tend to preserve craters formed just prior to periodic marine transgressions.

Could the periodicity in impact cratering be such an artifact? There are several strong arguments to the contrary. First, the thin cover of sediments and water that would have been present in stable platform areas could not have prevented the impacts from deeply affecting the basement rocks. For example, the bolide that created the 26 km diameter Ries Crater 15 Myr ago penetrated 600 meters of Mesozoic sediments and shocked the basement rocks to a depth > 1200 meters and most probably down to several kilometers (Hörz 1982). The preservation of large craters would thus be little affected by changes in sea level or sedimentation. Cumulative numbers of Phanerozoic impact craters summed over size also suggest that craters with $D > 20$ km will survive for at least a few hundred million years (Grieve 1982). More relevantly, if crater numbers are summed cumulatively over time, it appears that a high survival probability exists even for the smaller craters. For example, Grieve's list contains 11 craters with 2 km $< D <$ 10 km in the age range $t = 0$ to 250 Myr and 12 such craters in the age range $t = 251$ to 600 Myr.

If periodicity among the large craters is not a result of preservational bias, the same must be true of the smaller craters, because the latter are found to display essentially the same cyclical behavior over the time interval where they have survived. During the last 150 Myr, the 13 smallest ($D < 5$ km) craters show $P = 30 \pm 1$ Myr and $t_o = 4 \pm 4$ Myr. This implies that many of these craters are associated with the periodic showers of impactors (contrary to a claim by Alvarez and Muller 1984) and therefore that craters of all sizes have been properly included in our analysis of cratering periodicity.

IV. PERIODICITY OF MASS EXTINCTIONS

Raup and Sepkoski (1984) have recently analyzed a detailed time series of marine mass extinctions by families, with an emphasis on the last 250 Myr. Standard Fourier analysis of these data revealed a significant periodicity of about 30 Myr. Using the linear method of analysis mentioned in Sec. I, however, they found a stronger periodicity at 26 Myr by selecting only the 12 most conspicuous mass-extinction events; the most recent epoch of the mean cycle was $t_o = 13$ Myr. Following their analysis, we selected only the nine major mass-extinction events that consisted of more than 10% extinctions on a family level; a simple straight-line regression of the extinction date on the cycle number (assuming that all cycles are present) yielded $P = 30 \pm 1$ Myr and $t_o = 10 \pm 7$ (Rampino and Stothers 1984a). When the restriction on the cycle number was relaxed, however, the alternative period of 26 Myr was recovered. Sepkoski and Raup (1986) subsequently reduced the number of significant mass extinctions to eight; both Fourier analysis and the linear method yielded a best-fitting period of 26 Myr. The results were essentially independent of the geologic time scale used.

A serious difficulty with using mass extinctions for time-series analysis is that the extinctions must be placed into stratigraphical stages, which for the middle of the Mesozoic are arbitrarily assigned approximately equal length of ~ 6 Myr (Hallam 1984b; Hoffman 1985). We have analyzed with the linear method Palmer's (1983) dates of the stage boundaries during the Cenozoic and Mesozoic, and have found marked periodicities of 5 to 7 Myr. Since significant cycles longer than 7 Myr do not appear, the period of 26 or 30 Myr in the mass-extinction time series cannot be an artifact arising solely from the structure of the geologic time scale. However, the specific 26 Myr period may still be an associated artifact of poor resolution in the pattern of mass extinctions (Hoffman 1985). Therefore, the case for the 30 Myr period is correspondingly stronger. We believe that the true uncertainty can easily accommodate periodicities in the range 26 to 30 Myr, an opinion now shared by Raup (1985). Even this range is not sacrosanct: Kitchell and Pena (1984) have estimated by an elaborate statistical model a pseudoperiod of 31 Myr from Raup and

Sepkoski's data, while Fischer and Arthur (1977) previously inferred by inspection a 32 Myr periodicity in similar data.

The evidence for mass extinctions among nonmarine tetrapods (land vertebrates) is comparatively weak, according to Benton (1985). Nevertheless, by taking Benton's six post-Devonian apparent mass extinctions, adding his possible late Jurassic event, and adopting the Palmer (1983) time scale, we find from linear time-series analysis a double spectral peak at $P = 29 \pm 1$ and $P = 33 \pm 1$ Myr, with $t_o = 6 \pm 4$ Myr in both cases. These results refer to the last 350 Myr, and may support the case for a similar 30 Myr period in marine mass extinctions.

Several of the mass-extinction episodes of the last 600 Myr are now suspected to have coincided with large-body impacts, as was early conjectured by De Laubenfels (1956). Geochemical anomalies and/or microtektite layers have been found associated with extinctions in the Late Devonian (370 Myr; Playford et al. 1984), Late Cretaceous (66 Myr; Alvarez et al. 1980, 1982a), Late Eocene (37 Myr; Ganapathy 1982; Alvarez et al. 1982b; Sanfilippo et al. 1985), and Pliocene-Pleistocene (1 to 2 Myr; Glass 1982; Kyte and Brownlee 1985).

Raup and Sepkoski (1984) assumed that the most recent episode of mass extinction occurred in the Middle Miocene 11 Myr ago, although they noted that this episode was the most questionable in terms of amplitude and date. Their assumption has led some authors to believe that we are now about halfway between periodic extinctions. However, the rate of extinction in Sepkoski's (1982a) time series remains high in the Late Miocene (5 Myr) and Pliocene (2 Myr) (see also Stanley and Campbell 1981; Van Valen 1984b; Benson et al. 1984; Hoffman and Ghiold 1985). Raup (1985) now lists the most recent extinction event as occurring somewhere in the Middle Miocene to Pliocene (11 to 2 Myr ago). Fischer and Arthur (1977) placed this biotic crisis even closer to the present time. So recent an extinction event correlates very well with the increased rate of cratering from large impactors during the last 5 Myr (Rampino and Stothers 1984b).

Stratigraphic resolution and dating are relatively poor for the Paleozoic era. Nevertheless, Sepkoski and Raup (1986) have tentatively found a 37 Myr periodicity in their Paleozoic marine mass-extinction data.

V. PERIODICITY OF TECTONIC PHENOMENA

Geologists have long pondered the question of cyclicity in the tectonic record. A number of early 20th century workers claimed to perceive a periodicity of some 30 to 50 Myr in orogenic tectonism, sea-level fluctuations, and related sedimentation patterns (see, e.g., Joly 1924; Holmes 1927; Grabau 1936; for a review, see Umbgrove 1947). These ideas have been revived more recently in a new generation of tectonic studies (see, e.g., Damon 1971; for a bibliography, see Williams 1981).

TABLE II
Significant Periodicities in Tectonic Phenomena[a] (Myr)

Phenomenon	N	Time Range[a]	P_1	P_2[b]	$(t_o)_1$	$(t_o)_2$[b]
Low sea levels[c]	13	0–600	34 ± 1	260	−7 ± 5	40
Sea-floor spreading discontinuities	7	0–180	34 ± 2		13 ± 4	
Tectonic pulse maxima	18	0–600	33 ± 3	270	2 ± 8	40
	8	1200–3600		220		
Carbonatite intrusions	49	0–1840	34 ± 1	235	−5 ± 9	120
Kimberlite intrusions	38	0–420	35 ± 1	280	9 ± 6	100

[a] All times in Myr.
[b] Approximate.
[c] Dates of major episodes of Paleozoic low sea levels: 263, 306, 387, 468 and 570 Myr. Sources and other data are as found in Rampino and Stothers (1984b).

A number of recent authors (cited in Rampino and Stothers 1984b) have proposed that large-body impacts could have perturbed the Earth far beyond immediate atmospheric effects, through the transfer of energy and momentum to the Earth's interior. Urey (1973) calculated that a comet of mass 10^{18} g might release up to 10^{31} erg in a major impact. This release is 10^6 times the energy expended yearly by earthquakes. Perhaps a better comparison is made with the amount of heat energy escaping the Earth, since internal heat presumably fuels the motion of the Earth's plates. We find that the impact energy from a large comet equals 10^3 times the Earth's annual heat flux. Although this is clearly not enough to drive the mantle circulation, it might nevertheless have a perturbing effect (for example, by generating mantle plumes) that is large enough to be seen in the tectonic record in the form of sea-floor spreading discontinuities, sea-level fluctuations, mountain-building pulses, and magmatic intrusions.

In light of this possibility, we have culled from the published literature the best available data sets for all potentially relevant geotectonic phenomena, and have subjected them to time-series analyses using the linear method. A summary of our results is provided in Table II (mostly adapted from Rampino and Stothers 1984b). The geologic phenomena indeed appear to have an underlying period of 33 to 34 Myr.

To extend our work, Raup (1985) analyzed with the same method the record of 296 geomagnetic reversals going back 165 Myr (Harland et al. 1982). He found a marginally significant period of 30 Myr with $t_o = 10$ Myr, which can also be seen visually in his histogram of binned dates. His result essentially agrees with the earlier result of Negi and Tiwari (1983) who subjected to Walsh spectrum analysis an older data set covering the last 570 Myr;

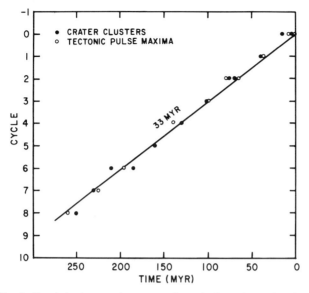

Fig. 2. Correlation between impact-cratering episodes and tectonic pulses.

that analysis yielded a significant period of 32 to 34 Myr. Negi and Tiwari found a number of other periodicities that we have simply interpreted as various harmonics of 33 Myr (Rampino and Stothers 1984b). By using the linear method of analysis on the 283 dates in a new geomagnetic-reversal time series covering the last 169 Myr (Kent and Gradstein 1985), we confirm the foregoing results in deriving $P = 30 \pm 1$ Myr with $t_o = 7 \pm 6$ Myr. By binning these dates in 5 Myr intervals and applying standard Fourier analysis to the histogram, we again find that $P = 30$ Myr with $t_o = 7$ Myr. However, the reliable time span covered by known geomagnetic reversals is only about five cycles long, which may be too short to yield a highly significant period (Lütz 1985).

Most of the present numerical results exhibit a considerable sharpness, despite the presence of dating errors, rounding of the dates in some cases, use of different geologic time scales, spurious events, omitted events, small numbers of events, and real geophysical phase lags. Thus, the statistical coincidence of the tectonic cycle with the impact-cratering cycle is quite striking (see Fig. 2). Consequently, the discovery that all the geologic phenomena under investigation show essentially the same underlying period and phase, strongly suggests that the phenomena are related physically (Rampino and Stothers 1984b).

VI. STABILITY OF THE PERIOD

Dividing a geologic time series into two segments allows us not only to check the reality of the derived period but also to determine its long-term stability. When the available Paleozoic dates are analyzed separately from the Cenozoic and Mesozoic dates, the formal difference in derived period turns out to be as follows: 33 Myr versus 31 Myr, from impact cratering; 35 Myr versus 32 Myr, from tectonic pulse maxima; and 34 Myr versus 35 Myr, from carbonatite intrusions (Rampino and Stothers 1984*b*). Less certainly, we mention 28 Myr versus 35 Myr, from geomagnetic reversal frequencies (Negi and Tiwari 1983); and 37 Myr versus 26 Myr, from mass extinctions (Sepkoski and Raup 1986). In view of the low accuracy of the Paleozoic dates, the differences are probably not statistically significant. If only the most reliable of the Paleozoic data sets are used (Rampino and Stothers 1984*b*), the period appears not to have drifted by more than ± 10% over Phanerozoic time. Long-term stability of the period is also indirectly suggested by the rough constancy of the cratering rate during the last 600 Myr, both on the Earth (Grieve 1982) and on the Moon (Baldwin 1985), and by the fact that the number of known large terrestrial impact craters per bombardment episode seems to have remained roughly constant over this time, according to the available cratering statistics.

VII. THE RAMPINO-STOTHERS GALACTIC MECHANISM

A possible explanation for the ~ 33 Myr terrestrial periodicity may be found in the dynamics of the Milky Way Galaxy (Rampino and Stothers 1984*a*). The essential driving mechanism is the 33 ± 3 Myr half-cycle of oscillation of the solar system perpendicular to the galactic plane, which can simply be viewed as the time interval from one plane crossing to the next. The galactic mechanism involves a gravitational interaction between the solar system and massive interstellar clouds of gas and dust. Hills (1981) has already shown that if comets in the solar system's inner comet reservoir, which is believed (see the chapter by Weissman) to feed the surrounding Oort (1950) comet halo, are gravitationally disturbed by a massive outside body, then some of these comets will lose energy and fall down to small perihelion distances, where a few will collide with the Earth. Such disturbances could arise by encounters with passing stars (Öpik 1932; Oort 1950; Hills 1981) or with passing interstellar clouds (Rampino and Stothers 1984*a*). Actual penetration of an interstellar cloud would lead to another consequence—the flooding of the inner solar system with particles of interstellar gas and dust, which might disturb the Earth's climate and biosphere (an idea first independently put forward by Nölke [1909] and Shapley [1921]). If the perturbing galactic objects are sufficiently concentrated toward the plane, encounters are more likely to occur at those times, like today, when the solar system lies near the plane. We find that intermediate-sized interstellar molecular clouds having mean density 10^2 to 10^3 particles cm^{-3}, radius 3 to 6 pc, and mass 10^3 to 10^4 $M_\odot$ are the most frequently encountered objects that could give rise to significant perturbations. During these encounters, most of the Earth-crossing comets will have initial semimajor axes of about 2×10^4 AU, which happens to be the dividing radius between Hills' inner comet reservoir and Oort's comet halo. If the inner reservoir does not actually exist, the lower shells of Oort's halo may be an adequate source of shower comets.

Our model for the encounters thus has two major components: (1) a periodic component due to the solar system's harmonic oscillation about the galactic plane; and (2) a stochastic component arising from the quasi-random distribution of interstellar clouds. The deciding factor is the relationship between the solar system's greatest vertical distance above and below the plane z_{max} and the vertical scatter of the clouds represented by the exponential scale height β. If the ratio z_{max}/β is large, the modulation should be easily detectable, while if z_{max}/β is small, encounters with the clouds will occur almost randomly throughout the solar system's orbit and so no 33 Myr signal would emerge.

In our original work, we adopted $z_{max} \sim 70$ pc and $\beta \sim 44$ pc, giving $z_{max}/\beta = 1.6$. An analysis of recently published observational data indicates that the best values for these parameters are probably $z_{max} = 88 \pm 13$ pc and $\beta = 50 \pm 5$ pc; therefore $z_{max}/\beta = 1.8 \pm 0.3$ (Stothers 1985). To test the ability of the galactic model to produce a 33 Myr signal in geologic time series, a large number of Monte Carlo simulations using pseudorandom time series have been made and analyzed by the same method that was used previously on the real geologic time series. On the assumption that molecular clouds are the dominant perturbers, the *a priori* probability of detecting the galactic signal in a record of 600 Myr length turns out to be approximately 50% (Stothers 1985). The *a posteriori* probability, of course, could be much higher.

These results contrast with those of Thaddeus and Chanan (1985; chapter by Thaddeus), who made their own Monte Carlo simulations with Fourier analysis, using different data, and concluded that the galactic signal would be undetectable. There seem to be two major flaws in their selection of data (Stothers 1985): (1) their value chosen for z_{max} was underestimated as it was based on an outdated value for the present z-velocity of the Sun; this flaw affects a similar study by Bahcall and Bahcall (1985); (2) their survey of the vertical distribution of molecular clouds used a galactic distance scale that was about 15% too large. Their survey was also heavily weighted by a large flare of interstellar material away from the galactic plane in the Carina spiral arm; consequently, their vertical scale height (Gaussian half-width at half-maximum $z_{1/2}$) for the clouds was overestimated. When corrections are made and the whole of Phanerozoic time is considered, their method of analysis predicts a signal-to-noise ratio of 1.5, which still might not be easily detectable by their approach. Therefore, our method, which is fully consistent with the way in which the observed time series were analyzed, appears to be preferable.

We emphasize that there is nothing special in our model about the actual crossing of the galactic plane. We do not expect comet showers to take place only, or exactly, at the times of plane crossing. The scatter of interstellar clouds above and below the plane guarantees a certain amount of randomness in the encounter times. Early criticisms that our galactic mechanism is out of phase with the mass extinctions have not considered that: (1) the most recent episode of mass extinctions and of impact cratering was most likely 1 to 5 Myr ago, not 11 Myr ago; and (2) some of these episodes ought to occur at times when the solar system is relatively far from the galactic plane. Other objections to our time-series analysis (Muller 1985; chapter by Muller; Stigler 1985) have been answered through further statistical testing (Rampino and Stothers 1985). A recently raised criticism has been that encounters of the solar system with *giant* molecular clouds would occur too rarely, on the order of every 10^2 to 10^3 Myr (Clube and Napier 1982*b;* van den Bergh 1982; Bailey 1983*a*), to account for disturbances of the comet halo every $\sim$ 33 Myr. Aware of this circumstance when we originally proposed our model, we argued that the main perturbers would probably be the more common intermediate sized interstellar molecular clouds (see also Stothers 1985).

VIII. OTHER GALACTIC MODELS

Early geologic investigations of the solar system's galactovertical oscillation considered both the half cycle and the full cycle of oscillation, even though the full cycle has no obvious physical significance. Tamrazyan (1957) and Steiner and Grillmair (1973), who discussed the half cycle, suggested that the large spatial variations of the galactic gravitational potential directly affected the strength of gravity in the solar system. They offered no real geologic or astronomical evidence to support their hypothesis, however. Crain et al. (1969) and Crain and Crain (1970) correlated their proposed 80 Myr geomagnetic cycle with the full oscillation of the solar system through the Galaxy's large-scale magnetic field, which lies moderately concentrated to the galactic plane. Magnetically trapped galactic cosmic rays could then cause a 80 to 90 Myr periodicity in mass extinctions, according to Hatfield and Camp (1970) and Meyerhoff (1973). The apparent 70 Myr regularity in the geologic periods has also been tentatively correlated with the full period of galactovertical oscillation (Innanen et al. 1978; Steiner 1979). Modern analysis of geologic time series, however, does not confirm a cycle of 70 to 90 Myr.

Napier and Clube (1979) and Clube and Napier (1982*b*, 1984*a*) have argued that cometary impacts on the Earth are responsible not only for mass extinctions but also for many other geologic phenomena. In their first two papers, they proposed that collisions between the solar system and giant molecular clouds led first to a stripping away of the Oort comet halo and then to the capture and infall of new comets manufactured in the giant molecular cloud itself; these episodes would take place when the solar system encountered galactic spiral arms on a quasi-periodic time scale of 50 to 150 Myr. Some doubt, however, exists as to whether a giant molecular cloud could either destroy (Hut and Tremaine 1985) or supply (van den Bergh 1982; Valtonen 1983) the Oort comet halo. In their more recent paper, Clube and Napier (1984a) acknowledged that the shower comets might come, instead, from Hills' inner reservoir of comets, which, if it exists, is probably mildly perturbed by close encounters with giant molecular clouds. Although they speculated that the solar system's galactovertical motion might modulate these encounters with a period of $\sim$ 30 Myr, any modulation by specifically giant molecular clouds is probably unobservable on account of the rarity of such clouds.

Davis et al. (1984), in a brief discussion that was in large part anticipated by Heylmun (1969), mentioned and rejected a simple galactic model in which interstellar comets, or other kinds of interstellar debris, are distributed within the plane of the Galaxy. As the Sun moves through the plane, these objects enter the solar system where some hit the Earth. This model is contrived since comets with hyperbolic orbits are unknown and, in any case, are unlikely to exist exclusively in the galactic plane (van den Bergh 1982). Besides, the observed cratering episodes are probably not regularly periodic. Goldsmith (1985) proposed a variant model where most of the missing mass (which has been inferred from stellar dynamical studies) lies within the galactic plane and gravitationally perturbs the Oort comet halo during solar system passages. This model, however, predicts almost strictly periodic encounters and assumes that the missing mass is much more concentrated to the plane than is actually allowed by stellar dynamical studies (Bahcall and Bahcall 1985). Other problems with missing mass as the main perturber have already been discussed (Stothers 1984).

Schwartz and James (1984), following Hatfield and Camp (1970), discussed the Sun's galactovertical oscillation in connection with a model of mass extinctions that relies on spatial variations in galactic cosmic-ray and soft X-ray intensity. Their model requires that mass extinctions occur at times when the solar system is farthest from the galactic plane. We believe that this phase requirement conflicts with existing paleontological data; the model also does not account for quasi-periodic impacts.

Scalo and Smoluchowski (1984) suggested that the compressional force exerted by the galactic disk induces a periodically varying gravitational perturbation of the Oort comet halo as the solar system oscillates about the galactic plane. This tidal force, however, is $\sim 10^2$ times weaker than the Sun's gravitational attraction at the outer edge of the inner cometary reservoir. Moreover, because the galactic force law is linear in z, the vertical compression of the outer comet halo probably remains constant in time and so no periodicity could result (Stothers 1984). Nevertheless, N-body calculations would be useful to test more fully this theory.

IX. SIGNIFICANCE OF THE PERIOD

Statistical significance of the 33 Myr geologic period can now be estimated in the framework of the proposed galactic models that consider impacts. Since all the data—galactic, tectonic and cratering—suggest a basic period close to 33 Myr, this value may be used as an *a priori* period. Monte Carlo simulations using the linear method of analysis on random time series then lead to confidence levels equal to 96% based on tectonic data, and 99.9% based on cratering data. An alternative approach is to calculate the probability that the galactic, tectonic and cratering periods are accidentally equal to each other within ± 3 Myr. This probability is less than 10^{-4} (Rampino and Stothers 1984b).

The galactic models also correctly predict that the geologic period should show no significant long-term drift. Owing to irregularities in the galactic mass distribution, however, a short-term period jitter of up to ± 10% ought to occur (Innanen et al. 1978; Stothers 1985) and may account for some of the imprecision seen in our derived results for the geologic period.

X. PHASE STABILITY

It is not possible to obtain accurate information about phase stability from any of the available geologic time series. The cycles covered are too few and too noisy, and seem to be intrinsically quasi-periodic. Numerical tests with periodic time series, however, indicate that the introduction of one, or even a few, abrupt phase shifts will change only slightly the two parameters, P and t_o, of the best-fitting mean cycle (Stothers 1985).

In our galactic model, the occasional close encounter of the Sun with a giant molecular cloud is capable of suddenly deflecting the Sun's ballistic trajectory in space. The Sun's present rms z-velocity averaged over a vertical orbit is ~ 6 km s^{-1}, which lies well below the ensemble mean of other solar-type stars, ~ 20 km s^{-1} (Wielen 1977). This may imply that the Sun's present low velocity is accidental. The impulse approximation can be used to calculate the possible change in the Sun's total velocity V on the assumption that a giant molecular cloud is at rest relative to the Sun and attracts like a point-mass M outside the cloud's mean radius R. Thus $\Delta V = 2\,GM/Vp$, where p is the impact parameter (Ogorodnikov 1965). For $M = 5 \times 10^5$ M$_\odot$ and $R = 20$ pc (Stark and Blitz 1978) together with $V = 20$ km s^{-1} and $p = R$, ΔV turns out to be ~ 10 km s^{-1}. Similarly, the angular deflection of the Sun's path amounts to $\psi = 2 \tan^{-1}(GM/V^2p) \approx 30°$. As a result of the total vector change of V, the z-component of velocity may easily change by more than 50% in a single close encounter.

Owing to the cancellation effect of several random encounters over the Sun's lifetime, it is possible that the Sun's rms z-velocity did average close to the ensemble mean observed for solar-type stars. If so, $z_{\max}$ would have been typically ~ 300 pc. This larger vertical trajectory could not have changed the half period of oscillation by a large amount, but it would have modulated very strongly the rate of random encounters with interstellar clouds of all sizes (Stothers 1985). A considerably smaller amount of phase jitter has been proposed by Bahcall and Bahcall (1985).

XI. A LONG PERIOD OF 260 MYR

Impact cratering on the Earth shows another periodicity of $P \approx 260$ Myr over Phanerozoic time; the most recent epoch of the mean cycle is roughly 40 Myr ago (Rampino and Stothers 1984b). The significance of this period is difficult to evaluate because the identifying spectral peak is rather wide, the number of cycles observed is small, and an *a priori* period is not available. However, a similar period of 250 ± 25 Myr (Hatfield and Camp 1970) or possibly ~ 300 Myr (Fischer 1979) has been inferred by inspection of modern data on mass extinctions. The existence of a long tectonic period has also been suspected since the early investigations of Holmes (1927) (see the review by Williams 1981). Modern time-series analyses that we have performed for several kinds of tectonic phenomena, shown in Table II, give best values of period and phase that are almost identical to those found for impact cratering. An analysis of older tectonic episodes suggests that this period has been roughly constant for at least 3600 Myr; the apparent trend with time in Table II is probably fictitious (Rampino and Stothers 1985b). In addition, time-series analyses of geomagnetic reversal frequencies during the Phanerozoic have yielded period estimates ranging from 250 Myr to 350 Myr (Crain et al. 1969; Crain and Crain 1970; McElhinny 1971; Ulrych 1972; Irving and Pullaiah 1976; Negi and Tiwari 1983).

Physical mechanisms to account for this long periodicity have been mainly concerned with explaining episodic glaciation, which is now thought not to follow, at least closely, the 260 Myr tectonic cycle (see, e.g., Christie-Blick 1982). Other suggested mechanisms, however, have dealt expressly with explanations of the strong cycle of global tectonism. Without worrying about the particular terrestrial phenomena invoked, we shall describe briefly the various proposed galactic mechanisms for a long-term periodicity. In all cases, this periodicity is linked, directly or indirectly, to the period of the solar system's quasi-elliptical orbit around the galactic nucleus, namely, 250 ± 50 Myr.

Holmes (1927) was probably the first author to suspect a connection with the Sun's motion through the Galaxy. Forbes (1931) subsequently suggested that an obscuring cloud, orbiting around the galactic nucleus inside the solar galactic circle, periodically blocks out the radiation coming from the nucleus. Lungershausen (1957) and Gidon (1970) pointed out that the radiation received from the nucleus will in any case vary because of the solar system's changing galactocentric distance. Brin (1984) postulated a radiating black hole which orbits the galactic nucleus inside the solar galactic circle and thus periodically alters its distance from the solar system. In all of these cases, however, the intensity of radiation reaching the Earth would almost certainly be too small to produce a noticeable climatic or biological effect. Even less credible, in our opinion, is the speculation that the spatially varying gravitational potential of the Galaxy somehow induces a significant change of the strength of gravity in the solar system and thereby affects tectonism and life on Earth (Tamrazyan 1967; Steiner 1967,1973,1974; Steiner and Grillmair 1973; Kropotkin 1970). On the other hand, the changing radial gravitational potential of the Galaxy will certainly lead to significant changes in the binding of distant comets to the Sun. Although these changes will alter periodically the overall size of the Oort comet halo, the inner comets of the halo should be affected very little.

Slow periodic interactions of the Earth with the galactic magnetic field (Steiner 1967; Crain et al. 1969; Crain and Crain 1970) or with magnetically trapped galactic cosmic rays (Hatfield and Camp 1970; Meyerhoff 1973) have been proposed as causing measurable geomagnetic and biological disturbances. However, the magnetic field strength of the Galaxy is only 10^{-5} times that of the Earth.

A very different suggestion makes use of quasi-periodic collisions between the solar system and dense complexes of interstellar dust clouds. Complexes will be encountered preferentially when the solar system crosses a spiral arm and other very dusty sectors of the Milky Way (Shapley 1949; McCrea 1975, 1981; Innanen et al. 1978) including the unwarped portion of the galactic disk (Williams 1975). The comet halo might then be lost to, and a new (highly perturbed) comet halo subsequently picked up from the dust clouds at those times (Napier and Clube 1979; Clube and Napier 1982*b*,1984*a* and their chapter herein). Alternatively, the gravitational attraction exerted by the dust clouds might merely perturb, without destroying, the comet halo and/or Hills' inner cometary reservoir (Rampino and Stothers 1984*b*). These are the only two ways that we can think of to explain the 260 Myr cycle in impact cratering. Passages through the spiral arms would probably serve as the basic clock. If the spiral pattern speed happens to be approximately half the Sun's galactic rotational speed and the Sun crosses only two spiral arms in a galactic circuit, the Sun will overtake a spiral arm every ~ 125 Myr.

Although none of these galactic mechanisms is a fully satisfactory explanation, a terrestrial mechanism is not obviously better. If the 260 Myr tectonic period were related purely to the turnover time of mantle convection for example, the identical period seen in impact cratering would have to arise from some unknown source of preservational bias.

XII. CONCLUSION

The search for long-term periodicities in the geologic record has made notable advances during the past few years. Accuracies of two significant figures are now being routinely obtained by the use of objective, nonparametric statistical methods. From a wide diversity of proposed long-term periodicities, only two periods have emerged as being probably significant: one at 33 ± 3 Myr and the other at 260 ± 25 Myr. The rest of the periods appear to be harmonics, integer multiples, and combinations thereof of the two basic periods, as well as accidental periods due to the occasional poor quality (or insufficient quantity) of data. What is most surprising is that mass extinctions have perhaps turned out to be the least suitable data set to use, primarily because the fossil record still has to be dated within a relatively coarse stratigraphic framework. The 33 Myr cycle in global tectonism is determined by the use of combined radiometric and stratigraphic dates, and appears to be intrinsically quasi-periodic like the sunspot cycle (although the underlying clock may be periodic); the most recent epoch of the mean cycle lies very close to the present time. Both cycles exhibit period stability within ± 10% over the past 600 Myr. (Phase stability cannot be determined accurately with any of the available data.)

Terrestrial impact cratering shows exactly the same two cycles. This coincidence has suggested a model in which comets quasi-periodically bombard the Earth, the timing being ultimately determined by the solar system's two components of cyclical motion through the Galaxy. It is not too much to say that the observed periodicity in impact cratering is the only real evidence for a connection to the Galaxy. Since, however, even the largest cometary impacts probably act only as triggers of terrestrial catastrophe, the timing as well as severity of the major geologic and biological crises must also depend on purely internal terrestrial stresses. Nevertheless, periodic impacts must have prevented the Earth from ever fully equilibrating—tectonically as well as biologically.

The galactic theory of geologic change, as outlined here, appears to be consistent with all the available astronomical and geologic data. Advantages of the theory are that it requires no special assumptions, while it supplies one (and perhaps two) *a priori* periodicities and satisfies the principle of unexceptionality (the solar system is not a special object in the Galaxy). The theory makes a number of substantial geologic and astronomical predictions that can be tested in the future through careful observation and theoretical calculations.

Acknowledgments. For valuable discussions, correspondence, and other assistance, we thank W. Alvarez, S. V. M. Clube, T. M. Dame, R. W. Fairbridge, E. B. Heylmun, A. Hoffman, D. V. Kent, R. J. Maddalena, G. L. Martin, W. M. Napier, D. M. Raup, C. K. Seyfert, H. R. Shaw, E. M. Shoemaker, S. M. Stigler, P. Thaddeus, P. R. Vail, T. Volk, P. R. Weissman, and G. E. Williams. This work was supported in part by a NASA Cooperative Agreement with Columbia University.

GIANT COMETS AND THE GALAXY: IMPLICATIONS OF THE TERRESTRIAL RECORD

S. V. M. CLUBE
University of Oxford

and

W. M. NAPIER
Royal Observatory, Edinburgh

We review various aspects of the theory of evolution involving terrestrial catastrophism in which the Oort cloud is disturbed quasi-periodically by close encounters with massive nebulae. Attention is drawn to aspects that differ from the stray meteorite thesis introduced to explain geochemical anomalies at the Cretaceous-Tertiary boundary and the more recent proposal requiring a hypothetical stellar companion. In general, Oort cloud disturbances generate bombardment pulses lasting a few million years, a pulse being dominated by a series of spikes of ~ 0.01 to 0.1 Myr duration at ~ 0.1 to 1.0 Myr intervals. Each spike reflects the arrival in circumterrestrial space of the largest comets and their disintegration into short-lived Apollo asteroids. Dense and sporadic dust veils are formed in the stratosphere during passage of the Earth through the cometary debris, leading, among other things, to rapid climatic deterioration and mass extinctions of life. Additionally, $\gtrsim 10$ planetesimals of $\gtrsim 1$ km dimensions will strike the Earth in the course of a bombardment episode comprising one or more pulses. Iridium bearing layers in the terrestrial record are thus unlikely to be simple impact signatures but are evidence of passage through the dusty environment created by these disintegrations. It is shown that the most recent pulse was probably induced 3 to 5 Myr ago following passage through Gould's Belt and that the most recent spike is likely to have been caused by debris from a Chiron-like progenitor of Comet Encke which has dominated the terrestrial environment for ~ 0.02 Myr. Geochemical evidence and climatic

GIANT COMETS AND THE GALAXY 261

variations relating to these events are discussed. It follows that the last major glaciation and subsequent interglacial periods exemplify a Galaxy-driven process which has effectively controlled terrestrial evolution through periodic destruction of the environment.

I. COMETS AND EXTINCTIONS

The catastrophic view of the Earth's history is often thought to have given way to the concept of more orderly evolution only during the nineteenth century. In fact, celestial missiles, the agents of catastrophe, had already become a drawing-room joke during the Age of Enlightenment ($\sim$ 1680–1710). Not everyone, though, went along with the popular view. Thomas Wright of Durham (1755), for example, who was perhaps the first to recognize the disklike structure of our Galaxy, noted that it was "not at all to be doubted from their vast magnitude, velocity and firey substance, that comets are capable of destroying such worlds as may chance to fall in their way." Sixty years later, Laplace (1816) was arguing from their nonplanar distribution for an interstellar origin of comets while claiming for their encounters with the Earth that "the seas would abandon their ancient positions . . . ; a great portion of the human race and animals would be drowned in the universal deluge, or destroyed by the violent shock imparted to the terrestrial globe; entire species would be annihilated. . . ." It is clear that a galaxy-solar system connection of the cometary kind is by no means a new idea.

Observed comets are of course in bound orbits so if they are interstellar, as Laplace imagined, they would have been captured some time ago. The unrelaxed $1/a$ distribution of long-period comets is consistent with this expectation (Yabushita 1979*a*,1983), but Oort (1950) argued that dynamically new comets with an excess of surface volatiles may be biasing an underlying population which is relaxed and primordial. Thus, a fading function was introduced whose purpose was to preserve the theoretical assumptions of the model. The required form of this function is quite specific but its adoption may well be arbitrary because there is as yet no physical model to explain it (Bailey 1984), and indeed it is not clear that new comets as such are physically exceptional (Oort and Schmidt 1951; Kresák 1977). An interstellar origin continues therefore to be a valid proposal although the $1/a$ observations imply, strictly speaking, only a disturbance of the comet cloud $\sim$ 3 to 9 Myr ago. The apparent clustering of long-period comet perihelia towards the solar apex and galactic plane likewise imply, *prima facie,* a recent ($\leq$ 10 Myr ago) disturbance. It is suggested from time to time that this effect too may represent some kind of observational bias (but see Bogart and Noerdlinger 1982); however, any such bias would have to generate the observed galactic alignments by chance. Further, it has been claimed that the system of new comets has recently acquired a general rotation (Delsemme 1985*a*), the equator of the system apparently containing the solar apex to within ± 3° (see the chapter by

Delsemme). It is conceivable that a few stars may be involved in producing these effects by moving relatively slowly through the Oort cloud and selectively removing comoving comets (Biermann et al. 1983). However, if this were the cause, the solar apex alignment would again have to be a chance effect. The perihelion clustering and rotation of new comets, if real, are thus more probably due to the distant and more rapid passage of appropriately massive bodies with a distribution similar to that of the Galaxy as a whole; i.e., lost comets are not likely to be due to approximately comoving stars but to approximately corotating massive clouds. It hardly seems fortuitous therefore that around 5 Myr ago, the Sun was close to the Scorpio-Centaurus association, part of a conspicuous massive structure in the solar neighborhood known as Gould's Belt (Clube 1967) which happens to be both flattened and inclined $\leq 20°$ to the galactic plane. Thus, unless some peculiar combination of observational biases is at work, the empirical evidence seems to point rather clearly to a long-period comet system that is unrelaxed and has undergone a disturbance a few Myr ago related to the Sun's passage through the immediate galactic environment. It should be noted, however, that in addition to this recent disturbance, the extent and mass of Gould's Belt probably imply a somewhat more extended period of Oort cloud disturbance for the last ~ 20 Myr, also involving a rather large deflection of the Sun's orbit relative to nearby stars (see Sec. III.B).

The possible connection between comets and extinctions was revived in recent times by Urey (1973) but without success because there are too few direct cometary encounters to explain the cratering record. The latter happens, however, to be broadly compatible with the fluctuating population of Apollo asteroids (Shoemaker et al. 1979). Dynamical simulations of Jupiter's role in deflecting bodies into short-period and Apollo orbits (Everhart 1973; Rickman and Froeschlé 1980) indicate that, if even a small proportion ($\sim 1\%$) of short-period comets could become asteroidal in appearance (Öpik 1963), then the comets would constitute an abundant source of Apollo asteroids. More specifically, given the mass distribution of comets (a power law with population index -1.75 ± 0.2 [Hughes and Daniels 1982]), the mass input to the inner planetary system will be dominated by the rarest, giant comets. There is no securely known upper limit to this distribution, but it is likely to extend beyond 10^{21} g (Fernandez and Jockers 1983). Inevitably therefore, the disintegration of the largest comets will yield large stochastic variations in dust input to the zodiacal cloud during a disturbance of the long-period comet population, such variation occurring at intervals of ~ 0.1 to 1 Myr and lasting ≤ 0.1 Myr. The disintegration will also produce stochastic increases in the number of Apollo asteroids. The actual number and physical and chemical state of these Apollos (i.e., their friability and composition) will be determined by the degree of differentiation in the progenitor giant comet. Each input of dust suffers progressive fragmentation and destruction as it evolves through its meteor stream and zodiacal cloud phases before being blown away by the solar wind. Much of the dust input on Earth will thus be in the form of μm-sized particles creating stratospheric dust veils of optical depths ~ 0.1 to 1. These veils will appear sporadically, depending on orbital commensurabilities with the disintegrating body, and have lifetimes $\sim 10^3$ to 10^4 yr. Hence, not only large impacts but successive climatic perturbations are in principle involved. The external causation of mass extinctions may therefore be a complex process, involving not only the prompt effects of large impacts, but also a variety of more prolonged and deleterious environmental conditions. The appropriate modern version of Laplace's theory therefore would appear to be one in which the course of terrestrial evolution is determined not by comets generally but by the meteor streams and asteroidal debris produced during the breakup of occasional giant comets that are deflected into Earth-crossing orbits during and after successive galactic disturbances of the (not necessarily primordial) Oort cloud.

II. TERRESTRIAL CYCLES

Despite the often crude nature of the data, an episodic or roughly cyclic character has often been claimed for biological and geologic evolution—recent suggestions that the cyclicity is a new discovery are quite misleading. Thus, over fifty years ago, Holmes (1927) argued for ~ 30 Myr cycles of orogeny correlated with sea-level fluctuations on which more intense episodes of orogeny at ~ 250 Myr intervals were superimposed (Fig. 1). Such claims, made before radiometric dating and the plate tectonic revolution, could of course be no more than suggestive. Nevertheless, the cyclicities, if real, were a fundamental, unexplained aspect of Earth's history of apparently galactic origin. Thus, if the time sequence of many terrestrial phenomena is stochastic with an underlying modulation at the above frequencies, the ~ 30 Myr cycle may be readily associated with the Sun's out-of-plane motion and generally weak encounters with significant mass concentrations close to the plane, while the ~ 250 Myr cycle may be related to the in-plane radial motion involving more intense passages through spiral arm concentrations (from which comets might also be captured). The qualitative picture that emerges from the terrestrial record therefore is of the quasi-regular distribution of young star-forming regions in the Galaxy producing quasi-regular disturbances of the Sun's comet cloud with corresponding quasi-regular deterioration of the terrestrial environment brought on largely by the evolution of giant comets. Evidently, if this picture is correct, the terrestrial record would require that the Oort cloud is currently disturbed, that we are currently within a bombardment episode and that we are currently immersed in an extended star-forming region.

In fact, the Oort cloud may well be currently disturbed, as we have seen in Sec. I, while the circumterrestrial missile population may be too high to match the time-averaged cratering rate both on the Earth (~ 500 Myr) and the Moon

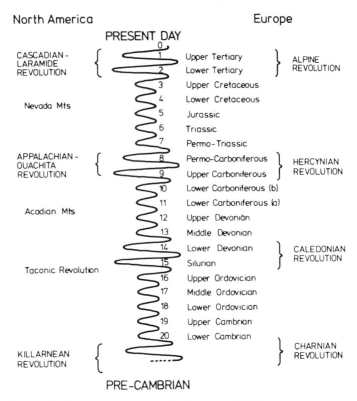

Fig. 1. Early qualitative representation of ~ 30 Myr and ~ 250 Myr sea-level cycles correlated with periodic geologic disturbances (figure after Holmes 1927).

(~ 3000 Myr) (Shoemaker 1983). Furthermore, there is good reason to believe that the Sun is in an extended star-forming region. Thus, although the Sun lies close to the central plane of a smooth stellar disk between two spiral arms, the immediate surrounding area is dominated by a number of stellar associations of which the nearby star-producing region in Orion is one and which form the prominent local feature known as Gould's Belt. The local distributions of H I and young stars (Olano 1982; Lindblad et al. 1984; Stothers and Frogel 1974) indicate that the surrounding complex has linear dimensions a few 10^2 pc, total mass ~ a few 10^6 M$_\odot$ and an age ~ 30 to 70 Myr, the whole system being perhaps a disjointed section of the Orion spiral arm. The internal motions of this complex point to a fairly rapid disruption consistent with its age, while its rather confused structure might be considered to resemble that of a well-evolved giant molecular cloud. Unless the complex is exactly comoving with the stellar disk, it and the disk would be expected to form two separate streams in local velocity space with ages ≤ and ≥ 100 Myr, respectively, and although such streams have been observed (Kapteyn 1905), they have not received much attention in recent years (Clube 1978,1983a; see chapter by Palouš). Inevitably though, there is a strong discontinuity in the parallactic motion of stars where lines of sight emerge from the local complex and such an effect has, for example, been demonstrated recently at $\sim$ 300 pc for $b \simeq -90°$ (Murray and Argyle 1984). Both spatial and kinematic evidence indicate therefore a rather large star-forming region in the wider solar neighborhood, while the Sun's orbit has probably crossed the Orion spiral arm during the last $\sim$ 50 to 100 Myr. Thus, the *prima facie* evidence places the Sun in an extended star-forming region in good agreement with the terrestrial record.

Now, the local complex includes many molecular clouds, particularly the massive concentration in Orion by which the Sun passed several Myr ago. The molecular clouds are themselves part of a Galaxy-wide system comprising a dense inner ring whose radial gradient may be in decline at the Sun's distance from the galactic center. It is a short step therefore to the general assumption that the system of dense molecular clouds in the Galaxy is responsible for the quasi-regular disturbances, that due to Gould's Belt $\sim$5 Myr ago, particularly the Scorpio-Centaurus concentration within it, being the most recent. In effect, the ($\sim$ 250, $\sim$ 30) Myr cycles of terrestrial evolution are directly attributable to the more or less regular features of the Sun's corrugated, quasi-elliptical orbit and the Galaxy's radial and vertical (R,z) distribution of molecular gas. The longer cycle essentially correlates with the Sun's periodic encounters with the outer parts of the molecular ring, and since it is conceivable also that the Galaxy is recurrently active, causing the molecular cloud system to be regenerated on time scales $\sim$ 100 to 200 Myr, the terrestrial record may be modulated by even larger cycles $\sim$ 500 to 1000 Myr. The existence of longer-period cycles has certainly been suspected (cf. McCrea 1981).

The basis for a quantitative theory of episodic terrestrial catastrophism has evidently been available in principle for some time (Clube 1978; Napier and Clube 1979), and following the discovery of siderophile trace elements at the Cretaceous-Tertiary boundary (Alvarez et al. 1980), there have been further developments of the theory both by Clube and Napier (1982a,b,1984a,b) and by Rampino and Stothers (1984a,b). Seyfert and Sirkin (1979) in effect rediscovered the shorter Holmes tectonic cycle and correlated it with impact episodes of $\sim$ 26 Myr separation (Table I), in accordance with the expectations of the galactic hypothesis. Although their arguments were qualitative, it is noteworthy that their impact epochs are virtually identical, to within the errors, with those found recently by Alvarez and Muller (1984) from a more rigorous statistical study. Other recent investigations of a wide variety of terrestrial phenomena (Tables II and III) not only confirm the presence of an underlying $\sim$ 30 Myr modulation intensifying at $\sim$ 250 Myr intervals but indicate by their correlation a single underlying driving force (cf. Clube and Napier 1982a).

That terrestrial evolution thus appears to be fundamentally controlled by

TABLE I
Impact Episodes and Mass Extinctions

Epoch	Impact Episodes[a] (Myr)	Impact Episodes[b] (Myr)	Orogenies[a] (Myr)	Mass Extinctions[c] (Myr)
1	1–2.5	—	2	—
2	15	15	15	11
3	42	40	42	37
4	70	75	70	66
5	100	100	100	91
6	135	130	136	144
7	160	160	159	176
8	180	185	185	193
9	205	210	205	217
—	—	—	—	245

[a] According to Seyfert and Sirkin (1979); [b] according to Alvarez and Muller (1984); [c] according to Rampino and Stothers (1984a,b).

the Galaxy through the agency of molecular clouds and comets is clearly a finding of considerable potential whose ramifications in astronomy, Earth science and biology have scarcely begun to be explored.

III. GALACTIC CONTROL

The most drastic perturbers of the long-period comet system appear then to be molecular clouds. Their effects on the comet population have been dis-

TABLE II
Short-term Fluctuations in Geophysical Record

Phenomenon	Period (Myr)	Source
Climatic (?) and sea-level variations	32	Leggett et al. (1981)
	30	Fischer and Arthur (1977)
Tectonic cycles	30	Holmes (1927)
Mass extinctions	32	Fischer and Arthur (1977)
	26–30	Raup and Sepkoski (1984)
Geomagnetically disturbed epochs	32–34	Negi and Tiwari (1983)
Ages of craters	~27	Seyfert and Sirkin (1979)
	31 ± 1	Rampino and Stothers (1984a,b)
	28	Alvarez and Muller (1984)

TABLE III
Long-term Fluctuations in Geophysical Record

Phenomenon	Period (Myr)	Source
Ice ages	~250	Holmes (1927)
	~200	Steiner & Grillmair (1973)
Major tectonic events	~200	Holmes (1927)
Climatic cycle	~300	Fischer and Arthur (1977)
Mixed magnetic intervals	285	Negi and Tiwari (1983)

cussed in a number of papers (Clube and Napier 1982b, 1983, 1984a; Napier and Staniucha 1982; Bailey 1983a); only a brief review is given here.

Let a comet receive a tangential impulse δv_A to its aphelion velocity v_A. Since

$$v_A^2 = \frac{\mu}{a} \cdot \frac{1-e}{1+e} = v_c^2 \frac{q}{Q} \tag{1}$$

it follows that

$$2 v_A \delta v_A \sim v_c^2 \frac{\delta q}{Q} \tag{2}$$

and the corresponding change in perihelion distance q is given by

$$\delta q \sim 2Q \frac{v_A}{v_c} \frac{\delta v_A}{v_c} \sim 2Q \frac{\delta v_A}{v_c} \left(\frac{q}{Q}\right)^{1/2} \tag{3}$$

where v_c represents the circular velocity at distance $0.5\,(Q+q)$ and Q represents the aphelion distance of the comet. Now, a dearth of comets with perihelia within the planetary system arises from planetary perturbations and comet disintegration. A bombardment episode will generally take place if a perturbation yields $\delta q \geq 50$ AU, so flooding the planetary system with comets from beyond the loss cone. A long-period comet with $Q = 40{,}000$ AU and $q = 50$ AU, for example, may be thrown into the planetary system if $\delta v_A \geq 2.5 \times 10^{-3}$ km s^{-1}. Such an impulse would be produced by the passage of a $10^{4.5}$ M$_\odot$ perturber within ~20 pc of the Sun at 20 km s^{-1} (cf. Clube and Napier 1982b). For a solar trajectory close to the galactic plane, weak encounters with molecular clouds ($M \geq 10^3$ M$_\odot$) producing impulses of about this magnitude (and hence bombardment episodes) will occur on average every

~ 15 Myr. The duration of a typical episode is dictated by the spread in half-period of disturbed comets and by the diffusion time of comets within the planetary system. A sharp onset is expected, with a decline to ~ 10% of maximum intensity within 5 Myr and a long tail, ~ 1% of maximum bombardment intensity being expected after 30 Myr (Clube and Napier 1984b). Within this episode, large random fluctuations in the zodiacal dust population and in a short-lived Apollo asteroid population are expected; these fluctuations, as we have seen in Sec. I, are dictated by the arrival of the largest comets (e.g., bodies of $\geq$ 50 km diameter arriving in the Apollo-Amor region every few 10^5 yr) and their disintegration into dense meteor streams containing dust, boulders and asteroids. A bombardment episode of a few Myr duration will yield, characteristically, an infall of ~ 10^9 to 10^{12} tons of dust on to the Earth, concentrated in a few discrete spikes of ~ 10^4 yr duration corresponding to the lifetime of the meteor stream. Even higher resolution would in general reveal the dust within a spike to be concentrated in a few short periods corresponding to close encounters with material from the prime disintegration. The dust within a bombardment episode will be interspersed with a broadly comparable mass of cosmic material from ~ 10 impacting planetesimals of $\geq$ 1 km dimensions.

The frequency of such bombardment episodes will of course vary with the solar trajectory and the distribution of molecular clouds; one expects an autoregressive time series, with galactic modulations of otherwise random events. Given the large stochastic element in the past solar motion, however, the relative strengths of the various periodicities is a matter for empirical discovery rather than revealed wisdom. We comment briefly on three cycles of possible interest.

A. Spiral Arm Penetrations

Cycles of 50 Myr or 100 Myr have often been claimed to occur in the terrestrial record (McCrea 1981). These geological cycles have yet to be statistically validated, but the intervals are about those expected for passage of the Sun through the spiral arms of the Galaxy, and such passages might create terrestrial disturbances through the mechanism reviewed here. However, whether or not molecular clouds strongly delineate spiral arms is a controversial question; different groups reach different conclusions. It now appears that warm molecular clouds, comprising about a quarter of the disk population, do indeed form patterns in the galactic plane, but that the cold molecular clouds are widely distributed in and out of the spiral arms (Solomon et al. 1985). In that case, spiral arm modulation might not be strong (e.g., signal-to-noise ratio $\leq$ 1 over 250 Myr).

B. Vertical Solar Oscillations

Because of their quasi-regularity and the probable ubiquity of molecular clouds in the galactic disk, it is not surprising that this 30 Myr half-cycle has been detected with the greatest confidence in the terrestrial record (Clube and Napier 1984a; Rampino and Stothers 1984a; but see Hoffman 1985). Thaddeus and Chanan (1985) and Bahcall and Bahcall (1985) have claimed that, because the current out-of-plane motion of the Sun ($Z_{1/2}$ (☉) ~ 35 pc) is comparable with the half thickness of the molecular cloud system, a significant 30 Myr modulation would not occur. However, the energy relaxation time T_E of main sequence stars is ~ 200 Myr (Wielen 1977), and whether or not giant molecular clouds (Lacey 1984) or transient spiral arms (Carlberg and Sellwood 1985) are the main perturbers, it is likely that at intervals of this order the Sun will receive a kick characteristically ~ 1 to 10 km s^{-1} during a major Oort cloud disturbance. The current $Z_{1/2}$ (☉) is anomalously low relative to that of main sequence stars of the same age ($Z_{1/2}$~ 250 pc: Wielen 1977); furthermore, as we have seen, the Sun has just emerged from encounters with the Orion spiral arm and Gould's Belt with, probably, a disturbed Oort cloud. The mass and location of Gould's Belt imply an immediate acceleration and deflection ~ 30° of the Sun's orbit. The Sun is therefore unlikely to have preserved much memory of the Z motion it had before it entered the arm. A modest ~ 30 Myr modulation (e.g., signal-to-noise ratio ≥ 2 over the last 250 Myr) is therefore a reasonable expectation of the theory, coupled with a phase shift of the current bombardment episode. The expected jitter in the period is consistent with that observed in the cratering record over the past 250 Myr (Rampino and Stothers 1984b).

C. The Longer Holmes Cycle

For galactic radial excursions of around one kpc, a variation in the cratering rate of order a few craters, with a ~ 250 Myr period, is expected. If the molecular ring of Solomon and Sanders (1980) is real, this modulation might become quite large and it would be further enhanced if the motion of the local standard of rest has to be seriously modified (see, e.g., Clube 1978; Clube and Pan 1985). However, unlike the other anticipated cycles, this one would be smooth rather than sharp. It is not to be expected that this periodicity like the others discussed here would be preserved for more than a few T_E (see Eq. 8 below).

In addition to the relatively numerous weak encounters producing the galactic modulation, about ten penetrations of giant molecular clouds have probably taken place within the lifetime of the solar system. The cumulative effect of the strong encounters is in fact to deplete the Oort cloud significantly during this period, with the probability that long-period comets survive *in situ* for 4.5 Gyr being ~ 10^{-2} to 10^{-3} (see chapter Appendix). The roughly constant time-averaged cratering rate over the last ~ 3 Gyr appears, however, to imply periodic replenishment. How the Oort cloud is replenished remains an unsettled question at present (Bailey 1986; Clube and Napier 1984a) but the issue is perhaps of lesser importance in the context of terrestrial catastrophism than it is in the context of cosmogony (Clube 1984). While not arguing against intermittent disturbance of the Oort cloud (crucial to the theory of terrestrial catastrophism), Hut and Tremaine (1985) have nevertheless disputed the need for replenishment. However, aspects of their analysis are open to question (see chapter Appendix).

IV. OTHER HYPOTHESES

Although stochastic and periodic disturbances of the Oort cloud by molecular clouds are inevitable (given their observed physical characteristics) and although the predicted and observed terrestrial records are in good agreement, it has also been suggested recently that the ~ 30 Myr extinction cycle may be due to a solar companion which disturbs an inner cloud of comets at these intervals (Whitmire and Jackson 1984; Davis et al. 1984), or that a planet beyond Pluto may periodically disturb a compact ring of comets (Whitmire and Matese 1985; chapters by Matese and Whitmire, by Delsemme and by Hut). The following comments seem relevant, beginning with the Nemesis hypothesis:

1. The orbit of Nemesis is likely to be unstable, with half-life ~ 50 to 100 Myr when molecular cloud perturbations are allowed for (Clube and Napier 1984c, 1985a). Given the recent extension of the known extinction cycle over the whole ~ 600 Myr of the Phanerozoic (Kitchell and Pena 1984), the probability of Nemesis surviving *in situ* for the required period, let alone holding its period constant to, e.g., ± 20% is vanishingly small.
2. The constancy of the cratering record is difficult to reconcile with the existence of such a companion. The cratering record should reflect the frequency of passage of the companion. If one starts it with an initial separation of 0.1 pc to keep it in a bound orbit (Hut 1984a), then the impact frequency over the Phanerozoic should have been an order of magnitude less than that at earlier epochs; in fact, it seems to have been significantly higher (Neukum et al. 1975). More generally, the approximately constant time-averaged cratering rate for the past ~ 3 Gyr (Hartmann 1975) is difficult to reconcile with the expected meanderings of both the period and eccentricity of the companion.
3. Binaries of the type postulated are very rare or unknown in the Galaxy (probably ≤ 0.1% of all binaries). The frequency of binaries falls off with increasing separation, only ~ 1% having periods ≥ 0.3 Myr: this seems to be a real phenomenon and not an artefact of the observations (Retterer and King 1982).
4. The hypothesis takes no account of the galactic alignments which appear to exist in the long-period comet system.
5. The inferred phase of the cycle is inconsistent with the evidence for recent disturbance of the Oort cloud and Earth, discussed herein.
6. The hypothesis is *ad hoc*. Whereas the galactic hypothesis is derived, Nemesis is invented, its ~ 30 Myr period being chosen to match a pattern

of bombardment which, on dynamical grounds, one expects that the molecular cloud system will generally impose in any case (Clube and Napier 1984b; Rampino and Stothers 1984a). In general, unless the binary marches in precise phase with the solar galactic orbit, the binary and galactic hypothesis are incompatible.

7. There is no explanation for the ~ 250 Myr cycle.

The Planet X hypothesis suffers from objections (4)–(7) and would apparently yield a strict periodicity of cycle, contrary to the observations. Both hypotheses also involve special requirements for the comet system: in the case of Nemesis, an inner cloud of $\geq 10^{13}$ comets; for Planet X, a flat ring of comets beyond Pluto. On the face of it, therefore, the Nemesis and Planet X hypotheses are neither consistent nor necessary in the current state of astronomical knowledge.

V. THE CONTEMPORARY GIANT COMET

From the empirical point of view the best prospect for unravelling the detailed astronomical interactions with the Earth is probably to examine recent events. The more remote geologic record is necessarily more obscure (see van Valen [1984a], for example, for a useful survey of problems and difficulties). An important line of research therefore is to test whether or not there is evidence directly connecting the most recent glaciation ~ 0.02 to 0.01 Myr ago with the dust from a recent giant comet. As we have seen in Secs. I and II, the Oort cloud may well have been disturbed ~ 5 Myr ago and the current circumterrestrial missile population may well be higher than average, implying a bombardment episode. While geologic mapping of the Earth's surface is incomplete and some craters that formed ≥ 1 Myr ago may be eroded or covered by sediments or ice (Shoemaker 1983), it is not possible to be certain that all recent impact structures have been recognized. Nevertheless, based on the current asteroid population, there is a 99% probability of at least one ≥ 18 km crater within the last 3.5 Myr. One such crater is known (Lake Elgygytgyn, Eastern Siberia) with a K/Ar age of 3.5 ± 0.5 Myr. In addition, the latest ice age began ~ 2 Myr ago with the appearance of mountain glaciers. Both records are apparently consistent with the delay in terrestrial response required by the half period of typical cometary orbits in the Oort cloud. There have also been very high rates of global volcanism during the last ~2 Myr (Kennett and Thunnell 1975). If the formation of large ice caps and corresponding reduction in sea level so alters the mantle loading as to induce vertical movements through viscous response, corresponding asthenospheric flows of a few cm yr^{-1} may result which superpose on deeper seated convection and enhance volcanic activity at tectonic boundaries (Clube and Napier 1984b). The Australite and Ivory Coast tektites are ≤ 1 Myr old, with an age gap of ~ 15 Myr between them and the Moldavites. Finally, the ~ 0.1 Myr intervals between occasional giant comets is typical of that between recent Pleistocene glaciations. Chiron, which is probably a comet to judge by its extremely short residence time in the planetary system, will with 80% probability enter the region of the inner planets in ~ 0.1 Myr (Oikawa and Everhart 1979; Kowal 1979). If Chiron is a comet (diameter ~ 350 km), its entry into the inner solar system will, on the present theory, produce a flood of dust and planetesimals adequate to induce a glaciation, mass extinctions and large tectonic upsets through mechanisms discussed in outline elsewhere (e.g., Clube and Napier 1982*a*, 1984*b;* Hoyle 1984; Olausson and Svenonious 1973; Rampino et al. 1979), and it will leave geochemical signatures.

There have been several other discoveries relevant to the general theory of extinctions discussed here. The discovery by the Infrared Astronomical Satellite (IRAS), for example, that Comet 1983 TB is part of the Geminid meteor stream gives obvious support to the comet-Apollo evolution required by the galactic theory (Fox et al. 1984). The discovery by IRAS of streams of dust in the asteroid belt is consistent with a recent asteroid encounter with a large comet (Gautier et al. 1984). However, probably the most significant new result in this context is the discovery of a stream of material whose mass is consistent with a giant comet in the circumterrestrial environment in comparatively recent times. Stohl (1983) has indicated that some 50% of the sporadic meteor flux can now be identified with two broad streams embracing the Taurid-Arietid stream that probably originated with a comet fragmentation in the asteroid belt. Although the masses of individual sporadic meteors are characteristically orders of magnitude smaller than ordinary meteors, the total sporadic flux is much greater than that of ordinary streams taken together. The massive Stohl streams are thus dominant in the circumterrestrial environment and their existence is confirmed by recent observations from space which have revealed an unexpectedly large flux of submicron particles in the Taurid-Arietid stream (Singer and Stanley 1980). Mutual collisions within and between the Stohl streams are thus probably feeding the zodiacal cloud (cf. Grün et al. 1984) which is likely to be increasing at the present time. Comet Encke and the boulder swarm that struck the Moon in 1975 (Dorman et al. 1978) probably also belong to the more eccentric of these streams. It seems in addition that an appreciable fraction of the Apollo asteroid population, say ~ 100 bodies ≥ 1 km in diameter, also belong to these streams (Clube and Napier 1984*b*). Such discoveries are consistent with the fragmentation history of the Comet Encke progenitor which has been inferred from a few meteor orbits by Whipple and Hamid (1952). It thus appears that a giant comet did indeed exist some 20,000 yr ago whose vast array of debris (Table IV) is still in circulation, periodically recharging the zodiacal cloud, and possibly inducing climatic effects.

Many years ago, for example, Bowen (1956; cf. Whipple and Hawkins 1956) correlated world-wide rainfall and noctilucent cloud observations with meteor streams in May–June and November–December. The meteor particles

TABLE IV
Probable Components of Taurid–Arietid Stream

Object	log m	a	e	i	$\bar{\omega}$	Comment
Meteor stream	$-4 \to +2$	2.2	0.85	6.0	162	
Boulder swarm	≤ 6					26–30 June 1975 (Dorman et al. 1978)
Boulder flux	≤ 7					April–June 1971–1975 (Dorman et al. 1978)
Tunguska	11					30 June 1908 (Kresák 1178)
Bruno	~15					26 June 1978 (Brecher 1984)
Comet Encke	15	2.2	0.85	11.9	160	
Unseen companion	—	2.4	0.86	—	160	(Whipple and Hamid 1952)
Oljato	~16	2.2	0.71	2.5	172	
1982 TA	~16	2.2	0.76	11.8	128	
1984 KB	~16	2.2	0.76	4.6	146	
Hephaistos	~18	2.1	0.83	11.9	258	$\sim 2 \cdot 10^4$ yr[a]

[a] The probability that of the ~ 50 known Apollos, three will by chance have $a,e,i,\bar{\omega}$ of Comet Encke to within 0.2 AU, 0.2, 10°, and 30°, respectively, is $\leq 10^{-6}$. Hephaistos, the largest known Apollo asteroid, could have separated from the Comet Encke progenitor $\geq 2 \cdot 10^4$ yr ago.

observed at the time did not seem adequate to explain the correlation but the discovery of the giant comet streams changes the situation. It now seems probable that the stronger concentrations in the stream precipitate simultaneous global rainfall. Moreover, global climatic recessions as measured by the long-term movement of the northern limit of forestation during the last 5000 yr (Nichols 1967) also show remarkably close correlation with the principal past fragmentations predicted by Whipple and Hamid (1952): see Fig. 2. (Such correlations would arise when the orbital nodes of the streams coincide with the terrestrial orbit.) In essence, the principal climatic variations during the Holocene seem to be under the control of the Taurid-Arietid stream. But perhaps the most telling test is the comparison of glacial dust a kilometer beneath the surface of Greenland with the fallout from the Tunguska event on 30 June 1908 believed to be associated with the β Taurid stream (Clube and Napier 1984b). Ganapathy (1983) has shown that random meteoritic sources produce considerable scatter in the Ni/Ir plane due to varying condensation histories, but samples from the above sources have an identical ratio suggestive of an origin from the same body. Although these sources are also predominantly chondritic, several volatile elements are overabundant by factors ~ 10^2 to 10^5 and the enrichment factors are identical within the errors (cf.

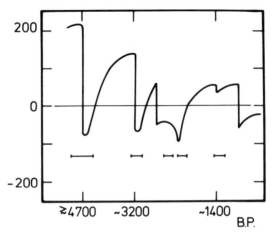

Fig. 2. Movement of northern limit of forestation (in miles) over the past 5000 yr (BP ≡ before present), showing possible correlations with fragmentation events in the Encke stream as deduced by Whipple and Hamid (1952). Horizontal bars correspond to periods of exceptional world-wide rainfall deduced from frequency distribution of radiocarbon dates (Gehy and Jakel 1974).

Golenetskii et al. 1982; La Violette 1983). Again a common origin seems to be involved and a body that apparently is of the most primitive meteoritic material known. A recent study of Brownlee particles also indicates a 10^3 enhancement of Bi (Mackinnon and Rietmeijer 1984) that is identical to Tunguska-like material. The evidence therefore appears to be consistent with the proposed single body, a comet ⩾ 50 km in diameter, dominating the terrestrial environment for the last ∼ 0.02 Myr. In particular, the greatly enhanced μm dust influx ∼0.02 to 0.01 Myr ago from this source is coincident with the duration of the last glaciation.

This new evidence seems to have a bearing on a major unsolved problem to do with ice ages, namely what determines their onset? As we have noted, the current Pleistocene ice age began only ∼ 2 Myr ago, which is too short a time for internal processes such as plate motion to have appreciably changed the continentality of the polar regions (Peltier 1982). It is, however, entirely consistent with a recently disturbed Oort cloud and a recent influx of dust from very large comets (Clube and Napier 1984b; Hoyle 1984). Steiner and Grillmair (1973) have suggested that ice ages were initiated multilaterally by, in order of importance: radial distance of the Sun from the galactic center; solar motion perpendicular to the galactic plane; Milankovič cycles; disposition of continents and oceans; altitude; and unknown causes providing 10^2 to 10^3 yr climatic variations. However, the heat and radiation budgets of a dust-loaded atmosphere are only just beginning to be studied (in a "nuclear winter" context: Turco et al. 1984) and the question of longer-term effects

GIANT COMETS AND THE GALAXY 275

(ice-albedo instabilities, etc.: Clube and Napier 1982a) has hardly been examined at all. Any discussion at this stage is necessarily phenomenological but the evidence in favor of giant comets playing a substantial if not dominant role in ice ages now seems to be rather strong.

Evidence in accord with the giant comet scenario also seems to be emerging at other extinction boundaries. Corliss et al. (1984) have argued for a rapid series of climatic coolings rather than impact to explain the Eocene-Oligocene extinctions. Although their argument does not exclude a role for impacts, glaciations correspond to sea-level changes (Vail et al. 1977); Hallam (1984) has recently emphasized the evidence for a general correlation between rapid sea-level falls and extinctions. Sea-level falls (Leggett et al. 1981; Fischer and Arthur 1977) and extinctions (Raup and Sepkoski 1984) independently correlate with the $\sim$ 30 Myr cycle. Smit and ten Kate (1982) have also found that the K-T iridium layer is enriched with various volatiles like antimony, similar to the contemporary input, so the source body could likewise have been a giant comet. Isotopic anomalies have also been detected in osmium (Luck and Turekian 1983) at this boundary, possibly indicating an interstellar origin.

To sum up, it seems that the most recent glaciation and its aftermath exemplify rather well a process, driven by the Galaxy, which has been effectively controlling terrestrial evolution through periodic destruction of the environment. Previously received ideas as to the cause of the climatic variations during this particular period have in fact been proving increasingly problematic, thus adding to the attraction of the present astronomical hypothesis. Indeed, the most recent mass extinction of large mammals occurred only $\sim$ 12,000 yr ago during the last major glaciation. It may also have an astronomical origin and thus have some bearing on the general process of extinction. The event clearly warrants further study.

VI. CURRENT CATASTROPHISM: AN APPRAISAL

The galactic hypothesis, reviewed here, is deductive in the sense that it has been developed in outline from new astronomical evidence, much of which became available throughout the 1970s, e.g., the molecular cloud and Apollo asteroid discoveries. It was thus developed independently of and indeed preceded the K/T trace element or cratering periodicity discussions which are currently attracting interest. In our view, the purely inductive approach of many geochemists and others, involving *ad hoc* theoretical modeling in response to such terrestrial facts while largely neglecting the astronomical ones, has led to numerous false trails in this field. Thus, it is likely that the geochemical anomalies at the Cretaceous-Tertiary extinction boundary have been wrongly interpreted by Alvarez et al. (1980) and many geochemists since: beyond a cosmic provenance, the "one-off meteorite" scenario (Öpik 1958) seems to have little relevance to the real astronomical environment

TABLE V
Some Geochemical and Other Discriminants between the Stray Meteorite (Alvarez/Öpik) and Galactic (Clube & Napier) Hypotheses

Stray Meteorite Hypothesis[a]	Galaxy/Giant Comet Hypothesis[b]
Simple impact ejecta; cosmic material diluted $\sim 10^{-3}$–10^{-4}	Complex deposition history includes layers too concentrated to be due to impact ejecta
Instantaneous deposition	Deposition over ~ 1 Myr, fine structure ~ 0.01–0.1 Myr within it
Chondritic (probably)	Roughly chondritic but with large over-abundances of Ag, Sb, Bi, Sn, etc, probable
Solar system isotope ratios	Isotope anomalies, especially $^{12}C/^{13}C$
Extinctions: prompt effects only	Extinctions: prompt effects, but also biotic trauma through prolonged and severe climatic deterioration, ocean regressions, etc. correlated with geologic noisy epochs
Random	Episodic

[a] According to Alvarez et al. (1980) and Öpik (1958).
[b] According to Clube and Napier (this chapter and the authors' earlier references given herein).

(some differing predictions of the Alvarez-Öpik and Clube-Napier hypotheses are listed in Table V). Hypothetical bodies like Nemesis or Planet X disturbing an equally hypothetical dense inner cloud of comets seem likewise to go well beyond the astrophysical requirements at the present time. One should also recognize the existence of a large community of Earth scientists who do not necessarily see sudden death at every mass extinction boundary, but rather consider a progressively disrupted environment, including sea-level recessions, as a prime mover (see, e.g., van Valen 1984a; Hallam 1984; Corliss et al. 1984; cf. Clube and Napier 1982a). In devising an astronomical model to explain *all* aspects of the terrestrial record, therefore, it is important to recognize the complexity of the induced phenomena (can impacts do it all?) and to take account of the known astronomical environment (do we need hypothetical agencies?). Specifically, much of the current theorizing in catastrophism seems to us to be missing several points (see subsections below).

A. Giant Comets

An essential element of any complete astronomical model is the need to explain recent as well as past events in the terrestrial record. These appear to indicate that giant comets play a dominant role. To understand geochemical anomalies or Earth history in a catastrophic framework, one must understand the disintegration history of these objects. Consider, for example, the disintegration of a 10^{21} g comet (cf. the Great Comet of 1729 or the progenitor of the Kreutz sun-grazers) in a short-period Earth-crossing orbit. This would enhance the mass of the zodiacal cloud by two or three powers of ten over its current value. A dust influx of $\sim 10^{-5}$ g cm^{-2} yr^{-1} to the stratosphere would occur, an appreciable proportion of which would be in the form of μm-sized particles. This would lead to a stratosphere with a semi-permanent dust veil of optical depth ~ 0.1 to 1.0 depending on the precise size distribution and properties of the dust; severe climatic effects would seem to be inevitable. This is true *a fortiori* when the Earth runs through the associated meteor stream or concentrations within it. For a chondritic relative abundance 5×10^{-7}, peak iridium deposition rates of up to 10^{-2} ng cm^{-2} yr^{-1} are expected for brief periods within the lifetime of the comet. There could be as many as 50 major climatic excursions during a bombardment episode, arising from brief dustings of the stratosphere due to close encounters with the debris from giant comets. Thus, in the present hypothesis, extinction boundary layers are not in general predicted to be simple impact ejecta. More often, they should be complex sandwiches over 1 Myr thick, comprising successive layers of dilute impact ejecta, cosmic dust from episodes of $\sim 10^3$–10^4 yr duration, and even layers dominated by local sedimentation.

B. Complexity of Effects

A corollary to the above is that prompt effects (e.g., dust obscuration, blast, ozone depletion: see Napier and Clube 1979) will compete with environmental deterioration caused by a series of sharp climatic coolings and the regression of seas from the continental margins. The hypothesis thus reveals several possible mechanisms of extinctions and it is not yet clear that it can discriminate between them. It is readily shown, however, that the angular momentum and energy inputs from giant comet debris must profoundly disturb the boundary conditions at the core, mantle and atmosphere of the Earth; prolonged volcanism, glaciations and magnetic field reversals are plausibly expected in consequence (Clube and Napier 1982*a*). As yet, however, the hypothesis lacks detailed modeling of such geophysical disturbances.

C. Episodicity

Even if the stochastic character of solar system interaction with molecular clouds were not responsible for random fluctuations of phase, amplitude and period in the various galactic cycles, as indicated in Sec. II, it is important to note that the finite relaxation time of the Oort cloud (~ 5 to 10 Myr) and the general unpredictability of the exact sequence of giant comets and their evolution are such that they impose a further stochastic element on any observed terrestrial cyclicity. To the extent that the relevant geophysical processes cannot yet be exactly modeled—either large body impacts or climate-induced variations in mantle-loading due to cometary debris, for example, produce magnetic reversals—exact one-to-one correlations between the various geophysical phenomena (e.g., extinction, orogeny, climate, magnetic field) based on the underlying galactic cycles are not expected. A study by Raup (1985) confirms this expectation. Unfortunately this author fails to note the earlier prediction (e.g., Clube and Napier 1982a) and seriously reduces the significance of the finding by invoking a new possibility, without physical foundation, that independent cyclicities may be operating with no causal connection (see van Valen [1985] for a comment on this style of statistical analysis). Admittedly, a stochastically varying cyclicity is less easy to test than a straightforward one but this does not render it any less real; it is for this reason that the present authors have frequently emphasized the essentially episodic rather than cyclic character of the terrestrial record as predicted by the galactic theory. The essentially stochastic or episodic character of the *observed* terrestrial record has of course long been noted; it is now obviously important also, so far as any objective judgement of the galactic theory is concerned, that two cyclicities with untutored estimates of ~ 250 Myr and ~ 30 Myr were originally culled from the terrestrial data without prior knowledge of any underlying physical process.

D. Oort Cloud Replenishment

Another important point to be noted is that the galactic theory, introduced by Clube (1978) and Napier and Clube (1979) probably implies more than mere disturbance of the long-period comet system to explain terrestrial catastrophism. There has been a tendency to overlook this factor in more recent discussions of the galactic mechanism (see, e.g., Rampino and Stothers 1984a; see also their chapter). Thus, since episodic catastrophism also implies depletion of the Oort cloud, the need to replenish the latter was emphasized and serious attention was given to the possibility that the Oort cloud may not be *primordial* but *captured*. The rationale for the proposal was that the observed constancy of the cratering record implies that the cometary reservoir replenishing the long-period population must reveal little or no significant decay over 3 Gyr. (Implicit in this is the further proposal that the long-period comets are a prime ultimate source of cratering bodies; the 30 Myr periodicity of the cratering record argues strongly in favor of this.)

Of course the very process that dissipates the Oort cloud might just as readily release comets into it from a concealed and necessarily primordial inner comet cloud, and indeed this is frequently suggested (see, e.g., Hills 1981; Shoemaker and Wolfe 1984; Bailey 1983a; Weissman's chapter). However, there are *prima facie* arguments to suggest that, to maintain a steady cratering rate over 3 Gyr, a dense compact cloud would have had an extremely large initial mass (Bailey 1986; Clube 1985; Napier 1985). Shoemaker's (see chapter by Shoemaker and Wolfe) Uranus-Neptune cloud, for example, requires 10^{14} comets when molecular cloud perturbations are allowed for, which, with a mean mass $\geq 10^{17}$ g, leads to a cloud $\geq 10^{31}$ g (Bailey 1986).

Likewise, Hills' inner cloud would require an initial mass of 10^{32} g. Such estimates exclude the H and He that are expected to be present in the primordial medium, and the implied initial masses are then prohibitively large. Of course, it is possible that the difficulties in these cases may be avoided by suitable adjustments to theories of solar system formation or to the mode of comet formation (Hills 1982). However, at the time of writing, an acceptable model of the inner cloud has yet to be constructed, and the only merit of the proposal is its maximal conformity to yesterday's orthodoxy (see, e.g., chapter by Weissman).

There are significant difficulties, therefore, with the *new* Oort cloud concept and it is necessary to examine other possibilities. It has been shown (Clube and Napier 1984*a*) that high-density aggregates of comets in star-forming regions are likely to exist, in general agreement with many aspects of the star-forming process (e.g., $10^{0\pm 1}$ comets AU^{-3} in regions of a few pc dimensions; cf. $10^{2\pm 1}$ AU^{-3} implicit in the primordial inner cloud scenario discussed above). The probable dissipation rates of molecular clouds are such that the number density of interstellar comets in the galactic disk may be 10^{-4} AU^{-3}. In these circumstances $\sim 10^{17}$ interstellar comets have passed within 50,000 AU of the Sun over 4.5 Gyr, about half of them during molecular cloud penetrations. If one in a million of such comets are captured, then one has a quasi-equilibrium between capture and dissipation and an Oort cloud population which will fluctuate within a power of ten around 10^{11} bodies. A capture mechanism of about this efficacy, essentially an inverse Fermi mechanism involving systematic deceleration of comets in the neighborhood of a molecular cloud, has been described by Clube and Napier (1984*a*). There may be other processes at work, however, as the topic is relatively unexplored (most studies have involved unrealistic assumptions about the astronomical environment, e.g., a fixed potential field in the solar neighborhood, yielding a capture cross section for the solar system $\sim 10^{-3}$ to 10^{-4} of the value above). The balance of evidence may therefore favor a capture rather than primordial origin for the Oort cloud; however, the issue is still quite open. It is of course a problem for celestial mechanics and the cratering record may also, in principle, constrain the possibilities.

No less important than the implications for terrestrial history and therefore biological evolution, the Clube-Napier galactic theory of episodic terrestrial catastrophism has potentially far-reaching consequences for astrophysics. Thus, if the theory is in fact Laplace's theory in a modern guise (i.e., the Oort cloud is replenished from recent star-forming regions), there are significant consequences for comet cosmogony which may also have some bearing on the evolution of the Galaxy itself (Clube and Napier 1985*b*) and several unresolved aspects of galactic dynamics (Clube 1978, 1983*a*). There is therefore a need for critical isotope tests to discriminate primitive solar nebula and genuinely interstellar origins for comets. The data existing so far (Golenetskii et al. 1982; La Violette 1983) seem to relate largely to the differentiation history of the most recent giant comet but if Brownlee particles mostly derive from this body, their isotope analysis in particular may be of fundamental importance.

VII. CONCLUSION

That there may be a catastrophic dimension to interaction with comets, introducing a galactic signature into the terrestrial record, has largely been seen so far for its revolutionary potential in the Earth sciences including biological evolution. In addition to these possibilities, the consequences, as we have noted, may be just as significant for astronomy. Ultimately this is because the less frequent giant comets have been isolated as the probable principal agent of terrestrial catastrophism. These bodies are unlikely to be ordinary large accreted comets. Thus, if giant comets in particular are copious sources of dust and asteroids because they have undergone significant internal differentiation, such bodies, like planetary moons of comparable dimensions, are more likely to have evolved through a hot phase brought on by very rapid collapse. Bodies of this character have often been favored in attempts to explain the properties of primitive meteorites. The results now emerging from the proposed Galaxy-solar system interaction therefore have important implications for the medium out of which such primitive bodies formed and hence for star and planet formation generally (Clube and Napier 1985b). This in turn provides us with a new handle on the state and origin of the preexisting spiral arm medium. The new understanding of the terrestrial record can thus lead to a new understanding of the state and evolution of the Galaxy itself.

APPENDIX

The cumulative effect of many encounters with molecular clouds is to cause Oort cloud comets to random walk in energy, the rms energy increment in units (km s^{-1})^2 at a distance Q from the Sun after time t (measured in units of 4.5 Gyr) being

$$\Delta E = 1/2\ \Sigma(\Delta V)^2 = 2.6\ \nu M^2 t\ (Q/R_c)^2\ [F+P(\varepsilon)]. \qquad (4)$$

Here ν represents the number of molecular clouds kpc^{-3}, M is a representative value for their mass in units of $2.2 \times 10^5 M_\odot$, and

$$R_c = \max\ (R, \bar{P}) \qquad (5)$$

with R being the radius of a molecular cloud and $\bar{P}$ the rms minimum impact parameter expected over time t. F and $P(\varepsilon)$ represent, respectively, the contri-

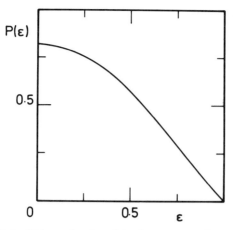

Fig. 3. Penetration factor $P(\varepsilon)$ as a function of the deepest penetration expected for the solar system over its lifetime, measured by $\varepsilon = \bar{P}/R$. It lies in the range $0 \leq P(\varepsilon) \leq 0.83$.

butions of flyby and penetrating encounters and are given in terms of a geometrical parameter x (see Clube and Napier 1983, Fig. 2) by

$$F = \int_1^\infty x^{-3} dx = 0.5 \tag{6}$$

$$P = \int_\varepsilon^\infty x^{-3}[1 - (1 - x^2)^{3/2}]^2 dx \tag{7}$$

where the latter is shown here in Fig. 3 in terms of a depth of penetration factor $\varepsilon = \bar{P}/R \sim 1/\sqrt{N}$, with N being the number of past penetrations of molecular clouds. The energy required to unbind comets at distance Q is of order $1/2\, v_c^2(Q)$, where $v_c(Q)$ represents the circular velocity at Q. Thus an unbinding time for the Oort cloud may be defined by

$$T_E = \frac{E}{\dot{E}} \simeq \frac{v_c^2(Q)}{5.6\, \nu M^2\, (Q/R_c)^2 [F + P(\varepsilon)]} \tag{8}$$

analogous to the relaxation time in stellar dynamics and proportional to $1/Q^3$. Converting νM to a density ρ $M_\odot$ pc^{-3}, and considering T_E at $Q = 50{,}000$ AU, one finds (in Myr)

$$T_E = \frac{2.4\,R_c^2}{[F + P(\varepsilon)]f_e f_z f_p f_g \left(\dfrac{\rho_o}{0.024}\right)\left(\dfrac{M}{2.2 \times 10^5}\right)} \qquad (9)$$

where $\rho = f_e f_z f_p f_g \rho_o$, ρ_o is the galactic midplane density of molecular clouds and the f_i are the efficiency factors introduced by Hut and Tremaine (1985). Here R_c is in pc, M is in $M_\odot$ and ρ_o is in $M_\odot$ pc^{-3}.

Suppose, for example, there have been 7 to 10 past penetrations of giant molecular clouds (GMCs) of radii 20 pc (see, e.g., Napier 1985). Then $\varepsilon \sim 0.35$ giving $F+P(\varepsilon) \sim 1.2$ and with $f_e = 1, f_z = 1/2, f_p = 3/2, f_g = 3/2, \rho_o = 0.024\,M_\odot$ pc^{-3} and $M = 2.2 \times 10^5 M_\odot$ Eq. 9 yields (in Myr)

$$T_E = 710. \qquad (10)$$

Estimates of GMC masses range from mean mass $5 \times 10^5 M_\odot$ to median mass $1.3 \times 10^5\,M_\odot$ (Wolfendale 1985; Bhatt et al. 1985). The latter corresponds to a mass-averaged mean mass of $\sim 1.7 \times 10^5\,M_\odot$ for a power-law distribution of masses with population index 1.5 or $\sim 2.5 \times 10^5\,M_\odot$ for index 1.65. The column density for the adopted parameters is $\sim 2 \times 10^{22}$ cm^{-2}. The effects of encounters with substructure are comparable or larger (Clube and Napier 1983), adding to the energy input and further decreasing T_E. The functional form of Eq. (9) differs substantially from that derived by Hut and Tremaine (1985), who adopt a point-mass assumption and then attempt to allow for penetration by introducing a finite size factor f_s which they put equal to 0.5. They find a half life in Myr

$$T_{1/2} = \frac{750}{f_z f_e f_p f_g f_s \left(\dfrac{\rho_o}{0.024}\right)}. \qquad (11)$$

Hut and Tremaine (1985) derive 3000 Myr from their formula. However they have taken an epicycle factor $f_e = 0.5$ and a gravitational focusing factor $f_g = 1$. The former value involves neglecting the past orbital diffusion of the Sun and the latter the strong preference for clouds to occur within massive complexes. The probable radial excursions of the Sun in the galactic plane amount to a few kpc (Wielen 1977) and so, depending on the vagaries of past motion and the assumed radial distribution of molecular clouds, we find $f_e \sim 1$ to 1.5 while making suitable allowance for the mass distribution of molecular clouds, we find $f_g \sim 1.5$ to 2.0.

A further difference between the Hut and Tremaine (1985) formula and ours relates to the effect of this finite size factor f_s. The functional forms of Eqs. (9) and (11) are quite different. This arises because Eq. (11) represents a limiting case (point-mass perturbers) wherein disruption is caused by one or two strong encounters. The other limiting case, wherein close encounters are softened by finite cloud size, and unbinding is due to the cumulative effect of many weak encounters, has the same functional form as Eq. (9). Hut and Tremaine attempt to evaluate f_s by equating the analytical $T_{1/2}$ for each of these extremes to yield a theoretical column density which they describe as critical, and then assuming that f_s is the ratio of actual to critical cloud column densities. However, this procedure involves hidden assumptions; it is more straightforward simply to evaluate $T_{1/2}$ directly from the two limiting cases. Thus, for point-mass perturbers ($f_s = 1$),

$$T_{1/2} = 670 \tag{12}$$

in Myr from Eq. (11), and for diffuse clouds

$$T_{1/2} = \frac{0.025 \, V \, M_\odot}{G \, \rho \, a^3 \, \Sigma_{av} \, C} \tag{13}$$

(Hut and Tremaine 1985, their Eq. 30). This, with flyby velocity $V = 20$ km s^{-1}, $\Sigma_{av} = 140$ M$_\odot$ pc^{-2} corresponding to number column density $\sim 2 \times 10^{22}$ cm^{-2}, clumping factor $C = 2.5$ (from Hut and Tremaine 1985) and $\rho = f_e f_z f_p f_g \rho_o$, yields in Myr

$$T_{1/2} = 780. \tag{14}$$

If the real half life lies between these two extremes,

$$670 \leq T_{1/2} \leq 780 \text{ Myr.} \tag{15}$$

Thus, contrary to any impression given by Hut and Tremaine, their semi-analytical approach strongly supports the short Oort cloud lifetime, in good agreement with Eq. (9) above and Clube and Napier (1983), and in good agreement also with the lifetimes which may be inferred from numerical simulations (Napier and Staniucha 1982) and other independent investigations (Bailey 1983a, 1986). It therefore seems very probable that the Oort cloud unbinds within the lifetime of the solar system, a conclusion which is further strengthened by the discovery of very fine structure within molecular clouds (Israel 1985).

Thus, if GMCs are relatively short-lived dissipating systems (~ 200 Myr) and it is necessary to take into consideration the existence of a more compact formative stage (~ 20 Myr), the influence of the latter towards enhancing the total number of GMC penetrations (e.g., ~ 20) cannot be overlooked, and it is easily shown that one or two cloudlets (~ 50 M$_\odot$, $\sim 10^{-1}$ pc) are then entered during the lifetime of the solar system, thereby reducing the zone of stability of the Sun's comet cloud to a radius ~ 5000 AU, well inside the nominal dimensions of the Oort cloud. In practice, the zone of stability may be even smaller because the 50 $M_\odot$ structure is probably not a natural limit but simply a matter of instrumental resolution; indeed structure at the level of ~ 1 $M_\odot$ is observed in at least one star-forming region (Kahane et al. 1985), not to mention the peak in the stellar luminosity function at or below this level. If ~ 1 $M_\odot$ structure in GMCs is ubiquitous, it is easily shown that stars or protostars encountering similar bodies in GMCs before they escape into the disk after ~ 100 Myr are unlikely to retain any primordial comet cloud larger than ~ 1000 AU, in agreement also with the general lack of wide binaries beyond this limit. Thus, apart from the question of Oort cloud survival in the presence of GMCs, its formation in GMCs is also problematical, a difficulty that extends moreover to any *inner* primordial comet cloud (see Sec. VI.D).

In an attempt to substantiate their figure of 3000 Myr for $T_{1/2}$, Hut and Tremaine also argue that the dissipation of the Oort cloud is difficult to reconcile with the high frequency of binaries with projected separations of 0.1 pc $\sim 20,000$ AU. However, unlike the primordial Oort cloud, binary stars have a source function: star formation is a continuing process. Let $N_b(t)$ and $N_f(t)$ represent respectively the formation rates of wide binaries and field stars t years ago. For simplicity, we may assume that these rates are constant. Let (T_b, T_o) represent, respectively, the half lives of wide binaries and the age of the galactic disk. Then the proportion of wide binaries (e.g., ≥ 0.1 pc separation) in relation to field stars at the present time is given by

$$p = \frac{\int_0^{T_o} N_b(t) e^{-t/T_b} dt}{N_f T_o} = \frac{N_b}{N_f} \frac{T_b}{T_o} (1 - e^{-T_o/T_b}). \qquad (16)$$

The distribution function of young wide binaries is not well known. Abt (1983) finds that in a sample of 114 B2 to B5 dwarfs, which are presumably young, the function is quite flat in log (period) whereas in a uniform sample of 123 F3 to G2 dwarfs (Abt and Levy 1976) of presumably mixed ages, the binary frequency declines as the period increases beyond $\sim 10^3$ yr, only two systems having periods ≥ 4 Myr corresponding to long-period comets. Abt (1983, his Fig. 2) suggests that $p \sim 0.02$ and $N_b/N_f \sim 0.1$; hence from Eq. (16) $T_b \sim 1800$ Myr. Since the G-type binaries considered have masses ~ 2 M_o per system, their binding energy is higher and a correction $1/\sqrt{2}$ must be applied to a comet lifetime; hence the half life of a comet at 20,000 AU becomes ~ 1200 Myr. This is an upper limit if the Sun is close to or within a star-forming region (cf. Sec. II) because the number of local binaries is then unrepresentatively high; however, because of the small-number statistics, the result is in any case not reliable to better than a factor of two or so. We conclude that the existence of wide binaries is consistent with and does not strongly constrain the rapid Oort cloud dissipation model.

To summarize, neither binary statistics nor the trend of increasing sophistication in the analysis of Oort cloud dissipation (Bailey 1983a; Hut and Tremaine 1985) and of increasing precision in the specification of molecular cloud parameters (see, e.g., Wolfendale 1985; Bhatt et al. 1985), lead to any very significant departure from earlier conclusions regarding the dissipation of the observed Oort cloud (Clube and Napier 1983). The new results are therefore consistent with the proposed Galaxy-solar system interaction and the need for Oort cloud replenishment, probably from an external source.

DYNAMICAL EVIDENCE FOR PLANET X

JOHN D. ANDERSON
and
E. MYLES STANDISH, JR.
Jet Propulsion Laboratory

Unmodeled gravitational forces in the outer solar system could possibly be detected using meridian transit observations of the outer planets or radio tracking data from interplanetary spacecraft beyond the orbit of Uranus. Indeed, there exists a general attitude that unexplained residuals in the motions of Uranus and Neptune suggest the presence of a Planet X of a few Earth masses ($M_\oplus$) with moderate eccentricity and inclination, just the conditions needed to produce a ~ 30 Myr modulation in the flux of short-period comets (chapter by Matese and Whitmire). However, notwithstanding the merits of the Planet X model, there is no clear-cut dynamical evidence for a planet of any type beyond the orbit of Neptune. This situation is reviewed. In addition, three years of radio tracking data from Pioneer 10, now at a distance of about 35 AU, can be fit to the noise level with no evidence for unmodeled accelerations at a level $> 5 \times 10^{-14}$ km s^{-2}. These considerations do not place severe limits on the Planet X model, but they do place a firm limit of 5 $M_\oplus$ on a hypothetical comet belt just beyond the orbit of Neptune. Continued acquisition and analysis of planetary observations as well as additional tracking data from Pioneer 10 and other spacecraft that may venture into the outer solar system are encouraged.

I. INTRODUCTION

Analysis of fossil data and certain geologic data indicates that the records are periodic, or at least quasi periodic. Particularly impressive are reports of ~ 30 Myr periodicity in mass extinctions of flora and fauna (Raup and

Sepkoski 1984; Sepkoski and Raup 1986), magnetic reversals (Raup 1986) and terrestrial cratering (Alvarez and Muller 1984; Rampino and Stothers 1984a; Sepkoski and Raup 1986; chapter by Shoemaker and Wolfe). Although the connection of these separate periodic phenomena by a common cause is not required by any rigorous cross correlation of the records to date (chapters by Shoemaker and Wolfe and by Tremaine), the possibility of such a connection is suggestive. Theoretical models have concentrated on cometary impacts. Mechanisms have included a solar companion star (Nemesis) that periodically disrupts the hypothetical Oort cloud of comets (Whitmire and Jackson 1984; Davis et al. 1984), a similar disruption by a periodic oscillation of the galactic orbit of the Sun along a direction normal to the galactic plane (Rampino and Stothers 1984a) and a trans-Neptunian Planet X that modulates the steady-state flux of short-period comets by means of the precession of its perihelion through a trans-Neptunian belt of comets (Whitmire and Matese 1985; Matese and Whitmire 1985; chapter by Matese and Whitmire).

In view of these extrasolar system explanations, the question arises whether any of them would have noticeable effects on the orbits of the outer planets or on spacecraft beyond the orbit of Uranus. It is apparent that such effects from galactic orbit oscillations or from Nemesis can be ruled out immediately as too small (see, e.g., Williams 1984). Unfortunately it is difficult to test the galactic oscillation model by any means, given the time scales (one might question whether it is a legitimate model in this regard), but it appears to be in theoretical difficulty anyway in that it fails to predict a significant ~ 30 Myr modulation of the comet flux (Thaddeus and Chanan 1985; see also chapter by Thaddeus).

Statistical examination of 126 cometary orbits by Delsemme (see his chapter) leads him to suggest that a slow massive body pierced the Oort cloud < 20 Myr ago. This fits in well with the Nemesis model, but the great distance to the Oort cloud rules out the possibility of detecting such a body by direct perturbations on planetary or spacecraft orbits. The discovery of Nemesis, if it exists, must await surveys of the sky in the visual and infrared, including the existing data from the Infrared Astronomical Satellite (IRAS).

Planet X, on the other hand, could be affecting orbits in the outer solar system. It was anomalies in the orbit fits of Uranus and Neptune that prompted P. Lowell early in this century to predict the existence of a ninth planet beyond Neptune (see the comprehensive history by Hoyt [1980]). However, the current situation on the orbits of Uranus and Neptune, as reviewed in the next section, does not necessarily suggest a new planet. A three-year analysis of radio tracking data from the Pioneer 10 spacecraft also fails to reveal the presence of a Planet X. Yet the search is worth pursuing in the future. There does seem to be an inconsistency between astrometric data on Uranus and Neptune taken in the last century and the modern observations after 1910. Although this inconsistency is not well enough defined to reveal anything about the orbit of Planet X, except to limit its mass to a few $M_\oplus$, an unmodeled force of some sort is a possibility. Similarly, though no unmodeled force is indicated by the three-year orbit of Pioneer 10, that spacecraft is sensitive to only a limited range of possible unmodeled forces. However, it will be joined in the next few years by Pioneer 11, Voyager 1 and Voyager 2, all at distances beyond the orbit of Uranus. These spacecraft are the first artificial probes of the gravitational field of the outer solar system. Although this region would not be expected to yield unusual gravitational features, perhaps the dynamical material reviewed here suggests otherwise.

II. ORBITS OF URANUS AND NEPTUNE

The basis for the more recent numerically integrated ephemerides at JPL has been described in detail elsewhere (Newhall et al. 1983). Weighted least-squares fits are made to optical meridian transit observations, radar ranges, planetary spacecraft positions and lunar laser ranges. The mainstay of the data for the orbits of Uranus and Neptune are the modern transit observations of the United States Naval Observatory (USNO) which extend from 1910 to 1982. The advantage of these observations is that they are all referenced to a single fundamental catalogue, the FK4. Similar transit observations have been received from the Royal Greenwich Observatory, extending up to 1982. Attempts to use other types of optical data, particularly photographic data referred to the Smithsonian Astrophysical Observatory (SAO) Catalogue, or the nearly equivalent reference system of the USNO Zodiacal Catalogue, have led to inconsistencies. The SAO Catalogue is a conglomeration of many star catalogues which lack a common reduction to the FK4. As a result, observations referred to the SAO Catalogue are often affected by systematic errors. Recent attempts to update the ephemeris of Uranus, in anticipation of the upcoming encounter with the planet by Voyager 2 on 24 January 1986, have only emphasized the inconsistencies.

From many photographic positions of Uranus recorded in 1983, only a few, those reduced with respect to the 1970 Perth Catalogue, are useful. Uranus has been in the southern hemisphere since 1970, and in that region it is the Perth Catalogue that is referenced to the FK4. Unfortunately, it does not contain enough stars to be useful for narrow field telescopes, thus reliable modern observations are severely limited in number to the few occasions where a wide-field Schmidt camera has been used. Differences between the SAO and Perth catalogues can range up to 2 arcsec or more for individual stars they have in common. It is nearly impossible to establish systematic differences over narrow fields because of the sparsity of Perth stars.

For the transit observations, the residuals in right ascension $\Delta\alpha$ and declination $\Delta\delta$ for Uranus and Neptune about the JPL ephemerides are shown in Figs. 1 and 2. These residuals, all referenced to the FK4, are obviously not ideally random, but neither is there a clear trend that would suggest the neces-

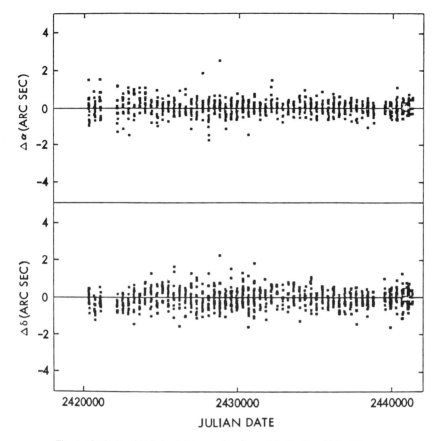

Fig. 1. Optical residuals in right ascension $\Delta\alpha$ and declination $\Delta\delta$ for Uranus.

sity of an unmodeled force acting on the outer solar system. The rms residual on the observations is ~ 0.5 arcsec.

Similar plots of residuals for Uranus and Neptune, but extending from 1832 to 1978 for Uranus and from 1846 to 1978 for Neptune, have been published for the research ephemerides of the USNO (Duncombe and Seidelmann 1980; Seidelmann et al. 1980). Systematic effects are apparent on a scale of ≥ 1 arcsec. Although the noise in the pre-1910 data is larger than in the modern data, there is a definite inconsistency. The year 1910 is significant in that it marks the introduction of the "impersonal micrometer" into transit observations, a greatly improved technique of observation. However, problems still remain with transit observations, even with the inner planets. The fact that many of these observations cannot be successfully fit with a dynamical ephemeris does not necessarily imply an unmodeled force. It seems possible that the discrepancies in the USNO fits could be the result of observational

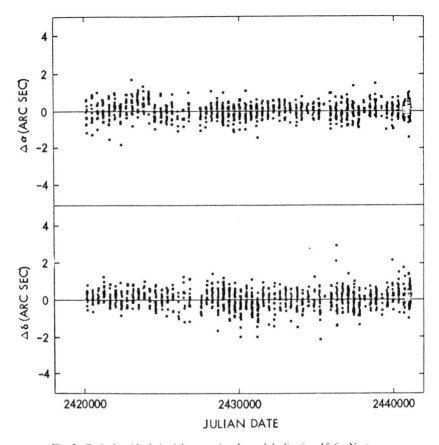

Fig. 2. Optical residuals in right ascension $\Delta\alpha$ and declination $\Delta\delta$ for Neptune.

errors. Yet, Seidelmann et al. (1980) conclude that "some unmodeled force is acting on bodies in the outer solar system." They further conclude that the patterns are inconsistent with simple explanations, such as instrumental or reference-system errors. More recently, Seidelmann et al. (1985) conclude that all of the Uranus data since 1830 cannot be fit with a single ephemeris, and that although the post-discovery data for Neptune can be fit, it is likely that the prediction ephemeris will fail to represent new observations after about 10 yr.

When considering the possibility of the existence of a Planet X as a source for inconsistencies in the orbits of Uranus and Neptune, it is important to recognize that there are two historically separate problems in the motion of Uranus that are often confused. The first and oldest involves a difficulty with trying to reconcile prediscovery observations of Uranus with the postdiscovery material of the 19th century. The elimination of this discrepancy was responsible for the prediction of Planet X by Lowell (1915) and Planet O by Pickering (1909) and it is the main source for the attitude that a planet of several $M_\oplus$ with a moderately high eccentricity and inclination is needed to remove unexplained residuals in the motions of the outer planets. In fact, the ancient observations of Uranus that were so important to Lowell have been largely discredited and ignored in more recent work, hence the first problem with the orbit of Uranus is not a serious consideration any longer (Abell 1982). Basically the prediscovery data are not reliable enough to conclude anything of interest about the motion of Uranus during that time.

It is the second more recent inconsistency that is not so easily dismissed. In particular, there is a secular trend in the declination residuals for Uranus extending from 1832 to 1978 (Seidelmann et al. 1980). Although it is possible that an undiscovered Planet X of a few $M_\oplus$ made its closest approach to Uranus in the 1700s as suggested by Harrington (1985b), and thus shaped the current orbit of Uranus, that is by no means the only explanation. Seidelmann et al. (1980) recall that "when Leverrier found the anomalous advance in the perihelion of Mercury and hypothesized an inter-Mercurial planet to explain it, the correct explanation had to await Einstein's theory of relativity."

There is a similar problem with the orbit of Neptune, although the number of prediscovery observations are limited to those of Lalande in 1795 and Galileo in 1613. The Lalande observations have been discussed by Rawlins (1970), while the Galileo observation was recognized more recently by Kowal and Drake (1980). Lalande was unaware that he had observed a new planet (Neptune) on two nights, 8 May and 10 May 1795; the confirmation was made a half century later (Mauvais 1847). These observations played an important role in the early postdiscovery orbits for Neptune as well as in the early determinations of Pluto's mass (Hoyt 1980). The modern reduction by Rawlins (1970) to the FK4 reference system shows that both observations differ in longitude by -7 arcsec from modern ephemerides. The latitude residuals (~ -1 arcsec) are not significant. The early attempts to reconcile Lalande's observations with postdiscovery observations of Neptune led to an erroneously large mass for Pluto, on the order of 1 $M_\oplus$ (Duncombe and Seidelmann 1980, and references therein). Such a large mass was not required after the Lalande observations were rejected as perhaps suffering from systematic errors (Duncombe et al. 1968a,b; Seidelmann et al. 1971; Ash et al. 1971). In addition, using the modern data, Seidelmann (1971) ruled out the possibility of E. C. Pickering's Planets P and S.

The approach to the Lalande observations at JPL has been to reject them, along with other pre-1910 observations, not because of any strong conviction that they are invalid, but because they do not add any information to ephemerides intended for the space program. The question of their validity is only important if one is attempting to gain information on forces acting on the outer solar system. Along these lines, Rawlins and Hammerton (1973) computed the positions and mass estimates of hypothetical planets beyond Neptune that would reduce the Lalande residuals and also fit the modern observations of Uranus and Neptune extending to 1966. Perhaps the most interesting aspect of their study was the indeterminancy of the solutions for the unknown disturbing planet. The determination of elements for an elliptical orbit was not feasible, so only three unknowns were introduced: the mean distance, the longitude at epoch (1973.0) for the circular orbit, and the mass. The data were shown to be insensitive to the inclination of the orbit. Even with these severe limits on a possible planet, a unique orbit could not be found. In fact, the size of the three-dimensional parameter space was relatively large. Without the Lalande observations it was so large that it revealed practically nothing about the position of Planet X on the sky. This study shows that a dynamical determination of the location of Planet X from unquestionably reliable data extending to 1966, is impossible. The best that can be done is to place limits on the mass. For a reasonable distance of less than 100 AU, it is limited to a few $M_{\oplus}$ (Rawlins and Hammerton 1973).

The situation with the 1613 observation by Galileo is similar to that of the Lalande observation in that the accuracy may be far worse than the ability of a modern ephemeris to predict that position. If Galileo's drawing in his notebook (see Kowal and Drake 1980) is assumed correct in scale, there is a residual (O − C) of −57 arcsec in longitude and 26 arcsec in latitude between the observed position (O) and the calculated position (C) from the JPL ephemeris DE 102. Yet the orbital plane of Neptune, especially in the region of the sky observed by Galileo, is established to within a small fraction of an arcsec by DE 102 (Standish 1981). It is unlikely that a Planet X will produce a change of 26 arcsec in latitude without a corresponding change a full order of magnitude larger in longitude (Standish 1981; Rawlins 1981). Further, if we discount only the scale of Galileo's drawing, his observation is completely consistent with the JPL ephemeris (Standish 1981). Although Galileo's observation may be of interest for other reasons, it tells us little or nothing about Planet X.

III. ORBITS OF PIONEER SPACECRAFT

Two Pioneer spacecraft, launched in the early 1970s for encounters with Jupiter on 4 December 1973 (Pioneer 10) and 3 December 1974 (Pioneer 11) are now receding from the solar system on trajectories that are essentially determined by gravitational forces. The only nongravitational forces of significance arise from solar radiation pressure on each spacecraft and from small orientation maneuvers performed every few months in order to keep the spacecraft's high-gain antenna pointed at Earth. The translational accelerations on the spacecraft from these two sources are modeled to sufficient accuracy in the equations of motion for the spacecraft. The spacecraft are spinning at about 5 rpm about an axis parallel to the axis of the parabolic antenna, thus the antenna can be pointed by applying attitude maneuvers over an interval of a few hours, thereby precessing the axis of the antenna to point at Earth. The small translational acceleration ($\sim 10^{12}$ km s^{-2}) imparted by attitude jets during this maneuver can be modeled to an accuracy of 10% or better. Of greater significance, the steady-state component of the acceleration from solar radiation pressure is about 10^{-13} km s^{-2} at 35 AU, the current distance to Pioneer 10, but it can be modeled to an accuracy of about 1%. Any remaining unmodeled time-varying component is no more than 2% of the steady-state contribution with a correlation time exceeding 4 hr (Anderson 1972).

Because of the dual advantage of being able to use spinning spacecraft with small, infrequent attitude control maneuvers, and distant spacecraft where the r^{-2} solar radiation pressure is small, it should be possible to detect unmodeled translational accelerations on the two Pioneer spacecraft at a level of 10^{-14} km s^{-2}, provided that the coherency time of the accelerations is at least 6 months. The radio data acquired by the Deep Space Network (DSN) are more than sufficient for a measurement of acceleration at this level of accuracy, even though the spacecraft carry a single-frequency 1960-vintage transponder that receives and transmits 13 cm (S-band) radio waves with no provision for ranging modulation.

The DSN utilizes large antennas, low-noise, phase-lock receiving systems, high-power transmitters, and atomic frequency standards to perform its basic functions, one of which is the acquisition of range-rate information from Pioneers 10 and 11. The DSN is physically located on 3 continents, with communication complexes in California, U.S.A.; near Canberra, Australia; and near Madrid, Spain. Each complex consists of one 64 meter diameter antenna, one 26 meter antenna, and either one or two 34 meter diameter antennas. A ground communication facility connects the 3 complexes to the DSN control center at the Jet Propulsion Laboratory in Pasadena, California.

Beginning with its first 26 meter antenna in 1958, the DSN has been managed as an evolving telecommunication and data acquisition facility. The 64 meter antennas are required for the tracking of the two Pioneers in the outer solar system. Given normal upgrades to existing facilities in the next few years, it is expected that it will be possible to track the Pioneers well into the 1990s when Pioneer 10 will be at a distance of 50 AU or more. Given the expected accuracy of the determination of unmodeled accelerations ($\sim 10^{-14}$ km s^{-2}), it may be possible to obtain new information on forces acting on the outer solar system.

The DSN tracking system measures Doppler and range, although the ranging assembly is useless for the Pioneer spacecraft which are unable to demodulate a ranging waveform. The frequency and timing system supplies stable references produced by a hydrogen maser. The 13 cm signal is amplified as high as 400 kW by a klystron amplifier and transmitted through a diplexer and microwave antenna to the spacecraft (Renzetti et al. 1982; Anderson et al. 1985).

The spacecraft receives the transmitted uplink signal and multiplies it by a rational fraction 240/221 to produce a coherent downlink carrier that is phase modulated with telemetry. The spacecraft radiates the downlink through a diplexer and antenna at a power of 8 W. The ground station receives the signal, amplifies it in a maser amplifier, and sends it to a triple conversion superheterodyne phase-locked loop receiver. The receiver reference frequency is supplied to the Doppler extractor. Because of the long round-trip light times involved with spacecraft in the outer solar system, practically all of the data beyond a distance of 30 AU are transmitted by one 64 meter antenna of the DSN and received by another 64 meter antenna on another continent. The stability of the hydrogen maser frequency standards in this regard is critical to maintaining the accuracy of the Doppler data.

Orbit fits for Pioneer 10 have been achieved with Doppler data extending over a period of 3 yr (Anderson and Mashhoon 1984). The fits are satisfactory with no indication that any unknown forces are acting on the outer solar system. Pioneer 10 is now moving slowly northeast on the celestial sphere, in the constellation of Taurus. Pioneer 11 is on the opposite side of the solar system, traveling southeast, near Libra. For now, the limit on unmodeled acceleration is 5×10^{-14} km s^{-2} out to a distance of 35 AU in the direction of Pioneer 10. In the future this limit will be improved by a factor of 5, the distance will be increased to ~ 50 AU, and the direction will be augmented to include the region of Pioneer 11.

It is meaningful to compare this limit with what can be achieved with the optical astrometric data on Uranus at the moment. Suppose that an unmodeled force perturbed Uranus significantly over a period of 50 yr in the recent past. The velocity change Δv from the unmodeled acceleration Δa on Uranus would amount to $\Delta v = \Delta a \times (50 \text{ yr})$. If practically all of this velocity perturbation went into a secular change in the latitude over a period of 75 yr (corresponding to the time interval of the modern data), it would be difficult to remove with the existing model of the solar system. The detection sensitivity in this case is about 1 arcsec for a total change $\Delta \beta$ in latitude. Using the expression $r\Delta \beta = \Delta v \times (75 \text{ yr})$, with $r = 20$ AU, the derived detection sensitivity for the unmodeled acceleration is $\Delta a = 4 \times 10^{-15}$ km s^{-2}. It is important to remember, however, that the signature of an unmodeled force would be quite different on the outer planets and on the Pioneer spacecraft, both from the viewpoint of orbital considerations and from the viewpoint of the respective measurements; thus data from planetary astrometry and from spacecraft are both important.

IV. LIMITS ON A TRANS-NEPTUNIAN BELT OF COMETS

The Planet X model, invoked to explain the periodicities observed in fossil and geologic records (Whitmire and Matese 1985; Matese and Whitmire 1985; chapter by Matese and Whitmire) requires a comet belt beyond the orbit of Neptune and a Planet X whose perihelion, driven by the action of the known planets, precesses through a primordial disk of material and leads to cometary showers. The existence of such a disk (of comets and planetesimals) has been suggested as a consequence of accretion theories of the origin of the outer planets (Kuiper 1951; Whipple 1972; Fernández 1980). The observations of thin disks of solid particles around Beta Pictoris (Smith and Terrile 1984) and Vega (Aumann et al. 1984) lend credance to these theoretical expectations (Weissman 1984). A similar disk surrounding the Sun is a possibility.

For purposes of placing limits on the mass of material in a solar disk, we approximate the cometary material by an annular disk of uniform density in the plane of the ecliptic with an inner radius of 35 AU and an outer radius of 56 AU. The inner radius is set by practical considerations of where a disk could possibly start, given the planetary orbits of the outer planets, and the outer radius is set by the canonical parameters of the Planet X model (chapter by Matese and Whitmire). Actually, that model does not require an inner disk; an outer one starting at a distance $\sim$ 100 AU would suffice, but whatever limits we can place on an inner disk is of some interest. If an inner disk of limited mass is allowed by the data, certainly an outer disk of comparable mass is allowed.

The accelerations produced by the assumed inner disk on an object with coordinates x,y,z can be evaluated using expressions given by Krogh et al. (1982). Because of symmetry, the spacecraft can be placed in the x–y plane with coordinates $x = r \cos \beta$ and $z - r \sin \beta$. The ecliptic latitude β for Pioneer 10 is 3.112 arcdeg, while the present distance is about 35 AU, near the inner boundary of the assumed disk. The acceleration on Pioneer 10 for a total disk mass of 1 $M_\oplus$ is shown as a function of heliocentric distance r in Fig. 3. Note that the acceleration is a maximum at the inner and outer boundaries of the disk, thus the effect on Pioneer 10 is most easily detected with data now being acquired. Unfortunately, the detection sensitivity is not particularly good. A total mass of 5 $M_\oplus$ or more would be required in order to exceed the current detection sensitivity of 5×10^{-14} km s^{-2}. Still, because we have seen no anomalous accelerations on Pioneer 10, we conclude from Pioneer data alone that if a belt of comets exists just beyond the orbit of Neptune, it has a mass < 5 $M_\oplus$.

A tighter bound on the mass has been set by Hamid et al. (1968) from the motions of periodic comets. They conclude that there is perhaps < 0.5 $M_\oplus$ of material to 40 AU and < 1.3 $M_\oplus$ to 50 AU, although they point out that

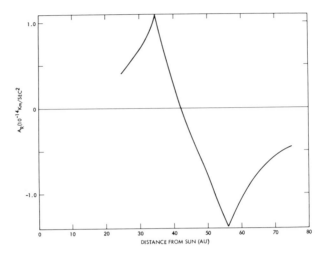

Fig. 3. Radial acceleration A_R on Pioneer 10 from an annular disk of cometary material beyond the orbit of Neptune in the ecliptic plane. Assumed total mass is 1.0 $M_\oplus$ distributed uniformly between an inner radius of 35 AU and an outer radius of 56 AU. The acceleration is directly proportional to the total mass in the disk.

nongravitational forces on comets leave some room for doubt. Within this context, the larger but firmer bound set by Pioneer is of value. The bound set by anomalous residuals in Neptune's latitude are even less stringent (Whipple 1964), so the best bounds are set by Pioneer and periodic comets. We can be reasonably certain, based on all available dynamical data sources, that the total mass of material in an inner disk is no more than 5 $M_\oplus$.

Acknowledgments. This work was supported by the Pioneer Project Office at NASA/Ames Research Center as part of the Heliospheric Mission on Pioneers 10 and 11. The work was performed at the Jet Propulsion Laboratory, California Institute of Technology, under contract with NASA. The manuscript was developed while one of us (J. D. A.) held a Visiting Lecturer position at Monash University, Victoria, Australia. Support provided by Monash staff and students, in particular A. J. R. Prentice, P. J. Thomas and R. Fok, is gratefully acknowledged. Helpful comments were received from D. P. Whitmire and from the referees P. K. Seidelmann and R. L. Duncombe.

PLANET X AS THE SOURCE OF THE PERIODIC AND STEADY-STATE FLUX OF SHORT PERIOD COMETS

JOHN J. MATESE

and

DANIEL P. WHITMIRE
University of Southwestern Louisiana

The cratering and fossil records suggest that impacts on the Earth have been modulated with a period of ≈ 30 Myr. In the Planet X model this modulation (comet showers) is the result of the interaction of Planet X with a primordial disk of planetesimals lying beyond the orbit of Neptune. The peak in the flux of short-period comets occurs when Planet X's precessing perihelion and aphelion points make their deepest penetration into the primordial disk every 30 Myr. Here we briefly review some independent observational and theoretical evidence supporting the existence of Planet X and the comet disk. We then consider the Planet X model in more detail and summarize how the model restricts the orbital parameters and mass. An attractive feature of this model is that these parameters are consistent with those independently deduced from an analysis of the residuals in the motions of Uranus and Neptune. Within the context of several simplifying approximations, the model makes a unique prediction of the relation between Planet X's semimajor axis, inclination and eccentricity. Independent of explaining periodic comet showers and the Uranus/Neptune residuals, Planet X, if it exists, can readily account for the present observed flux of short-period comets over a wide range of planet and disk parameters. The ability of Planet X to prepare the required comet density profile (a radial scale length which is ≲ 20 AU) in its primordial and inclined orbits is discussed. The required inclination (≳ 25°) may explain the failure of previous surveys to discover the planet as its present latitude is not likely to be near the ecliptic. If it exists, the best immedi-

ate hope of finding Planet X is the full-sky optical search by Shoemaker and Shoemaker. In the infrared the on-going IRAS search in the 100 μm band may in principle be capable of detecting the planet.

I. INTRODUCTION

A comprehensive and entertaining historical account of the various predictions of and searches for Planet X over the past century (through 1980) can be found in Hoyt's (1980) book *Planets X and Pluto*. Although the emphasis of the book is on the searches carried out at Lowell Observatory by Lowell and his successors, the discoveries of Uranus and Neptune as well as other less extensive Planet X searches and predictions are also chronicled.

The planet Neptune was discovered in 1846 on the basis of an approximately 100 arcsec discrepancy in the motion of Uranus. A subsequent reanalysis of Uranus' orbit indicated the possibility of a 5 arcsec residual in the longitude which led Percival Lowell (1915) to predict orbits for a ninth planet (Planet X) beyond Neptune. When Pluto was discovered by Clyde Tombough in 1930 near one of the predicted longitudes, it was initially identified with Lowell's Planet X. To account for the Uranus residuals, a planet at Pluto's distance would need a mass of $\approx 1\ m_\oplus$. However, subsequent occultation measurements placed restrictions on Pluto's size, and it became apparent that a $1\ m_\oplus$ Pluto would require an unrealistically high density. Over the years, Pluto's mass was gradually revised downward until its moon, Charon, was discovered thus yielding the first accurate mass determination (Christy and Harrington 1978). The total mass of the system was found to be $2 \times 10^{-3} m_\oplus$, which is far too low to have any measurable effect on Uranus. Pluto is therefore not Lowell's Planet X, and there is now a renewed interest in identifying the origin of the observed small residuals.

The residuals that led Lowell to predict his Planet X are now considered unreliable. Nonetheless, there are smaller but apparently real discrepancies between modern ephemerides and the historical observations of Uranus and Neptune (Seidelmann et al. 1980,1985; Rawlins and Hammerton 1973). For Uranus, the U.S. Naval Observatory (USNO) (Seidelmann et al. 1985) and the Jet Propulsion Laboratory (chapter by Anderson and Standish), ephemerides fit to post-1900 data do not require an unmodeled force for agreement during this century. However, using the USNO's most recent ephemerides for Uranus and Neptune, Seidelmann et al. find that no single ephemeris fits all the data for Uranus, and that although Neptune's ephemeris can be fit to the present, past experience indicates that the O-C plot will run-off within ten years. If the 1795 prediscovery observations of Neptune by Lalande are accepted, then the Neptune longitude residuals are ≈ 10 arcsec. Assuming that the observed residuals are due to Planet X, most estimates placed it at 50 to 100 AU with nontrivial eccentricity and inclination and a mass of ≈ 1 to

5 $m_\oplus$. Anderson and Standish (see their chapter) discuss the less restrictive dynamical constraints based on Pioneer 10 tracking data.

In addition to the dynamical constraints imposed on Planet X by the residuals in the motions of Uranus and Neptune, independent and more specific predictions of the orbital elements can be made if Planet X is the origin of the reported periodicity in terrestrial cratering (Alvarez and Muller 1984; Rampino and Stothers 1984*a*; Sepkoski and Raup 1985; also see chapters by Muller and by Rampino and Stothers), mass extinctions (Raup and Sepkoski 1984; Sepkoski and Raup 1986) and possibly other geophysical phenomena (Rampino and Stothers 1984*b*). The cratering record suggests an impact periodicity of $\approx$ 30 Myr, comparable numbers of shower and steady-state impacts and a shower duration of $\lesssim$ 15 Myr. A critical discussion of the statistical significance of the cratering and extinction periods is given in chapters by Shoemaker and Wolfe and by Tremaine.

II. THE PLANET X MODEL

In the Planet X model (Whitmire and Matese 1985; Matese and Whitmire 1986) periodic comet showers are associated with the precession (advance or regression) of the perihelion and aphelion points through a primordial disk of comets believed to lie beyond the orbit of Neptune. The existence of a residual disk of comets (planetesimals) is a natural consequence of accretion theories of the origin of the outer planets (Kuiper 1951; Whipple 1972; Fernández 1980,1984*a*; Shoemaker and Wolfe 1984). The observation of a thin disk of solid particles around β Pictoris (Smith and Terrile 1984), and possibly other stars (Aumann 1984), strongly support these theoretical expectations (Weissman 1984; see also his chapter). The disk in β Pictoris is largely free of particles inside 30 to 40 AU. It extends out to 400 AU, has an angular thickness of $\approx$ 10° and obeys an r^{-3} density variation beyond 100 AU.

During the lifetime of the solar system, Planet X will have tended to clear a gap in the solar comet disk (i.e., a region of reduced density, bordered by transition zones). If Planet X has always been in its present orbit, the gap boundaries correspond to the present values of perihelion and aphelion distance. Transition-zone comets can be resupplied by diffusion due to Planet X itself and/or by internal diffusion driven by massive comets (Fernández 1980). However, Planet X may have been perturbed from a primordial ecliptic orbit to its present inclined orbit in which case the gap boundaries may be the primordial perihelion and aphelion distances.

If the orbit of Planet X is eccentric and inclined to the planetary plane, the flux of comets scattered by Planet X will maximize twice during each precession period as the perihelion and aphelion points make their deepest penetration into the transition zones. This is illustrated schematically in Fig. 1 for the extreme case of $i_x = 90°$. In this figure only aphelion showers are illustrated for clarity. Perihelion showers, if relevant, will be additive as dis-

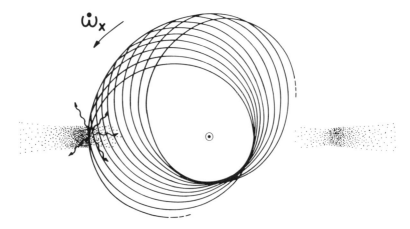

Fig. 1. Schematic illustration of the proposed shower mechanism. The extreme orbital inclination illustrated, $i_x = 90°$, is chosen for clarity only. The shower at perihelion is also omitted for clarity.

cussed in Matese and Whitmire (1986) and Whitmire and Matese (1985). Comets whose perihelia are scattered directly into the zones of influences of Saturn and Jupiter will constitute the shower flux as their subsequent dynamical lifetimes are $\lesssim 4$ Myr (Everhart 1977). Comets scattered into the zones of influence of Uranus and Neptune, some of which ultimately become Jupiter family short-period (SP) comets, have lifetimes of ~ 100 Myr (Muller, personal communication 1985; Weissman, personal communication 1985). These comets will therefore contribute to (and in the context of the model, dominate) the steady-state flux of SP comets. We emphasize that independent of the question of periodicities, the existence of Planet X and the comet disk can readily explain the origin of SP comets over a wide range of planet and disk parameters. The case against captured long-period comets as the source of the SP flux is reviewed by Fernández (1980).

The semimajor axis required for a specified precession period T_ω is given by (Kozai 1959)

$$a_x = \left[\tfrac{3}{2}T_\omega\left(\sum_p m_p a_p^2\right)\frac{|1 - \tfrac{5}{4}\sin^2 i_x|}{(1 - e_x^2)^2}\right]^{2/7} \quad (1)$$

where i_x and e_x are the inclination and eccentricity of Planet X, p specifies the known planets, masses are in solar units, semimajor axes are in AU, times are in years and $T_\omega = 6 \times 10^7$. This equation includes only the first (quadrupole) term in a multipole expansion for the precession rate. In this approximation the outer planets are treated as rings with radii equal to their semimajor axes

PLANET X AND PERIODIC COMET SHOWERS 301

a_p. The effects of the disk itself and possible commensurabilities are neglected. Within these limitations, the Planet X model predicts a unique relationship between a_x, i_x and e_x.

To obtain the time dependence of the comet flux scattered by Planet X at the aphelion transition zone, we model the local comet number density by

$$\mathcal{N}(r,z) = \mathcal{N}(r_d,0)\exp\left[-\left(\frac{r_d - r}{\sigma_r}\right)^2 - \left(\frac{z}{\sigma_z}\right)^2\right]. \qquad (2)$$

Here σ_r is the scale length for radial density variation which serves to define the transition zone at the aphelion distance $Q_x = a_x(1 + e_x)$, σ_z is the scale length for vertical density variation which defines the angular dimensions of the disk edge, $<i(\text{comet})> \approx \sigma_z/Q_x$, and $r_d \equiv Q_x + \sigma_r$ defines the nominal disk edge. Planet X orbits with coordinates

$$\begin{aligned} r &= \frac{a_x(1 - e_x^2)}{1 + e_x \cos f} \\ z &= r \sin i_x \sin (f + \omega_x) \end{aligned} \qquad (3)$$

where f is the true anomaly and $\omega_x = 2\pi t/T_\omega$ is the perihelion angle. We assume that the scattered flux, averaged over an orbit, is proportional to the average comet density sampled. Thus, the time dependence of the scattered flux is determined from the ω_x dependence of the orbital average of Eq. (2). Matese and Whitmire (1986) found that the numerical results of the averaging could be adequately reproduced by the analytical estimate that the flux has a time scale given by

$$\tau/\tfrac{1}{2}T_\omega \approx 0.6 \max\left\{\left[\frac{\sigma_r(1 - e_x)}{Q_x e_x}\right]^{1/2}, \left[\frac{\sigma_z}{Q_x \sin i_x(1 - e_x)}\right]\right\}. \qquad (4)$$

In the above, τ is the time scale for the scattered comet flux to rise from e^{-1} of its maximum value and return to e^{-1} of its maximum value.

In addition to the requirements for predicting the correct shower period, Eq. (1), and the correct shower time scale, Eq. (4), the model must also predict the requirements for an adequate number of impacts on the Earth per shower interval. We estimate the number of comets scattered into the zone of influence of a specified planet p during a single shower to be

$$N_{x \to p} = \mathcal{N}_o u \Sigma_{x \to p} f \tau \qquad (5)$$

where $\mathcal{N}_o \approx e^{-1}\mathcal{N}(r_d, 0)$ is the comet number density sampled by Planet X at aphelion, u is the relative velocity between Planet X and a comet, $\Sigma_{x \to p}$ is the cross section for scattering comet perihelia into the specified zone, and f is the fraction of the orbit time that Planet X is in the disk

$$f = 2\sigma_z/\sin i_x \, v_x \, T_{\text{orbit}}. \tag{6}$$

The number density of comets $\geq 10^{15}$g is deduced from the estimates of Fernández (1980) (see Matese and Whitmire 1986) to be

$$\mathcal{N}_o \approx 10^5 \, (\sigma_z/10 \text{ AU})^{-1} (Q_x/100 \text{ AU})^{-2} \text{AU}^{-3}. \tag{7}$$

This value is obtained by using a single power-law distribution $m^{-1.75}$ with a maximum comet mass of 10^{23}g. The inferred areal mass density at 100 AU is within the limits implied by the absence of disk perturbations on comets. However, the actual value of $\mathcal{N}_o$ could be larger by a factor $\lesssim 10$ if other reasonable distributions are chosen. For example, Greenberg et al. (1984) have suggested that a two-power mass distribution law may be required to explain the Oort cloud and primordial planet growth.

The relative velocity u is obtained by assuming the comets to be in near circular orbits at aphelion since any significant noncircular motion would imply a long shower time scale as can be seen from Eq. (4).

To calculate the cross sections in Eq. (5), we adopt the zones of influence from Everhart (1977). Jupiter (J) controls comets with perihelia $q \leq 5.8$ AU, Saturn (S) those with $5.8 \text{ AU} \leq q \leq 10.7 \text{ AU}$, Uranus ($U$) those with $10.7 \text{AU} \leq q \leq 22$ AU and Neptune (N) those with $22 \text{ AU} \leq q \leq 34$ AU. The calculation of $\Sigma_{x \to p}$ proceeds in a straightforward manner (see Matese and Whitmire 1986) once the parameters of Planet X (a_x, e_x, i_x and m_x) are specified consistent with the constraint of Eq. (1). Since $\Sigma_{x \to p}$ scales with m_x^2, one need only to investigate the cross section dependence on i_x and e_x.

To predict the number of Earth impactors in a single shower cycle, we assume that every comet which ultimately becomes a Jupiter family SP comet has a probability of 0.1 of becoming an Earth crosser (Fernández 1984b) and that these in turn have a probability of 7×10^{-9} per orbit of colliding with the Earth during their dynamical lifetime of 10^5 orbits (Weissman 1982b). Thus, the net collision probability during the lifetime of a Jupiter family SP comet is adopted to be

$$P_{\text{coll}} = 7 \times 10^{-5}. \tag{8}$$

Comets scattered to Saturn have a pass-on time to Jupiter of ≤ 4 Myr. Thus, comets directly scattered to Saturn or Jupiter can be considered to be shower comets, and the net number of shower impacts in a single cycle is given by

$$N_{\text{shower}} = [N_{X \to S} f_{S \to J} + N_{X \to J}] P_{\text{coll}} \tag{9}$$

where $f_{S \to J} = 229/500$ is the pass-on fraction from Saturn as estimated by Everhart (1977).

Comets scattered to Uranus and Neptune have a pass-on time scale of $\sim 10^8$ yr such that the fractions ultimately becoming Jupiter family comets ($f_{N \to J} = \frac{1}{2} f_{U \to J}$, $f_{U \to J} = 40/69 \, f_{S \to J}$) can be considered to be steady-state comets. Therefore, the number of steady-state (ss) impacts in a single cycle is given by

$$N_{ss} = [N_{X \to U} f_{U \to J} + N_{X \to N} f_{N \to J}] P_{coll}. \tag{10}$$

Finally, the net flux of Jupiter family SP comets is given by

$$\Phi_{sp} = [N_{X \to N} f_{N \to J} + N_{X \to U} f_{U \to J} + N_{X \to S} f_{S \to J} + N_{X \to J}]/\tfrac{1}{2} T_\omega. \tag{11}$$

III. LIMITS ON THE MODEL PARAMETERS

The basic equations which provide limitations on the Planet X model are then Eqs. (1), (4), (9) and (11). We now discuss the constraints on the orbit parameters of Planet X and the disk parameters which must be satisfied to ensure consistency with observational limits. These two sets of parameters are sufficiently interrelated that a rigorous set of limits is not readily obtainable. In the following, the results given should be taken as self-consistent estimates.

A. Limits on σ_r and σ_z

The scale parameters are the most difficult to estimate as they depend on the interaction between Planet X and the comets, comet-comet interactions and on the past history of the system. As an initial estimate we shall assume that the solar system disk is similar to that in β Pictoris in which it is found that the full angular thickness is $\approx 10°$. Thus, we adopt as an estimate

$$\frac{\sigma_r}{Q_x} \approx \frac{\sigma_z}{Q_x} \approx \tfrac{1}{2}\sin 10° \approx 0.1. \tag{12}$$

We shall assume that the actual scales lie in the range between 0.05 and 0.2. Ultimately, we shall discuss the consistency of this estimate with that implied by diffusion due to Planet X.

B. Limits on e_x

The shower time scale, Eq. (4), shows that e_x must be intermediate in value in order that $\tau \lesssim 15$ Myr,

$$\sigma_r/Q_x \lesssim e_x \lesssim 1 - \sigma_z/Q_x \sin i_x. \tag{13}$$

We adopt a value of $e_x = 0.3$. The choices of $e_x = 0.1, 0.5$ are discussed in Matese and Whitmire (1986).

C. Limits on a_x

If one considers the behavior of $a_x(e_x, i_x)$ from the period formula, Eq. (1), one observes that the quadrupole approximation breaks down near the critical angle of $63°.4$. Including the next term in the expansion (Kozai 1959) merely shifts the critical angle. In the vicinity of the critical angle the motion changes from precession to libration of the perihelion (Garfinkel 1960; Allan 1970). Although an acceptable 60 Myr libration orbit might still occur, we shall assume that the inclination is not near the critical angle. In Fig. 2 we show the one- and two-term results for $e_x = 0.3$ (the curves are relatively insensitive to e_x). Requiring that Planet X not overlap the orbit of Neptune, we deduce that 50 AU $\lesssim a_x \lesssim$ 100 AU.

D. Limits on i_x

The shower time scale, Eq. (4), restricts $\sin i_x \gtrsim \sigma_z/Q_x (1-e_x)$. More stringent restrictions arise from the requirement that the cross section for scattering comets to Saturn and Jupiter be nontrivial. In order to scatter a comet to the deep interior of the solar system, the comet's velocity must be signifi-

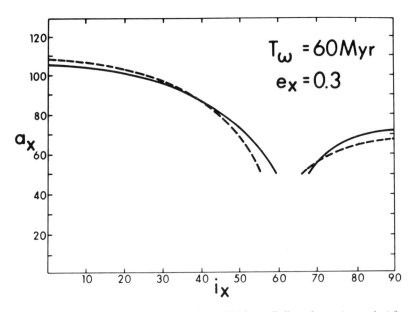

Fig. 2. The semimajor axis required to produce a 60 Myr perihelion advance (regression) for $e_x = 0.3$ as a function of orbital inclinations i_x. The solid curve is for the quadrupole term only while the dashed curve includes the octupole term. The region excluded is the interval where omitted higher multipoles are not negligible.

cantly changed. This requires that the relative velocity between Planet X and the comet, u, be a significant fraction of the local Kepler velocity $v_o \equiv (Gm_\odot/Q_x)^{1/2}$. Since the comets themselves must be more nearly circular than Planet X in order to obtain an acceptable shower time scale, the large relative velocity u must be due to Planet X itself.

In addition to requiring a large u, we also need to scatter the direction of **u** through a large angle θ

$$\tan\frac{\theta}{2} = \frac{Gm_x}{bu^2} = \frac{m_x}{m_\odot} \frac{Q_x}{b} \left(\frac{v_o}{u}\right)^2 \qquad (14)$$

where b is the impact parameter. Thus one observes that so long as $u^2 \approx \frac{1}{2}v_o^2$, the scale of the cross section for scattering to the inner solar system is $\Sigma_{x\rightarrow p}$ is of the order of $(m_x Q_x/m_\odot)^2$. Matese and Whitmire (1986) have shown that

$$u^2 = [2 - e_x - 2(1 - e_x)^{1/2} \cos i_x]v_o^2. \qquad (15)$$

Inserting $e_x \approx 0.3$ and $u^2 = \frac{1}{2}v_o^2$ yields $i_x \approx 45°$. A more detailed discussion of the i_x, e_x dependence of $\Sigma_{x\rightarrow p}$ is given in Matese and Whitmire (1986) where it was concluded that a minimum inclination of $i_x \gtrsim 25°$ is required to scatter comets in circular orbits to the inner solar system. Adopting a canonical value of inclination to be 45°, we then find from Fig. 2 that the related canonical value of a_x is 80 AU.

In all subsequent discussions the results given will be appropriate to the adopted values of $a_x = 80$ AU, $e_x = 0.3$ and $i_x = 45°$.

E. Limits on m_x, $\mathcal{N}_o$

In the above we discussed the limits on a_x, e_x and i_x implied by the constraints $T_\omega \approx 60$ Myr, $\tau \lesssim 15$ Myr and $\Sigma_{x\rightarrow p} \approx$ order $[(m_x Q_x/m_\odot)^2]$. Limits on m_x and $\mathcal{N}_o$ are deduced from the requirement that the number of shower and steady-state impacts per cycle (Eqs. 9 and 10) be significant. For the adopted orbit parameters, the scattering cross sections were found to be $\Sigma_{x\rightarrow p} = (5.9, 5.8, 18, 31) \times (m_x Q_x/m_\odot)^2$ for $p = (J, S, U, N)$, respectively. Inserting these results in Eqs. (9) and (10), we find the number of impacts of comets scattered at aphelion to be

$$N_{ss} \approx N_{shower} \approx 30 \, (\mathcal{N}_o/10^5 \text{AU}^{-3})(m_x/5m_\oplus)^2. \qquad (16)$$

The observed cratering rate over the last 300 Myr of craters ≥ 20 km in diameter corresponding to comets of mass $\geq 10^{15}$g (Weissman 1982b) is $\approx 50 \pm 20$ per cycle (Grieve and Dence 1979). In Matese and Whitmire (1986) a mass of $m_x = 5m_\oplus$ was adopted as the canonical value. Given the uncertainties in $\mathcal{N}_o$ discussed in Sec. II as well as the exact contribution of Planet X to the net cratering rate, we allow for $m_x \gtrsim 1m_\oplus$. An upper limit on the mass will come from limits on observed planetary perturbations which depend in turn on a_x. In addition, self-consistency with the transition zone scales will restrict the value of m_x.

The dependence of the transition zone scales σ_r and σ_z on the mass of Planet X inevitably requires a specification of the history of Planet X. It is unlikely that protoplanet X could have grown to its present size if it had always been in its present orbit without requiring an excessive amount of matter in the disk. This follows because the large relative velocity u would preclude efficient accretion. Thus we must assume that Planet X was essentially completely formed in the disk before being perturbed to its present orbit.

The perturbation could have been due to a close approach with Neptune (Harrington and van Flandern 1979) or, perhaps more likely, with another trans-Neptunian planet of mass comparable to Planet X that was subsequently ejected when its orbit overlapped the known planets. A close approach of this type could have been caused by the migration of the semimajor axes of Planet X and its perturber as they exchanged angular momenta with planetesimals in the manner described by Fernández and Ip (1984). An initial inward migration from a radius a_{max} is caused by the acceleration mechanism acting on the near circular comets (Öpik 1976) while a later outward migration results when comets are eventually transferred to the inner planets. Thus we envision protoplanet X and its perturber migrating in opposite directions, dispersing and accreting the intervening belt of comets, until their orbits are close enough to allow the mutual acceleration mechanism to random walk their relative velocity leading to an eventual strong perturbation of their orbits. Such orbit crossings did in fact occur in some of the simulated Uranus/Neptune cases considered by Fernández and Ip (1984).

In this scenario Planet X would find itself in its present orbit with the disk efficiently cleared of comets out to a distance of a_{max}. Thus the primordially prepared transition zone would be $\sigma_r(t=0) = Q_x - a_{max}$. This region would then be further diffused by subsequent interactions with Planet X. The latter effect can be approximated as follows.

Using the conventional estimate (Wetherill 1980) of the diffusive spread in the disk due to Planet X, we have

$$\frac{\Delta\sigma_z}{Q_x} \approx \frac{\Delta\sigma_r}{Q_x} \approx \frac{u_c}{\sqrt{3}\,v_0} \qquad (17)$$

where u_c is the rms relative velocity of the comets to the Kepler circular standard. In Matese and Whitmire (1986) the adopted orbit parameters were used to obtain

$$\frac{u_c}{\sqrt{3}\,v_0} \approx 0.2\left(\frac{m_x}{5m_\oplus}\right) \qquad (18)$$

in the lifetime of the solar system. This result is in agreement with the estimate of Tremaine (see his chapter).

Of more direct use are the Monte Carlo calculations of Shoemaker and Wolfe (personal communication). They have considered the interaction of $\approx$ 500 planetesimals with a 10 $m_\oplus$ Planet X having the orbit parameters adopted here. The method used is based on the Öpik analysis and neglects three-body effects as well as comet-comet relaxation effects. The planetesimals are initially distributed in a narrow band with $a \approx 90$ AU, $e \approx 0.1$ and $i \approx 5°$. In the lifetime of the solar system, the survivors have acquired an average inclination $\approx 10°$. Therefore $\sigma_z/Q_x \approx \sin 10° \approx 0.2$ which is approximately ½ the analytical estimate of Eqs. (17) and (18). This implies that the σ_z scale height limit requires that $m_x \lesssim 10\ m_\oplus$.

More stringent limits are inferred from the diffusion of planetesimal perihelia and semimajor axes. The initially localized band of planetesimals has the peak of its q and a distributions shifted outward as the distributions broaden. Within the gap the radial density of planetesimals is largest at $Q_x = 104$ AU and decreases by e^{-1} over a distance of ≈ 40 AU. This result is in agreement with the analytical estimate of Eqs. (17) and (18) and suggests that the distribution has little memory of its primordial preparation as long as the planetesimals are initially beyond a_x. If we assume the mass dependence of Eq. (18), the requirement that $\sigma_r/Q_x \lesssim 0.2$ implies that $m_x \lesssim 5\ m_\oplus$ and the adopted value of 0.1 is consistent with $m_x \approx 2$ to 3 $m_\oplus$.

Tremaine (see his chapter) has argued that a mass $m_x \gtrsim 30\ m_\oplus$ is required to significantly deplete the gap region of comets. In contrast, Shoemaker and Wolfe (personal communication) have found that a 10 $m_\oplus$ Planet X in its present orbit will have its encounter rate with comets decreased by a factor ≈ 10 in the lifetime of the solar system. If one further considers that the formation stage of Planet X in an ecliptic orbit must have substantially contributed to gap depletion, we conclude that his estimate is in error. The reason for the discrepancy is that Tremaine has assumed that gap clearing occurs predominantly because of close collisions ($\theta \approx 1$ rad), whereas it was argued in Matese and Whitmire (1986), and verified by Shoemaker and Wolfe, that diffusion is the dominant gap clearing mechanism subsequent to perturbation to its present orbit.

IV. SUMMARY AND DISCUSSION

If Planet X is the origin of the apparent 30 Myr periodicity in cratering and mass extinctions, then the allowable range of parameters was found to be: $a_x \approx 50$ to 100 AU, $e_x \approx 0.1$ to 0.5, $i_x \gtrsim 25°$, $m_x \approx 1$ to 5 $m_\oplus$. The adopted self-consistent values chosen were $a_x = 80$ AU, $e_x = 0.3$, $i_x = 45°$ and $m_x = 3\ m_\oplus$. Neglecting contributions to the precession rate from the disk itself and from commensurabilities, the Planet X model predicts a unique relation between a_x, i_x and e_x given in Eq. (1) and Fig. 2. The disk transition-zone scale lengths required are $\sigma/Q_x \lesssim 0.2$. Our adopted value of 0.1 corresponds to a e^{-1} shower duration of 9 Myr.

In the context of the model, the cratering data require only that the density of sampled comets vary by a factor of $\gtrsim 3$ over a radial distance of $\lesssim 20$ AU, a seemingly modest requirement. We have suggested that Planet X may itself have created such a density gradient, either in its present or primordial ecliptic orbit. A full three-body statistical analysis including the effects of comet-comet relaxation on diffusion is required to determine if Planet X can indeed create a disk density profile with the required modulation. These results, in turn, will depend on the evolutionary scenario for perturbing it to its present orbit.

One scenario has already been suggested (Harrington and van Flandern 1978) in which a single close encounter between Planet X and a primordial Neptune system of satellites simultaneously ejected Pluto-Charon, left Triton in a retrograde orbit, and placed Nereid in its present highly eccentric orbit. These authors conclude that the planet mass was 2 to $5m_\oplus$, that its present semimajor axis is 50 to 100 AU, and that its present eccentricity is $\lesssim 0.6$. An alternative scenario was suggested (Sec. III.E) in which Planet X was perturbed to its present orbit in a close encounter with another planet of comparable mass as the semimajor axes of both planets migrated due to planetesimal scatterings (Fernández and Ip 1984).

We note here a distinction between the Planet X, Nemesis and galactic oscillation models. The Nemesis model (Whitmire and Jackson 1984; Davis et al. 1984; also see chapter by Muller) predicts a short ($\lesssim 4$ Myr) shower duration with a larger (~ 5 Myr) period variation (Whitmire and Jackson 1984; Davis et al. 1984; Hut 1984a). The Planet X model predicts a longer ($\gtrsim 5$ Myr) shower duration with no significant period variation. The galactic oscillation model predicts a long shower duration and variable period (Rampino and Stothers 1984a). At present, it is unclear if the cratering or extinction data can discriminate between these predictions.

If Planet X's albedo is similar to that of Neptune, then its apparent brightness at 80 AU would be reduced by a factor of $(30/80)^4 = 0.02$ compared to Neptune. Assuming a density similar to that of Neptune, then the surface areas scale as $(m_x/m_N)^{2/3} \approx 0.31$. Thus, with all these assumptions, Planet X would be ≈ 160 times fainter than Neptune. Neptune's apparent magnitude is 8 and so Planet X's magnitude at 80 AU may be ≈ 13.5 which is about one magnitude brighter than Pluto. If it exists, the failure of past optical surveys to identify Planet X may be due to its large ecliptic inclination which suggests a significant present latitude. The best immediate hope of finding Planet X optically is the proposed full sky survey (Shoemaker and Shoemaker, personal communication) which is expected to require about three years to complete. In the infrared the best immediate prospect is in the IRAS (infrared astronomical satellite) data. This survey covered most of the sky and may have recorded a 3 $m_\oplus$ planet within 150 AU in the 100 μm band (Reynolds et al. 1980). These data are currently being searched for unknown bodies in the outer solar system. The parallax of Planet X can easily be measured at IRAS resolutions.

Acknowledgments. We thank E. Shoemaker and R. Wolfe for providing us with the results of their Planet X Monte Carlo calculations prior to publication. We also wish to thank J. Anderson, A. Jackson, IV and R. Reynolds for useful suggestions and assistance. This work was supported in part by a grant from the National Science Foundation.

PART VI
Existence and Stability of a Solar Companion

EVOLUTION OF THE SOLAR SYSTEM IN THE PRESENCE OF A SOLAR COMPANION STAR

PIET HUT
Institute for Advanced Study

A review is presented of the dynamical implications of a companion star in a wide orbit around the Sun, with a semimajor axis of about half a parsec, as proposed recently to explain the periodicity found in mass extinctions on Earth. Three main topics are discussed. First, the full theoretical problem is formulated: the gravitational dynamics of the complicated network of interactions between Sun, planets, comets, field stars and a companion star; and the evolution of the solar system under these combined interactions. A comparison is made with the traditional calculations of the evolution of the Oort comet cloud in the absence of a companion star. Second, four crucial tests of the solar companion theory are addressed, which can be summarized by the following four questions: what is the expected lifetime of such a wide binary system; how stable is the orbital period; will a companion star cause enough comets to impact on Earth to offer an explanation for periodic mass extinctions; and will a companion star not overly deplete the comet reservoir with so many strong perturbations. Detailed numerical orbit calculations are discussed which answer these questions, showing the Nemesis theory to be viable. Third, related research in other fields in astronomy is reviewed briefly. One application of the calculations mentioned above concerns the stability of wide binaries in the solar neighborhood; a similar application uses these binaries as probes to study the character of dark matter in the disk of the Galaxy; in another study the influence of interstellar gas clouds, both atomic and molecular, on cometary orbits as well as on wide binary orbits has been investigated in detail.

I. INTRODUCTION

This chapter is organized as follows. Section I introduces the motivation behind the recent hypothesis of a solar companion star and also gives alternate hypotheses. Section II gives a discussion of the general problem of solar system dynamics with a solar companion star. In Secs. III through VI, four principal questions are posed, and answered, concerning the consistency of the solar companion theory in its role as providing the required modulation in comet arrival times. In Sec. VII some applications outside the solar system are discussed, involving wide binaries, interstellar clouds and dark matter in the galactic disk. The discussion in Sec. VIII contains a critical assessment of the viability of the solar companion theory.

A. A Solar Companion Star?

Most solar-type stars are part of a binary or multiple system (Abt and Levy 1976; Abt 1983). It is therefore a natural question to ask whether the Sun really is exceptional in being a single star. Or could it be that the Sun, too, has a companion star faint enough and at a large enough distance that it may have gone unnoticed so far? The answer is that this is quite possible for orbital semimajor axis values in the range 10,000 to 100,000 AU (van de Kamp 1961; Davidson 1975). Systematic searches for nearby stars rely heavily on those stars having a high proper motion. Therefore, a solar companion star in a wide orbit may easily have escaped detection, simply because it has a negligible proper motion, one or two orders of magnitude lower than that of field stars passing close to the Sun.

Recently it has been suggested that the apparent periodicity in mass extinctions observed in the fossil record (Raup and Sepkoski 1984; Fischer and Arthur 1977) might be driven by an astronomical clock in the form of a distant companion star to the Sun (Davis et al. 1984; Whitmire and Jackson 1984). Each perihelion passage of the companion star would result in an enhanced rate of arrival of comets in the inner planetary system, some of which could collide with the Earth and perturb the atmosphere strongly enough to cause a catastrophic extinction of many, if not most, species (for a short review, cf. Alvarez [1983]; many details can be found in Silver and Schultz [1982]). A similar periodicity has been observed in the cratering rate on Earth (Alvarez and Muller 1984).

Various other chapters (see especially those by Shoemaker and Wolfe, by Tremaine, and by Muller) discuss the evidence for periodicities in extinctions and cratering. This review therefore is limited to a survey of the consequences of a periodicity in both, without trying to evaluate the likelihood that these periodicities are merely a statistical coincidence. In my opinion, the case for a significant periodicity is strong enough to take seriously, and only time will tell which of us have been too enthusiastic and which too cautious in our initial reactions to these new ideas.

1. Various Names for the Solar Companion. Although the name "Nemesis" has been chosen as the favorite description of our theory in the popular press (with "Death Star" a close second), I wish to add a historical note here. It is not widely known that several other names such as "George" and "Kali" were proposed simultaneously with the name "Nemesis" in the preprint version of Davis et al. (1984). Unfortunately, these did not survive the editorial pruning of the published version in *Nature*. The name "George," however, did make it into *Scientific American* ("A star named George," vol. 250, p. 68, 1984). The name "Kali" I find to be the most appropriate, for much the same reasons as given by Gould (1984) who argues for the name "Shiva." Shiva, the Hindu God of destruction, and his consort, the Mother Goddess Kali, share similar traits, but Kali's name "the black one" seems to make her even more appropriate. She generally appears in threatening forms, with a garland of skulls, as the destroyer of life. In her grim aspects she carries other names as well, such as Durga "the inaccessible" and Candi "the fierce." But she also appears as a beautiful young woman, symbolizing life-giving qualities, under the names Parvati, Mahadevi, Sati, Gauri and Annapurna (cf. Basham 1967). These dual aspects seem to cover the implications of mass extinctions better than the association with the Greek Goddess Nemesis. However, in the following I will use the name Nemesis since that name seems to have found the widest recognition at present.

B. Alternative Hypotheses

If significant periodicities in impacts and extinctions are with us to stay, then we have to find an explanation for the unexpected phenomenon of impacts being modulated in time. It seems extremely difficult to produce a periodicity of tens of millions of years in the impact of asteroids; no mechanism is known to operate on that particular time scale which can drastically affect asteroids. Perhaps some caution should be expressed because we do not really know what the effects of long-term resonances in the planetary system are, since the motion of the planets are only beginning to be modeled for such long time scales (cf. chapter by Tremaine). However, it seems *a priori* extremely unlikely that one particular long-term resonance should stand out sufficiently strongly from the others in modulating the number of Earth-crossing asteroids.

Comets are the alternative candidates for producing impacts on Earth, and for them it is easier to find a clockwork mechanism. Three different types of proposals have been made since Raup and Sepkoski's (1984) findings became available. One of these, a galactic modulation of comet arrival times by passages of the Sun through molecular clouds (Rampino and Stothers 1984*a*) is virtually ruled out. Thaddeus and Chanan (1985; see also the summary in their chapter) have shown that the motion of the Sun perpendicular to the galactic plane cannot produce any detectable periodicity in encounters with molecular clouds. The reason is that the distribution of molecular clouds is known to have a scale height comparable with that of the present amplitude of the solar motion perpendicular to the galactic plane. This argument against a galactic explanation of periodic comet showers is the most convincing one because it does not involve any assumptions about the mechanism of comet perturbations by passages near or through molecular clouds.

Another proposal for generating periodic comet showers in the inner planetary system involving an (as yet) unseen companion star was put forward independently by Davis et al. (1984) and Whitmire and Jackson (1984). The strengths and weaknesses of this theory are discussed in the following sections: the likelihood of both the presence and persistence of a solar companion star in its present orbit, and its ability to perform the required modulation of comet arrival times on Earth.

Recently a third mechanism has been proposed to explain periodicities in comet showers, involving a tenth planet (Whitmire and Matese 1985; see also chapter by Matese and Whitmire). This suggestion is interesting as well as original, and its consequences have not as yet been worked out well enough to allow a detailed evaluation, although there seems to be a serious problem to clear out the needed gap in the comet distribution around the tenth planet (cf. Tremaine's chapter).

II. THE GENERAL PROBLEM

A. Dynamics of a Comet Cloud around a Single Star

The existence of a large cloud of more than 10^{11} comets surrounding the Sun, out to distances of ~ 1 pc, was inferred by Oort (1950) from the characteristics of observed long-period comets. This theory has found general acceptance, and is in good agreement also with the present body of comet observations (Marsden et al. 1978) which is much more extensive than the material on which Oort founded his original theory. Not only the present state of the Oort cloud can be understood, but also the formation of such a comet cloud can be modeled in a satisfactory way; the perturbations of the outer planets can have pumped up some of the debris left over after planet formation to the presently observed wide comet orbits (for a review, see, e.g., Weissman 1982a and references therein). However, there still seem to be some problems in deriving a consistent picture for the formation of short-period comets (for a review, see, e.g., Everhart [1982] and references therein).

Most results mentioned above have been obtained under a variety of simplifying assumptions for the interactions between the different players in the cometary drama: the Sun, the planets, the comets, passing stars and interstellar clouds. An accurate and detailed description of the total complicated network of interactions between all these objects is beyond the power of present-day computers. Fortunately, such an exhaustive description is not needed. First, comets are so light that they can be treated as massless test particles, even when interacting with planets. Second, the total number of comets is so large that a statistical treatment is appropriate, for which most of the detailed knowledge of individual orbits is irrelevant.

The statistical description of the evolution of the Oort cloud involves a diffusion of comet orbits, that are perturbed at the inner boundary of the cloud by the planets and at the outer boundary by passing stars and interstellar clouds. Several interesting conclusions have been obtained under rather simplified assumptions for the diffusion coefficients governing comet cloud evolution (cf. Weissman 1982a).

B. **Dynamics of a Comet Cloud around a Wide Binary**

The complexity of the network of interactions mentioned above is considerably increased when we allow a solar companion star to appear on the scene. Now the players interacting with the comets can no longer be neatly divided into locals (planets) and passing foreigners (field stars and interstellar clouds). A new character arises in the form of a periodic visitor (Nemesis) who stirs up the inner part of the comet cloud during every perihelion passage. The addition of this new element complicates the computation of cometary orbit perturbations, because the relatively slow passage of Nemesis through the inner comet cloud can no longer be treated in an impulsive approximation, as is appropriate for rapidly passing field stars.

A more fundamental change of the standard models concerns the formation of the comet cloud in the presence of a companion star. The initial stages of pumping up the orbits of the comets, left over as unused building blocks at the outskirts of the region of planet formation, are unlikely to be altered much by the addition of a companion star. Although the separation between Sun and Nemesis has probably been smaller in the past (see Sec. VI below), it is unlikely that Nemesis ever significantly affected the planetary system. However, during the later phases in which comets were scattered into the present more isotropic distribution (which at least for the outer comet cloud is the most likely one), the presence of a companion star must have drastically altered cometary orbit evolution.

The relative merits of the two pictures sketched above, of comet cloud evolution with and without a solar companion star, cannot be easily evaluated. While the latter has been modeled in considerable detail, little quantitative work has been done with respect to the former type of evolution. Some results have been obtained regarding the question of the stability of planetary orbits in relatively close binary systems (cf. Pendleton and Black 1983 and earlier references therein), but I am not aware of any studies of planetary and cometary systems around members of wide binaries.

It is conceivable that presently unsolved puzzles, such as the dynamical history of short-period comets, will find their solution in a Nemesis-type evolutionary history. Alternatively, a scenario of comet evolution under the influence of a solar companion might turn out to be inconsistent. I expect that more or less plausible explanations can be constructed for the formation and early evolution of comets in either scenario, with or without Nemesis. At present there are still many uncertainties in our understanding of star and planet formation, probably enough to obscure any evidence which could settle the question of the existence of a companion star.

Instead, a different approach will be followed in the remainder of this chapter. No speculations about the formation and early evolution of Sun, comets and companion star will be put forward. More to the point, four crucial questions will be discussed (Secs. III–VI), each of which could have ruled out the Nemesis theory on dynamical grounds, by leading to inconsistencies for the *present* rate of change of parameters for the solar system.

III. QUESTION 1: WHAT IS THE EXPECTED LIFETIME OF A SOLAR COMPANION?

On Earth, the dating of mass extinctions as well as the determination of crater ages is subject to observational uncertainties. In the heavens, too, we should not expect a double-star clock to be perfect, due to the influence of the perturbations by stars and interstellar clouds. The net galactic perturbation can be conveniently divided into two parts: a slowly changing background field and a rapidly fluctuating contribution by individual passing stars and clouds.

A. Slowly Varying Perturbations

Let us first consider the background perturbing forces. These include the tidal field of the Galaxy, which is strongest in the direction perpendicular to the galactic plane, and the Coriolis forces exerted in the solar rest frame which rotates around the center of the Galaxy. These forces limit the types of orbit which are allowed for comets and companion stars in bound orbits to the Sun. The edge of the region of dominance of the Sun within the Galaxy lies around 1 pc, allowing bound orbits with semimajor axes at most not much exceeding 0.5 pc (Smoluchowski and Torbett 1984). The assignment of a value for the semimajor axis of an orbit at the verge of instability is, however, not very well defined because the orbits deviate significantly from Kepler orbits (Hut 1984a; Torbett and Smoluchowski 1984).

The conclusion from the calculations mentioned above is that the present Nemesis orbit is still bound to the Sun. Had the period been significantly longer, e.g., 50×10^6 yr, then Nemesis would have left the Sun before even completing one revolution. For orbits with periods of 26 to 30×10^6 yr, the orbits are more clearly periodic if they are oriented parallel to the galactic plane than if they are perpendicular to the plane (Hut 1984a, Fig. 1). This suggests that Nemesis is somewhat more likely to be found at lower galactic latitudes. However, this should not discourage any search at higher galactic latitudes (where, incidentally, a search is easier). First, the periodicity can still be quite pronounced for hundreds of millions of years at intermediate latitudes (Torbett and Smoluchowski 1984, Fig. 1). Second, the orbit is not frozen in space, but rather tumbles around and deforms continuously on a time scale of hundreds of millions of years; therefore the possibility cannot be excluded that a past orientation at intermediate inclination to the galactic plane has increased to a higher value at present. The conclusion by Torbett and Smoluchowski (1984) that only low inclinations to the galactic plane allow orbits to be formally stable over the age of the solar system does not constrain Nemesis' present orbit, since the initial orbit has most likely been much less wide than the present one, as is discussed below (Sec. III.B.2).

B. Rapidly Varying Perturbations

Encounters with field stars and interstellar clouds cause a slow net outward drift of the orbits of comets and wide binaries. Because these two types of perturbations have a rather different character, they will be treated separately.

1. Encounters with Field Stars. Close passages between a star and either the Sun or Nemesis will continuously change the parameters of the relative orbit of Sun and Nemesis. Most changes will be small, as can be estimated in the impulsive approximation which neglects the motion of Nemesis during the quick passage of a field star. Whenever such a star passes on a random trajectory close to, or even through, the Sun-Nemesis system the damage done is relatively minor and can be simply estimated as follows for encounters closer than the Sun-Nemesis separation. The relative change in the orbital velocity v_{orb} of Nemesis during such a passage is proportional to the length of time during which the perturbation can act. This duration is of order $\Delta v_{orb}/v_{orb} \sim v_{orb}/v_{pass}$, where v_{pass} is the speed of the passing field star.

With a typical field star passing at 40 km s^{-1} and Nemesis moving at 0.1 km s^{-1}, it would take some 400 of these passages to significantly change Nemesis' orbit and unbind it, if all perturbations would add coherently. However, the velocity vector of Nemesis' orbit will of course follow a random walk instead, since the passages are not correlated. In a random walk, the distance traversed scales roughly with the square root of the number of steps. Thus more than 10^5 such passages seem to be required to unbind Nemesis.

With interstellar distances of one or two parsecs, and velocities of a few tens of km s^{-1}, passages of field stars through or near the Sun-Nemesis system occur at a rate of a few times per million years. Thus the naive estimate above leads to a lifetime of order a few tens of billions of years. This is rather an overestimate, for two reasons. First, the contribution of passages close to either the Sun or Nemesis contribute significantly to the overall effect, adding a logarithmic factor to the efficiency of binary disruption. This factor is similar to the logarithmic factor in the standard treatment of stellar orbit diffusion in the computation of relaxation times (Chandrasekhar 1942) and has been discussed by Retterer and King (1982). Bahcall et al. (1985) determined this factor to be ~ 10, modeling the diffusion process through explicit numerical orbit calculations. This correction brings the expected lifetime of Nemesis down to a few billion years.

The second correction results from the fact that Nemesis only needs to be perturbed into an orbit just outside the region of influence of the Sun. Just as in atomic physics, where the ionization potential of an atom is lowered by the interactions with other nearby electrons, also for the ensemble of Nemesis orbits there is a depression of the continuum, due to nearby stars and the galactic tidal field. This again lowers the lifetime, and indeed the final value found by detailed modeling is $\sim 10^9$ yr (Hut 1984*a*). This result is the median lifetime, obtained by numerical orbit calculations where several hundred replicas of the Sun-Nemesis system were followed in time, starting at the present separation, until final dissolution.

In the numerical calculations of Nemesis' fate, each Sun-Nemesis system was continuously bombarded by passing field stars, with impact parameters out to several parsecs. These field stars were drawn randomly from the observed distribution of star densities, represented by ten different stellar mass groups. The velocity of each field star was drawn randomly from a Maxwellian distribution with the observed velocity dispersion for stars of the corresponding mass class (Hut 1984*a*).

As expected from the above crude estimates, each run in which a Sun-Nemesis system was dissolved required on the order of a hundred thousand interactions with passing field stars. Such a large number might seem to indicate that an analytic diffusion treatment would in principle be a good approximation. However, in this particular case analytic approximations cannot be trusted to be accurate to better than a factor of three or so. One problem is the competition between slow diffusion and instantaneous ionization of the orbit (due to a very close encounter, a very heavy mass of a passing field star, a very slowly moving field star, or a combination thereof). Another problem is the addition of complicated galactic tidal and Coriolis forces. Nevertheless, the expected lifetime of a billion years is in reasonable agreement with the Fokker-Planck treatment of the diffusion process of wide binaries given by Retterer and King (1982).

2. Encounters with Interstellar Clouds. The time scale on which interstellar clouds can disrupt wide binaries is of the same order of magnitude as the time scale on which stars dissolve those binaries (this was first pointed out for the specific case of comets around the Sun by Biermann and Lüst [1978] and Biermann [1978]). The process of dissolution is different: a single close passage near a dense molecular cloud can disrupt a wide binary instantly, while little damage is done in the intervening times. Although this simplifies the theoretical description, in practice an accurate estimate of the dissolution time scale is much harder to give for the interstellar cloud case than for the previous stellar case. In the latter case, the distribution of stars in the solar neighborhood is known accurately. In the former case there are significant uncertainties, *both* in the total amount of interstellar matter at the Sun's distance from the center of the Galaxy, and in the internal structure of the interstellar clouds, especially of the molecular clouds which are the densest and most damaging.

Some estimates have resulted in very short disruption times of binaries by passages through molecular clouds. For example, Napier and Staniucha (1982) find that only comets with semimajor axes of < 8000 AU can survive for the age of the solar system. A most direct objection to this result follows from the fact that such short lifetimes would predict the rapid dissolution of wide binaries as well. Instead, observations show a significant abundance of wide binaries, with separations up to 0.1 pc (Bahcall and Soneira 1981; Latham et al. 1984; cf. Bahcall et al. 1985). These observations are in good agreement with our recent analytical estimates (Hut and Tremaine 1985), which yields lifetimes which are an order of magnitude larger than those found by Napier and Staniucha (1982), but only somewhat larger than those found by Bailey (1983a).

An interesting result reported by Hut and Tremaine (1985) is the unexpectedly simple form for the dependence of the lifetime of binaries perturbed by passages of interstellar clouds; independent of the details of the mass spectrum and distribution of cloud radii, the main parameters of importance are the average mass density of the clouds (in the Galaxy at the distance of the Sun from the center of the Galaxy); and the surface density of the clouds and their subclumps. The approach of Hut and Tremaine (1985) differs from previous work in that they introduced a series of six efficiency factors, to facilitate detailed comparisons between different theoretical treatments as well as the input of observational constraints. These six factors correct for the motion of the Sun perpendicular to the galactic disk; its radial motion in an epicyclic orbit; deviations from the impulse approximation; effect of finite-size of gas clouds (not point masses); higher densities of gas clouds in the past; and gravitational focusing of the Sun's orbit in encounters with heavy molecular clouds.

With an estimate of the average density of molecular material in the Galaxy at the distance of the Sun from its center of ~ 0.013 $M_\odot$ pc^{-3} (Dame and Thaddeus, preprint, 1985), and a scaling to a semimajor axis of 0.4 pc, the analysis by Hut and Tremaine (1985) yields a half-life $t_{1/2}^{\rm ISM} \approx 10^9$ yr. Longer half-lives, up to 2×10^9 yr, as well as shorter half-lives, down to $1 \sim 2 \times 10^8$ yr, cannot be excluded given the present uncertainties in both the average density of molecular material in the Galaxy and the degree of subclumping within individual molecular clouds (cf. Hut and Tremaine 1985). The shortest values for the half-life would be uncomfortably small and would require that we live in a special period of time, just before Nemesis will become unbound. The longer values would not pose that problem, and therefore the evidence against Nemesis posed by molecular cloud perturbations remains inconclusive.

In the previous section, the half-life under perturbations by stars (st) and by the galactic field was found to be $t_{1/2}^{\rm st} \approx 10^9$ yr. The combined effects of molecular clouds (ISM), passing stars (st) and tidal galactic fields can be found by adding the rates

$$t_{1/2} = [(t_{1/2}^{ISM})^{-1} + (t_{1/2}^{st})^{-1}]^{-1} \approx 5 \times 10^8 \text{ yr} \qquad (1)$$

subject to the uncertainties mentioned above. Such a short half-life implies that Nemesis probably formed in an orbit with a semimajor axis, say, 2 to 5 times smaller than the present value, and thus with an orbital period of only a few million years. The half-lives of these narrower orbits are comparable to the age of the solar system, and a binary starting in such orbits could easily be close to dissolution at present. The important conclusion of this subsection is that the life expectancy of Nemesis is long enough, so that it does not place us in a very special time, just at the brink of dissolution. Instead, Nemesis can be expected to complete another 20 or 30 orbits in the near future. If Nemesis exists, then the bulk of its destructive work has been done and retirement is approaching, but it can still look forward to a future which is estimated statistically to last $\sim 10\%$ of the duration of its past life span.

Two preprints have recently appeared which address questions similar to the ones discussed above. M. E. Bailey (preprint, 1985) and M. D. Weinberg et al. (preprint, 1985) study the effects of encounters with stars and molecular clouds on comets and on wide binaries, respectively. Bailey reports results similar to those by Hut and Tremaine (1985), although different in some details. Weinberg et al. obtain a somewhat faster disruption rate for Nemesis' orbit than Hut (1984a) found, but conclude that the two results are not inconsistent.

IV. QUESTION 2: HOW STABLE IS THE ORBITAL PERIOD?

With a life expectancy of < 1 Gyr, Nemesis' orbital period cannot have been constant over the geologic period of 250 Myr in which Raup and Sepkoski (1984) found a periodicity in mass extinctions. One might fear that no clear periodic signal would remain observable, and that the Nemesis hypothesis would fall victim to criticism similar to that given by Thaddeus and Chanan (1985) with respect to the galactic oscillation theory.

However, one might also argue that the orbital periodicity will remain recognizable until just before the final disruption of the Sun-Nemesis system. The galactic tidal forces, which will eventually disrupt the system, drop off strongly with decreasing distance. Although they will hide any periodicity in the last few orbits before dissolution, they might not disturb the period very much as long as Nemesis is not yet on the brink of dissolution. However, such a qualitative argument is not very convincing either, and it is not easy to test this conclusion without numerical orbit calculations. Indeed, this problem was not addressed in any detail in the original papers that proposed the Nemesis hypothesis (Davis et al. 1984; Whitmire and Jackson 1984).

Soon afterwards the question was settled satisfactorily: the Nemesis orbit is likely to remain stable over the geologic period of interest of 250 Myr to within 10 to 20%, according to a variety of independent detailed numerical orbit calculations. Hills (1984a) investigated the perturbing effects of passing field stars; Torbett and Smoluchowski (1984) studied the influence of the galactic tidal field on Nemesis orbits of different orientations; and Hut (1984a) combined both effects in the most realistic set of calculations which included field stars from ten different mass groups.

The conclusion of a 10 to 20% drift in orbital period over 250 Myr is valid for the median in an ensemble of the Sun-Nemesis system, but individual replicas can of course show a wide variety. A rare passage of a single (slow, heavy or close) star can cause an instantaneous change in orbital period much larger than 20%. A close encounter with a molecular cloud can similarly cause a sudden large change in period. Both types of effects are unlikely to occur over a 250 Myr period (although the effects of molecular clouds are somewhat uncertain; see Sec. III.B.2).

Another important outcome of the numerical calculations concerns the sensitivity of Nemesis' period stability on orbital orientation. Hut (1984a) and Torbett and Smoluchowski (1984) have shown how the period stability increases with decreasing inclination with respect to the galactic plane. Although this effect again points to a somewhat higher probability of discovering Nemesis at lower galactic latitude, the differences are not large enough to discourage a search at higher galactic latitudes (cf. the discussion in Sec. III.B.2). Figure 1 of Torbett and Smoluchowski (1984) shows clearly how the semimajor axis (and therefore the period) fluctuates only to a small degree until just before the final dissolution, for an orbit with a significant inclination of 40° with respect to the galactic plane.

The most important implication of the present discussion concerns the astronomical explanation for mass extinctions observed on Earth. Even an idealized, completely accurate dating of mass extinctions and crater impacts should not be expected to yield perfectly sharp peaks in the power spectrum of a Fourier analysis of those dates. The present data are barely accurate enough to distinguish periodicities, and therefore cannot determine how stable these periodicities are. It is hoped that the dating of craters and extinctions will improve to such a point that such an observational stability analysis will become feasible.

V. QUESTION 3: DOES A SINGLE PERIHELION PASSAGE OF A SOLAR COMPANION PERTURB ENOUGH COMETS?

The answer given by Davis et al. (1984) is, "Yes, the numbers work out surprisingly well." Their analytic arguments are summarized below (with some detailed improvements added which result in raising their number by a factor of about 3). Further on (Sec. V.B.1), I report more accurate results based on recent numerical orbit calculations.

The estimates below can only determine the relative number of comets taking part in comet showers, and we therefore need to assume a normalization factor. The total number of comets in the Oort cloud was at first estimated conservatively by Oort (1950) to be 2×10^{11}. Oort added a warning that the total number could be significantly larger because of our lack of knowledge about the population of cometary orbits with semimajor axes $a < 20,000$ AU. These inner comets are only perturbed occasionally when a single star happens to pass close to the Sun, and are therefore presently not observable. If the comets formed as debris left over outside the planet-forming region in the protosolar nebula, their density would be expected to drop significantly with increasing semimajor axis. This would imply that the observed comets form only the tip of the iceberg; the majority of the comets may simply have gone unnoticed.

Subsequent detailed improvements in theory as well as observations have resulted in estimates which are one or two orders of magnitude higher. Each estimate of the total number of comets is of course sensitive to the cut-off in cometary brightness, below which the observation of comets cannot be expected to have been complete. How well such a cut-off correlates with a cut-off in cometary mass is not clear, because of substantial uncertainties in cometary albedos.

Hills (1981) estimates the total number of comets to be a few times 10^{13}, most of which reside in the (presently unobserved) inner Oort cloud. A detailed study by Weissman (1983b) yields an estimate of 1.4×10^{12} comets brighter than $H_{10} = 11$ (H_{10} is defined as the brightness of a comet with coma corrected to unit heliocentric and geocentric distance), and a total mass for those comets of roughly 2 $M_\oplus$. A detailed discussion of these and many other estimates for the total number of comets, made since the Oort cloud was first proposed, is given by Weissman (1985a) including an improved mass estimate of about 5 $M_\oplus$ for the comet cloud.

Davis et al. (1984) illustrated their estimates with a total number of comets taken to be 10^{13}, although we are about to see that even a conservative estimate of 10^{12} suffices to produce comet bombardments that are heavy enough to explain periodic mass extinctions.

A. An Analytic Estimate

Before discussing recent numerical calculations, I review the Davis et al. (1984) analytical estimates of the number of comets expected to hit the Earth during a comet shower induced by a passing companion star. Let us start with the simplest model for the distribution of comet orbits in the inner Oort cloud, in which all comets have the same semimajor axis $a = 10,000$ AU. I assume the simplest distribution of positions and velocities of comets: a homogeneous distribution of points in phase space, which implies an isotropic distribution of velocities and transforms to a distribution function

$$f(e) = 2e \tag{2}$$

for the eccentricity e (Heggie 1975).

A preference for high eccentricities follows naturally in a random distribution of positions and velocities in a set of Kepler orbits. In other words, there is proportionally more phase space volume available for higher eccentricity orbits, a fact which can be derived in two ways. The hard way is to determine the full phase space volume by explicit integration over the other five classical orbital elements of the full family of elliptic Kepler orbits. A much simpler derivation starts with a quantum-mechanical description of a double star, similar to that of a hydrogen atom. In the classical limit (i.e., highly excited states) the eccentricity of a Kepler orbit is related to the quantum number l by

$$1 - e^2 \propto L^2 \propto l^2. \tag{3}$$

Here L denotes the orbital angular momentum, and l and m the usual quantum numbers for the total and projected orbital angular momentum. For every value of l, the hydrogen atom admits m values in the range $m = -l, \ldots, +l$ and therefore the distribution function for l takes the form

$$g(l)dl \propto l \, dl \tag{4}$$

in the continuum limit. Substituting Eq. (3) in Eq. (4) gives Eq. (2). See Hut (1985) for further use of the correspondence principle between atoms and double stars.

Only those comets that cross the Earth's orbit can cause impacts and lead to mass extinctions. This requires the perihelion distance q to obey

$$q = a(1-e) < 1 \text{ AU}. \tag{5}$$

Starting with a completely undisturbed, isotropic distribution of comets, we thus find that a fraction f_{cr} of all comets, with

$$f_{cr} = \int_{(1-1/a)}^{1} f(e)de = \frac{2}{a} - \frac{1}{a^2} \approx \frac{2}{a} = 2 \times 10^{-4} \tag{6}$$

will cross the Earth's orbit within one cometary orbital time (10^6 yr for the present choice of orbits).

With a total number of comets of order 10^{13}, we obtain 2×10^9 Earth-crossing comets. In a steady-state situation, most of these comets would be quickly removed from the solar system by the perturbations of the planets, mainly Jupiter and Saturn, after a few cometary orbital periods. The reason is that Jupiter strongly perturbs the orbital energy of any comet crossing its orbit. The sign and magnitude of the energy perturbation depends sensitively on the orbital parameters of the comet, as well as on the position of Jupiter in its orbit during the perihelion passage of the comet. To study the average behavior of Jupiter's (and Saturn's) perturbations, one can first determine the average and rms of the energy perturbations for a large ensemble of comet orbits; and later draw random values for the perturbations from a Gaussian distribution with the same moments, to speed up the evolutionary calculations in a Monte Carlo approximation (Weissman 1982a). The typical Jovian perturbation in the orbital energy of a comet is comparable to the binding energy of a comet with a semimajor axis of about 2000 AU. All comets with a semimajor axis much larger than this value have about a fifty-fifty chance of being ejected from the solar system altogether in a hyperbolic orbit, and a comparable chance of being captured into a much smaller orbit (from which they disappear in a million years typically, either by disintegration, capture in a short-period orbit, or subsequent scattering into a hyperbolic orbit). Comets of interest for the present discussion, having a semimajor axis of order 10,000 AU (see below), satisfy the above condition, and therefore are effectively removed from the Oort comet cloud soon after they are perturbed into an orbit crossing the orbits of Jupiter or Saturn.

At any point in configuration space, the unfilled region in velocity space (directionality space) resulting from this depletion by Jupiter and Saturn is sometimes called a loss cone (cf. Hills 1981), in analogy with plasma physics, where a mirror machine has a cone-shaped region in velocity space from which the plasma can escape. The term loss cone is convenient, but not altogether accurate, in the gravitational case; gravitational focusing by the Sun causes a depletion of comets within a region bounded by a hyperboloid (Cohn and Kulsrud 1978). In fact, the hyperboloid has a shape somewhat in between that of a cone and that of a cylinder, which is the term used by Oort (1950) to describe the region of planetary depletion of comet orbits. However, the term loss-hyperboloid-of-one-sheet, although more accurate, is less euphonious, so we stick to the term loss-cone.

If Nemesis is heavy enough and approaches the Sun closely enough, it will perturb the comets sufficiently to fill the entire loss cone. This will result in a comet bombardment involving billions of comets, as shown above, arriving over a time interval of a million years. This will cause the appearance of more than one new Earth-crossing comet per day, and a fantastic interplay of comet tails lighting up the night sky. Davis et al. showed that in a simple impulsive approximation a Nemesis with a mass of 0.05 to 0.1 $M_\odot$ could indeed fill the loss cone in a single perihelion passage, even for typical eccentricity values of $e = 0.7$ (the median value, as well as the rms value in an isotropic distribution; see Eq. 2).

The geometric cross section of the Earth is only 1.8×10^{-9} of that of the Earth's orbit. Gravitational focusing will enhance the cross section a little, with a correction factor (cf. Wood 1961)

$$f_g = 1 + (v_e/v_{rel})^2 \tag{7}$$

where v_e is the escape velocity from the Earth, and v_{rel} is the relative velocity in an encounter between a comet and the Earth. Typical encounter velocities of order 20 to 70 km s^{-1} result in $f_g \approx 1.03 \sim 1.3$, with a proper average $<f_g> \approx 1.1$. A comet crossing the Earth's orbit, i.e., approaching the Sun to within 1 AU, will have two chances to hit the Earth, once on its way in and once on its way out. The total effective cross section for a collision with the Earth, during a single perihelion passage of a comet, is thus $2 \times 1.1 \times 1.8 \times 10^{-9} = 4 \times 10^{-9}$. Therefore, at the first arrival of the comets in the bombardment following a perihelion passage of a companion star, the expectation value for the number of direct hits on the Earth's surface is about $2 \times 10^9 \times 4 \times 10^{-9} = 8$.

The total number of impacts on Earth will be significantly higher than the above estimate. Within 2 Myr after the start of the bombardment, a small fraction of the newly incoming comets will have been captured into smaller orbits, by the perturbations of (mainly) Jupiter and Saturn. These comets repeatedly visit the planetary system, until they are either ejected on a hyperbolic orbit by further planetary perturbations or disintegrate by a close passage near the Sun. Although only a small fraction of newly incoming comets are trapped into frequently returning orbits, the cumulative probability that they impact on the Earth is significantly higher than the impact probability of all of the first arrivals combined. Davis et al. took this repeated-planetary-scattering enhancement factor to be $f_{pl} \approx 4$, following Hills (1981). More detailed calculations give preliminary values which are at least twice as high. For the present discussion we therefore adopt $f_{pl} \approx 8$ to 10.

This multiple scattering correction brings the number of impacts during and soon after a companion-induced comet bombardment to about 70. A more conservative assumption for the total number of comets to be of order 10^{12} would have resulted in an expected number of 7 impacts on Earth per comet bombardment. Both estimates imply a very high probability for at least one impact to occur (for a Poisson distribution with an average value of 70 or 7), and also a high probability for several impacts to occur over a period of order 2 Myr.

It is important that the above simple estimates provide the right order of magnitude for the number of comets hitting the Earth during a single perihelion passage of a companion star; no extra assumptions or alterations of the standard Oort comet cloud model are needed. Either we have here an unfortunate numerical coincidence, where extremely large numbers cancel each other fortuitously to result in a small but nonzero number of comets predicted to impact on Earth, or we have found a valid hint for the existence of a companion star. We think the latter is more likely, and it was exactly this simple calculation which convinced us of the viability of our theory (viability of, not evidence for our theory, because closely passing stars may also fill the loss cone, and produce similar comet showers, cf. Hills [1981]; the evidence hinges on the reality of the *periodicity* in mass extinctions and cratering, which of course cannot be produced by randomly passing stars).

However, Davis et al. (1984) were well aware of one weak point in this original estimate. They determined the perturbation of a companion star on a cometary orbit using an impulsive approximation, strictly valid only for an instantaneous perturbation. In fact, a typical comet in the inner Oort cloud will describe a large fraction of an orbital revolution during the perihelion passage of the companion, which invalidates the assumptions underlying their calculation. Unfortunately, an accurate determination of the perturbation cannot be derived without lengthy numerical orbit calculations, and therefore they decided to stick with their simple estimate, arguing that at least the order of magnitude of the result was unlikely to be affected by the shortcomings in the derivation.

B. Numerical Orbit Calculations

To settle the remaining doubts, referred to above, concerning the validity of the analytic estimates, I have recently embarked on an extensive set of orbit calculations for a variety of orbital parameters for both the companion and a typical comet. This work is still in progress, but the statistics based on preliminary results are already accurate enough to offer significant improvements over the original estimate (see above). I summarize these new results below.

1. Average Eccentricity Perturbation. Let us focus on a companion orbit with a semimajor axis of 0.4 pc and an eccentricity of 0.7 (the rms value; see Eq. 2). For these values of the perturbing orbit, the average change in eccentricity $|\Delta e|$ of a comet orbit with initial eccentricity e and semimajor axis a, during a single perihelion passage of the companion, is found from numerical orbit calculations to be

$$|\Delta e| = (0.0058 \pm 0.0008)\left(\frac{1-e}{0.01}\right)^{1/2}\left(\frac{a}{10{,}000 \text{ AU}}\right)^{3/2}\left(\frac{m}{0.1 \text{ M}_\odot}\right). \quad (8)$$

Here m is the mass of the companion star, and the indicated uncertainty is an estimate of the 1σ error. The functional form of Eq. (8) is a slightly simplified version of an equation which has been derived analytically by Heggie (1975, Eq. 5.66) for a similar case, involving three-body scattering. In Heggie's case the orbit of the perturber was hyperbolic rather than elliptic as in the case of a solar companion star, but in both cases the orbits are sufficiently close to being parabolic that the same approximations suffice.

Heggie's functional dependence on eccentricity and semimajor axis of the comet, as well as on the mass of the perturber, are accurately reproduced by my numerical orbit calculations. As is often the case with analytic estimates, the best fit to the numerical data gives a slightly different coefficient, with some weak dependence on the parameter domain of interest. In the present case, and for an a and e ranging around the fiducial values given above, the coefficient in Eq. (8) is within 20% of the corresponding value which can be derived from Heggie's (1975, Eqs. 5.31, 5.33 and 5.66) equations after averaging his expressions over all encounter orientations. This is yet another tribute to Heggie's remarkable analytical achievements (for other comparisons between his analytic estimates and experimental results from detailed numerical orbit calculations, cf. Hut and Bahcall [1983], Hut [1983c, 1984b, 1985]). In the notation of this chapter, and with the coefficient derived from Eq. (8), Heggie's result reads

$$|\Delta e| = (1.7 \pm 0.2) e (1 - e^2)^{1/2} (1 - e_c)^{-3/2} \left(\frac{a}{a_c}\right)^{3/2} \left[\frac{m}{(M_\odot(M_\odot + 1))^{1/2}}\right]$$

(8a)

where a_c and e_c are the semimajor axis and the eccentricity of the orbit of the solar companion star, respectively.

Hills (1984a) has reported orbit calculations which approximately agree with Eq. (8) for some parameter values, but differ for other values. The largest deviations between Hills' and my results concern the dependence of the perturbation in cometary orbit eccentricity on the mass of the companion star. Hills (1984a, Eq. 2) finds a quadratic, rather than linear, dependence on companion star mass, in contradiction with Heggie's analytic estimates. To test Hills' result, I have performed orbit calculations for two extreme choices of companion star masses: 0.1 $M_\odot$ and 0.01 $M_\odot$. I found the average Δe to be (9.4±0.9) times stronger for the first choice than for the second, in excellent accord with Eq. (8) and Heggie's (1975, Eq. 5.66) equation, but in disagreement with Hills' (1984a, Eq. 2).

Starting from the average change in eccentricity Eq. (8), we can estimate which fraction of the total comet population will be perturbed into an orbit that can cause an impact on Earth. First we determine the minimum initial eccentricity e_{min} required for a comet to be perturbed into an Earth-crossing orbit, under a typical perturbation. Second, we determine the fraction of comets which start out with an eccentricity higher than e_{min}, taking into account that we start with an empty loss cone, and a corresponding maximum value $e_{max} < 1$. Not all of these perturbed comets will become Earth-crossers, however; they will be spread out over a range of perihelion distances q, and only those with $q < 1$ AU are actually able to hit the Earth. The determination of this last fraction constitutes the third and last step.

The estimate sketched above cannot be expected to be much more accurate than to within a factor of two, since Eq. (8) contains no information about the detailed spectrum of the perturbations. A complete determination of this spectrum would involve a much more time-consuming investigation of the dependence of the perturbations on all cometary orbit parameters. While the present work, based only on the average behavior of the perturbations, can thus be improved upon, it already constitutes a major advance over the previous estimates by Davis et al. (1984).

2. Eccentricity Range for Earth-crossing Comets. For the first step mentioned above we have to isolate those comet orbits that can be perturbed strongly enough to intersect the Earth's orbit. A simple criterion is the requirement that the initial eccentricity be close enough to unity $(1-e) \leq |\Delta e|$, which guarantees that the comet can even hit the Sun, and therefore certainly can reach the Earth. With Eq. (8), this results in the condition $e > e_{\min}$, with the minimum eccentricity $e_{\min}$ given by

$$1 - e_{\min} = (0.0033 \pm 0.0009)\left(\frac{a}{10{,}000 \text{ AU}}\right)^3 \left(\frac{m}{0.1 \text{ M}_\odot}\right)^2. \tag{9}$$

The second step involves the determination of an upper limit on the eccentricity of surviving comet orbits, posed by the requirement that they initially lie outside the loss cone. By far the strongest planetary perturbations are those by Jupiter and Saturn, while the perturbations by Uranus and Neptune are too small to clean out a region in velocity space (cf. Hills 1981; Weissman 1982a and references therein). For the edge of the loss cone, we adopt a perihelion distance of 12 AU, as a compromise between the radius of Saturn's orbit of 9.6 AU and the distance 1.5 times larger beyond which Saturn's influence becomes negligible (Everhart 1968).

The maximum eccentricity $e_{\max}$ allowed for a comet with semimajor axis a under the condition that the initial perihelion distance obeys

$$q = a(1-e) > 12 \text{ AU} \tag{10}$$

is $e < e_{\max}$, with

$$1 - e_{\max} = 0.0012\left(\frac{10{,}000 \text{ AU}}{a}\right) \tag{11}$$

The fraction of comets in the Oort cloud, at a given semimajor axis a, with eccentricity $e_{\min} < e < e_{\max}$, follows from Eq. (1) as

$$\int_{e_{\min}}^{e_{\max}} 2e \, de = e_{\max}^2 - e_{\min}^2 \approx 2(e_{\max} - e_{\min}) = 0.0024\left(\frac{10{,}000 \text{ AU}}{a}\right) \times$$

$$\times \left[\left(\frac{m}{0.06\ M_\odot}\right)^2 \left(\frac{a}{10{,}000\ AU}\right)^4 - 1 \right]. \quad (12)$$

It is interesting to note the mass scale of 0.06 $M_\odot$ appearing in this expression. A companion star with a mass smaller than that is ineffective in filling the loss cone with comets which have a semimajor axis of 10,000 AU.

3. *Relative Probability for Earth-crossing, in the Eccentricity Range.* The condition $e_{min} < e < e_{max}$ is necessary, but not sufficient for a comet to be perturbed into an Earth-crossing orbit; it merely implies that the perturbations are large enough to allow such an outcome. Many of the comets satisfying the above condition will have their perihelion perturbed from one position near, but outside, the Earth's orbit to a very different position, also outside the Earth's orbit. In general, only a small fraction of those selected by Eq. (11) will actually approach the Sun to within 1 AU. To determine this smaller fraction, we first have to determine the spread in perihelion distance q of comets implied by the average change in eccentricity, given by Eq. (8):

$$\Delta q = a\Delta e = 580\sqrt{(1-e)}\left(\frac{a}{10{,}000\ AU}\right)^{5/2}\left(\frac{m}{0.1\ M_\odot}\right) AU. \quad (13)$$

Within this range Δq, any q value is equally likely to occur. This is a consequence of gravitational focusing by the Sun of near-parabolic cometary orbits (cf. Weissman 1980a). We thus arrive at an expression for the fraction of those comets which *can* approach the Sun within 1 AU and those which actually *will* approach the Sun this closely, given by

$$\frac{1\ AU}{\Delta q} = 0.0017(1-e)^{-1/2}\left(\frac{10{,}000\ AU}{a}\right)^{5/2}\left(\frac{0.1\ M_\odot}{m}\right) AU. \quad (14)$$

Equation (14) is valid for a fixed value of e. To determine the overall effect of this correction, we have to average it over the allowed interval $e_{min} < e < e_{max}$. Note that Eq. (14) is valid as long as $\Delta q > 1$ AU. This condition is automatically fulfilled whenever $e < e_{max}$.

4. *Percentage of Earth-crossing Comets in a Comet Shower.* We can now determine the final fraction $f_{cr}(a,m)$, within a group of comets which all have the same initial semimajor axis a, and which will become Earth crossers after a single perihelion passage of a companion star with mass m. For every value of the cometary eccentricity, we have to take into account the correction

given by Eq. (14). This modifies the derivation leading to Eq. (12), and we get instead, using Eqs. (9, 11, 14):

$$f_{cr}(a,m) = \int_{e_{min}}^{e_{max}} 2e \frac{1 \text{ AU}}{\Delta q} de \approx 2 \int_{e_{min}}^{e_{max}} \frac{1 \text{ AU}}{\Delta q} de = 4 \times 10^{-4} \left(\frac{10{,}000 \text{ AU}}{a} \right)$$
$$\times \left[1 - \left(\frac{0.06 \text{ M}_\odot}{m} \right) \left(\frac{10{,}000 \text{ AU}}{a} \right)^2 \right]. \quad (15)$$

This expression is only valid when the right-hand side is positive; a negative value would indicate a perturbation which is too small to fill the loss cone, leading to a small number of impacts on Earth (zero impacts in our approximation).

The fractional error in the final expression above is unlikely to exceed a factor two, as discussed in the previous subsections. Most of the error stems from the fact that we have approximated the distribution of changes in eccentricity by a step function. However, the result above is considerably more accurate than the extrapolation of a simple approximation beyond its realm of application as was done by Davis et al. (1984). A comparison with Eq. (6) shows that the analytic estimate turns out still to be close to the numerical determination above, as so often happens when one (moderately) exceeds the range of validity of an analytic approximation—although even a relatively close agreement is never guaranteed *a priori*.

Another significant improvement of Eq. (15) over the previous arguments given by Davis et al. (1984) concerns the lower limit on the mass of the companion star. They argued that the loss cone will not be completely filled if the mass of the companion is much smaller than $0.1 \text{ M}_\odot$, in which case the Earth will be relatively safe in the eye of the comet storm, with most of the comets falling in at the edge of the loss cone, at perihelion distances larger than 1 AU. This argument has now been made more quantitative in the above analysis, and we see that a mass smaller than $\sim 0.06 \text{ M}_\odot$ does not suffice to fill the loss cone for comets with initial semimajor axes of $\leq 10{,}000$ AU.

We are now in a position to give a more accurate answer to the question "Does a single perihelion passage of a solar companion perturb enough comets?"—the remaining uncertainty being mainly the unknown distribution of comets. Let us take 10,000 AU as a typical value for the semimajor axis of a comet orbit in the inner Oort cloud, and let us assume the total number of comets to be 10^{13}, following Davis et al. (1984); see the beginning of Sec. V for a discussion about the uncertainty in this number. With a companion star mass of $0.1 \text{ M}_\odot$ we find from Eq. (15) that a perihelion passage of a companion star produces a total of 1.6×10^9 new Earth-crossing comets. This result is only slightly smaller than the estimate of 2×10^9 given by Davis et al. (1984), and indeed constitutes a deviation of less than an order of magnitude, as conjectured by them.

With a companion star mass of 0.06 $M_\odot$, Eq. (15) gives zero impacts, a consequence of the square approximation made for the representation of the distribution of perturbations in eccentricity. A more careful analysis would instead show a finite tail in the fraction of Earth crossers well below 0.06 $M_\odot$. However, the efficiency factor in this tail, which will replace the factor in square brackets in Eq. (15), will quickly drop with decreasing m, probably by an order of magnitude for m values lower than about 0.04 $M_\odot$. This is likely to result in too small a number of cometary impacts to explain mass extinctions on Earth. A more accurate determination of the small-m tail of Eq. (15) would be useful, but would require a substantially more detailed investigation than the present one.

However, it is not clear whether a significant improvement over Eq. (15) would improve our predictions of the number of cometary impacts on Earth during a comet shower. The large uncertainties in the total number of comets, as well as in their distribution with respect to semimajor axis, introduce an uncertainty which in most cases of interest is much larger than the remaining uncertainties in Eq. (15). In summary, Davis et al.'s (1984) conclusion is confirmed: for a companion star mass of 0.1 $M_\odot$, comet showers indeed have the required strength; for $m = 0.05$ $M_\odot$ comet showers become much less effective; and for significantly smaller values the comet showers will be orders of magnitude weaker.

VI. QUESTION 4: DO REPEATED PERIHELION PASSAGES OF A SOLAR COMPANION PERTURB TOO MANY COMETS?

Having seen that a single passage of a companion star will perturb enough comets to cause several impacts on the Earth, the next question is: are there not too many comets being perturbed? Let us first analyze the long-term effects of the companion if its orbit would have remained unchanged over the lifetime of the solar system; afterwards we will correct for the orbital evolution which will increase the damage done to the Oort cloud by the companion.

With the present orbital period of 26×10^6 yr, and an age for the solar system of 4.5×10^9 yr, a first estimate implies 173 perihelion passages. During each passage most of the comets crossing Saturn's orbit will be lost from the Oort comet cloud; some will be captured in short-period orbits, while others will escape from the solar system in hyperbolic orbits (see the discussion following Eq. 6). An upper limit for the number of comets lost in the loss cone per perihelion passage is given by the assumptions that the loss cone is completely filled, with the same density in phase space as the regions in velocity space far outside the loss cone. In this case we can use Eq. (11) to find the fraction of comets lost per companion passage, as

$$\int_{e_{max}}^{1} 2e \, de = 1 - e_{max}^2 \approx 2(1 - e_{max}) = 0.0024\left(\frac{10{,}000 \text{ AU}}{\alpha}\right). \quad (16)$$

For the fiducial value $a = 10,000$ AU, the cumulative effects of 173 perihelion passages simply gives a survival fraction of the total number of comets of $\exp(-173 \times 0.0024) = 0.66$. Thus we find that at most one third of all comets will be lost down the loss cone, during the past history of the solar system, for the simplest case in which the companion star has an unchanging orbit. This fraction will become significantly larger when we take into account the orbital evolution of the companion.

In the previous section we have seen that the companion probably has formed in an orbit significantly less wide than the present one. This implies a shorter initial orbital period and therefore a larger number of perihelion passages. Moreover, each perihelion passage is likely to generate a stronger comet bombardment because the loss cone is wider at distances closer to the Sun. The combined effect may substantially increase the total number of comets lost. The increase in the fraction could be as large as an order of magnitude, over and above the simple estimate above. In this case, a linear estimate would break down, and the number of comets required initially would indeed be significantly higher than the number of comets presently left over in the comet cloud.

A number of remarks are in order. First, the original comet cloud could easily have been one or more orders of magnitude more populous than the present one. Even a total mass of 0.01 $M_\odot$ for the comet cloud cannot yet be excluded observationally (Whipple 1975; Hills 1981; Bailey 1983d; but cf. Sekanina 1976). With a typical cometary mass of 10^{15} to 10^{16}g, the upper limit translates into 10^{15} to 10^{16} comets, two or three orders of magnitude larger than the number used here as a more realistic estimate for the inner Oort cloud.

Second, we have no idea *how* close to the Sun the solar companion was formed. We can only compute the average rate of orbital growth induced by passing stars and interstellar clouds, but the variations between individual realizations of this type of random walk are enormous (Hut 1984a). An initial period of 26×10^6 yr is quite unlikely, though. It is much more likely that the initial orbital period of the companion lay somewhere in the range 1 to 5×10^6 yr. However, since we are postulating the existence of only one companion, a statistical estimate can be misleading, and we cannot definitely exclude larger values. Therefore, we have to keep in mind that the total amount of damage done to the comet cloud could have been significantly less than the average estimate for an ensemble of solar systems.

Third, the previous analysis has completely neglected the evolutionary history of the comets. If the companion star actually exists, it must have played an important role in the process of shaping the present Oort cloud out of planetesimals left over after the formation of the planetary system.

VII. APPLICATIONS OUTSIDE THE SOLAR SYSTEM

The research concerning the dynamics of a companion star in a wide orbit around the Sun has stimulated research in related fields, concerning perturbations of comets and wide binaries in the Galaxy (these two being very similar dynamically, differing only in the fact that a comet can be treated as a test particle whereas the stars in a wide binary can have comparable masses). We review briefly these and related recent papers below.

A. Stability of Wide Binaries

The detailed calculations by Hut (1984a) concerning the stability of a solar companion star can be readily applied to wide binaries in the Galaxy as well. They confirm the picture of a gradual diffusion towards larger orbits as described in a Fokker-Planck treatment by Retterer and King (1982). We have to conclude that wide binaries with a semimajor axis > 0.1 pc are destroyed by the continual perturbing influences of passing field stars on time scales significantly shorter than the age of the Galaxy. This is in good agreement with observations, which show a clear cut-off at a projected separation of 0.1 pc (Bahcall and Soneira 1981; Latham et al. 1984).

B. Limits on the Constituents of Dark Matter in the Galactic Disk

Two corollaries follow from the agreement between observations and the theory of the rate of decay of wide binaries, described in the previous subsection. The first concerns the nature of unseen matter in the galactic disk in the solar neighborhood, which has a density of ~ 0.1 $M_\odot$ pc^{-3} (Oort 1932; Bahcall 1984b), comparable to that of observed stars and interstellar matter combined. The existence of this matter has been inferred indirectly from their gravitational attraction which influences the observed distribution of stars perpendicular to the galactic plane. Although no direct observations of this material exist, the unseen disk objects cannot, on average, be heavier than ~ 2 $M_\odot$, since they would have increased the destruction of wide binaries beyond what is observed (Bahcall et al. 1985). This conclusion was established accurately on the basis of hundreds of dynamical calculations, each following the dissolution of a wide binary, under the combined influence of tens of millions of encounters with dark objects.

C. Limits on the Distribution of Interstellar Clouds in the Solar Neighborhood

The second corollary concerns the influence of interstellar clouds on wide binaries and comets. Since stellar encounters can already explain the existence of the observed cut-off in the distribution of separations of wide binaries (Bahcall et al. 1985), it is clear that molecular clouds cannot be much more damaging than passing stars. Similarly, it is clear that molecular clouds cannot significantly have depleted the observed Oort comet cloud, with typical values for the semimajor axis of a cometary orbit of $\sim 25{,}000$ AU (Marsden et al. 1978), comparable to that of the widest of the observed wide binaries (Latham et al. 1984). A detailed analysis, and a comparison with earlier results by other authors reaching different conclusions is given by Hut and

Tremaine (1985). Related preprints have recently appeared by M. E. Bailey (preprint, 1985) and M. D. Weinberg et al. (preprint, 1985), as discussed in Sec. III.B.2.

VIII. DISCUSSION

What are the odds that the Nemesis hypothesis will prove to be correct? We have heard a variety of opinions, ranging from extremely unlikely to rather likely. Most of the arguments against the theory fall into one of three categories, arguments from: (1) those who prefer alternative theories involving the Sun's galactic motion; (2) those who argue that the Nemesis hypothesis leads to contradictions; and (3) those who feel that the *a priori* likelihood for the Sun to have a companion star with the required properties is rather small.

The first argument can be dismissed, since Thaddeus and Chanan (1985) have shown that galactic modulation is far too weak to produce an observable modulation in passages between the Sun and interstellar clouds. A last-ditch attempt to salvage the galactic modulation theory by proposing a very recent large change in the solar velocity is artificial and implausible; also it does not remove the additional objection that strong perturbations by interstellar clouds are too infrequent (Hut and Tremaine 1985).

The second argument has been countered in detail in Secs. III-VI above. An important conclusion is that Nemesis cannot be expected to remain near its present orbit for more than 10^9 yr; the original orbit probably had a period of 1 to 5 Myr. This raises the question of whether it will be possible to find a consistent scenario for the formation of the solar system, including Nemesis and comets. For example, will Nemesis cause too much cratering in the planetary system, due to intense early comet showers?

Interesting though such questions are, they are unlikely to provide any clear criterion for deciding whether or not the Nemesis hypothesis is viable. After all, for many *observed* objects we have not yet found a unique, quantitative, consistent description of their formation and early evolution, and comets are no exception. Attempts to rule out objects by lack of a consistent theory of their early history are therefore ill-founded.

This does not mean that we cannot construct scenarios to counter, e.g., the above objection. For example, we might conjecture that the early heavy bombardment of the Moon might be connected to a specially close passage of Nemesis, when still in a tight orbit around the Sun. However, the approach taken in this chapter is a less flamboyant one; I have set out to test the viability of the Nemesis hypothesis by formulating four crucial questions, each of which could rule out the Nemesis hypothesis on dynamical grounds. These questions all concern the *present* rate of change of parameters for the solar system, and are discussed in detail in Secs. III through VI. The answers found there are encouraging; the Nemesis theory has passed all four crucial tests unscathed.

With the first two arguments dismissed, the plausibility of the Nemesis hypothesis hinges on the third, i.e., the likelihood of the Sun having a companion with exactly the choice of orbital parameters needed. Questions about *a priori* probabilities are much more difficult to discuss than those concerning the first two points. Here "plausibility" or "implausibility" are often in the eye of the beholder.

As an example, Shoemaker and Wolfe and Tremaine in their chapters have argued that a companion star with a semimajor axis of 0.4 pc is a rather extreme postulate, since all observed double stars have smaller separations. In this chapter it is assumed that the original separation between Sun and Nemesis most likely was ~ 0.1 pc, as discussed in Sec. VI, which is not an unusual separation for a double star. That few of those have yet been observed is caused solely by the difficulty to detect such wide pairs against the confusion of background stars (Bahcall and Soneira 1981; Latham et al. 1984).

Actually arguments such as the above hardly touch the question of the viability of the Nemesis hypothesis. The final answer has to come from a direct detection of a companion star; or lack thereof after a detailed search down to objects of ~ 0.05 $M_\odot$. In the meantime, the largest uncertainty seems to be the question of whether the observed periodicity in mass extinctions and cratering are not just statistical coincidences. We have seen an interesting mix of skeptics (see chapter by Tremaine) and optimists (such as the present author) with regards the significance of the periodicities found. Clearly, a more detailed set of data concerning mass extinctions and cratering is needed to settle this question.

In conclusion, it seems that an agnostic attitude towards Nemesis is the most reasonable, pending further evidence of periodic extinctions and periodic cratering. Should such evidence be presented, it would strengthen the solar companion hypothesis.

Acknowledgments. I happily acknowledge interesting discussions with many colleagues, especially L. W. Alvarez, W. Alvarez, J. N. Bahcall, M. E. Bailey, L. Blitz, M. Davis, F. J. Dyson, I. R. King, R. A. Muller, E. M. Shoemaker, A. A. Stark, P. Thaddeus, A. Toomre, S. D. Tremaine and P. R. Weissman. This work was supported in part by a grant from the National Science Foundation.

MASS EXTINCTIONS, CRATER AGES AND COMET SHOWERS

EUGENE M. SHOEMAKER
and
RUTH F. WOLFE
U.S. Geological Survey

Six strong mass extinctions have occurred in the last ~ 250 Myr, but only three of these are accurately dated. The apparent best-fit period is 31 Myr. If mass extinctions are actually randomly distributed in time, there is about a 10% probability that the two time intervals separating the three well-dated strong extinctions would be as nearly equal as observed. The formation of large (≥ 5 km diameter) impact craters in the last 250 Myr also appears to be periodic. The period and phase of the cycles yielding the best fit to the crater ages match fairly closely the best-fit cycle obtained from strong extinctions. This apparent periodicity may also be due to chance. Sharp pulses of impact events at ~ 1 Ma and ~ 35 Ma are indicated by strewn fields of impact microspherules. These pulses coincide approximately with the last two strongest peaks in the crater-age distribution and rather precisely with two mass extinctions. The pulses are best explained by mild comet showers. Various astronomical mechanisms that have been invoked to explain periodic comet showers either are improbable or cause only weakly periodic modulation of the comet flux. The mild comet showers that appear to be recorded in the Earth's impact history probably have been produced by the nearly random close passage of stars through the Sun's comet cloud.

A now famous report by Raup and Sepkoski (1984) on the apparent periodicity of mass extinctions of families of organisms in the last 250 Ma has led to the suggestion of various astrophysical mechanisms that might produce periodic or quasi-periodic extinction of life on Earth (Davis et al. 1984; Rampino and Stothers 1984*a*; Whitmire and Jackson 1984; Schwartz and James

1984; Whitmire and Matese 1985). Most attention has been focused on the possibility that the mass extinctions were caused by impact of extraterrestrial bodies, as is suggested by considerable evidence at the Cretaceous-Tertiary boundary, about 65 Myr ago (see, e.g., Alvarez et al. 1980, 1982a). Accordingly, the ages of known terrestrial impact structures have been scrutinized by means of a variety of statistical tests, and are reported to be periodically distributed (Alvarez and Muller 1984; Rampino and Stothers 1984a; Sepkoski and Raup 1986). Periodic fluctuation in the bombardment of solid bodies has been interpreted, in turn, as the consequence of the modulation of the flux of comets in the inner solar system. This modulation might take the form of discrete pulses or "showers" of comets or of a smoother, low-amplitude variation in the comet flux.

In this chapter, we review the underlying evidence upon which the claims of periodicity of mass extinctions and crater ages have been based and compare in detail the observed geologic record of impact events on the Earth with the paleontologic record of mass extinctions. Our conclusion is that there have been pulses in the impact rate on Earth, some of which are correlated with mass extinctions, and some probably are due to comet showers. Finally, we examine briefly the efficacy or likelihood of the existence of some of the astronomical clocks purported to modulate the comet flux.

I. MASS EXTINCTIONS

Although paleontologists have tended to regard the history of changes in the biota of the Earth in terms of gradual transition (Raup 1986), there is moderately strong agreement that this history has been marked by occasional episodes of rapid loss of taxa referred to as mass extinctions (see, e.g., Newell 1967). Five or six mass-extinction events are generally acknowledged, including those near the ends of the Permian (Guadalupian and Djulfian Stages), the Triassic (Norian and Rhaetian Stages) and the Cretaceous (Maestrichtian Stage). Other, lesser extinction events have been recognized by various authors (see, e.g., Fischer and Arthur 1977). The identification and precise determination of the time and duration of mass extinctions, however, are fraught with difficulties. These difficulties arise from incompleteness in the stratigraphic record, in the record of fossils within the preserved strata, and in the study of the contained fossils—as well as from problems of global correlation of the preserved beds (see, e.g., Newell 1982; Signor and Lipps 1982; Hoffman and Ghiold 1985).

Perhaps the most exhaustive attempts to define mass extinctions on a quantitative basis have been made by Sepkoski and Raup, who used a large catalog (Sepkoski 1982a) of the observed stratigraphic ranges of various taxa. Working at the taxonomic level of families of organisms, Raup and Sepkoski (1982) and Sepkoski (1982b) reported three high- to intermediate-intensity extinctions within the last $\sim$ 250 Myr (Djulfian, Norian and Maestrichtian

Stages) and five "lesser," possibly regional, extinctions (Toarcian, Tithonian (?), Cenomanian, late Eocene, and late Pliocene Stages). Raup and Sepkoski then expanded the list of possible mass extinctions in this ~ 250 Myr interval to include 12 events; the list formed the basis of their 1984 analysis of periodicity. Although this analysis has been superseded, to some extent, by two more recent papers (Raup and Sepkoski 1986; Sepkoski and Raup 1986), it has stimulated great interest and a number of hypotheses concerning possible astronomical driving mechanisms for extinctions. Therefore, we will review the 1984 analysis in some detail.

The extinction events recognized by Raup and Sepkoski in 1984 were found by the following procedure (1984 Raup-Sepkoski algorithm):

1. Sepkoski's catalog was screened to eliminate families whose stratigraphic range is poorly known;
2. All living families were subtracted from the screened list;
3. The number of families that are now extinct was tabulated for each stratigraphic stage (comprising 40 in all, from the Djulfian Stage of the Permian to the end of the Tertiary), and the ratio of the number of families that fail to appear in the following stage to the total number of extinct families known from each stage was used to calculate "percent extinction";
4. The time of all extinctions for each stage was plotted at the upper boundary of the stage, although the precise stratigraphic level of extinction of a given family commonly is not known, and extinctions may be distributed through a stage.

The fluctuation of "percent extinction" from one stage to the next gives rise to a series of 12 peaks (Fig. 1). Each peak found by this procedure was included in the statistical analysis for periodicity, although Raup and Sepkoski recognized explicitly that not all peaks may be significant.

Some indication of the inherent difficulty in identifying "lesser" mass extinctions is given by changes in identification by Raup and Sepkoski between 1982 and 1984. An extinction event suggested by Sepkoski (1982*b*) to occur at the end of the Toarcian Stage of the Jurassic was later resolved into two peaks, one in the Pliensbachian and the other in the Bajocian; the Toarcian extinction level, in the 1984 analysis, falls in a valley between these two peaks. The extinctions per stage at the Pliensbachian and Bajocian peaks, 15% and 11% respectively, are below the overall average level of extinction per stage from the end of the Permian to the middle Miocene (19% of extinct families per stage). A lesser mass extinction near the end of the Pliocene (at ~ 2 Ma),[a] which had been described by Stanley and Campbell (1981) and re-

---

[a]Ma is used here to denote the *age* of a geologic event 10^6 yr before the present. In contrast, Myr indicates the *duration* of a remote interval of time lasting 10^6 yr.

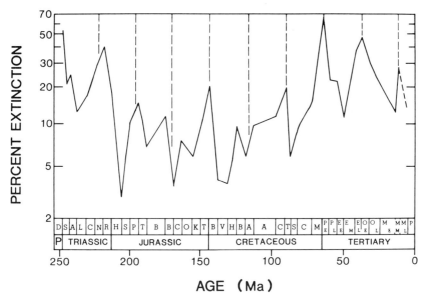

Fig. 1. Distribution of percent extinction of families with respect to time for the past 250 Myr, as calculated by the algorithm of Raup and Sepkoski (1984). Letters in small boxes along the abscissa are the stages adopted in the analysis of Raup and Sepkoski; the boundaries of the stages are plotted primarily according to the Harland et al. (1982) geologic time scale. (There is an unexplained expansion of the duration of the Bathonian in the middle of the Jurassic, as plotted by Raup and Sepkoski [1984], which does not correspond to the time scale of Harland et al. [1982].) Positions of peaks for a best-fit cycle having a 26 Myr period, determined by time series analysis by Raup and Sepkoski, are shown as dashed lines.

ported in Sepkoski's 1982 paper, was dropped in the 1984 analysis, and a peak in the middle Miocene (at 11.3 Ma according to Harland et al. [1982]) was somewhat tentatively introduced.

A peculiarity of the extinction rate calculated by the 1984 Raup-Sepkoski algorithm is that extinction rates in the relatively recent geologic past are greatly exaggerated by subtraction of living families. Families that became extinct in late Tertiary time actually constitute a very small fraction of the extinct families. The middle Miocene mass extinction, which represents a sharply defined loss of only 6 out of a total of about 700 living and extinct families was shown as questionable by Raup and Sepkoski (1984), but it has been affirmed in their later papers. The evidence cited by Sepkoski (1982b) for a mass extinction at the end of the Pliocene is not contravened by the 1984 analysis.

Some of the deficiencies of the 1984 Raup-Sepkoski algorithm have been rectified in their 1986 papers. First of all, percent extinction has been calculated on the basis of standing taxonomic diversity rather than on the basis of taxa now extinct. Secondly, extinction has been examined at the level of genera as well as families. The analysis at the genus level reinforces the identification of mass extinctions found to be significant at the family level. Most important, the statistical significance of the extinction peaks was examined; several of the peaks identified in the 1984 analysis have been rejected.

Among many factors that affect the apparent extinction rate are differences in the lengths of the stages (see Hoffman 1985). The durations of stages in the Late Cretaceous and Tertiary differ by as much as a factor of 3 or more; the durations of most of the stages in the Mesozoic are virtually unknown. If the actual extinction rate were perfectly uniform, it would appear to fluctuate by at least a factor of 3, and probably much more, when analyzed by the 1984 Raup-Sepkoski algorithm. Whether small peaks in the percent extinction per stage correspond to actual peaks in extinction rate cannot be determined until the apparently high rates of extinction have been verified by detailed stratigraphic studies. Peaks that lie below the average level of extinction per stage should be regarded as indicators of unusual biological events only with caution; peaks below 10% extinction per stage probably represent nothing more than statistical noise.

Sepkoski and Raup, in their 1986 paper, explicitly calculated extinction *rates* on the basis of estimated stage lengths. However, for stages older than the Albian (older than $\sim$113 Ma) their calculated rates (25 of their total of 44 calculated rates) are virtually meaningless. In general, the uncertainties of the ages of the boundaries of the pre-Albian stages exceed the estimated durations of these stages. These uncertainties are described in detail in Harland et al. (1982, ch. 3). No pre-Albian stage duration is known within a factor of 10. We wish to stress here the inadvisability of any investigator attempting to use first derivatives of published geologic time scales without first assessing with great care the available chronometric calibration. Large errors in first derivatives are likely even for the best calibrated part of the time scale (the last 100 Myr). As an example, a duration for the late Eocene Stage of 4.0 Myr can be obtained from the ages of stage boundaries (42.0 Ma and 38.0 Ma) estimated by Harland et al. (1982), yet a more recent calibration by Montanari et al. (1985) suggests that the late Eocene lasted for only 0.7 Myr. The possible error for the duration of the late Eocene derived from Harland et al. (1982) is greater than a factor of 5.

Because the reported periodicity of Raup-Sepkoski extinction peaks was the starting point for several papers in which various astrophysical mechanisms were proposed, we review here the status and timing of each of these peaks. As the principal time scale used by Raup and Sepkoski in their 1984 and 1986 analyses is that of Harland et al. (1982), we will examine the uncertainties in this time scale as well as the inherent limits to the resolution of the time of extinctions of taxa.

Guadalupian-Djulfian Extinction (End of the Permian)

There is universal agreement that a great extinction at both the genus and family levels occurred near the end of the Permian Period. This extinction marks the change from Paleozoic to Mesozoic life. Sea level was low near the end of the Permian and beginning of the Triassic, however, so an apparently continuous record of deposition across this boundary has been found in fossiliferous rocks in only a few regions of the world. How extinctions were distributed in the Djulfian Stage (the last stage of the Permian) is not known. A major loss of taxa also took place in the preceding Guadalupian Stage of the Permian (Raup and Sepkoski 1986; Sepkoski and Raup 1986). Hence, the great terminal Paleozoic extinction was definitely distributed through at least two stages. The range of uncertainty in the estimated age of the Permian-Triassic boundary, roughly determined from chronograms presented by Harland et al. (1982), is 20 Myr. The error is not symmetrical about the 248 Ma age given for the boundary by Harland et al. (1982), which was derived by interpolation, but the age of the boundary is estimated as between about 234 and 254 Ma. The beginning of the latest Permian stage (referred to as Tartarian by Harland et al. [1982]) is also estimated to lie in this interval. Thus the interval in which the Djulfian (Tartarian) extinction is estimated to have occurred extends from 234 to 254 Ma, and the next oldest stages (referred to as Kazanian and Ufimian by Harland et al. [1982] roughly and equivalent to the Guadalupian) also fall in this interval. The great episode of extinction near the end of the Permian was distributed in an unknown way in this ~20 Myr time interval.

Olenekian Extinction

This minor extinction peak (about 2% amplitude) is on the flank of the great extinction peak near the Permian-Triassic boundary. It was recognized only by splitting the Scythian Stage and was regarded by Raup and Sepkoski (1984) as probably not significant. It has been dropped in their 1986 analyses.

Norian-Rhaetian Extinction

A major mass extinction near the end of the Triassic Period has been recognized by most specialists. Some have thought that a separate stage, the Rhaetian, should be designated in the latest Triassic, others that the Rhaetian should be included with the Norian Stage. Extinction appears highest in the Norian at the family level and highest in the Rhaetian at the genus level (Raup and Sepkoski 1986). Evidently the mass extinction is distributed over at least these two stages. The range of the estimated age of the Norian-Rhaetian boundary is from about 204 to 222 Ma (Harland et al. 1982). Both the Norian and Rhaetian Stages are estimated to range from 222 Ma or later to 204 Ma or earlier, so that the late Triassic extinction is distributed somewhere in this 18 Myr interval.

Pliensbachian, Bajocian and Callovian Extinctions

These three Jurassic extinction peaks are of low amplitude and also of very uncertain age. The Pliensbachian extinction has been accepted as significant in the 1986 Raup-Sepkoski papers and has been independently identified by biostratigraphic studies of Hallam (1976, 1977). The Bajocian and Callovian extinctions, however, have been rejected in the 1986 Raup-Sepkoski papers as not statistically significant. The end of the Bajocian is estimated to fall somewhere between 165 and 200 Ma and the end of the Pliensbachian somewhere between 167 and 199 Ma (Harland et al. 1982). The beginning and end of the Callovian fall somewhere in the interval from 150 to 165 Ma. Until the time scale for the Jurassic is improved, paleontological data for this period cannot be used for tests of periodicity of extinctions or correlation with isotopically determined crater ages.

Tithonian Extinction

This extinction at or near the end of the Jurassic is a significant mass extinction on a statistical basis and is also recognized in the detailed biostratigraphic studies of Hallam (1976, 1977). The ending date of 144 Ma for the end of the Tithonian given by the Harland et al. (1982) time scale and cited by Raup and Sepkoski was obtained by interpolation, however, and does not correspond to the best-fit age of 135 ± 3 Ma derived from reported isotopic dates (see Harland et al. 1982, ch. 3). For the beginning of the Tithonian, the best-fit age lies between 140 and 149 Ma. The Tithonian extinction probably occurred sometime between 132 and 149 Ma, most likely near the later end of this time interval.

Hauterivian Extinction

An extinction peak found by Raup and Sepkoski (1984) in the Hauterivian Stage of the Early Cretaceous rises only 3% above that in the following Barremian Stage and is no greater than the extinctions per stage in all but one of the later stages of the Cretaceous. This extinction has been found to be not statistically significant in the 1986 Raup-Sepkoski papers. Best-fit ages for the beginning and end of the Hauterivian are 124 ± 4 Ma and 122 ± 5 Ma.

Cenomanian Extinction

The time near the Cenomanian-Turonian boundary was a period of definite biological crisis, as shown not only by the Raup-Sepkoski 1984 and 1986 analyses but also by detailed biostratigraphic studies. A mass extinction occurred in five discrete steps extending from the middle late Cenomanian to the middle early Turonian (Kauffman 1984; Elder 1985; Hut et al. 1986). The following Coniacian Stage has a normal level of extinction per stage, but the stage appears to be very short (~1 Myr); the high *rate* of extinction in the Cenomanian-Turonian boundary may have continued into the Coniacian.

The end of the Cenomanian has been dated at 91.0 ± 1.5 Ma (Harland et al. 1982); the stepwise mass extinction that straddles the Cenomanian-Turonian boundary probably spanned about 2.5 Myr (Hut et al. 1986).

Maestrichtian Extinction

The disappearance of many forms of living organisms near the end of the Mesozoic Era is one of the best documented mass extinctions in the paleontological record. As summarized by Kauffman (1984), this extinction took place in about five distinct steps over a probable interval of about 2.0 to 2.5 Myr, extending from the middle-upper Maestrichtian boundary into the early Paleocene. In recent papers, the end of the Maestrichtian has been variously estimated at 65 to 66.4 Ma (Harland et al. 1982; Palmer 1983).

Late Eocene Extinction

Another well-documented stepwise mass extinction extended from latest middle Eocene to the Eocene-Oligocene boundary (Keller 1983; Corliss et al. 1984; Hut et al. 1986). According to the time scale of Harland et al. (1982) the duration of the mass extinction can be estimated at about 4 Myr, with the last step occurring at about 38 Ma. New isotopic ages presented by Montanari et al. (1985), however, indicate that the duration was only about 1 Myr and that the last step of the extinction occurred at 35.7 ± 0.4 Ma. Given the probable short duration of the late Eocene, the average rate of extinction may have been as high or higher than that of the Maestrichtian extinction.

Middle Miocene Extinction

Loss of families at this extinction peak was only about 1% of combined extinct and living families. Sepkoski and Raup (1986) showed a calculated extinction rate for the middle Miocene that appears to be much higher than average for the Tertiary, but the actual time of extinction is well defined for only half the families lost at about this time. The end of the middle Miocene is fairly accurately dated at 11 Ma.

In summary, six relatively strong mass extinctions have occurred in the last ~ 250 Myr. The last three, the Cenomanian, Maestrichtian and late Eocene, are fairly accurately dated. The range of uncertainty in timing of the three earlier strong extinctions (Guadalupian-Djulfian, Norian-Rhaetian and Tithonian) is 17 to 20 Myr. The Guadalupian-Djulfian and Norian-Rhaetian mass extinctions are separated from the poorly dated Tithonian by about 70 to 100 Myr and from the well-dated Cenomanian by about 120 to 150 Myr. Hence, they provide only relatively weak control in testing cycles fitted to the younger, well-dated extinction events.

If we choose the apparent best central values for the ages of the last four strong mass extinctions, cycles can be fitted to their ages by a least-squares procedure (Table I). If only the three relatively well-dated strong extinctions are used, the period of the best-fit cycle is fairly sharply defined at 28 Myr; the

TABLE I
Cycles Fitted to Best Estimated Ages of the Relatively Strong Mass Extinctions in the Last 250 Myr

Mass Extinction	Estimated Age (Ma)	Cycles Fitted by Least Squares							
		25-Myr Period	28-Myr Period[a]	31-Myr Period[b]	32-Myr Period[b]	33-Myr Period[b]	34-Myr Period[b]
Late Eocene	36.2[c]	38	36	35	34	32	31
Maestrichtian	65[c]	63	64	66	66	65	65
Cenomanian	91[c]	88	92	97	98	98	99
Tithonian	135[d] (132–149)	138	148	128	130	131	132
Norian-Rhaetian	204–222	213	204	221	226	197	200
Guadalupian-Djulfian	234–254	238	232	252	258	230	234

[a]Fitted only to last three comparatively well-dated strong mass extinctions.
[b]Fitted to last four strong mass extinctions, assuming the sequence from Tithonian to late Eocene is complete.
[c]Adopted best estimated age of the midpoint of stepwise mass extinction.
[d]Adopted best estimated age of the end of the Tithonian.

last peak of this fitted cycle is at 8 Ma. Extrapolation of the 28 Myr cycle back in time yields a peak at 148 Ma, close to the earliest estimated bound for the beginning of the Tithonian. Peaks are also predicted at 204 and 232 Ma, fairly close to the latest bounds estimated for the ends of the Norian and Djulfian Stages.

On the other hand, the stratigraphic studies of Hallam (1976,1977) show that the Tithonian extinction was distributed through the Tithonian Stage (end of the Jurassic Period); if one chooses the most probable age of 135 ± 3 Ma for the end of the Tithonian from the chronogram by Harland et al. (1982) and takes this as the culmination of the distributed extinction, then the best-fit period for the last four definite mass extinctions is 32 Myr; the last predicted peak then falls at 2 Ma. This fit is based on the assumption that the sequence of four extinctions from the Tithonian to the Eocene is complete. As the residuals are fairly large, the period is not sharply defined; a 33 Myr period whose last peak is at 32 Ma is nearly as good a fit. Neither the 32 nor the 33 Myr cycles yield peaks that fall in the very broad age ranges for either the Norian-Rhaetian or Guadalupian-Djulfian mass extinctions. However, a 31 Myr cycle, whose residuals are slightly higher, predicts peaks within the age ranges of the Norian and Djulfian, and a 34 Myr cycle yields peaks at 201 and 235 Ma, which jointly fit the latest bounds for the Norian-Rhaetian and Guadalupian-Djulfian extinctions about as well as the 28 Myr cycle. Interestingly, Fischer and Arthur (1977) originally estimated the period at 32 Myr; Kitchell and Pena (1984) found a best-fit apparent periodicity of 31 Myr for the entire set of the 1984 Raup-Sepkoski extinctions of the last ~ 250 Myr and Rampino and Stothers (1984a) found a best-fit period of 30 Myr, from the Raup-Sepkoski extinctions.

As the Tithonian extinction evidently extended to the end of the Tithonian Stage, we find that no cycles with periods > 28 Myr fit the ages of more than four of the six strong mass extinctions particularly well. The period of the best-fit cycle for all six extinctions is 25 Myr, which is close to the 26 Myr period determined by Raup and Sepkoski (1984) and Sepkoski and Raup (1986); the 25 Myr cycle predicts a mass extinction between the Tithonian and Cenomanian, which Raup and Sepkoski (1984) identified with the low-amplitude Hauterivian peak that they have now rejected. This cycle also fits the low-amplitude middle Miocene extinction, but is out of phase with the late Pliocene extinction identified by the detailed studies of Stanley and Campbell (1981).

Because the observational base from which all the cycles listed in Table I have been derived includes, at most, six events, the case for periodicity is no longer compelling. Indeed, the evidence for periodicity rests chiefly on the relatively secure ages of just three strong mass extinctions. For a random distribution of three events over the last 91 Myr, there is about a 10% probability that the two intervals of time separating these events would be as nearly equal as observed. Kitchell and Pena (1984) calculated best-fit models for both the times and amplitudes of the 1984 Raup-Sepkoski extinctions on the assumptions of (1) a deterministic periodic impulse, (2) a deterministic cycle of sinusoidal wave form, and (3) a stochastic dynamic system. They found the stochastic model with a 31 Myr pseudocycle to provide a superior fit to the 1984 Raup-Sepkoski extinctions, taking into account the amplitudes obtained by the 1984 algorithm. In our view, this is the result that should be expected, inasmuch as the times of two-thirds of these extinctions are not known to within one-half the cycle length and, therefore, can be considered random, and the amplitudes of most of the 1984 Raup-Sepkoski extinctions depend ultimately on virtually unknown durations of the stages and can also be considered random.

II. AGES OF KNOWN TERRESTRIAL IMPACT EVENTS

If most mass extinctions are related to the impact of comets or asteroids, it is reasonable to search for correlation between the times of mass extinction and the times of known impact events. Our knowledge of the geologic record of impact events is, at present, only fragmentary, however, and the determinations of the ages of these events have diverse and, in many cases, large uncertainties. Moreover, because the geologic record of impact structures is incomplete, a well-known bias exists in the observed sample of the ages of these structures; this bias is due to a decrease in the probability of both the preservation and the discovery of impact structures with an increase in age. Iridium anomalies and strewn fields of impact glass provide additional information on the timing of large impact events that is partly independent of the recognized impact craters. In order to assess the correlation between impacts and mass extinctions, we will draw upon information both on impact structures and on these more widespread markers of impact events.

Impact Craters and Structures

In order to reduce biases in the statistics of impact crater ages due to actual losses by erosion as well as to failure to detect degraded or buried impact structures, it is useful to restrict the statistics to craters and structures above some limiting size. Most known impact craters smaller than 1 km in diameter, for example, are of late Quaternary age (Shoemaker 1983), whereas the distribution of ages of known very large impact structures is much more uniform. The discovery of structures > 20 km in diameter formed in the last 125 Myr may be nearly complete for the North American and European cratons (Grieve 1984). A compromise must be made, however, between reducing the observational bias and retaining sufficient data to obtain useful statistics; therefore, we have chosen a diameter of 5 km as the lower limiting crater size for our age distribution study. This size is somewhat above the threshold at which craters are produced by most or nearly all extraterrestrial bodies entering the atmosphere (Shoemaker 1983).

TABLE II
Impact Structures 5 km or Greater in Diameter Whose Reported or Derived Ages are 250 Myr or Less

Impact Structure	Diameter (km)	Method of Dating	Age (Ma)[a]	Reference[b]
Bosumtwi, Ghana	10.5	Fission-track	1.04 ± 0.11*	(1)
		K/Ar	1.3 ± 0.2	(2)
Zhamanshin, USSR	7	Fission-track	1.07 ± 0.05*	(1)
Elgygytgyn, USSR	19	K/Ar	3.5 ± 0.5	(3)
		Fission-track	4.5 ± 0.1*	(4)
Karla, USSR	12	Stratigraphic (late Miocene–early Pliocene)	5 ± 3	(5)
Ries, W. Germany	27	Fission-track	14.7 ± 0.4*	(1)
		K/Ar	14.8 ± 0.7	(6)
Haughton, Canada	20	Stratigraphic (Miocene)	20 ± 5	(7)
Popigai, USSR	100	Fission-track	30.5 ± 1.2*	(4)
		K/Ar	39 ± 6	(8)
Wanapitei, Canada	8.5	$^{40}Ar/^{39}Ar$	32 ± 2*	(9)
		K/Ar	37 ± 2	(10)
Mistastin, Canada	28	$^{40}Ar/^{39}Ar$	38 ± 4 mean	(11)
		K/Ar	37.8 ± 0.9 38.5 ± 2	(11)
		Fission-track	39.6 ± 4.4	(1)
Goat Paddock, Australia	5	Stratigraphic (early Eocene)	55 ± 3	(12)
Kara, USSR (includes Ust Kara)	60	K/Ar	57 ± 9	(13)
Kamensk, USSR	25	Stratigraphic (early Paleocene)	65 ± 3	(5)
Manson, USA	35	Fission-track	61 ± 18	(14)
		$^{40}Ar/^{39}Ar$	≤70*	
Lappajärvi, Finland	14	$^{40}Ar/^{39}Ar$	78 ± 2	(15)
Steen River, Canada	25	K/Ar	95 ± 7	(16)
Boltysh, USSR	25	K/Ar	88 ± 17	(17)
		Fission track	100 ± 5*	(17)
Dellen, Sweden	15	$^{40}Ar/^{39}Ar$	100 ± 2	(18)
Carswell, Canada	37	$^{40}Ar/^{39}Ar$	117 ± 8*	(18)
		K/Ar	485 ± 50	(20)
Mien, Sweden	5	Fission-track	92 ± 6	(1)
		$^{40}Ar/^{39}Ar$	119 ± 2*	(19)
Gosses Bluff, Australia	22	Fission-track	130 ± 6	(21)
		K/Ar	133 ± 3*	(21)

TABLE II (*Continued*)

Impact Structure	Diameter (km)	Method of Dating	Age (Ma)[a]	Reference[b]
Vyapryai, USSR	8	Stratigraphic (late Jurassic)	150 ± 16	(17)
Rochechouart, France	23	K/Ar	160 ± 5*	(22)
		Fission-track	198 ± 25	(1)
Puchezh-Katunki, USSR	80	K/Ar	183 ± 3	(23)
		Stratigraphic (early Bathonian)	183 ± 17	(17)
Oboloń, USSR	15	Stratigraphic (Bajocian)	183 ± 17	(24)
Manicouagan, Canada	70	K/Ar	210 ± 4 mean	(25)
		Rb/Sr	214 ± 3 212 ± 3	(26)

[a] Age dated with a precision of 20 Myr or better. Preferred age used in construction of Figs. 2 and 3 is indicated by asterisks.
[b] References: (1) Storzer and Wagner (1977); (2) Gentner et al. (1964); (3) Gurov et al. (1978); (4) Storzer and Wagner (1979); (5) Stratigraphic age from Masaitis et al. (1980). Age based on Harland et al. (1982) geologic time scale; (6) Gentner and Wagner (1969); (7) Robertson et al. (1985); (8) Masaitis et al. (1975); (9) Bottomley et al. (1979); (10) Winzer et al. (1976); (11) Mak et al. (1976); (12) Stratigraphic age from Harms et al. (1980). Age based on Harland et al. (1982) geologic time scale; (13) Masaitis et al. (1980). Although these authors cite the K/Ar age given above, they state, on the basis of regional stratigraphic evidence, that "the most probable time of formation of the crater is the interval between late Eocene and early Oligocene"; (14) Hartung et al. (1986). On the basis of the published ^{39}Ar release diagram, an age of 67.5 ± 2.5 Ma is here assigned for construction of Figs. 2 and 3; (15) Jessberger and Reimold (1980); (16) Carrigy and Short (1968); (17) Masaitis et al. (1980); (18) Bottomley (1982); (19) Bottomley et al. (1978); (20) Currie (1969); (21) Milton et al. (1972); (22) Lambert (1974); (23) Firsov (1965); (24) Val'ter et al. (1977); (25) Wolfe (1971); (26) Jahn et al. (1978).

Twenty-five terrestrial impact structures are known whose mean diameters are greater than or equal to 5 km for which ages less than 250 Myr have been reported or can be estimated with a precision thought to be better than 20 Myr (Table II). Their ages have been determined by a wide variety of methods. In many cases, measurements of age of impact glass or crystallized impact melts have been made by standard isotopic techniques, including the K/Ar, the ^{40}Ar/^{39}Ar plateau, and Sr/Rb methods. Some of the most precise ages have been obtained by the fission-track method. We have derived other estimates of age for several structures on the basis of stratigraphic evidence and the Harland et al. (1982) geologic time scale.

As ages have been obtained by multiple methods for 60% of the structures listed in Table II, it is instructive to compare the results obtained by different methods. One uncertainty in K/Ar dating of impact glasses or recrystallized impact melts is the degree to which old radiogenic argon was lost from the melt. These melts commonly contain unmelted clasts of older rocks that retain some preimpact argon. The inherited argon generally leads to anomalously high age estimates, particularly in very young impact glass. Most of the K/Ar ages listed for structures younger than 40 Myr old are greater than the corresponding fission-track ages, probably chiefly as a result of inherited argon. A particularly severe case is the K/Ar age of 39 ± 6 Ma listed for the Popigai structure in the Soviet Union, which has been cited in several previous studies (see, e.g., Grieve 1982; Alvarez and Muller 1984). This age is based on the average of age determinations from six different samples. A large spread of K/Ar ages was obtained from these samples, almost certainly as a consequence of inherited argon (Masaitis et al. 1975); a seventh sample yielded a much higher age. Although Soviet investigators found a fission-track age consistent with the reported mean K/Ar age (Masaitis et al. 1980), Storzer and Wagner (1979) obtained a fission-track age of only 30.5 ± 1.2 Myr, based on techniques that correct for track fading and that control other sources of error. The age found by Storzer and Wagner is close to the youngest K/Ar age obtained from Popigai impact glasses, which is probably the least affected by inherited argon. The fission-track and the youngest K/Ar ages, therefore, may be most accurate. Fission-track ages, on the other hand, are less reliable for more ancient glasses because of problems of track fading. For example, a relatively precise 119 ± 2 Myr $^{40}Ar/^{39}Ar$ plateau age reported for the Mien structure in Sweden (Table II) is more stable and is preferred to a 92 ± 6 Myr track age reported by Storzer and Wagner (1977).

Estimates of precision of the ages given in Table II are chiefly estimates of analytical precision expressed as greater than one standard deviation. For ages assigned from stratigraphic evidence, the precision listed expresses the estimated uncertainty in chronostratigraphic position, as well as the uncertainties in the calibration of the geologic time scale. The latter uncertainties are relatively small for ages < 70 Ma, but they are the dominant uncertainties for ages based on stratigraphic evidence for Jurassic and older impact structures. For almost all ages listed in Table II, the true uncertainties probably are larger than indicated by the estimated precision, owing to diverse sources of systematic error such as inherited argon, argon loss, fission-track fading, and uncertainties in the identification and the chronological range of fossil taxa. The possible hazards of systematic error, particularly in the case of K/Ar ages for impact-metamorphosed or shock-melted rocks, are indicated by the extreme difference between a relatively well-defined $^{40}Ar/^{39}Ar$ plateau age and a previously published K/Ar age for the Carswell structure, Canada (Table II).

The ages listed in Table II are portrayed graphically in Fig. 2, where the probability distribution of age for each impact structure is taken simply as a box of 2 σ width. Where more than one estimate of age is available for a given impact structure, the preferred age (indicated by asterisk in Table II) has been plotted. A somewhat more easily evaluated picture of the composite age probability distribution is obtained by smoothing the distribution of Fig. 2 with a 6-Myr running mean (Fig. 3). Because of the unknown magnitude of the systematic errors, it does not appear that a more sophisticated representation

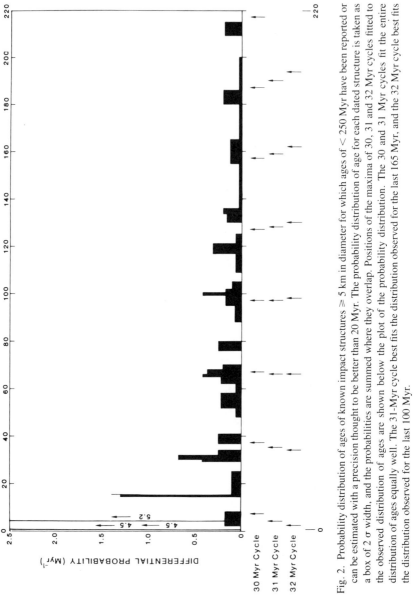

Fig. 2. Probability distribution of ages of known impact structures ⩾ 5 km in diameter for which ages of < 250 Myr have been reported or can be estimated with a precision thought to be better than 20 Myr. The probability distribution of age for each dated structure is taken as a box of 2 σ width, and the probabilities are summed where they overlap. Positions of the maxima of 30, 31 and 32 Myr cycles fitted to the observed distribution of ages are shown below the plot of the probability distribution. The 30 and 31 Myr cycles fit the entire distribution of ages equally well. The 31-Myr cycle best fits the distribution observed for the last 165 Myr, and the 32 Myr cycle best fits the distribution observed for the last 100 Myr.

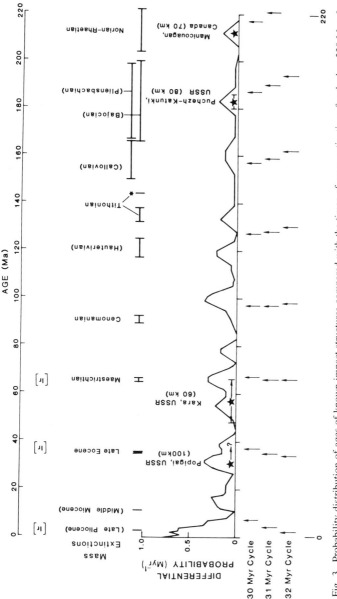

Fig. 3. Probability distribution of ages of known impact structures compared with the times of mass extinctions for the last 220 Myr. In this figure, the probability distribution of ages shown in Fig. 2 has been smoothed with a 6 Myr running mean in order to facilitate comparison with times of mass extinction. Best-fit 30, 31 and 32 Myr cycles are shown below the distribution of ages, as in Fig. 2. Isotopically determined ages of the four largest known impact structures formed during the last 220 Myr are shown with stars. The range of uncertainty of the time of each mass extinction recognized by Raup and Sepkoski (1984) is shown by error bars above the probability distribution of ages of impact structures. These times are derived from the geologic time scale of Harland et al. (1982). Weak to moderate and questionable mass extinctions are indicated in parentheses (see text). A mass extinction in the late Pliocene reported by Stanley and Campbell (1981) and recognized by Sepkoski (1982) is also shown. Two times are shown for the Tithonian mass extinction: (1) the range of uncertainty of the end of the Tithonian obtained from the chronogram of Harland et al. (1982); and (2) the age (shown with *) of the end of the Tithonian obtained by linear interpolation between chronometric tie points at 113 and 238 Ma.

of the data (such as modeling the probability distributions of age for individual structures with Gaussian functions, as was done by Alvarez and Muller [1984]) is warranted.

Inspection of Figs. 2 and 3 and Table II reveals that the number of relatively well-dated impact structures decreases with increasing age, as expected from known observational and geologic selection effects. The total number of dated structures listed in Table II decreases, on average, by about 44% in each successive interval of 75 Myr. When the decrease is fit to an exponential function, the number decreases by half per 90 Myr increase in age (characteristic time for decrease is 130 Myr). Of course, the structures listed are only a fraction of the known impact structures younger than 250 Myr. At least 16 other impact structures greater than 5 km in diameter are known whose ages probably lie in this interval but whose dates do not yet meet our rather broad precision limits. A plot of the cumulative size-frequency distribution for the impact structures with ages in the intervals 0–75 Ma, 75–150 Ma, and 150–225 Ma (Fig. 4) shows that the decrease in the number of recognized structures with increase in age applies chiefly to structures < 50 km in diameter. As many craters > 50 km in diameter are recognized in the 150–225 Ma age range as in the 0–75 Ma age range, although none this size has yet been dated in the age range 75–150 Ma (log 50 ≃ 1.7).

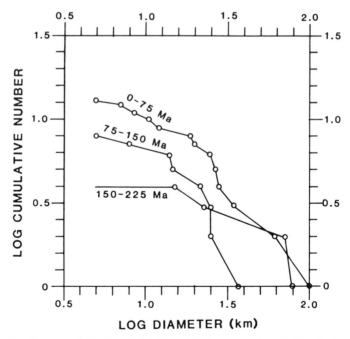

Fig. 4. Size-frequency distribution of dated terrestrial impact structures ⩾ 5 km in diameter in the age ranges 0 to 75, 75 to 150 and 150 to 225 Ma.

The probability distribution of observed ages shown in Fig. 3 consists of a pattern of peaks somewhat irregularly distributed over the past 220 Myr. Many of the peaks correspond to relatively precise individual ages; for the most part, the occurrence of these peaks is simply a reflection of the low density of age data in the time interval investigated. Four peaks, centered at about 2, 32, 65 and 99 Ma, rise above the rest of the peaks shown in Fig. 3. Each represents a cluster of three or more ages. This type of clustering resembles that generally found in random time sequences of events commonly referred to as "shot noise." Most distributions of 25 ages drawn at random from a uniformly distributed population of ages will contain several similar clusters. Hence, the presence of these peaks does not, by itself, indicate surges or modulation of the underlying cratering rate.

The highest peak shown in Fig. 3, which is centered at about 2 Ma, represents four impact events in the last 8 Myr, at least three of which occurred within a 4 Myr interval. There is about a 29% probability that three precisely determined ages out of 25 that are randomly distributed over 225 Myr would fall in the last 4 Myr and about a 4% chance that a fourth age would also fall in this interval. However, with a characteristic time of 130 Myr for the decrease in average probability of observing ages of impact structures, ages are expected to be observed about 5.5 times more frequently near the present than at 225 Ma, and twice as frequently at the present as at 90 Ma. After correction for the selection effects, even the strongest peak in the observed crater-age distribution is not statistically significant at the 95% confidence level. There is about a 50% chance that four crater ages drawn from 25 ages distributed randomly over 225 Myr would fall in a cluster spanning only 4 Myr.

Our primary purpose here is to test whether the ages of impact structures are correlated with mass extinctions. This question is partly independent of the tests for random or periodic surges in the cratering rate. Mass extinctions might arise from single large impact events or from real randomly occurring clusters of large impacts. Single observed ages of impact structures may also be samples of age clusters, particularly at times earlier than 100 Ma, as our knowledge of the actual cratering history is especially incomplete for these times.

In Fig. 3, one can see a moderate correlation between the relatively accurately dated mass extinctions in the last 100 Myr and the four strongest peaks in the probability distribution of crater ages. The stepwise mass extinction spanning the Cretaceous-Tertiary boundary is centered on the crater-age peak at $\sim$ 65 Ma, and the late Eocene stepwise mass extinction occurs on the shoulder of the $\sim$ 32 Ma crater-age peak. The precise position of the $\sim$ 32 Ma peak is uncertain, owing to the problem of interpreting the age of the Popigai crater; the true peak of crater ages might be closer to the Eocene mass extinction than shown.

Two of the four largest known impact structures of Phanerozoic age may be associated with the ~ 32 Ma and ~ 65 Ma peaks. Within the uncertainty of the K/Ar age determination, the 60 km diameter Kara crater might be correlated with the Cretaceous-Tertiary mass extinction. On the other hand, if the stratigraphic age assignment given by Masaitis et al. (1980) is correct (see footnotes to Table II), the Kara crater, along with the 100 km diameter Popigai crater, might be associated with the late Eocene mass extinction. As described below, iridium anomalies and other stratigraphic tracers of large impact events are precisely correlated with individual extinctions at the Cretaceous-Tertiary boundary and in the late Eocene. Hence, these extinctions are unequivocally correlated in time with large impact events, and the observed crater-age peaks at or near these times probably reflect or are at least close in time to real surges in the cratering rate.

The Pliocene mass extinction reported by Stanley and Campbell (1981) and by Sepkoski (1982b) is centered on the ~ 2 Ma crater-age peak. Although Sepkoski considered this extinction to be only regional, and although it was not included in the 1984 Raup-Sepkoski mass extinction list, it is fairly closely associated in time with a number of known impact events, including at least one major event, as shown below. Curiously, a local iridium anomaly has been discovered in late Pliocene deep-sea sediments (age ~ 2.3 Ma) near Antarctica (Crocket and Kuo 1979; Kyte et al. 1981); the anomaly is very close in age but apparently slightly younger than a significant extinction of radiolarian species that is estimated by Hays and Opdyke (1967) to have occurred at ~ 2.5 Ma. The anomaly is associated with definite meteoritic particles (Kyte and Brownlee 1985), and it is considered to have been produced by impact in the ocean of a basaltic asteroid 100 to 500 m in diameter. This event may be merely part of the background-level asteroid bombardment of the Earth, but it is intriguing that it occurred during an apparent surge in the impact rate that is fairly closely correlated with the late Pliocene extinction (see below).

Two Raup-Sepkoski extinction events in the last 100 Myr do not correlate well with crater-age peaks. The relatively weak mass extinction in the middle Miocene, if it occurred at the end of the middle Miocene, as assumed by Raup and Sepkoski (1984,1986), is about 3.5 Myr younger than the accurately dated Ries crater in West Germany and the correlative tektite (moldavite) strewn field and of the order of 10 Myr younger than the Haughton crater of early Miocene age in Canada. In any case, there is neither a strongly defined crater-age peak nor a strong mass extinction late in the middle Miocene.

The strong and relatively accurately dated stepwise mass extinction spanning the Cenomanian-Turonian boundary does not coincide with a crater-age peak; the nearest is the ~ 99 Ma peak, which is offset from the center of the extinction by about 8 Myr. This appears to be a clear miss, although the relatively poorly dated Boltysh structure in the USSR, as well as the Steen River structure in Canada, might have been formed at about the time of the Cenomanian-Turonian boundary. No iridium anomalies or other stratigraphic tracers of large impacts have yet been found at any of the Cenomanian-Turonian extinction steps, despite searches for them and a false report of an iridium anomaly (see Hut et al. 1986). There is no strong evidence that links the Cenomanian-Turonian extinction to impacts.

The dating of mass extinctions older than 100 Myr is so uncertain that, in most cases, detailed comparison of the timing of these events with crater-age peaks older than 100 Myr cannot be made. However, a crater-age peak occurs within the possible time range of each of the six Raup-Sepkoski extinctions identified between ~ 100 and 225 Ma. Each of these peaks corresponds to a single age determined with relatively high precision. No precise crater ages are available for the time between 225 Ma and 254 Ma (the earliest likely time of the Permian-Triassic boundary), although the age of the 23-km-diameter St. Martin structure of Canada, cited by Grieve et al. (1985) to be at 225 ± 25 Ma, probably falls in this interval. It is of interest that the age of one of the largest known Phanerozoic (i.e., the Paleozoic, Mesozoic and Cenozoic eras taken together) impact structures (Manicouagan, Canada) is nearly centered in the estimated range of possible age of the strong Norian-Rhaetian mass extinction.

Three of the mass extinctions older than 100 Myr shown in Fig. 3 (Bajocian, Callovian and Hauterivian) are not considered to be statistically significant, as described above. However, the age of the Puchezh-Katunki structure in the USSR is estimated from stratigraphic evidence to be early Bathonian (Table II). This great impact event could have coincided with the possible extinction event placed by Raup and Sepkoski at the end of the Bajocian (11% extinction). The 15 km diameter Obolon' structure in the USSR, estimated from the stratigraphic evidence to be of Bajocian age (Table II), might be contemporaneous with the tentative Bajocian extinction, if the Bajocian extinction preceded the end of the Bajocian Stage.

As a number of claims have been made concerning the periodicity of the ages of impact structures, it is pertinent to reexamine these claims on the basis of our updated list of ages presented in Table II. Following procedures of least-squares fitting similar to those employed by Rampino and Stothers (1984*a*), one can find best-fit cycles for the entire sequence of ages given in Table II or for various parts of it. A choice can be made either to rigidly fit a cycle to the entire sequence, which extends to 212 ± 3 Ma, or to consider the period somewhat variable, so that fits can be made to different parts of the sequence. Here we examine several cases.

For the entire sequence of 25 ages, periods of 30 and 31 Myr fit equally well; the last maxima in the cratering rate fall at 7 and 4 Ma, respectively. The mean squared deviation of observed ages from the idealized 30 and 31 Myr cycles is 55% of that expected for purely random distributions of age. Hence, there is a suggestion of periodicity in the distribution of crater ages, but there is also clear evidence of a strong nonperiodic component. If we consider ages that extend to only 165 Ma, which include 84% of the observations, the best-fit period is 31 Myr, but a 32 Myr cycle fits almost as well, with the time of its last maximum falling at 2 Ma. The mean squared deviations of the observed ages for these fits are 54% and 56% of that expected for random distributions. If one considers only the last 100 Myr interval, which includes 68% of the observations, the best-fit cycle is sharply defined with a period of 32 Myr.

The four highest peaks in the probability distribution of crater ages appear fairly strongly periodic. However, a number of accurately determined ages in the last 100 Myr interval are well separated from these peaks. If the crater-age clusters are periodic, an unequivocal second component exists that is probably random. The distribution of the various best-fit cycles is illustrated in Fig. 3. The period and phase of the best-fit cycle to the crater ages for the last 100 Myr match those of the best-fit cycle for the last four strong mass extinctions, when fitted with the assumption that the sequence of these four mass extinctions is complete.

The statistical significance of the apparent periodicity of the crater ages remains open to question. Grieve et al. (1985), using a list of 26 crater ages that is similar but not identical with ours, tested for periodicity by time-series analysis. On the basis of various criteria of data selection, they found periods of ~ 29, ~ 21, ~ 18.5 and ~ 13.5 Myr, with diverse power and phase. They also found that periodicities equivalent in power to those determined from the observed ages could be obtained in about 25% of their Monte Carlo runs when sets of 20 random numbers chosen from 1 to 250 were used. The apparent periodicity of the crater ages, considered by themselves, probably is not significant at the 95% confidence level.

Impact Glasses, Microspherules and Iridium Anomalies

Widely dispersed glass, produced by shock melting of rock, and iridium anomalies detected at discrete horizons in the stratigraphic column constitute an important record of impact events in the last 65 Myr. This record supplements the evidence from impact craters (Table III). At least 12 and possibly 13 impact events in the past 65 Myr can be identified from strewn fields of tektites, from microtektites and related glassy or cryptocrystalline microspherules, from other impact glasses, such as Libyan desert glass and Darwin glass, and from iridium anomalies discovered in sediments of late Pliocene and late Eocene age and in a thin claystone at the Cretaceous-Tertiary boundary. For five of these events, the associated or apparently associated impact crater has been identified; however, only three of these craters—Zhamanshin, Bosumtwi and Ries—are independently dated and listed in Table II. One of the associated craters, the Darwin crater, is smaller than our 5-km-diameter cutoff; the Darwin glass strewn field evidently records a relatively minor impact event. On the other hand, impacts that produced very widely distributed microspherules associated with an iridium anomaly in the late Eocene, and microspherules, shocked mineral grains and the associated iridium anomaly in

TABLE III
Ages of Strewn Fields of Impact Glass and Stratigraphic Iridium Anomalies

	Method of Dating	Age (Ma)	Reference[f]
Indochinites and philippinites (includes microtektites)	$^{40}Ar/^{39}Ar$	0.690 ± 0.028	(1)
	Fission-track	0.693 ± 0.025	(2)
Darwin glass[a]	K/Ar	0.70 ± 0.08	
	Fission-track	0.74 ± 0.04	(3)
Australites	Fission-track	0.830 ± 0.028	(2)
	K/Ar	0.86 ± 0.06	(4)
	$^{40}Ar/^{39}Ar$	0.887 ± 0.034	(1)
Irgizites[b]	Fission-track	1.07 ± 0.06	(5)
Ivory Coast tektites[c] (includes microtektites)	Fission-track	1.08 ± 0.10	(6)
Iridium anomaly and meteoritic particles in late Pliocene deep-sea sediments	Stratigraphic	2.3 ± 0.1	(7)
South Australian high-Na tektites	Fission-track	≥8.35 ± 0.90	(8)
Moldavites[d]	Fission-track	14.7 ± 0.4	(6)
Libyan Desert glass[e]	Fission-track	29.4 ± 0.5	(6)
North American tektites (includes microtektites)	Fission-track	34.6 ± 0.7	(6)
	Stratigraphic	36.0 ± 0.5	(9)
Clinopyroxene-bearing spherules and microtektites in *Globorotalia cerroazulensis* zone. Associated with strong iridium anomaly	Stratigraphic	36.0 ± 0.5	(9)
Cryptocrystalline spherules and microtektites in *Globigerapsis semiinvoluta* zone	Stratigraphic	36.4 ± 0.5	(9)
Iridium-bearing clay layer at Cretaceous-Tertiary boundary containing microspherules of diverse mineral composition and shocked grains of quartz and feldspar	Stratigraphic	65.0 ± 1.0	(10)

[a] Associated with "Darwin" crater, Tasmania (undated).
[b] Associated with Zhamanshin crater, USSR.
[c] Associated with Bosumtwi crater, Ghana.
[d] Associated with Ries crater, W. Germany.
[e] Apparently associated with Oasis crater, Libya (undated).
[f] References: (1) Storzer et al. (1984); (2) Storzer and Wagner (1980); (3) Gentner et al. (1973); (4) McDougall and Lovering (1969); (5) Storzer and Wagner (1979); (6) Storzer and Wagner (1977); (7) Age determined from paleontology and magnetostratigraphy of deep-sea core (Hays and Opdyke 1967); (8) Storzer (1985); (9) Age assigned on the basis of stratigraphic position and a new K/Ar–Sr/Rb calibration of the geologic time scale in the late Eocene by Montanari et al. (1985); (10) Age assigned from Harland et al. (1982) geologic time scale.

the Cretaceous-Tertiary boundary claystone may have released more kinetic energy than any of the impacts that formed the largest known Phanerozoic craters.

Most ages of impact events recognized from the sources of evidence listed in Table III are grouped into two well-defined clusters, one centered at about 35 Ma and the other at about 1 Ma. These clusters correspond fairly closely to the two youngest peaks recognized from the distribution of crater ages. If ages of impact glasses that are derived from well-dated craters are eliminated, four ages remain in the ~ 35 Ma cluster and four in the 1 Ma cluster. These clusters represent a set of observations that are completely independent of crater-age data. Only two of the remaining ages in Table III fall outside these clusters, one at the Cretaceous-Tertiary boundary and one in the late Miocene. The Cretaceous-Tertiary boundary event falls on the ~ 65 Ma crater-age peak. Hence, in the last 65 Myr, the ages of nearly all impact events recognized from the independent evidence of impact glass, microspherules and iridium anomalies are correlated closely with the three principal peaks observed in the crater-age distribution.

The relations of the best-fit cycles derived from the crater ages to the ages of tektites and related glass, the microtektites and related spherules, and the known associated iridium anomalies are shown in Fig. 5. Two cycles derived from the crater ages fit the ages in Table III rather well: the 31 Myr cycle, whose last maximum was 4 Ma, and the 32 Myr cycle whose last

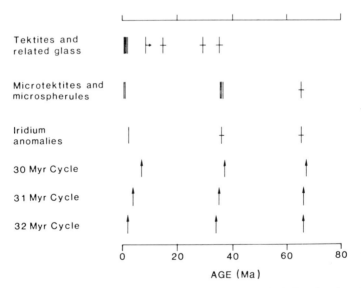

Fig. 5. Age distribution of impact events recorded by tektites and related glass, by microtektites and microspherules, and by iridium anomalies in the stratigraphic column. Uncertainties in the ages are indicated by horizontal bars. Cycles fitted to the distribution of ages of known impact structures are also shown.

MASS EXTINCTIONS, CRATER AGES, COMET SHOWERS 361

maximum was at 2 Ma. In fact, the best-fit cycle derived directly from the ages of Table III has a period of 31 Myr and the last maximum was at 4 Ma; observed ages depart from a 32 Myr cycle whose last maximum was at 3 Ma with only a slightly higher mean squared deviation. The mean squared deviations of the entire set of ages from the 31 and 32 Myr cycles are < 25% of those expected for a random distribution; for the 31 Myr cycle, the mean squared deviation of ages of only the microspherules and iridium anomalies is < 10% of that expected for a random distribution. Thus, the apparent periodicity found in the distribution of crater ages over the last 165 Myr is reflected much more strongly in the ages of impact glass and iridium anomalies in the last 65 Myr.

The distribution of microspherules and iridium anomalies in the stratigraphic column permits a much more precise test of the time relation between impacts and mass extinctions than does the distribution of crater ages. The Cretaceous-Tertiary boundary claystone that contains anomalous abundances of iridium and other noble metals, as well as microspherules and shocked mineral grains has now been identified in the stratigraphic section at about 70 sites around the world. It coincides with the most thoroughly studied and documented extinction in the entire stratigraphic record. This extinction event marking the end of the Cretaceous (end of the Maestrichtian stage) is only one, perhaps the most severe of about five sharp extinctions that occurred in an interval of a few million years, however, and a search for evidence of major impacts at the times of the other extinction steps should be undertaken.

Three horizons of microtektites and microspherules in late Eocene strata lie within the stratigraphic sequence spanned by the late Eocene stepwise mass extinction. One of these microspherule horizons in the *Globorotalia cerroazulensis* foraminiferal zone, which coincides with a relatively strong iridium anomaly (Table III), is correlated with an extinction of radiolaria (Hut et al. 1986). This microspherule horizon and associated iridium anomaly have been identified over a very large area extending from the Carribean to the central Pacific; they clearly record a major impact event, possibly one that produced a large, as yet undiscovered, crater in the sea floor. The stratigraphic distribution of radiolaria in the late Eocene is best defined in sections on Barbados, where the iridium anomaly is found to coincide with a sharply defined extinction of about 25% of the identified extant radiolarian taxa (Sanfillipo et al. 1985). At these sites, the microsperules associated with the iridium anomaly have been almost entirely lost by solution corrosion (Hut et al. 1986), but they are well preserved in deep sea cores obtained in the Caribbean and Gulf of Mexico (Glass 1986).

In the Caribbean and Gulf of Mexico, a microtektite horizon that is correlated with the North American tektites occurs a few tens of centimeters above the stratigraphic position of the iridium anomaly (Glass et al. 1982; Glass et al. 1985). No extinctions have been recognized at the North American microtektite horizon, but here the foraminifera show evidence of environmental stress (Hut et al. 1986; Keller et al. 1986).

The stratigraphically lowest microspherule horizon of late Eocene age is found in the *Globigerapsis semiinvoluta* foraminiferal zone (Table III). It coincides with an episode of strong environmental stress marked by an abrupt permanent decline of the *Globigerapsis* group from about 50% to 1% of the planktonic foraminiferal faunas (Hut et al. 1986; Keller 1983; Keller et al. 1985,1986). As noted by Hut et al. (1986) and Keller (1986), the time of extinction of species is not necessarily the best indicator of environmental stress, as species commonly are rare at the time of their final demise, which may occur as a result of minor ecological perturbations. Hence, the impact-induced environmental crises in the late Eocene may be closely related to the observed stepwise mass extinction of the late Eocene, even though the microspherule horizons do not, in every case, coincide with the extinction steps. Several of these steps do coincide with abrupt cooling events in the ocean (Keller et al. 1983), which conceivably could have been triggered by major impacts not yet recognized.

In this context, it is of interest that the ~ 2 Ma crater-age peak and associated tektite- and microtektite-producing impact events occurred at about the time of a major change in global climate near the end of the Pliocene that resulted in the great Pleistocene glaciations. In all likelihood, the late Pliocene mass extinction was related to this climate change (Stanley 1984), but it is pertinent to ask whether the late Pliocene-Pleistocene glaciations were, themselves, brought about by the apparent surge in the impact rate. Discrete short cooling events caused by global veils of dust from large impacts might have been responsible for the inception of major continental glaciation, provided that the pattern of oceanic circulation in the already cool ocean and an existing ice sheet in Antarctica had set the stage. Many large impacts occurred during the Pleistocene, including at least one that produced a widespread microtektite horizon in the Indian Ocean correlative with Southeast Asian tektites (Table III). An extinction of one radiolarian species at the time of the Jaramillo magnetic event at 1.0 Ma (Hays 1971) appears to be nearly coincident with the formation of the Zhamanshin and Bosumtwi craters and the strewn field of Ivory Coast tektites and microtektites associated with Bosumtwi.

In summary, there is clear-cut evidence of a broad correlation in time between apparent pulses in the impact rate and the major stepwise mass extinctions near the Cretaceous-Tertiary boundary and in the late Eocene. Specific extinction steps in each of these stepwise mass extinctions are coincident with major impact events. Not all recognized impact events are precisely correlated with extinction steps, however. Circumstantial evidence suggests that the ~ 2 Ma peak in the cratering rate may be related to late Pliocene-Pleistocene continental glaciation and to the late Pliocene mass extinction. Episodes of relatively frequent impact-induced global environmental crises may have been more effective in producing mass extinctions than individual very large impact events. Moreover, multiple large impacts and an accompanying increase in the infall rate of cometary dust may have caused climatic instability or may have triggered long-lasting climate changes, and thereby led indirectly to mass extinctions. Hence, not all extinction steps in a time-extended mass extinction should necessarily coincide with large impact events, even if a surge in the impact rate is the ultimate or a contributing cause.

III. EVIDENCE FOR COMET SHOWERS

The combined data on ages of craters and strewn fields of impact glass and associated iridium anomalies indicate at least three and possibly four pulses or surges in the terrestrial impact rate in the last 100 Myr. These surges might be due to a variety of causes. First of all, some of the largest and comparatively rare impact events should have occurred in random clusters in time. This type of surge, which is a perfectly real phenomenon (as opposed to an apparent surge due to sampling error), may be thought of as true "shot noise" in the impact history. The surge of large impacts would not be expected to be accompanied by a surge in the much more numerous small impact events, however, and should not show up in statistics based chiefly on relatively small craters (i.e., craters < 20 km diameter). Thus, the ~ 2 and ~ 99 Ma crater-age peaks are unlikely to represent surges of this type, but, on the basis of crater ages alone, the ~ 32 and ~ 65 Ma peaks might be examples of "shot noise" surges.

Secondly, a surge in the impact rate on Earth can be produced by catastrophic collisional breakup of a large (100 to 200 km diameter) asteroid in the main asteroid belt (Shoemaker 1984*b*). Depending on the position of the asteroid in orbital element phase space, collision fragments can be delivered to one or more resonances and subsequently perturbed to Earth-crossing orbits. Appreciable surges due to breakup of large asteroids should occur once every few hundred million years, and they can be expected to have a characteristic decay time of a few tens of million years (Shoemaker 1984*b*). The ~ 65 Ma crater-age peak, or perhaps the ~ 99 Ma peak, could be related to surges of this type.

A third type of surge expected in the Earth's cratering history is produced by a shower of comets caused by close passage of a star through the comet cloud that surrounds the Sun (Hills 1981). The strength of this type of surge depends on the mass, velocity and impact parameter (miss distance) of the star relative to the Sun and on the space density and distribution of comets in a theoretical inner reservoir of comets, where comet semimajor axes range between about 500 and 10,000 AU. As there are both theoretical and fairly strong empirical grounds to infer a fairly massive inner cloud or reservoir of comets (Shoemaker and Wolfe 1984; chapter by Weissman), there are good reasons to think that comet showers have, in fact, occurred. Hills (1981) calculated that very strong comet showers ($\sim 10^4$ comets passing perihelion per year) have a frequency of about once per 500 Myr. On the basis of a theoretical model of the inner comet cloud (Shoemaker and Wolfe 1984, and unpublished analysis) and the present observed flux of stars in the solar neighborhood, we calculate that significant, but much milder, comet showers occur as frequently as once every few tens of million years, on average. All four of the observed crater-age peaks of the last 100 Myr might reflect comparatively mild comet showers.

A critical test for surges in the cratering record due to comet showers is the decay time of the surges. Comets injected by stellar perturbation into Earth-crossing orbits are fairly quickly ejected from the solar system by planetary perturbations, chiefly those of Jupiter. The characteristic decay time for comet showers produced by stellar perturbations has been found from very extensive Monte Carlo simulations to be close to 1 Myr (Hut et al. 1986). If the perturbation is due to a relatively distant passage of a giant molecular cloud, the duration of the comet shower may be extended for about 1 Myr more. For simplicity of calculation, we will adopt a half-life of 1 Myr as reasonably representative of the decay of comet showers.

If comet showers are periodic, one can calculate the cumulative frequency distributions of deviations of ages of the dated impact structures listed in Table II from the various best-fit cycles (Fig. 6). The observed distributions of deviations of age from the 30, 31 and 32 Myr cycles lie, on average, about midway between the theoretical distribution (1 Myr half-life) for comet showers and for purely random distributions of age. The distribution of deviations from the 32 Myr cycle is closest to that predicted for comet showers, up to about 50% cumulative frequency, but the remainder of the distribution approaches that expected for randomly distributed ages. In part, this approach to the random distribution line reflects the fact that the apparent 32 Myr periodicity is best defined for the last 100 Myr and tends to break down in the time interval of 100 Ma to 225 Ma.

A better fit to the theoretical distribution for comet showers can be obtained if the time interval between showers is considered to be somewhat variable, but, even with the most flexible fits that maintain the 30, 31 and 32 Myr average periods, the "random" component remains at the level of several tens of percent. Of course, some of the deviations could be due to error of age determination, but it should be borne in mind that the cycles have been fitted to the observations. In this type of comparison, therefore, perfectly random distributions of 25 ages tend to be biased away from the random-distribution lines shown in Fig. 6. Within the observational errors of age determination, it appears plausible that at least half of the relatively precisely dated impact structures could be associated with quasi-periodic comet showers.

Given the errors of age determination, identification of impact surges due to comet showers must rest, in the final analysis, on clusters of dated impact events that have the highest relative precision. In general, only the ages of

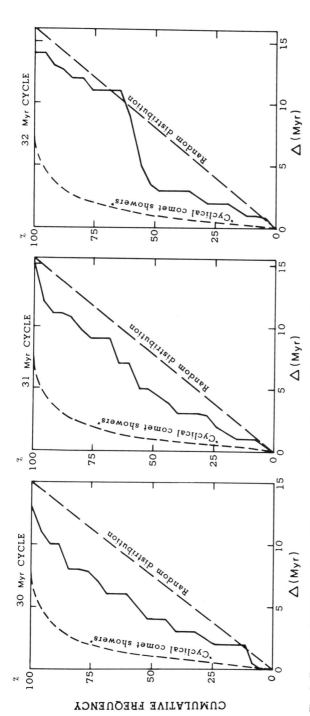

Fig. 6. Frequency distributions of the deviations of ages (Δ Myr) of known impact structures from the best-fit 30, 31 and 32 Myr cycles. Deviations of observed ages for structures ≥ 5 km in diameter formed in the last 250 Myr are shown; only ages determined with a precision better than 20 Myr have been included. For comparison, the cumulative frequency of deviations for perfectly periodic comet showers with 1 Myr half lives and for random distributions are also shown.

craters formed in the last 15 Myr and the relative ages of impact glass horizons in deep-sea sediments are dated with a precision better than 1 Myr; only these ages provide strong tests for the comet shower hypothesis. In this regard, the clustering of ages of known microtektites and other glassy microspherules into two tight groups at $\sim$ 1 Ma and $\sim$ 35 Ma is a striking observational result. In each cluster, the spread of ages is less than the probable duration of comet showers.

The stratigraphic spacing of two microspherule horizons in the upper Eocene, in particular, appears indicative of a comet shower. The spherule-bearing layers are so close, in fact, that at one stage of investigation they were thought to be at the same stratigraphic horizon (see, e.g., Glass et al. 1979). After discovery of an iridium anomaly just below the North American microtektite horizon in a deep-sea core in the Caribbean (Ganapathy 1982), a reexamination of the core revealed a second horizon of chemically distinct microspherules, many of which are crystal bearing, that is coincident with the iridium anomaly. This lower spherule horizon and the associated iridium anomaly have proven to be areally much more extensive than the North American microtektites (Glass et al. 1985). The geographic distributions of the two horizons overlap in the Caribbean region and Gulf of Mexico. On the assumption that average rates of sedimentation during the late Eocene apply to the $\sim$30 cm of clay that separates the two horizons, the time interval between deposition of the two spherule layers has been estimated to be as short as 13,000 to 14,000 yr (Sanfilippo et al. 1985). The clay separating the two spherule horizons has a very low carbonate content everywhere, however, due either to leaching of carbonate or to low biological productivity, either of which may be the consequence of an impact-induced environmental crisis. Thus, the actual time interval represented by the $\sim$ 30 cm of sediments between the two horizons is not precisely known. It may be several tens of thousands of years or even a hundred thousand years.

A still lower, third horizon of glassy microspherules that occurs in the *Globigerapsis semiinvoluta* zone in the western Pacific and probably in the Indian Ocean has been recognized through high-resolution foraminiferal biostratigraphy (Hut et al. 1986; Keller et al. 1983,1986; Keller 1986). These spherules can be distinguished from the higher upper Eocene spherules on a statistical basis by their chemistry (D'Hondt et al. 1986; Keller et al. 1986). On the basis of the Montanari et al. (1985) recalibration of the late Eocene time scale, the spherule horizon in the *G. semiinvoluta* biozone is very roughly estimated by us as about 0.5 Myr older than the two spherule horizons in the overlying *G. cerroazulensis* zone.

If the ages of the five confirmed horizons of glassy microspherules at $\sim$ 1 Ma and $\sim$ 35 Ma were distributed randomly over the past 35 Myr, the chances that two of them would be deposited in the same 0.1 Myr time interval are about 3%. The joint probability that the age of a third horizon would lie within 0.5 Myr of two horizons separated by 0.1 Myr is about 0.12%. Finally, the joint probability of obtaining both a cluster of three ages like that observed at

~ 35 Ma and a second cluster in which the remaining two ages are separated by less than 0.4 Myr (Table III), is 0.012%. Similar odds for obtaining one tight cluster of three ages and one cluster of two ages are found if we include the nonglassy microspherule horizon at the Cretaceous-Tertiary boundary (as shown at 65 Ma in Fig. 5), and if we consider the ages of six microspherule horizons to be distributed randomly over the last 65 Myr.

The odds of finding the tight clusters at ~ 35 and ~ 1 Ma by chance are impressively low, but the possibility of observational bias should be kept in mind. The three upper Eocene microspherule horizons were discovered, in part, as a result of work initially aimed at tracing out just one microtektite horizon thought to be correlative with the North American tektites. The microtektite horizons of the ~ 1 Ma cluster were found partly by deliberate search for microtektites correlative with previously known tektite strewn fields. Hence, the search for glassy microspherules of impact origin has not been carried out systematically for the last 65 Myr of the stratigraphic column. Keller et al. (1983) reported a few possible glassy microspherules at the Eocene-Oligocene boundary (~ 36 Ma) and in the middle Oligocene (~ 30 Ma). The Eocene-Oligocene boundary objects did not turn out to be glassy microspherules, however, and classification of two objects from the middle-Oligocene has yet to be confirmed. It is hoped that both G. Keller and other micropaleontologists, who will have the opportunity to examine thousands of samples from deep-sea sediments, will look for new horizons of impact microspherules.

Surprising results on the dating of tektites have emerged recently that appear to strengthen the evidence for a tight cluster of impact-event ages near ~ 1 Ma. A great strewn field of tektites that extends from Tasmania to southern China, commonly referred to as the Australasian strewn field, has long been thought to be the product of a single large impact event. New fission-track ages reported by Storzer and Wagner (1980), which appear to be confirmed by new $^{40}Ar/^{39}Ar$ ages by Storzer et al. (1984), suggest that the Australasian field actually consists of two strewn fields of tektites separated in age by about 14×10^4 yr (Table III). Tektites found in Australia (australites) may belong chiefly to the older strewn field, whereas tektites found in Southeast Asia (indochinites and philippinites) may belong chiefly to the younger strewn field. A reexamination of deep-sea cores from the Indian Ocean by Glass (1986) has revealed only one microtektite horizon of about the same age as that reported for Southeast Asian tektites, although a few microtektites were observed at a level that might be correlative with the 0.83 to 0.86 Ma age reported for australites. If the reported age difference between australites and indochinites is firmly demonstrated by further work, two (not one) large tektite-producing impacts on the continents are indicated in the middle Pleistocene. So far, no source crater for these tektites has been identified, although two craters of substantial size and nearly the right age have been found (Table II).

The tight cluster of ages of three microspherule horizons at ~ 35 Ma and the equally tight cluster of ages of microspherule horizons, tektites and impact craters at $\sim$ 1 Ma provide fairly strong circumstantial evidence for comet showers centered at $\sim$ 35 and $\sim$ 1 Ma. Evidently the $\sim$ 1 Ma comet shower has nearly decayed, as the present flux of comets can be explained as the consequence of the background level of stellar perturbation of the outer comet cloud (Oort 1950; Hills 1981) and also by the perturbations due to galactic tidal forces (Morris and Muller 1986; Heisler and Tremaine 1986; chapter by Torbett). The observed frequency distribution of semimajor axes of long-period comets can be accounted for by an equilibrium distribution of comet orbits that results from planetary perturbation of comets from the outer cloud arriving at about the present from the outer cloud flux (Weissman 1978).

The cluster of impact-structure ages and the associated Cretaceous-Tertiary boundary impact event at $\sim$ 65 Ma might also be related to a comet shower, but, on the basis of present evidence, they might also reflect a major catastrophic collision in the asteroid belt. Strong evidence for another comet shower at $\sim$ 65 Ma would be provided if most of the steps in the distributed mass extinction at the Cretaceous-Tertiary boundary were found to coincide with large impacts. We have no evidence, at present, to link the $\sim$ 99 Ma crater-age peak to a comet shower. This peak may just be "shot noise" in our sample of crater ages. The $\sim$ 99 Ma peak is chiefly of interest because it fits the apparent $\sim$ 32 Myr cycle of crater ages rather closely.

Three objections have been raised to the interpretation of two or three comet showers in the observed impact record of the last 100 Myr:

1. The present estimated flux of Earth-crossing asteroids already accounts for the known impact events of this time interval (Grieve et al. 1985);
2. Projectile contamination of investigated impact melt sheets, particularly as revealed by siderophile trace-element abundances, indicates that the impactors were chiefly differentiated asteroids (Grieve et al. 1985; Weissman 1985c);
3. No large excursions in the background iridium abundance have been observed in a continuous set of samples from a deep-sea core spanning the interval of 33–67 Ma (Kyte and Wasson 1986).

We examine the merits of each of these objections below.

The present collision rate of Earth-crossing asteroids has been estimated by Shoemaker et al. (1979) at about 3.5 asteroids brighter than or equal to absolute visual magnitude 18 per million years. This estimated rate was shown by Shoemaker (1983) to correspond to a probable production on the continents of about four craters $\geq$ 10 km in diameter per million years. The probable error of the derived asteroid impact cratering rate was estimated as a factor of 2; chief sources of error, of about equal importance, are uncertainties in (1) the Earth-crossing asteroid population, (2) the distribution of albedo and compositional types among the Earth-crossing asteroids, and (3) the crater-scaling relations.

The observed record of large impact events in the past ~ 1 Myr is about that predicted from observations of Earth-crossing asteroids, if the Australasian tektite strewn field represents two events. However, there is an $\sim 25\%$ chance that the production of craters $\gtrsim 10$ km in diameter by asteroid impact is less than half as great as estimated, and there is a good chance that the identification of large impact events that have occurred on the continents in the last million years is still incomplete. The combined astronomical and geological observations certainly permit a sharp pulse of ~ 1 Myr duration in the impact rate that is at least twice and perhaps several times above background. It is unlikely, on the other hand, that this pulse could have been 10 times above background.

The contribution of the background flux of long period comets must also be considered. The present cratering rate due to comet impact has been variously estimated at about 10% to 50% of the present cratering rate due to asteroid impact (Shoemaker and Wolfe 1982; Weissman 1982b). It is not yet clear, however, whether the present comet flux reflects the tail of a pulse. Inclusion of the background flux of comets, in any event, reduces to some extent the plausible amplitude of a comet shower at ~ 1 Ma.

On a longer time scale, Shoemaker et al. (1979), Shoemaker (1983) and Grieve (1984) have suggested that the present flux of Earth-crossing asteroids is consistent with the record of large impact structures in North America, Europe and the USSR for the past hundred or several hundred Myr. It was also recognized, however, that the cratering rate estimated from the present asteroid flux is at least twice the average cratering rate on the Moon over the last 3.3 Gyr.

If a mild comet shower occurred at ~ 1 Ma, one of the consequences would have been an increase in the number of extinct comets in the present Earth-crossing asteroid population. Perhaps half or more of the present Earth-crossing asteroids are either extinct or are dormant, very short period comets (Shoemaker 1984b). A moderate to strong comet shower may produce a step increase of a factor of 2 or more in the Earth-crossing asteroid population, which then decays with a characteristic time of a few tens of million years.

In short, not only is there a probable error of a factor of 2 in the estimate of the present cratering rate by asteroid impact, but the present population of Earth crossers may be substantially higher than average, owing to the postulated comet shower at ~ 1 Ma. The integrated contribution of comet showers to the observed record of impact structures of the last 100 Myr could readily amount to about 50%, within the error of estimation of the long-term average population of Earth-crossing asteroids. If comet showers occurred, on the average, about once every 32 Myr, and if each lasted about 1 Myr, the average cratering rate during a shower could be as much as ~ 30 times the background cratering rate.

The composition of the impacting bodies has been estimated from the relative abundances of siderophile trace elements and, in some cases, from the occurrence of taenite or kamacite in the impact melt rocks, for about half the structures listed in Table II. (See Grieve [1982] and Grieve et al. [1985] for a tabulation.) Five of the impactors are classified as iron or iron (?) bodies, one as an iron or achondrite (?), one as an iron (?) or chondrite, one as an achondrite (ureilite), one as a chondrite or achondrite (?), two as chondrites, and one as stone. Of the craters made by these impactors, the ages of nine fall in the clusters that peak at ~ 2, ~ 32, ~ 65 and ~ 99 Ma. From study of the composition of meteoritic dust (Brownlee 1985), most workers think that the nonvolatile component of comets is similar in elemental composition to carbonaceous chondrites. Thus, only four impactors classified in the above list have siderophile element-abundance patterns suggesting that they might have been comet nuclei; two of these do not fall on the crater-age peaks. The impactors at the 1 Myr old craters Zhamanshin and Bosumtwi are both classified as iron, and the impactors at the three structures of the ~ 32 Ma crater-age peak are classified as an achondrite, as an iron or achondrite (?), and as an iron. Hence, the trace-element evidence appears to contravene the comet-shower hypothesis for the origin of the ~ 2 Ma and ~ 32 Ma crater-age peaks and, indirectly, the comet showers at ~ 1 Ma and ~ 35 Ma postulated from the microspherule distribution.

An apparent anomaly in the identification of impactors, however, is that over half are identified chiefly as iron or as differentiated stony meteorite bodies, and another two as possible differentiated bodies. These indicators stand in contrast to the proportions of differentiated meteorites recovered from observed meteorite falls, among which only about 10% are either iron or differentiated stony meteorites (Mason 1962). Moreover, no more than one-quarter to one-half of the bodies that produce very bright meteoritic fireballs are strong enough to reach the ground in the form of recoverable meteorites (Ceplecha and McCrosky 1976; Wetherill and ReVelle 1981). The objects that break up readily in the atmosphere probably are chiefly friable carbonaceous meteorites. Just a few percent of incoming meteoroids are likely to be irons or differentiated meteorites. On the basis of spectrophotometric and broadband photometric observations, no more than 5% to 10% of observed asteroids are likely to be iron objects (Zellner 1979); only two good candidate iron asteroids (Tedesco and Gradie 1986) have been identified among about 30 well-observed Earth-approaching asteroids.

The difficulty with identifying the impacting body from the abundance pattern of siderophile trace elements in impact-melt rocks is that the trace elements, including noble metals that are used principally as a fingerprint for compositional type, may have been fractionated in the shock-melted and partly shock-vaporized material. The observed abundances of noble metals are highly variable in impact-melt rocks and, in our opinion, generally are unreliable guides to the composition of the impacting bodies.

Finally we come to the objection of Kyte and Wasson (1986) that comet showers in the time interval of 33 to 67 Ma are precluded by the observed iridium content in slowly deposited deep-sea sediments. Their objection is based on the assumption that the intensity of a comet shower that might produce mass extinctions is about 200 times the background comet flux. This intensity of shower certainly appears to be precluded by their observations.

The combined astronomical observations of the present Earth-crossing asteroid flux and the geologic record of cratering events suggest that comet showers in the last 100 Myr have been no greater than about 30 times the combined background flux of comet nuclei and asteroids; the "typical" comet shower may be about 10 times background. Kyte and Wasson's observations of a rather broad iridium anomaly near the Cretaceous-Tertiary boundary are entirely consistent with a comet shower of 1 Myr duration peaking at 30 times background, although they attributed the strong anomaly observed at this stratigraphic position to a single event. Other minor peaks that they observed in the iridium distribution in the Eocene could correspond to showers as strong as 10 times the background comet flux. Indeed, Kyte and Wasson found a relatively high abundance of iridium (1 to 2 parts per billion) throughout the middle and upper Eocene, which they attributed to a slow sedimentation rate in the section sampled. In the core that they studied, the general abundance of iridium is so high that the iridium anomaly known to occur in the *G. cerroazulensis* zone is not recognizable. Comet showers whose intensities were 30 times background at ~ 65 Ma and 10 times background at ~ 35 Ma are compatible with the results of their detailed iridium survey.

IV. ASTRONOMICAL CLOCKS

As the four apparent pulses in cratering rate at ~ 2, ~ 35, ~ 65, and ~ 99 Ma appear periodic, it is appropriate to review in some detail the various astronomical mechanisms that have been suggested for periodic modulation of the comet flux. These include:

1. A hypothetical companion star of the Sun on a highly eccentric orbit that perturbs the comet cloud at each perihelion passage (Davis et al. 1984; Whitmire and Jackson 1984);
2. A hypothetical tenth planet (Planet X) that perturbs an inner reservoir of comets when the line of apsides lies close to the line of nodes on the proper plane of the solar system (Whitmire and Matese 1985; Matese and Whitmire 1986; chapter by Matese and Whitmire);
3. Vertical z oscillation of the Sun through the galactic plane, which leads to modulation of the rate of encounter of stars and molecular clouds with the Sun (Rampino and Stothers 1984a; chapter by Rampino and Stothers);
4. Passage of the Sun through the galactic spiral arms, which also leads to modulation of the encounter rate of stars and molecular clouds with the Sun (Napier and Clube 1979; Clube and Napier 1982a).

The first two of these proposed "astronomical clocks" are speculative, as they involve hypothetical members of the solar system for which there is no direct observational evidence. The second two "clocks" certainly exist, but the amplitude of the periodic or quasi-periodic modulation in the comet flux that they produce is in question.

Solar-Companion Hypothesis

To produce cyclical showers with a period of 30 to 32 Myr, corresponding to the apparent period of pulses in cratering rate, the semimajor axis a_s of an undiscovered solar-companion star must be 97,000 to 101,000 AU. This is approximately the size of the orbit that has long been considered to be the outer bound of the comet cloud (see, e.g., Oort 1950). The outer boundary of the region beyond which a small body would be unbound and would simply drift away from the Sun is an ellipsoid with semiaxes of x = 293,000 AU, y = 196,000 AU and z = 152,000 AU, where x is in the direction toward the galactic center and z is the direction perpendicular to the galactic plane (Antonov and Latyshev 1972). A companion star must have a perihelion distance q_s of $\sim$ 16,000 AU to perturb the inner comet cloud and produce a comet shower, if its mass is $\sim$ 0.1 $M_\odot$ (Hills 1984a); its eccentricity must be $\gtrsim$ 0.84 and its aphelion distance Q_s must be $\gtrsim$ 185,000 AU (for $a_s \approx$ 100,000 AU). The eccentricity could be much smaller only if the mass of the companion is much closer to the solar mass. If the mass is 0.2 $M_\odot$, the perihelion distance required to produce a comet shower is $\lesssim$ 30,000 AU and the aphelion must be $\gtrsim$ 170,000 AU. Torbett and Smoluchowski (1984) and Torbett (see his chapter) have shown that, if the inclination of the orbit of the companion star to the galactic plane exceeds about 30°, a small companion capable of producing comet showers would become unbound over solar system time by the action of the galactic tidal forces alone.

At aphelion distances of 170,000 to 190,000 AU, a companion star would be very weakly bound to the Sun, even if the orbital inclination lay in the stable region. Perturbations by passing stars and molecular clouds would tend to detach the companion over a period of time shorter than solar system time. Therefore, it comes as no surprise that, although binary stars with separations up to $\sim$ 20,000 AU are fairly common (Bahcall and Soneira 1981; Latham et al. 1984), none has been recognized with a separation remotely approaching 100,000 AU. On strictly empirical grounds, it appears unlikely that the Sun is accompanied by a star with the semimajor axis and eccentricity required to produce cyclical comet showers with a period of 30 to 32 Myr.

In order to evaluate the probability that a distant solar companion exists, we have carried out a series of Monte Carlo calculations that began with the orbital evolution of the companion at various distances from the Sun. We treated the companion simply as a very large planet (i.e., a body having a small mass compared with the mass of the Sun). We studied two different cases, one in which perturbations by passing stars alone were considered, and one which also included perturbations by giant molecular clouds.

For the case involving perturbations by passing stars alone, 100 trials were run at each of the following starting semimajor axes: 10,000, 20,000,

30,000, 40,000, 50,000, 70,000, 80,000, 87,764 and 100,000 AU. Initial eccentricity e_o of the solar companion was taken as 0 and its orbital plane was taken as initially coincident with the proper plane of the solar system as determined from the known planets. A mass distribution for passing stars down to 0.1 $M_\odot$ was adopted from the present mass distribution in the solar neighborhood as estimated by Miller and Scalo (1979). A total mass density for observable stars in the solar neighborhood of 0.1 $M_\odot$ pc^{-3} (see reviews by Krisciunas [1977] and Bahcall [1984a]) and a uniform speed of 20 km s^{-1} for all stars with respect to the Sun, appropriate for the weighted average of various classes of stars (Delhaye 1965), were adopted. Although a more realistic distribution of speed, dependent on stellar mass, could have been used, the assumption of uniform speed is considered conservative, in that it eliminates rare strong perturbations due to slow encounters. The motion of the companion about the Sun was taken to be Keplerian until perturbed by a passing star. Perturbations were calculated by the impulse approximation; impulses were derived from the approximate formula used by Oort (1950) and determined for the time of closest approach of the star to the Sun or to the solar companion. Corrections were applied for the finite durations of the perturbations, as derived by Hut and Tremaine (1985). New orbital elements after each perturbation were computed on the basis of Öpik's (1951) equations, and the motion after each perturbation was again considered Keplerian. The motion of the companion was followed until the orbit passed beyond the envelope of stability or for 4.5 Gyr. The results are illustrated in Fig. 7.

In the case of perturbations by passing stars alone, the companion star survived over solar system time (4.5 Gyr) in 91% of the trials, when started with a semimajor axis a_o of 10,000 AU. The fraction surviving for 4.5 Gyr dropped rapidly with increasing a_o, however. At a_o = 50,000 AU, the fraction dropped to 25%; at a_o = 100,000 AU, it dropped to 4%. In most cases of survival with large initial orbits, we found that the final semimajor axes a_f were much smaller than a_o. Only 8% of the trials started at 50,000 AU yielded survivors with a_f larger than the starting orbit, and the 100 trials started at 100,000 AU yielded only one survivor with $a_f \geq$ 50,000 AU. For $a_f \geq$ 50,000 AU, a maximum of 12% survivors was found at a_o = 40,000 AU.

It could be argued that we used too large a flux for the passing stars in our Monte Carlo calculations, as the Sun is now near the galactic plane and the present flux may be near a maximum. If the Sun had a much larger z oscillation in the past, the mean flux of stars might have been lower by about a factor of 2. On the other hand, we neglected objects smaller than 0.1 $M_\odot$, which might account for about half the local mass density of the Galaxy; the whole mass is now estimated to be about 0.19 $M_\odot$ pc^{-3} (Bahcall 1984a,1986). Moreover, passing stars are not the principal cause of unbinding of wide binary systems. The much more massive giant molecular clouds, with masses up to ~ 10^6 $M_\odot$, are more effective than stars in detaching distant companion stars over solar system time.

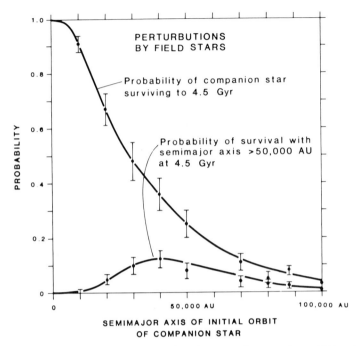

Fig. 7. Probability of survival for 4.5 Gyr of a solar-companion star whose orbit is subject to impulsive perturbations only by passing field stars in the Galaxy. The upper curve shows total probability of survival; the lower curve shows combined probability of surviving and also having a semimajor axis > 50,000 AU at 4.5 Gyr from the beginning of the perturbation history. Estimates of probability are based on a Monte Carlo procedure described in the text. All Monte Carlo runs begin with the companion star in a circular orbit that is in the proper plane of the solar system as determined from the observed planets.

In the second series of Monte Carlo runs, we added perturbations by giant molecular clouds to the perturbations by stars. A smooth mass-frequency distribution of molecular clouds, from 10^3 $M_\odot$ to 10^6 $M_\odot$, fitted to a distribution suggested by van den Berg (1982), was adopted; the mean density of molecular clouds through the region traversed by the Sun was taken as 0.0128 $M_\odot$ pc^{-3}. This mean density corresponds to a density at the galactic midplane of 0.026 $M_\odot$ pc^{-3}, which is nearly the same as that adopted by Hut and Tremaine (1985) and somewhat lower than that estimated by Sanders et al. (1984). The perturbations by the molecular clouds were also modeled by the impulse approximation, with corrections for the durations of the perturbations and for the finite sizes of the giant molecular clouds derived from the work of Hut and Tremaine (1985). Trials were started, as before, with orbits for the hypothetical solar companion at $e_o = 0$ and the orbit plane in the proper plane of the solar system. At $a_o = 100,000$ AU, there were no survivors out of 100 trials; the maximum time to detachment of the companion star was ~ 1.7 Gyr, and the mean time was 196 Myr. Giant molecular clouds accounted for about one-third of the total perturbations and were responsible for three-fourths of the final perturbations in which the companion was lost. Trials were also run at a_o = 10,000, 20,000, 30,000, 40,000 and 87,764 AU. The fractions surviving were 56% at 10,000 AU, 22% at 20,000 AU, 9% at 30,000 AU and 3% at 40,000 AU (Fig. 8). It can be fairly readily understood from these results why no binary star systems have been discovered with separations much greater than 20,000 AU (see also Bahcall et al. 1985). A maximum of ~ 1% survivors with a_f > 50,000 AU at 4.5 Gyr was obtained in the range a_o = 10,000 to 30,000 AU.

After we presented these results at the conference *The Galaxy and the Solar System,* Tucson, Jan. 1985, a paper by Weinberg et al. (1986) appeared that gave an analysis of the stability of wide binaries based on a different method of calculation and somewhat different canonical values of the number

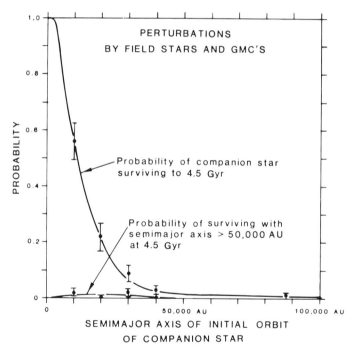

Fig. 8. Probability of survival for 4.5 Gyr of a solar-companion star whose orbit is subject to perturbations both by field stars and molecular clouds. The upper curve shows total probability of survival; the lower curve shows combined probability of surviving and also having a semimajor axis > 50,000 AU at 4.5 Gyr from the beginning of the perturbation history. Estimates of probability are based on a Monte Carlo procedure described in the text. All Monte Carlo runs began with the companion star in a circular orbit that is in the proper plane of the solar system as determined from the observed planets.

density, masses and velocities of the perturbing bodies. They found even shorter lifetimes for wide binaries than ours.

We conclude that the odds are no greater than $\sim 1\%$ that the Sun has a companion star with a semimajor axis in the range necessary for the production of comet showers with any period > 11 Myr ($a > 50,000$ AU). Indeed, 1% may be a generous upper bound. Moreover, to produce comet showers with a period of 30 to 32 Myr, the solar companion must also have an eccentricity of $\gtrsim 0.7$, if its mass is relatively large, and $\gtrsim 0.9$ if its mass is small. There is a $\lesssim 50\%$ chance that the eccentricity of the companion is $\gtrsim 0.7$, and $\lesssim 20\%$ that the eccentricity is $\gtrsim 0.9$ (see, e.g., Hills 1981, for the theoretical frequency distribution of eccentricities). Finally, the orbit of the companion must be fairly stable for ~ 100 Myr in order to produce the apparent periodicity observed in the crater ages. From our Monte Carlo trials at $a_o = 100,000$ AU, we found that there is a chance of $< 50\%$ that the orbit remains sufficiently stable over 100 Myr. When giant molecular cloud perturbations are included, the solar companion is lost before 100 Myr in $\sim 40\%$ of the trials. The overall odds that a companion star produces comet showers with a period of about 30 to 32 Myr appear to be no better than about 1 in 1,000.

V. PLANET X HYPOTHESIS

Whitmire and Matese (1985) suggested that an undiscovered tenth planet orbiting in the region beyond Pluto might produce comet showers in the vicinity of the Earth with a very stable frequency. In their hypothesis, Planet X has a substantial orbital eccentricity e_x and inclination i_x; it revolves within a gap, cleared by the planet, in a thin disk of comet nuclei. Showers of comets occur when the perihelion and aphelion of Planet X pass near the edges of the gap in the plane of the comet disk, as a result of the steady advance of the argument of perihelion ω_x. Two passages of the aphelion and perihelion through the plane of the comet disk occur during every 360° advance of ω_x. The period T_ω is determined chiefly by the size of the major axis of Planet X, $2a_x$, which precesses under the well-known perturbing influence of the other planets. From the formula adopted by Whitmire and Matese (1985), the required value of a_x is about 105 AU for $0.5\,T_\omega = 31$ Myr, on the assumption that $e_x \approx 0.25$ and $i_x \approx 15°$; T_ω is relatively insensitive to moderate variations in e_x and i_x.

For comet showers to occur, it is essential that Planet X clear a gap or greatly reduce the areal density of comets in the region between the perihelion distance q_x and aphelion distance Q_x of the planet. Moreover, the edge of the gap or depleted region must be fairly sharply defined. The comets might either be lost entirely from the neighborhood of Planet X or be piled up just sunward of q_x and beyond Q_x (Matese and Whitmire 1986). Comets in the relatively undepleted disk on both sides of the gap are envisioned as diffusing toward the gap edge as a result either of perturbations by large comet nuclei embedded in the comet disk (Whitmire and Matese 1985) or of the intermediate-range perturbing influence of Planet X itself (Matese and Whitmire 1986). The origin of the gap is not discussed in detail by Whitmire and Matese (1985), although it is treated indirectly by Matese and Whitmire (1986). We consider the putative gap to be the Achilles heel of this intriguing hypothesis.

In order to test whether a gap is a stable feature of the Planet X-comet disk system postulated by Whitmire and Matese, we conducted a number of Monte Carlo simulations of the system. Trial runs were made for various assumed values of i_x and for the mass of the planet m_x with 506 comets each, starting with comet orbits near the aphelion and perihelion of Planet X. All comets were initially assigned an eccentricity e_c of 0.1 and an inclination i_c of 0.05 radians (2°865) to the central plane of the comet disk (assumed to be the proper plane). Comet semimajor axes a_c were chosen near q_x and Q_x such that all comets could encounter Planet X. The semimajor axes were distributed to simulate steep diffusion gradients at the gap edges. Cases were investigated for $i_x = 5°$, 15° and 45° and corresponding values for $a_x = 106.0$, 104.5 and 80.0 AU; e_x was assumed to be 0.25 when $i_x = 5°$, 15°; $e_x = 0.3$ when $i_x = 45°$ (cf. canonical value of $e_x = 0.3$ adopted by Matese and Whitmire 1986). Masses of 5 $m_\oplus$ and 10 $m_\oplus$ were adopted for Planet X (cf. 5 $m_\oplus$ adopted by Matese and Whitmire 1986), and a few runs were made with $m_x = 1$ $m_\oplus$.

The evolution of the comet orbits was followed by a Monte Carlo procedure derived from that described by Arnold (1965), which we have used extensively to investigate the dynamical history of the Uranus-Neptune planetesimal swarm (Shoemaker and Wolfe 1984). Perturbations due to encounters with Planet X and the known planets, as well as the times between encounters, were based on Öpik's (1976) equations for single encounters; changes in orbit due to all encounters within Öpik's (1951) radius of action were calculated for each comet. Perturbations due to encounters of stars with the solar system were also included in the Monte Carlo code for all orbits that started with or evolved to semimajor axes in excess of 100 AU. The effective impulses due to stellar encounters were calculated as described above in our investigation of the hypothetical solar-companion star. In general, the contribution of stellar perturbations was very small, except when the orbits evolved to large size ($a_c > 10^3$ AU). The history of each comet orbit was followed until the comet was ejected from the solar system or until it collided with a planet or the Sun; if neither event occurred, the orbit was followed for 4.5 Gyr. A statistical summary of the results is given in Table IV.

Early trials quickly showed that a planet of 1 $m_\oplus$ to 5 $m_\oplus$ would be ineffective in clearing a gap in the comet disk at any distance appropriate for the Planet X hypothesis. Therefore, we concentrated our investigation on the effects of a 10 $m_\oplus$ planet, in order to elucidate the general pattern of orbital evolution of comets in this region. Even a 10 $m_\oplus$ planet does not clear a gap, although it can reduce significantly the number of comets in the region of close encounters in the cases that we studied, the fraction of comets remaining in the neighborhood of Planet X after 4.5 Gyr ranged from 68% to 82%, for Monte Carlo runs in which $i_x = 5°$, 15°; the fraction remaining in runs with i_x

TABLE IV
Statistical Summary of Results from Monte Carlo Simulations of the Orbital History of Comets in the Neighborhood of Planet X

Parameters of Planet X			Fraction of Comets Surviving in Neighborhood of Planet X After 4.5 Gyr		Mean Semimajor Axes of Comet Orbits				Mean Eccentricities of Comet Orbits			
					Comets Initially Near Perihelion of Planet X (q_x)		Comets Initially Near Aphelion of Planet X (Q_x)		Comets Initially Near Perihelion of Planet X (q_x)		Comets Initially Near Aphelion of Planet X (Q_x)	
Mass	Orbital Inclination	Semimajor Axis	Comets Initially Near q_x	Comets Initially Near Q_x	Initial Mean Semimajor axis	Geometric Mean Semimajor Axis After 4.5 Gyr	Initial Mean Semimajor Axis	Geometric Mean Semimajor Axis After 4.5 Gyr	Initial Eccentricity	Mean Eccentricity After 4.5 Gyr	Initial Eccentricity	Mean Eccentricity After 4.5 Gyr
	(deg)	(AU)	(%)	(%)	AU	AU	AU	AU				
5 $m_\oplus$	15	104.5	89.4		72.00	163.15			0.100	0.445		
10 $m_\oplus$	5	106.0	67.79	70.5	74.17	412.10	145.46	360.00	0.100	0.652	0.100	0.619
10 $m_\oplus$	15	104.5	72.13	83.2	73.15	279.90	143.37	225.32	0.100	0.570	0.100	0.529
10 $m_\oplus$	45	80.0	22.7	96.6	54.36	475.77	113.80	168.35	0.100	0.710	0.100	0.406
10 $m_\oplus$	45	80.0	44.6	81.8	62.23	349.70	89.77	198.38	0.100	0.656	0.100	0.509

Parameters of Planet X			Mean Inclinations of Comet Orbits					Ratio of Comets Perturbed Directly to Jupiter- and Saturn-Crossing Orbits to Total Comets Lost Out of 506 Trial Runs	
			Comets Initially Near Perihelion of Planet X (q_x)		Comets Initially Near Aphelion of Planet X (Q_x)			Comets Initially Near Perihelion of Planet X (q_x)	Comets Initially Near Aphelion of Planet X (Q_x)
Mass	Orbital Inclination (deg)	Semi-major axis (AU)	Initial Inclination (deg)	Mean Inclination After 4.5 Gyr (deg)	Initial Inclination (deg)	Mean Inclination After 4.5 Gyr (deg)			
5 $m_\oplus$	15	104.5	2.865	7.71					
10 $m_\oplus$	5	106.0	2.865	17.07	2.87	14.45		53/61	74/77
	15	104.5	2.865	12.79	2.87	9.98		65/76	28/57
	15	104.5	2.865	23.81	2.87	8.33		250/269	8/8
	45	80.0	2.865	18.78	2.87	10.46		162/172	56/59

= 45° ranged from 23% to 97% (Table IV). Comets removed from the strong-encounter region were not piled up on both sides of the depleted region, as suggested by Matese and Whitmire (1986). Instead, nearly 90% of the comets removed from the neighborhood of Planet X were perturbed directly to Saturn- and Jupiter-crossing orbits and were ejected from the solar system. The remainder were lost chiefly by ejection due to encounters with Neptune or Uranus.

In every set of Monte Carlo runs, the semimajor axes, eccentricities, and inclinations of comets remaining in the neighborhood of Planet X tended, on average, to increase with time, and the distribution of all three orbital elements became widely dispersed. The geometric mean of the semimajor axes generally increased to several hundred AU, the mean eccentricity increased to about 0.5 to 0.7, and the mean inclination to about 10° to 19°. Except in the cases of comets near q_x and Q_x when $i_x = 45°$, the trends and the amounts of orbital evolution were relatively insensitive to the starting orbits of the comets. In most cases, the final distribution of orbits for comets starting near the perihelion distance of Planet X, q_x, was not statistically discriminable from the final distribution for comets starting near the aphelion distance Q_x.

Our results show that, while the number density of comets in the neighborhood of Planet X can be reduced if the planet is very massive, and while the probability of encounter with Planet X for the remaining comets is also reduced as a result of dispersion of their orbital elements, it is impossible for Planet X to form a gap with sharp edges. In fact, if a sharp-edged gap in the comet disk were formed initially by some other mechanism, Planet X would disperse the comets near the gap's edges. If comets diffused toward the edges, the gradients of comet density near the original edges would be low at the present time, and the gap would be partly filled with the dispersed comets. If no gap were formed by other, unspecified mechanisms, a 10 $m_\oplus$ Planet X with high inclination would significantly deplete the density of comets near q_x. The result would be that the minimum rate of comet perturbation would now occur when the apsides lie near the central plane of the comet disk. This is precisely the opposite effect from that predicted by Whitmire and Matese.

Because the perturbations by Planet X would tend to fill a sharp-edged gap, rather than create one, and would smooth out the edges, we consider it very unlikely that an undiscovered planet is responsible for comet showers. On the other hand, if a relatively massive planet exists in the region beyond Pluto, and if this region is fairly densely populated with comets, there is no question that this planet could perturb comets fairly efficiently into Jupiter-crossing orbits. Therefore, a tenth planet could contribute to the capture of Jupiter-family comets and account for part of the background population of these comets, as suggested by Matese and Whitmire (1986).

There are three principal reasons, however, for doubting the existence of a relatively massive (2 $m_\oplus$ to 5 $m_\oplus$) planet beyond the orbit of Pluto. First of all, the reported residuals in the motion of Uranus and Neptune used to deduce the existence of Planet X (Rawlins and Hammerton 1973) may be spurious. We now know the present position and motion of Uranus with great accuracy as a result of the Voyager 2 encounter; comparison of the position and motion of Uranus derived from spacecraft-tracking data with those derived from conventional astrometric observations reveals systematic errors in the relatively recent astrometry (M. Standish, personal communication, 1986). In our judgment, the reported residuals in the motion of both Uranus and Neptune may be due largely or entirely to systematic errors of both the old as well as the more recent astrometric measurements (see also the chapter by Anderson and Standish).

Secondly, a planet whose mass is in the range 2 $m_\oplus$ to 5 $m_\oplus$ probably would have an apparent B magnitude in the range 13 to 15 (cf. Matese and Whitmire 1986). This magnitude is significantly brighter than the conservatively estimated limit of completeness of the survey for additional planets conducted by Tombaugh (1961). If a planet as bright as magnitude 13 to 15 exists, its orbital inclination must be high and it must have been hiding in the southern skies below the declination limits of Tombaugh's survey.

Finally, the time scale of planetary accretion at distances beyond the aphelion of Pluto is long in comparison with solar system time. It is unlikely that a planet as massive as $\gtrsim 1$ $m_\oplus$ could have formed in this region. It is also unlikely that a planet of large mass (i.e., ~ 10 $m_\oplus$) could have been deflected into this region by encounter with Neptune (cf. Harrington and Van Flandern 1979) without leaving Neptune in a much less regular orbit than what we observe today. On the other hand, it is entirely plausible that planets about the size of Pluto remain to be found at distances of 50 to 100 AU; systematic searches for faint distant planets are appropriate and timely (Kowal 1979; C. S. Shoemaker, unpublished proposal).

VI. OSCILLATION OF THE SUN PERPENDICULAR TO THE GALACTIC PLANE

The hypothesis of Rampino and Stothers (1984a) that the vertical z oscillation of the Sun through the galactic plane leads to periodicity of comet showers has two attractive features. Both the period and phase of the cyclical comet showers predicted by this hypothesis fit the apparent pulses in cratering rate over the last 100 Myr. The Sun's last crossing of the galactic plane occurred at about the time of the crater-age peak centered at ~ 2 Ma. The height of the Sun above the galactic plane $z_\odot$ is variously estimated at 8 ± 12 pc, from observations of H I clouds, to 20 to 30 pc, from the distribution of young stars and star clusters (Stothers 1986). As the Sun's z component of velocity $Z_\odot$ is fairly well determined at about 8.6 ± 1 km s^{-1}, (8.8 ± 1 pc yr^{-1}) (Stothers 1986), the last plane crossing probably occurred in the last ~ 3 Myr, perhaps at 1 ± 1 Ma, if H I observations provide the best frame of reference for $z_\odot$ (see also Bahcall and Bahcall 1985). Taking Bahcall's (1984a) latest estimate of 0.185 ± 0.02 $m_\odot$ pc^{-3} for the local mass density at the galactic plane $\rho_\odot$, we can roughly estimate the half period $P_{1/2}$ of the z oscillation of the Sun from the formula for a harmonic oscillator:

$$P_{1/2} = (\pi/4G\rho_\odot)^{1/2} = 30.7 \pm 0.2 \text{ Myr}. \tag{1}$$

This period matches very closely the best-fit period derived from the ages of impact structures over the last 165 Myr and the best-fit period for impact glass, microspherule horizons and iridium anomalies. Bahcall and Bahcall (1985) have evaluated the period and height of the Sun's oscillation on the basis of a variety of models of the vertical distribution of mass in the galactic disk and found periods ranging from $\sim$ 26 Myr to $\sim$ 36 Myr. The most conventional model, based on approximately equal amounts of unobserved and observed material, yields periods of 30.8 to 32.1 Myr, which again match very closely the period obtained from the dates of terrestrial impact events. Perturbations of the Sun's motion due to encounters with a massive object typically produce a jitter in the phase of oscillation of about 6 to 9%.

The principal difficulty with the hypothesis of Rampino and Stothers is that stars and molecular clouds, whose encounters with the Sun produce comet showers, are not sharply concentrated at the galactic plane. The scale height of observable stars, at the present galactocentric radius of the Sun, is variously estimated at 46 pc to 57 pc; somewhat similar estimates are made for the vertical distribution of molecular clouds (see Stothers [1986] for a review). An amplitude of z excursion of the Sun from the galactic plane z_{max} consistent with the values of $z_\odot$, $Z_\odot$ and $P_{1/2}$ used above can be roughly estimated from the harmonic oscillation formula by

$$z_{max} = (z_\odot^2 + P_{1/2}^2 Z_\odot^2/\pi^2)^{1/2} = 86 \pm 4 \text{ pc}. \tag{2}$$

As shown by Bahcall and Bahcall (1985), however, this formula tends to overestimate z_{max} by about 5% to 30%, depending on the actual vertical distribution of mass in the galactic disk. For the most conventional model, they find $z_{max} \simeq 70$ pc. As z_{max} is comparable to or possibly slightly greater than the probable scale height of most perturbing bodies, a fairly strong fluctuation should occur in the probability of occurrence of a comet shower over a half cycle of z oscillation. Depending on the model of vertical mass distribution, the present amplitude of the variation probably is about a factor of 2 to 5. The Sun passes rather quickly through the region of highest density of perturbing bodies, however, so that a substantial fraction of comet showers should occur at times when the Sun is far from the galactic plane.

Stothers (1986) has shown that the periodicity of comet showers should be detectable in the existing terrestrial impact record for the last $\sim$ 600 Myr, provided that the z distribution of perturbing bodies in the galactic disk is not dominated by a dispersed component near z_{max}. For a much shorter interval of time, the detection of periodicity is problematical. To first order, the distribution in time of the last 4 comet showers should be more nearly random than periodic.

In the limiting favorable case that the distribution of all the mass near the present galactocentric radius of the Sun is distributed with a scale height of 50 ± 5 pc, Stothers (1986) found that about 74% of the encounters of perturbing bodies with the Sun should occur during the 50% of the cycle of z oscillation when the Sun is closest to the galactic midplane. Under these conditions, there are *a priori* probabilities of $\sim 40\%$ that a comet shower occurred within ± 0.5 $P_{1/2}$ of each of the last 3 midplane crossings and $\sim 30\%$ that comet showers occurred within ± 0.5 $P_{1/2}$ of the last 4 crossings (assuming unit probability of a comet shower during each one-half cycle of z oscillation). The apparent pulses in cratering rate at ~ 2, ~ 35, ~ 65 and ~ 99 Ma are, however, much closer to the estimated times of midplane crossing than ± 0.5 $P_{1/2}$. Deviations in time from perfectly periodic crossings corresponding to the harmonic oscillator solution for $z_\odot = 8$ pc, $Z_\odot = 8.6$ km s^{-1} and $P_{1/2} = 30.7$ Myr are no greater than $\sim 0.2\ P_{1/2}$. The probability of a shower-producing perturbation occurring within ± 0.2 $P_{1/2}$ of the midplane crossing is only ~ 0.3. Hence the *a priori* odds that the last three comet showers should have occurred this close to the estimated time of crossing are no greater than $\sim (0.3)^3 \simeq 3\%$; the odds that the last 4 showers occurred this close to the midplane crossing are $\sim 1\%$. If periodicity of large body impact on Earth is attributed largely or entirely to the modulation of comet showers by the z oscillation of the Sun, it is difficult to escape the conclusion that the near coincidence of the last 4 apparent pulses in cratering rate with a likely sequence of times of galactic midplane crossings is an improbable statistical fluke.

Thaddeus and Chanan (1985) have analyzed the probability of detecting periodicity in the terrestrial cratering record produced solely by perturbations of the comet cloud by molecular clouds. They found the scale height of molecular clouds to be somewhat greater than z_{max}, and they concluded that observations of the impact effects of more than 300 comet showers would be required to detect even a relatively modest periodic signal. On independent grounds, however, it is unlikely that molecular clouds are responsible for the apparent 30 to 32 Myr periodicity in the impact record. Encounters with molecular clouds are insufficiently frequent, and long-range encounters with giant molecular clouds are ineffective in perturbing the inner comet reservoir, where comet showers must originate. Most sharply defined comet showers (half life ~ 1 Myr) must be produced chiefly by stellar perturbations of the inner comet reservoir.

VII. PASSAGE OF THE SUN THROUGH THE GALACTIC SPIRAL ARMS

It has been realized for many years that the Sun must travel through the spiral arms of the Galaxy; in an extended series of papers, W. M. Napier and

S. V. M. Clube have discussed how the encounter of the Sun with a giant molecular cloud might produce a strong comet shower. In most of their papers, they have argued that the Oort cloud of comets would be stripped from the Sun during a giant molecular cloud encounter and that there would be a shower of comets derived from the molecular cloud. They have further pointed out that the probability of encounter with a molecular cloud is enhanced during passage through a galactic spiral arm. As spiral-arm passages are quasi-periodic, the comet showers produced would be quasi-periodic also (see, e.g., Napier and Clube 1979; Clube and Napier 1982a). Recent investigations show that the Sun's comet cloud is fairly stable against stripping by giant molecular cloud encounters (Hut and Tremaine 1985), a result that is confirmed by our study of the stability of a solar-companion star given above. The outer bound of the comet cloud is established primarily by molecular cloud encounters; comets stripped from the outer edge probably are replaced by outward diffusion of comets from the massive inner part of the cloud. The vast majority of comets that have entered the inner solar system probably are original members of the system and were derived from the cloud of comets that has remained bound to the Sun. Nevertheless, the quasi-periodic passage of the Sun through the galactic spiral arms almost surely modulates, to some extent, the flux of comets at the Earth.

The spiral arms of the Galaxy are controlled by waves of stellar density that revolve about the galactic center at about half the rate of revolution of the stars themselves at the present galactocentric distance of the Sun (Lin et al. 1969). There is some debate as to whether our Galaxy has two spiral arms, as observed in most other spiral galaxies and assumed by Lin and his coworkers, or whether it has four spiral arms (Blitz et al. 1983). The average period between spiral-arm passages of the Sun is approximately equal either to half the galactic year (~ 125 Myr), or to one-quarter of the galactic year. As both the orbit of the Sun and the spiral arms are somewhat irregular, the time between spiral-arm passages should also be somewhat irregular. In any event, the mean period is about two to four times greater than the apparent period of pulses in terrestrial cratering.

The increase in encounter rate with stars during spiral-arm passages probably is on the order of 5% to 10% (cf. Lin et al. 1969). Molecular clouds may be somewhat more concentrated in the arms, where the clouds are evidently shock compressed. This compression accounts for the birth of massive young stars that make the arms visible (Shu et al. 1972). It is certain that the rate of encounter with these stars is highest in the arms, which increases the rate of perturbation of the inner comet cloud during passage through an arm. Overall, the probability of perturbation of the inner comet cloud may increase by about 10% during spiral-arm passages. The resulting quasi-periodic modulation of the terrestrial cratering rate would be detectable only with a very long and much more complete record of impact structures than is available to us now.

VIII. SUMMARY

The formation of large impact craters in the last 220 Myr appears to have been moderately cyclical, with a best-fit period of 30 to 31 Myr; four apparent pulses in the cratering rate, broadly centered at ~ 2, ~ 35, ~ 65 and ~ 99 Ma, are fairly strongly cyclical, with a best-fit period of 32 Myr. Sharp pulses of impact events at ~ 1 Ma and ~ 35 Ma are indicated by strewn fields of impact microspherules. The period and phase of the impact-event cycles match fairly closely the best-fit cycle obtained from strong well-dated mass extinctions; impact pulses at ~ 35 Ma and ~ 65 Ma, in particular, are correlated with strong stepwise mass extinctions. The sharp pulses of impact events are best explained by comet showers stimulated by close encounters of stars or molecular clouds with the Sun.

Neither the solar-companion star hypothesis nor the Planet X hypothesis appears to offer a likely mechanism for producing periodic comet showers. While the existence of a star revolving about the Sun on an orbit of the size, eccentricity and short-term (~ 100 Myr) stability required to produce the four apparent pulses in impact rate cannot be disproved, the odds that a star with the required orbit exists appear to be no better than 1:1000. An undiscovered tenth planet whose apsides precess with respect to the node with a period of 30 to 32 Myr might well exist, but we have found that such a planet, even if it is as massive as 10 $m_\oplus$, would not clear a sharp-edged gap in a thin disk of comet nuclei. Therefore, the postulated Planet X cannot produce detectably cyclical comet showers.

Oscillation of the Sun normal to the galactic plane appears to have very close to the right frequency and phase to explain the apparent pulses of impact events in the last 100 Myr. However, the correlation in time between apparent impact pulses and the last four galactic midplane crossings of the Sun, calculated on the basis of Bahcall's (1984a) estimate of the mass density in the solar neighborhood, is much stronger than can be expected from any plausible scale heights of stars and molecular clouds above the galactic plane. The *a priori* odds of finding a correlation as high as that observed are $\sim 3\%$ for the last three apparent pulses and $\sim 1\%$ for the last four.

Quasi-periodic passage of the Sun through the galactic spiral arms probably has resulted in very weak ($\sim 10\%$) modulation of the comet flux. If the Galaxy has four arms (Blitz et al. 1983), the mean period between passages of the Sun through the arms is close to twice the period of passages through the galactic plane and might lead to reinforcement of the modulation due to z oscillation, provided the passages are approximately in phase.

To sum up, the z motion of the Sun has probably resulted in periodic modulation of the comet flux at the orbit of the Earth sufficiently strong that the period should be detectable in the long term (Phanerozoic) cratering record. This is especially true if, during most of the Phanerozoic, z_{max} were closer to the average height expected for a star of the Sun's mass and age. The very good fit of the last four apparent impact pulses evidently is a statistical fluke. The odds of occurrence of such a fluke, on the other hand, are at least an order of magnitude better than the odds that a solar companion is responsible for the impact pulses. Most comet showers are generated by close passages of stars, and the estimated strength and frequency of possible showers in the last 100 Myr is roughly consistent with the estimated flux of stars in the solar neighborhood.

If most strong mass extinctions are related to comet showers, a periodicity probably should also be detectable in mass extinctions over Phanerozoic time. However, the present chronometric control for the geological time scale is inadequate to demonstrate periodicity of the known mass extinctions. A crucial direction for future research is to identify and accurately date a much larger fraction of the terrestrial impact events that have occurred in the last half-billion years and to determine accurately the correlation between these events and the recognized biological crises. If comet showers have been the primary trigger for mass extinctions, most mass extinctions probably have occurred in discrete steps distributed over intervals of no more than a few million years. Therefore, the detailed history of each mass extinction should be examined by means of intensive stratigraphic studies. Evidence for large impact events should be looked for both at the stratigraphic position of each extinction step and in the general stratigraphic interval of the mass extinctions.

Acknowledgments. It is a pleasure to acknowledge detailed discussions, correspondence and exchange of information over the past few years with L. Alvarez, W. Alvarez, F. Asaro, J. Bahcall, V. Clube, S. D'Hondt, W. Glass, R. Grieve, J. Hills, P. Hut, G. Izett, A. Jackson, E. Kauffman, G. Keller, F. Kyte, J. Matese, S. Montanari, R. Muller, C. Orth, M. Rampino, D. Raup, J. Sepkoski, R. Smoluchowski, M. Standish, S. Stanley, D. Storzer, R. Stothers, M. Torbett, J. Wasson, P. Weissman and D. Whitmire. We thank D. Raup, in particular, for very helpful critical review of an early draft of the manuscript. Finally, we are indebted to D. Weir, S. Bounds, and C. Shoemaker, without whose special help the manuscript could not have been brought to a satisfactory editorial state in time, and to M. S. Matthews and R. Smoluchowski for their encouragement and patience.

EVIDENCE FOR NEMESIS: A SOLAR COMPANION STAR

RICHARD A. MULLER
University of California at Berkeley

The evidence that the Sun has a companion star "Nemesis" responsible for periodic mass extinctions is reviewed. A Gaussian ideogram of the rates of family extinctions in the oceans shows periods of 26 and 30 Myr. Analysis of impact cratering on the Earth shows a period of either 28.4 or 30 Myr, depending on the crater selection. Models which attempt to explain these periods with either oscillations through the galactic plane, or through the effects of a tenth planet, are seriously flawed. If the periods seen in the data are real (and not a spurious result of a statistical fluctuation), then the Nemesis hypothesis is the only suggested explanation that has survived close scrutiny. The Nemesis model predicts that the impacts took place during brief storms of several million years duration, perhaps accounting for the extended nature of the mass extinctions. A search for Nemesis is underway at Berkeley.

We have recently seen the status of the Alvarez et al. (1980) claim, that an asteroid or comet hit the Earth at the time of the Cretaceous extinctions, changed from that of a disputed hypothesis to that of standard dogma. Although there are still some who dispute that the impact caused the extinctions, there is virtually nobody now who disputes that an impact took place. But at the same time that the Alvarez model was becoming the standard picture, two new and equally startling claims have arisen to keep the skeptics busy.

The first is the discovery by Raup and Sepkoski (1984; Sepkoski and Raup 1986) that the mass extinctions were not isolated events, but that they have occurred regularly with a period of 26 Myr. The original papers of Raup and Sepkoski show that an alternative period of 30 Myr fits the data almost as well. In Fig. 1, I have plotted the data of Raup and Sepkoski in a way some-

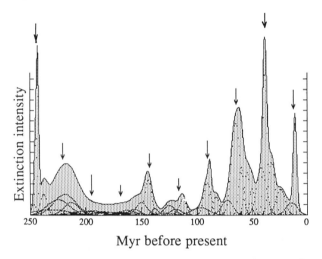

Fig. 1. The data of Raup and Sepkoski (1984) plotted as a Gaussian ideogram. For each extinction, an error estimate ß was calculated from the uncertainty published in the Harland time scale (Harland et al. 1982), taken in quadrature with the duration of the preceding stage. Each point was turned into a Gaussian curve, with width equal to ß and area equal to the percentage of family extinctions published by Raup and Sepkoski. The Gaussian curves for all the 39 stages were added to produce the figure. The arrows indicate the 26 Myr periodicity found by Raup and Sepkoski.

what different from that in their paper. The percentage of family extinctions at each of the geologic stages has been represented by a Gaussian curve, with area equal to the percentage, and half width equal to the uncertainty in the time of the extinctions. In this plot the regularity of the extinctions is somewhat more evident to the eye than in the original plot presented in their paper, but more importantly the curve in this plot represents an estimate of the extinction rate in the last 250 Myr. (We believe the statistical significance of the Raup and Sepkoski periodicity is more evident than in the logarithmic plot originally published by them; we suspect that much of the skepticism about the significance of their discovery arises because of the relatively unconvincing way in which they plotted the data.) The Fourier power spectrum of this curve is shown in Fig. 2. As in the original time series analysis of Raup and Sepkoski (1984), the two dominant periods at 26 Myr and 30 Myr are evident. One should not assume that both of these periods are necessarily physical. One physical period can give rise to two apparent Fourier peaks if a few of the cycles are missing, as explained by Alvarez and Muller (1984), who showed that a 28.4 Myr period could fit the extinction data of Raup and Sepkoski by slipping phase during the missing cycles. A single frequency with modulated amplitude is mathematically indistinguishable from the sum of two pure sine waves:

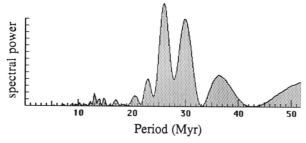

Fig. 2. Fourier transform of the curve shown in Fig. 1. The existence of significant periodicity at 26 or 30 Myr is indicated by the two peaks at these periods.

$$2 \cos[\Omega t] \cos[\omega t] = \cos[(\Omega+\omega)t] + \cos[(\Omega-\omega)t]. \qquad (1)$$

We do not claim that the two peaks seen in the data must be interpreted in this way, but likewise it is premature to conclude either that there are two physical mechanisms responsible for the two peaks, or that at most only one of the two peaks in the Fourier transform is significant. The truth could lie in between.

The second startling discovery was the claim made independently by Alvarez and Muller (1984) and by Rampino and Stothers (1984) that a similar periodicity existed in the dates of impact craters on the Earth. Both of these groups had read the Raup and Sepkoski paper, but they were inspired by different models: Alvarez and Muller by the Nemesis hypothesis of Davis et al. (1984), and Rampino and Stothers by their own theory of galactic oscillations. (The model of Whitmire and Jackson (1984) is equivalent in most respects to the Nemesis model, but requires an orbit of eccentricity $e = 0.9$; it is not clear whether such a high eccentricity is sufficiently stable to describe the extinction data [Hut 1984a].) Alvarez and Muller found a dominant period of 28.4 Myr with an uncertainty of ±1 Myr. Rampino and Stothers saw a broad peak in the Fourier transform at 31 Myr with a full width at half maximum of 4 Myr, but did little statistical analysis and did not offer an uncertainty limit. Shoemaker (see the chapter by Shoemaker and Wolfe) made a reanalysis of the crater periodicity, using the craters selected by Alvarez and Muller, supplemented with a few additional craters, and with his own selection of the best dates for each one, based on a search of the original literature. (Shoemaker preferred dates obtained by fission-track dating, for example, over dates obtained from the potassium-argon method.) He concluded that there was a periodicity present but it might not be statistically significant. If one was present, its dominant period was 30 ± 1 Myr. The Fourier transform of a Gaussian ideogram that we made from Shoemaker's crater list is shown in Fig. 3.

These two discoveries, if real, significantly alter the original model of Alvarez et al. (1980). The most important change is that the impacts must come during showers or storms when it is possible that there may be several

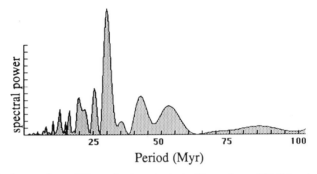

Fig. 3. Fourier transform of Shoemaker (see chapter by Shoemaker and Wolfe) crater ideogram, showing a single broad peak at a period of 30 Myr.

impacts in a relatively short time. The idea of comet showers triggered by a passing star was originally due to Hills (1981). The Nemesis model (Davis et al. 1984) predicts the duration of these storms to be from 1 to several Myr, depending on the eccentricity of the companion star orbit. Thus, impacts are not expected to be isolated events, but should occur in short bursts. The significance of this change is that it offers the possibility of explaining the claim of some paleontologists that the extinctions were not abrupt, but took several Myr. The gradual mass extinction would, upon closer scrutiny, be due to a series of impacts closely spaced in time.

I. NEW IRIDIUM DETECTOR

F. Asaro and collaborators at Berkeley are making a careful search for the multiple iridium layers that are predicted in this model. A new detector invented by L. W. Alvarez is being constructed that will enable a complete scan of rock for the last 250 Myr. The instrument uses two germanium detectors to look for coincident gamma rays at 316 and 468 keV from the decay of activated Ir-192; a mineral oil scintillation counter surrounding the germanium gamma detectors will be used to reject cosmic ray events and background from Compton-scattered gammas. The system will be capable of detecting iridium levels as low as 50 parts per trillion in 7 minutes of counting, using raw rock samples that have not been chemically purified.

II. NEMESIS AND THE SOLAR SYSTEM

The Nemesis hypothesis which offers a solar companion star as the possible trigger for the comet showers, has held up quite well after a year of close scrutiny. As stated in the original paper by Davis, et al. (1984), the orbit required for a 26 to 30 Myr period is unstable against perturbations from passing stars, with a lifetime of about 10^9 yr. Thus, it is extremely unlikely that the star has been in this orbit since the beginning of the solar system; it must have begun in a much more tightly bound orbit. Hut (1984a) has shown that, on the average, the effects of passing stars will drive the binding energy of the Nemesis-Sun system linearly towards zero. Thus, with 1 Gyr remaining, we can extrapolate into the past and conclude that Nemesis was probably 4.5 times as tightly bound at the time of the creation of the solar system. At this time Nemesis would have been 4.5 times closer, with a semimajor axis of 20,000 AU and a period of 3 Myr. The reader is referred to Hut's chapter for details of the effect on the original solar system. One important note of caution: the average behavior is a very poor indicator of actual behavior, as Hut's Monte Carlo calculations have shown. Thus, we cannot really conclude that we know anything about the past history of Nemesis in the case of interest. Nevertheless, it is still somewhat reassuring that the extrapolated separation of the Sun and Nemesis at the creation of the solar system is only 20,000 AU, a value at which other binary stars have been found.

As stated in Davis et al.'s (1984) original paper, the orbit is sufficiently unstable that we do not expect a strict periodicity, but expect to see fluctuations of 10 to 15% over the past 250 Myr. It is possible that the fluctuations in the orbit, both in semimajor axis (affecting the intensity of the storms) and in period, give rise to the double peak structure seen in the Fourier analyses.

Soon after the original Nemesis hypothesis was published, objections were raised that passing molecular clouds would have a much bigger effect than passing stars in disrupting the orbit. Hut has shown that this is not true, by extrapolation from the relatively high abundance of double stars with semimajor axis of 10,000 to 20,000 AU. He found that the lifetime against breakup for this separation was comparable to the ages of the stars, and a simple extrapolation to the Sun-Nemesis system then indicates that the lifetime could not be much less than our calculated 1 Gyr. Therefore, at present (1985), we believe that there is no theoretical difficulty in postulating the existence of Nemesis with the orbit originally assigned to it.

III. OSCILLATIONS ABOUT THE GALACTIC PLANE?

Other theories to account for periodic comet showers have not fared so well. The model of Rampino and Stothers (1984a) had the showers caused by molecular clouds concentrated in the galactic plane. Their original paper claimed a 99% correlation between their calculated plane crossings for the Sun, and the Raup and Sepkoski (1984) mass extinction times. Unfortunately this was in error; the correlation in the two data sets is easily shown to be zero, within statistics. Rampino and Stothers had made a simple mathematical mistake, as was pointed out by S. M. Stigler (1985): they had compared the correlation cofficient which they had obtained to the correlations expected between unordered numbers, when in fact the numbers in their two lists were ordered. And finally, in a careful analysis done by Thaddeus and Channan (1985; also see the chapter by Thaddeus), it was shown that the mechanism proposed by Rampino and Stothers, perturbations by molecular clouds, could not possibly account for the observed showers. In fact, the Thaddeus and Channan method can be used to show that it is impossible to trigger the observed showers from any mass concentration in the galactic plane.

IV. GEOLOGIC RHYTHMS?

Rampino and Stothers (1984b) have proposed that a period of 33 ± 3 Myr can be seen in many geological and biological upheavals. However, I do not think that their conclusion follows from their data because most of the phenomena referred to have no statistically significant periodicity. One can always find a peak in the Fourier power spectrum of any data, and if the peaks from many data sets always fell at the same period that would then be of interest; this is not the case. I have plotted a histogram of all the frequencies that they found in Fig. 4; in addition, I have added two periods from publications to which they refer but do not mention explicitly (26 Myr from Raup and Sepkoski [1984], and 28.4 Myr from Alvarez and Muller [1984]). As can be seen in the figure, the distribution is relatively flat, as expected from the fact that most of the periods plotted (with the exception of those two that I added) are from data with no statistically significant periodicity. Rampino and Stothers incorrectly conclude that a period of 33 ± 3 Myr is present, and to do so they ignore all periods in the range 11–14 Myr because they are third harmonics of 33; they ignore the periods from 15 to 16 because they are approximately second harmonics of 33; they ignore the periods from 18 to 23 because they are 3/2 harmonics of 33, and as mentioned previously they ig-

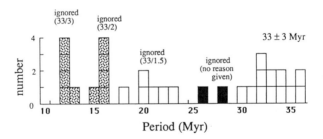

Fig. 4. Histogram of periods found by Rampino and Stothers (1984). The clear boxes represent periods listed in Table 2 of their paper. In their text they mention other periods that they found in the literature; these are plotted as the speckled boxes. Two other periods are plotted in solid black from Raup and Sepkoski (1984) and Alvarez and Muller (1984), respectively; these papers also appear in the list of references of Rampino and Stothers but the periods plotted are not explicitly mentioned in their paper. The flat nature of their complete histogram suggests that no real periodicity is present.

nore the periods of 26 and 28.4 with no reason given. Then they conclude that there is statistically significant clumping around 33 Myr. There is no valid justification for their procedure, particularly since in doing such work it is extremely important, as they say in their paper, "to avoid any possibility of subjective bias in selecting the data to be analyzed." In summary, the data they present in their paper does not justify their conclusion of statistically significant "geologic rhythms."

V. PLANET X?

Another clever trigger for comet showers was suggested by Whitmire and Matese (1985) and Matese and Whitmire (1986; see also their chapter and chapter by Anderson and Standish) who postulate the existence of a tenth planet, Planet X, with mass of 1 to 5 $M_\oplus$, orbiting 70 to 100 AU from the Sun. They must postulate that there is an inner part of the Oort comet cloud that has maintained a disklike shape in the plane of the inner planets. Planet X periodically scrapes the edge of this disk, as its perihelion advances around its own orbit which is tilted at an angle of perhaps 45 degrees to the ecliptic. The major problem with this model is possibly the stability of the disk of comets near Planet X, as was first pointed out by D. Morris (personal communication). Perturbations from Planet X itself will cause these comets to leave the ecliptic and fill the postulated gap in a time comparable to the 26 to 30 Myr perihelion advance time. The comets would not then arrive in storms, but sprinkled out over the entire period.

VI. MAGNET REVERSALS

Raup (1985) has recently made another discovery of potentially great significance: he has found a statistically significant periodicity in the rate of reversals of the Earth's magnetic field. His conclusions are difficult to accept simply because the impact of a comet conveys little energy compared to that stored in the field of the Earth. Morris and I have found a possible model (Muller and Morris 1985) that can account for the correlation. Briefly, we assume that an impact can trigger a climate change that persists for at least a few hundred years. As ice is deposited at polar latitudes, the sea level drops. (Sudden sea-level drops of 10 m or more are known to have occurred at least 41 times in the last 65 million years.) The redistribution of the mass of the water affects the moment of inertia of the crust and mantle to cause a sudden increase in its motion relative to the liquid and solid core. The resulting shear in the liquid core, $\geq 10^{-2}$ cm s^{-1}, is sufficient to disrupt the dynamo. When the dynamo regenerates itself it has a 50% probability of creating a field opposite to the original one, thus causing a magnetic reversal. In the other 50% of the cases, we predict a magnetic excursion or aborted reversal.

Some paleontologists have noted the correlation between sudden sea level drops and mass extinctions, and concluded that the drops *caused* the extinctions. We see, however, that they may have both been caused by the same agent, a comet or asteroid impact. And, we cannot rule out that the sea-level drop was important in killing some of the species that survived the more immediate effects of the impact.

VII. TULIP ORBIT

While looking for alternative models for periodic perturbations, I found one that is particularly interesting and pretty. In the end the idea was not sufficient by itself to explain the periodic mass extinctions, but the physics of the problem is very relevant for understanding the stability of the Nemesis orbit, as well as that of comets. Suppose we postulate an object orbiting the Sun in a moderately eccentric orbit, with major axis initially perpendicular to the galactic plane. Let its orbital period be $t_N \ll t_g$, where t_g is the period of oscillation of the Sun up and down in the galactic plane. ($t_g \cong 66$ Myr.) Due to the nearly constant gravitational gradient in the galactic plane, the perihelion of the orbit will precess. As the major axis develops a component parallel to the galactic plane, the gradient will put a torque on the Sun-object system, and remove angular momentum from it. Gradually the eccentricity will increase, and the object's distance of closest approach to the Sun r_{min} will decrease. When the instantaneous value of the orbit parameter r_{min} passes through zero (usually not when the object is near the Sun) the angular momentum of the orbit will reverse sign, and the orbit will begin to precess in the opposite direction. The magnitude of the angular momentum will increase, until the major axis oscillates all the way to the other side of the normal to the galactic plane. A computer simulation of this orbit is shown in Fig. 5; for obvious reasons I have come to call this orbit the "tulip orbit".

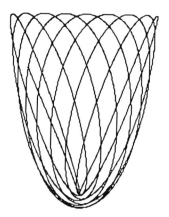

Fig. 5. The "tulip orbit," the path taken by an object in the presence of both a central gravitational force $F_r = 1/r^2$ and a uniform gravitational gradient $F_z = k\,z$.

The eccentricity of the tulip orbit changes with a period t_T given roughly by

$$t_T = \frac{t_g^2}{t_N}. \tag{2}$$

I arrived at this formula by guess and checked it by using a Monte Carlo simulation (written by J. Kare); it was proven analytically and independently by M. Rosenbluth (personal communication). It implies that the eccentricity of the obit will change cyclically in a number of cycles n given by

$$n = \frac{t_T}{t_N} = \left(\frac{t_g}{t_N}\right)^2. \tag{3}$$

Note that for Nemesis this value is $n = (66 \text{ Myr}/30 \text{ Myr})^2 = 5$ cycles. Thus, we expect the distance of closest approach to vary with this period.

VIII. THE BERKELEY SEARCH FOR NEMESIS

There is no need to assume that Nemesis is an exotic object, such as a brown dwarf or black hole. If it were a red dwarf, the most common known star type in our Galaxy, then it could have apparent magnitude between 8 and 12, dim enough to have been missed in full-sky parallax surveys. If the last comet shower was 13 Myr ago, then Nemesis would be at its greatest distance, about 3 lightyears; if the shower was 5 Myr ago, then Nemesis would be only half that distance. Its proper motion, due to its orbital velocity, would be < 0.01 arcsec yr^{-1}. Our group at Berkeley, including C. Pennypacker, J. Kare, F. Crawford, S. Perlmutter and R. Williams, are making a search for Nemesis. We are currently taking electronic images of 5000 red stars (M3 or later) in the Dearborn Catalog of the northern hemisphere, using a 512×320 element CCD (charge-coupled device). To save expense and tape, the images have been recorded on betamax videotape; pairs of images are analyzed with software we have developed for our PDP-11/44 computer. At 3 lightyears, the expected 6 month peak-to-peak parallax of Nemesis is nearly 3 arcsec, and we are making a crude (± 0.2 arcsec) measurement of the parallax of each star. Stars that we identify as candidates will be studied with great care, and have their proper motion and radial velocity measured. If we fail to find Nemesis in the northern hemisphere, we hope to do a full-sky survey of the southern hemisphere using Schmidt plates measured on the Minnesota "Starcruncher," a technique suggested by J. Kare.

IX. SUMMARY

We believe that the periodicity found by Raup and Sepkoski is made particularly evident when their data is plotted as a Gaussian ideogram. There are two strong periods present, at 26 and 30 Myr, although this may be due to a modulated single periodicity. Iridium layers have been found at at least two of the cycles (the Cretaceous/Tertiary and the Eocene/Oligocene), and there is unconfirmed evidence of iridium at a third (the Permian/Triassic), indicating the impact of a comet or asteroid. By making the natural assumption that all the cycles are due to impacts, we are drawn to the conclusion that the Earth is periodically (or quasi-periodically) subjected to storms of comets or asteroids. Evidence of multiple iridium peaks within a few Myr of the boundaries will confirm or deny this conclusion. The existence of multiple impacts during a storm could account for the extended periods of extinctions reported by some paleontologists.

The only model that has been proposed that is both self-consistent and compatible with all the known facts of astronomy and paleontology, is the Nemesis hypothesis which postulates a small star (mass from 0.3 to 0.05 $M_\odot$) orbiting the Sun in a moderately eccentric orbit ($0.7 < e < 0.9$) with a period of 26 to 30 Myr. Despite early claims to the contrary, it has been shown that the orbit of Nemesis is not highly unusual, and it is sufficiently stable against perturbations by passing stars and molecular clouds to account for the observed periodic extinctions and periodic impacts.

DEFLECTION OF COMETS AND OTHER LONG-PERIOD SOLAR COMPANIONS INTO THE PLANETARY SYSTEM BY PASSING STARS

JACK G. HILLS
Los Alamos National Laboratory

If the solar system were isolated, we would know nothing about comets. Comets in orbits that pass through the planetary system would rapidly be ejected into hyperbolic orbits while the remaining comets would never pass through the planetary system. All currently observed comets have been perturbed into planet-crossing orbits by passing stars and the tidal field of the Galaxy. Perturbations by molecular clouds may also be important when the Sun passes through galactic spiral arms. The probability of a comet entering the planetary system is strongly dependent on its semimajor axis. The probability is highest for comets in the inner edge of the classical Oort (steady-state) comet cloud at a semimajor axis of 20,000 AU. Objects in the classical Oort cloud are perturbed frequently enough by passing stars that they enter the planetary system in a constant stream (or trickle). Comets with semimajor axes < 20,000 AU only enter the planetary system in brief, intense showers which are activated by the exceptionally close passage of a perturbing star. Although the duration of each shower is very short compared to the mean time between showers, it is likely that most comets enter the planetary system in intense showers from the inner comet cloud (semimajor axes < 20,000 AU) rather than in the steady-state trickle from the classical Oort cloud. No comets with semimajor axes < 500 AU are likely to have been deflected into the planetary system by passing stars. Comets with perihelia beyond the orbit of Neptune are not greatly perturbed by the planets. These comets should exhibit the full range of semimajor axes of all comets in the solar comet cloud and not just the range spanned by comets in the classical Oort cloud which have semimajor axes > 20,000 AU. The total comet flux in this region should be many orders of magnitude greater than the comet flux from the classical Oort cloud alone because most comets are expected to have semimajor axes < 20,000 AU, their orbit periods are much shorter than those in the classical Oort cloud, and a higher fraction of them are in orbits that take them through the outer planetary system.

I. INTRODUCTION

If the solar comet distribution extends smoothly from the outer edge of the planetary system to the outer part of the Oort cloud at a semimajor axis of 10^5 AU, the bulk of the long-period comets with perihelia inside the orbit of Jupiter would usually come from the classical Oort cloud where comets have semimajor axes > 20,000 AU, as is currently the case. Only the rare close passage of a star to the solar system would bring comets with semimajor axes < 20,000 AU into Jupiter-crossing orbits. The observed semimajor axis distribution of long-period comets which cross the orbit of Jupiter is a result of the interplay between the perturbations of the major planets which tend to eliminate such comets and passing stars which tend to resupply them. The observed fact that new comets are continuously being sent into Earth-crossing orbits is a manifestation of the nonisolation of the solar system.

On crossing the orbit of Jupiter, comets from the inner edge of the classical Oort cloud, which have semimajor axes on the order of 2×10^4 AU, suffer an average perturbation in their orbital binding energy on the order of 15 times their preencounter orbital binding energy (Everhart 1969,1972; Yabushita 1979a; Hills 1983). About half the comets that cross the orbit of Jupiter are ejected into hyperbolic orbits (prompt ejectors) and half are deflected into more tightly bound shorter period orbits. The comets that are perturbed into these shorter period orbits make more frequent passes across the orbit of Jupiter, so most are ejected into hyperbolic orbits or lost through other means within a time which is short compared to their original orbital periods in the solar comet cloud. Weissman (see his chapter) finds that only about 5% of these comets eventually return to the Oort cloud. In summary, of the long-period comets crossing the orbit of Jupiter, all prompt ejectors and almost all those brought into shorter period orbits are lost in a time less than the original orbital periods of these comets in the solar comet cloud.

Based on the frequency at which the perturbations by passing stars force the solar comet cloud to resupply the comets lost by planetary perturbations, the comet cloud can be divided into three zones according to the semimajor axes of the comets. The outer zone comprises the classical Oort cloud comets which have semimajor axes > 2×10^4 AU. The comets in this zone are frequently enough perturbed by passing stars that the rate at which they enter the planetary system is at the maximum allowed by statistical equilibrium; the rate is not determined by the passage of an individual star through the solar comet cloud. The rate at which comets in the middle zone, which have semimajor axes between 500 AU and 20,000 AU enter the planetary system is

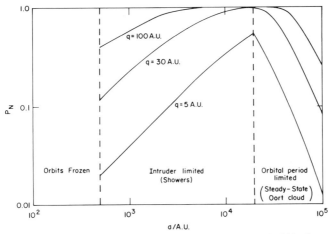

Fig. 1. The fraction of the comets of semimajor axis a which have passed within distance q of the Sun during the lifetime of the solar system.

determined by the frequency of individual, close stellar encounters. The comets in the inner zone, which have semimajor axes $\leq$ 500 AU, have never had their orbits significantly altered by passing stars.

The initial Jeans radius of the protosun was about 5000 AU (Hills 1981) which is well inside the middle zone. It is likely that the comets formed somewhere within the Jeans radius. We expect that most comets which formed in the inner part of the middle zone and in the inner zone to have remained in these zones. As shown in Fig. 1, which we discuss later in this chapter, only a small fraction of such comets have been perturbed by passing stars into orbits of sufficient eccentricity to bring them into the planetary system where perturbations by the major planets can increase their orbital energies to the point where they enter the classical Oort cloud or go into hyperbolic orbits. Direct perturbations by passing stars are even less effective in changing the semimajor axes of these comets (Hills 1975). Only a small fraction of the comets in the inner and middle comet zones are likely to have made their way to the classical Oort cloud, which constitutes the outer zone.

In the following sections, I discuss each of these zones in turn.

II. THE CLASSICAL (STEADY-STATE) OORT CLOUD

The classical Oort cloud constitutes the comets with semimajor axes between about 2×10^4 AU and 1×10^5 AU. This comet zone was found empirically by Oort (1950) from his study of the orbits of the long-period comets. More recent studies of the orbits of long-period comets indicate that comets at the inner edge of this zone may actually have semimajor axes between 10,000 and 20,000 AU (Marsden et al. 1978). The largest semimajor axis of the comets in this zone is determined by the tidal field of the Galaxy (Hills 1981). A recent discussion of the effect of the tidal field of the Galaxy on distant solar companions is given in Torbett and Smoluchowski (1984).

Weissman (1982a; also see his chapter) has reviewed the dynamics of the classical Oort cloud (outer zone) as perturbed by passing stars and the planets. I will only emphasize some points not covered by Weissman in order to relate the Oort (steady-state) Comet cloud to the rest of the solar comet cloud.

A principal characteristic of the classical Oort cloud is the fact that the eccentricity distribution of its comet orbits must be in statistical equilibrium due to the randomization imposed by passing stars. In the inner part of the classical Oort cloud, the tidal field of the Galaxy is not important and the comet orbits are nearly Keplerian. In this case the fraction of the comets of semimajor axis a which have pericenter distances of q or less is given (Hills 1981) by

$$F_q = \frac{2q}{a}\left(1 - \frac{q}{2a}\right) \cong \frac{2q}{a}. \tag{1}$$

The rate at which the comets of semimajor axis a in the classical Oort cloud enter a sphere of radius q centered on the Sun is given by

$$\dot{N}_c = \frac{dN_c}{dt} = \frac{N_c F_q}{P} = \frac{N_c}{\pi}\left[\frac{q(GM_\odot)^{1/2}}{a^{5/2}}\right] \tag{2}$$

where P is the orbital period and N_c is the total number of comets of semimajor axis a.

Equation (1) assumes that each comet-Sun pair can be treated as an isolated two-body system with only occasional, abrupt perturbations by passing stars. For comets in the outer parts of the Oort cloud where the galactic tidal field becomes important, this assumption breaks down. The statistical-equilibrium distribution of orbital eccentricities for a two-body system in the presence of a strong tidal field has not been worked out. An examination of the plot given in Fig. 3 of Hut (1984a) of the long-term evolution of a Nemesis-like object in the outer part of the Oort halo where the tidal field is important seems to suggest that the tidal field may result in a smaller percentage of high-eccentricity orbits in statistical equilibrium than is the case for isolated two-body systems. In this case, F_q in the outer part of the Oort cloud would be less than that given by Eq. (1). The effect of the tidal field is very small for comets in the inner part of the classical Oort cloud and in the inner cloud, so Eq. (1) would certainly be applicable to them.

An integration of Eq. (2) from the inner boundary of the Oort cloud at 20,000 AU and beyond relates $(N_c)_{OC}$, the total number of comets in the classical Oort cloud, to $(\dot{N}_c)_{AU}$, the average number of comets in the classical Oort cloud that pass inside 1 AU each year (Hills 1981)

$$(N_c)_{OC} = 9 \times 10^{10} (\dot{N}_c)_{AU}. \quad (3)$$

This result is not very model dependent because most comets in the classical Oort cloud are concentrated towards its inner boundary where $a = 2 \times 10^4$ AU. With the usual value of $(\dot{N}_c)_{AU} = 2$ to 3 yr^{-1} for the number of newly discovered classical Oort cloud comets which cross inside 1 AU per year (cf. Shoemaker and Wolfe 1982), this gives the number of comets in the Oort cloud as $(N_c)_{OC} = 2 \times 10^{11}$.

Everhart (1967b) and Shoemaker and Wolfe (1982) have emphasized that it is likely that only a fraction of the fainter long-period comets are observed by present-day techniques. The discovery of several new comets by IRAS is illustrative of the potential for finding fainter comets. Improvements in observational techniques will result in the quoted number of comets in the classical Oort cloud increasing monotonically with time. This is to be expected and is not very interesting since it does not affect our conceptual understanding of the classical Oort cloud. It would be better if model makers refrained from quoting a value for the number of comets in the classical Oort cloud, and instead quote and attempt to refine the coefficient given in Eq. (3), which is independent of the mass function of comets in the Oort cloud.

Equation (3) only applies to comets in the current classical Oort cloud. This can be evaluated from the rate at which its members cross the orbit of the Earth because their eccentricity distribution has come to statistical equilibrium. It does not apply to the comets in the inner comet cloud (semimajor axes < 20,000 AU). It also does not apply to the long-term history of the comets in the classical Oort cloud which may occasionally be stripped from the Sun by molecular clouds to be repopulated from the inner cloud, as discussed in the chapters by Torbett and by Shoemaker and Wolfe. There is still considerable theoretical work and new observational tests needed to understand the structure of the complete solar comet cloud and its history, but the relationship between the comet distribution in the classical Oort (steady-state) cloud and the rate at which these comets cross the orbit of the Earth as summarized by Eq. (3) should be well determined or at least determinable.

Let us consider the distribution of comet velocities at some point in space far from the Sun. If there were no perturbations by passing stars, there would be a hole in velocity space occupied by comets with perihelia inside the orbits of Jupiter and Saturn. This hole is usually called the loss cone from an analogous physical situation which occurs in plasma physics. There is no net depletion of comets in the loss cone if, within an orbital period, passing stars and other perturbers produce a change in the velocity directions of the comets which exceeds the angular size of the loss cone in velocity space. If the loss-cone comets are a small minority of the comets of a given semimajor axis, then the amount of perturbation required to fill in the loss cone is a very small fraction of the perturbation needed to completely randomize the orbits of the comets. This is the situation in the solar comet cloud.

In the classical Oort cloud (semimajor axes > 20,000 AU), the perturbations by passing stars are frequent enough, compared to the lifetime of the comets in the loss cone, for there to be no net depletion of comets in the loss cone. The fraction of the comets of a given semimajor axis in the loss cone is just that given by Eq. (1). The comets in the loss cone are flushed out in one orbital revolution, so the number of independent loss-cone fillings that have occurred in the lifetime of the solar system for the comets in the classical Oort cloud is just the number of orbital revolutions they have made,

$$N = \frac{t}{p} = 1626 \left(\frac{20,000 \text{ AU}}{a} \right)^{3/2} \qquad (4)$$

(Hills 1981) for $t = 4.6 \times 10^9$ yr. Because of the steady-state distribution of eccentricities of comets in the classical Oort cloud, the number of these comets perturbed out of the loss cone by passing stars just equals the number perturbed into it, so the number of these comets entering the planetary system in each orbital period is just the steady-state number in the loss cone.

The fraction of these comets which have passed within distance q of the Sun is given by

$$P_N = 1 - (1 - F_q)^N = 1 - \left(1 - \frac{2q}{a} \right)^N. \qquad (5)$$

This equation is valid if the average accumulated change in the pericenter distance of a comet per orbital revolution due to the perturbations of passing stars exceeds q. This is true for all $q \leq 5$ AU at the inner edge of the Oort cloud at $a = 20,000$ AU and for increasingly larger values of q for semimajor axes > 20,000 AU.

The fraction of the original Oort cloud comets which have passed inside the orbit of Jupiter, i.e., those which have $q = 5$ AU, drops monotonically from a peak of $P_N = 56\%$ at the inner edge of the classical Oort cloud where $a = 2 \times 10^4$ AU to $P_N = 1.4\%$ at $a = 10^5$ AU.

It is interesting to consider the example of the hypothetical stellar companion Nemesis (Davis et al. 1984; Whitmire and Jackson 1984). With an orbital period of 2.6×10^7 yr this object should have a semimajor axis of about 9×10^4 AU. This places it within the classical Oort cloud. The change in its q per orbital period due to stellar perturbations exceeds 100 AU (Hills 1984a), so all loss cones with q less than this value are refilled within each orbital period. The number of orbital revolutions made by this object in the age of the solar system, assuming its semimajor axis to be constant, is $N = 170$. This is also the number of independent loss-cone fillings. (Here we invoke the ergodic hypothesis that a time average for one representative object is equivalent to an ensemble average over many objects.) Equation (5) gives $P_N = 14\%$ for $q = 40$ AU and $P_N = 2\%$ for $q = 2$ AU. If Nemesis should pass inside the planetary system, it would greatly increase the eccentricity of the orbit of any planet which it crossed and it would have a high probability of ejecting such a planet into a hyperbolic orbit (Hills 1985). If Nemesis has a mass of 0.05 $M_\odot$, the average post-encounter eccentricity of any planetary orbit which it crosses would exceed 0.15. Such visits are not encouraged.

III. MIDDLE COMET ZONE AND COMET SHOWERS

This comet zone has been discussed by Hills (1981) and Bailey (1983a). In this zone, in which the semimajor axes go from about 500 AU to about 20,000 AU, comet loss cones are filled, and a comet shower begins whenever a star passes sufficiently close to the Sun. Hills (1981) estimated analytically that this occurs for comets of semimajor axis a when a star passes within distance a of the Sun. Computer simulations (Hills, work in progress) indicate that the loss cone of these comets is filled if a passing star more massive than 0.25 $M_\odot$ passes within distance $2a$ bf the Sun. The filling of the loss cone of comets of semimajor axis a causes a shower of them to enter the planetary system for a time comparable to their orbital periods. Within an orbital period, all the comets in the loss cone have been through the planetary system and most have been ejected into hyperbolic orbits by Jupiter and Saturn.

The outer boundary of the middle (comet-shower) zone is determined by the condition that the comets have an orbital period equal to the mean time between their loss cone fillings. The semimajor axes of these comets can be expressed analytically (Hills 1981) as

$$R_{\max} = a_c = \left[\frac{(GM_\odot)^{1/2}}{8\pi^2 n_s V_s}\right]^{2/7}. \tag{6}$$

Here we assume that the comet loss cones are filled if a star passes closer to the Sun than twice the semimajor axes of the comets, and the characteristic duration of a shower equals the orbital period of these comets. Here n_s is the number of perturbing stars per unit volume and V_s is their average velocity with respect to the Sun. Using $V_s = 30$ km s^{-1} and $n_s = 0.1$ pc^{-3}, we find that $a_c = 20,000$ AU. If the value of $n_s V_s$ were uncertain by an order of magnitude, which is not the case, the uncertainty in a_c would be less than a factor of two. It is clear that a_c is well defined.

From his empirical study of the long-period comets, Oort (1950) found the inner edge of the observed comet cloud to be at a semimajor axis of about 20,000 AU, which is just the value we have found for the inner edge of the steady-state comet cloud from theoretical considerations. More recent work by Marsden et al. (1978) indicates that the semimajor axis of the observed inner edge of the Oort cloud may actually lie between 10,000 and 20,000 AU. While this is still close to the theoretical value of the inner edge of the steady-state cloud it may indicate a relatively recent stellar encounter and a current recovery from a comet shower. The observed edge of the comet cloud will be at $0.75\,a_c = 15{,}000$ AU or less about 35% of the time due to a recent stellar passage, so the present situation is not unusual. (This number comes from Eq. [7], which is discussed below.)

The inner boundary of the middle comet (shower) zone occurs at about $a_{min} = 500$ AU. For a stellar space density of $n_s = 0.1$ stars pc^{-3}, the closest approach of any star to the Sun was about 1000 AU (cf., Hills 1984b). The smallest semimajor axis of comets which have had their loss cones filled by passing stars is about half this distance or 500 AU.

The fraction F_{shower} of the time during which comets of semimajor axis a are engaged in a shower is just the number of loss-cone fillings suffered by these comets per unit time multiplied by the duration of the showers. From these considerations it is evident that for comets of semimajor axis a, F_{shower} is given by

$$F_{shower} = \left(\frac{a}{a_c}\right)^{7/2} \qquad (7)$$

where $a_c = 20{,}000$ AU is again the semimajor axis of the comets at the inner edge of the Oort (steady-state) comet cloud. The power of $7/2$ comes from the fact that the cross section for filling the comet loss cone scales as a^2 and the orbital period (shower duration) scales as $a^{3/2}$.

The number of loss-cone fillings within the age of the solar system for comets of semimajor axis a_c is given by

$$N = 1626(a/a_c)^2 \qquad (8)$$

as a result of the cross section for loss-cone fillings scaling as a^2. The numerical coefficient comes from Eq. (4).

We note that Eq. (5) determines the fraction P_N of the comets of semimajor axis $a/a_c < 1$ which have passed within distance q of the Sun if the value of N used in this equation is given by Eq. (8). For $q = 5$ AU, P_N drops from 56% for $a = a_c = 20{,}000$ AU to 2% for $a = 500$ AU. Figure 1 shows P_N plotted as a function of semimajor axis a. The comets which have the highest probability of entering the planetary system are just those with $a = a_c = 20{,}000$ AU. For $a > a_c$, the loss cone is always filled but the fraction of the comets of semimajor axis a in the loss cone scales as a^{-1} and the effective number of independent loss-cone fillings scales as $a^{-3/2}$, so there is a rapid decrease in P_N with increasing a. For $a < a_c$, the loss cone is only filled when a perturbing star comes sufficiently close to the Sun, but the infrequency of such encounters at small values of a is compensated in part by the increase in the fraction of the comets in the loss cone. This results in P_N dropping less rapidly for $a < a_c$ than for $a > a_c$.

The number of comets in the middle (shower) comet zone can be estimated from the observed number in the classical Oort cloud by assuming that these comets originated in the middle comet zone (Hills 1981; Bailey 1983a). This analysis indicates that the number of comets in the middle zone between semimajor axes 500 AU and 20,000 AU is up to two orders of magnitude greater than that of the comets in the classical Oort comet cloud. The work further indicates that up to 10 times as many comets may have entered the planetary system in showers as have entered it in a steady trickle from the classical Oort cloud.

These comet models indicate that the peak intensity during a comet shower produced by filling the loss cone of all comets with semimajor axes $> a_{min}$ in units of the integrated steady-state flux of comets from the classical Oort cloud is given by

$$\frac{I_{shower}}{I_{Oort}} = \left(\frac{a_c}{a_{min}}\right)^{7/2}. \tag{9}$$

For a_{min} = 3000 AU, $(I_{shower}/I_{Oort}) \sim 10^3$ while for a_{min} = 500 AU, $(I_{shower}/I_{Oort}) \sim 10^5$. For a_{min} = 3000 AU, this corresponds to a major comet crossing inside the orbit of the Earth every four hours while for a_{min} = 500 AU, this corresponds to a comet crossing the orbit of the Earth every half minute. We contrast these fluxes to the observed flux from the classical Oort cloud which amounts to two or three major comets crossing the orbit of the Earth per year. (We use "major" to indicate the fact that only the larger Oort cloud comets are observable, but the ratio of the intensities of comet showers to the steady-state trickle from the Oort cloud as given by Eq. [9] is independent of this observational selection effect.)

One of the characteristcs of the inner-cloud which has not been pointed out previously is the relatively high frequency of comet collisions. We can estimate the number of comet collisions per year using a crude model. Suppose there are N comets in a volume of radius S around the Sun. We assume the comet velocity distribution is isotropic and the comets are uniformly distributed in this volume. The mean time between collisions for an individual comet is

$$\tau = \frac{1}{\sigma V n} \tag{10}$$

where V is the relative velocity of the comets with respect to each other during collisions. This is on the order of the orbital velocity, or

$$V \sim \left(\frac{GM_\odot}{S}\right)^{1/2}. \tag{11}$$

The number of comets per unit volume is given by

$$n = \frac{N}{\frac{4}{3}\pi S^3}. \tag{12}$$

The collision cross section is

$$\sigma = \pi(R_1 + R_2)^2 \tag{13}$$

where R_1 and R_2 are the radii of the two comets involved in the collision. The average time between any two comet collisions in the inner cloud is given approximately by

$$\tau_c = \frac{\tau}{N} = \frac{(4/3)S^{7/2}}{(R_1 + R_2)^2(GM_\odot)^{1/2}N^2}. \tag{14}$$

Most of the collisions will occur in the inner part of the middle comet cloud where the comet density is highest and the relative comet velocities are the largest. Hills (1981) estimated that there are $N = 10^{13}$ comets in the middle comet zone (semimajor axes $\geqslant$ 500 AU). The density distribution he found is such that half these comets have semimajor axes < 1000 AU. If we let $R_1 = R_2 = 2$ km, $S = 10^3$ AU and $N = 5 \times 10^{12}$ comets, then $\tau_c = 0.38$ yr which corresponds to about 2.5 comet collisions per year. This crude calculation may exaggerate somewhat the collision rate beyond $S = 10^3$ AU, but it may underestimate the collision rate inside $S = 10^3$ AU. The calculation illustrates the potential importance of collisions.

While these collisions would not have destroyed an appreciable fraction of the comets in the inner cloud they would produce much debris which builds up within the inner cloud. Some of this debris would enter the planetary system when the innermost part of the comet cloud is perturbed by a close stellar encounter. Because of the higher surface-to-mass ratio of this debris compared to that of the original comets, the debris will suffer sublimation more readily than the comets. This will produce a larger ratio of zodiacal light and meteor flux to comet flux during a comet shower than is the case under present circumstances where comets, but no significant collisional debris, enter from the classical Oort cloud. The dust from comet debris may also contribute to the infrared cirrus found by IRAS (cf., Bailey 1983a).

IV. REGION OF FROZEN ORBITS, THE INNER COMET ZONE

Comets with semimajor axes < 500 AU are not likely to have been perturbed enough by passing stars ever to have had their loss cones refilled. However, a substantial fraction of these comets may originally have formed in planet-crossing orbits unless angular momentum constraints produced a highly rotationally flattened distribution of comets in this zone (which is certainly possible). In this case, the average angular momentum of the comet orbits in this zone is quite large, and there will be far fewer comets in planet-crossing orbits than given in Eq. (1). It is often assumed that the comets formed in this zone and then were ejected into the middle comet zone and the classical Oort cloud by planetary perturbations. This mechanism is inefficient in ejecting comets into the classical Oort cloud because the large masses of the outer planets (particularly Jupiter and Saturn) produce large perturbations which tend to eject most of the comets into hyperbolic orbits. (However, comets may have formed in the middle comet zone as well as in the inner comet zone [Hills 1982].) One factor, which may make the ejection mechanism more efficient, is the likelihood that the masses of the outer planets grew on a time scale which was long compared to the orbital periods of the comets in this zone. The smaller perturbations by the lower mass planets may have allowed substantially more of the comets to make their way to the weakly bound orbits of the classical Oort cloud. A Monte Carlo model illustrating such a possibility is given by Fernandez and Ip (1981).

V. COMET FLUX IN THE OUTER PLANETARY SYSTEM

The average perturbation imposed on a comet with a perihelion, which is a fixed fraction of the semimajor axis of a planet, is directly proportional to the mass of the planet and inversely proportional to the semimajor axis of the planet (Hills 1983). The larger semimajor axes of Uranus and Neptune and their smaller masses result in their perturbations on comets being much smaller than those of Jupiter and Saturn.

Computer simulations show a very rapid decrease in the perturbations imposed on a comet by a planet-star system if the closest approach of the comet to the star is > 1.5 times the semimajor axis of the planetary orbit (Everhart 1968; Hills 1983). This indicates a rapid decrease in the perturbations on comets having perihelia > 1.5 times the semimajor axis of Saturn or about 14 AU. As a result, these comets can make many passages through the outer planetary system before they are ejected into hyperbolic orbits or perturbed into Saturn and Jupiter crossing orbits by the lower-mass planets beyond Saturn. This causes a decrease in the value of the critical semimajor axis a_c separating the Oort (steady-state) comet cloud from the middle (cometshower) zone as q increases beyond 14 AU; i.e., the mean semimajor axes of those comets which enter the planetary system in a steady stream rather than in showers drops rapidly with increasing distance beyond Saturn so that, at about the distance of Pluto, comets will enter in a steady stream from the middle comet zone as well as from the classical Oort cloud.

If comets only entered the planetary system from the classical Oort cloud (with semimajor axes $a > 20{,}000$ AU), there would be 30 times as many comets per unit time entering a sphere of radius $q = 30$ AU as entering a sphere of radius $q = 1$ AU (see Eq. 1). Because the area of the larger sphere is $30^2 = 900$ times that of the inner sphere, the number of comet hits per unit area and unit time at 30 AU should be 30 times less than at 1 AU if only comets from the classical Oort cloud are considered. We see from the estimates of peak comet shower intensities given by Eq. (9) that the comet flux in the outer planetary system may be many orders of magnitude greater than one would expect from considering only comets that enter from the classical Oort cloud because comets from the entire solar comet cloud would enter this region in a steady stream. The peak shower intensity given by Eq. (9) would apply to the steady-state flux in the outer parts of the planetary system. This should produce a noticeably enhanced comet cratering rate on the planetary satellites beyond Saturn. There also should be more micron-sized dust and larger debris entering this region (due to the collisional breakup of comets in the inner comet cloud) than a simple extrapolation based on classical Oort cloud comets alone would suggest. (Dust grains with diameters less than about a micron are blown out by radiation pressure.)

IS THERE EVIDENCE FOR A SOLAR COMPANION STAR?

SCOTT TREMAINE
Massachusetts Institute of Technology

This chapter contains a brief review of the major astronomical explanations for a possible 25 to 35 Myr period in the cratering and extinction records. The only viable theory is that the Sun has a distant companion star; however, the required orbit for this companion is an improbable one, since the companion will probably escape in a time short compared to its present age. The most likely resolution of this uncomfortable situation is that there is no periodicity: in both the cratering and extinction records, the statistical evidence for periodicity is very weak.

In this chapter I review, as requested, the evidence for the existence of an unseen solar companion, the so-called death star Nemesis, which may be responsible for periodic comet showers which are thought to have triggered many of the major extinction events of the last 200 Myr. There are three separate issues to be addressed: is there a periodicity in the cratering and/or extinction record; if it exists, can it be the result of comet showers induced by a companion of this sort; and, are there any other viable explanations of the periodicity. I will discuss these three issues in reverse order.

Explanations for a Periodicity

Several different explanations for the periodicity have been proposed. For example, Rampino and Stothers (1984*a,b;* see also their chapter) have suggested that the periodicity is associated with the interval between successive passages of the Sun through the galactic plane (the agreement between the times of the Sun's crossings of the galactic midplane and the boundaries of major geologic periods was pointed out by Innanen et al. [1978]), and arises because direct collisions and tidal encounters with interstellar clouds are more frequent when the Sun is in the plane. In these encounters, the gravitational field of the passing interstellar cloud perturbs the Sun's comet cloud, causing a shower of comets to rain into the planetary system and onto the Earth, with disastrous biological consequences. Assuming that the amplitude of the solar oscillation (around 70 pc) is small compared with the thickness of the dominant components of the disk, this interval is simply $P_g = 0.5(\pi/G\rho)^{1/2}$, where ρ is the local density. For a reasonable range in ρ, say 0.15 to 0.20 $M_\odot pc^{-3}$ (cf. Bahcall 1984a), P_g is between 29.5 and 34 Myr, tantalizingly close to the periods of 26 to 35 Myr which have been found in various biological and geological records.

The most incisive criticism of this possible source of periodicity is due to Thaddeus and Chanan (1985; see also chapter by Thaddeus). They show that the fractional variation in density of molecular clouds due to the Sun's vertical oscillation is simply too small to produce a significant period signal in the nine extinction events used by Rampino and Stothers. To save the Rampino-Stothers hypothesis, one would have to invent some unseen component of the disk with a thickness much smaller than that of molecular clouds (whose half-thickness at half-density $z_{1/2} = 85$ pc is already the smallest of any known component of the Galaxy); this component would have to be either dissipative or young (so that it was not puffed up from encounters with molecular clouds), and dense enough to cause substantial perturbations to the comet cloud, yet not so dense as to be gravitationally unstable. Because of all these complications, I think that it is extremely unlikely that the periodicity arises from variations in the rate of encounters with interstellar clouds—or any other galactic objects—due to the Sun's motion perpendicular to the galactic plane. Other galactic environmental effects which might be modulated with the period P_g were discussed by Rampino and Stothers (1984a) and Schwartz and James (1984), but all of them seem to have similarly serious problems in explaining the observed periodicities.

The galactic tidal field also varies with the period P_g as the Sun oscillates through the plane. The tidal force is proportional to the local density, which is smaller when the Sun is at its maximum height. A. Toomre and I have wondered whether comets with periods close to P_g might be trapped into resonance with this oscillating force field. If a substantial population of comets were in this resonance, and if they had small libration amplitudes, so that they were always near aphelion when the tidal force was weakest, then periodic comet showers could occur when the Sun was passing through the plane. We have done a number of orbit integrations and have found resonant orbits of this type. However, it is our impression that these resonant orbits occupy a relatively small fraction of phase space, and, in addition, there is no obvious dissipative mechanism which could reduce the libration amplitudes to a point where well-defined showers occurred.

Thus, there is no promising mechanism which can connect the Sun's vertical oscillations to the cratering and extinction records, and I am forced to agree with Innanen et al. (1978) that the rough consistency of P_g with the apparent period in the terrestrial record is no more than "an interesting coincidence."

Another interesting possibility is that the periodicity might be due to some unrecognized oscillatory behavior in the known solar system. Important long-term periodicities are common, even though the orbital periods of the planets are comparatively short. For example, the Earth's eccentricity varies between 0.003 and 0.057 with a period of ~ 0.1 Myr, and this variation is believed to be responsible for a substantial fraction of the climatic variation over the last 1 Myr (Hays et al. 1976; Berger 1977). The eccentricity of Pluto varies between 0.21 and 0.27 with a period of 4.0 Myr (Williams and Benson 1971). Are there still longer periodicities? The answer is that we do not know, because the integrations have not been done. The longest direct integration of the outer solar system was for 1.0 Myr and was done over a decade ago (Cohen et al. 1973). Fortunately, the construction of fast hard-wired integrators (Applegate et al. 1984) should lead to very accurate long-term integrations in the near future. Of course, even if a strong 25 to 35 Myr period is found, we would still need a mechanism to deliver bursts of comets or asteroids to the Earth. This will probably not be easy to arrange; so my best guess is that the answer does not lie in this direction either.

Comet Showers to Explain Periodicity

Whitemire and Matese (1985) have suggested that comet showers might be induced at regular intervals by a hypothetical planet X rather than by a distant companion like Nemesis. They argue that an unknown planet of mass $M_X = 1$ to 5 $M_\oplus$ at semimajor axis $a_X = 100$ AU would have a perihelion precession period of about 56 Myr, so that if its orbit were inclined, the intersections of its orbit with the ecliptic would oscillate in radius with a period of 28 Myr. Some theories of the origin of comets suggest that there is a disk of comets extending outward from around 40 AU, and Whitmire and Matese suggest that planet X clears out an annular gap in this disk between its perihelion and aphelion, with comet showers occurring when its perihelion lies in the ecliptic so that it passes close to the inner edge of the gap. One serious flaw in this proposal is that planet X is probably not able to clear out such a well-defined gap. The typical velocity impulse received by a comet in an encounter with impact parameter p and relative velocity v is $\Delta v = 2GM_X/pv$. This formula may overestimate the impulse in close encounters, but in any case the bound $\Delta v \leq 2v$ is imposed by energy conservation. If the encounter is to remove the comet from the disk, then we require $\Delta v \gtrsim v_c$, where $v_c = (GM_\odot/r)^{1/2}$ is the circular speed; thus we need $v \gtrsim v_c/2$ and $p \lesssim p_c = 4GM_X/v_c^2 = 4r(M_X/M_\odot)$. In time t planet X sweeps out a volume $\pi p_c^2 v_c t$, and if the comet disk has thickness h, then the fractional volume swept free of comets is

$$f = Q\frac{p_c^2 v_c t}{h(r_a^2 - r_p^2)} = \frac{4Q}{e}\left(\frac{M_X}{M_\odot}\right)^2 \frac{v_c t}{h}$$

where r_a and r_p are the aphelion and perihelion distances, e is the eccentricity of planet X, and Q is the fraction of time when the orbit of planet X lies in the disk. Taking parameters suggested by Whitmire and Matese (1985), $e = 0.3$, $h = 35$ AU (based on a typical comet inclination of $10°$ and a typical radius of 100 AU), $M_X = 5 M_\oplus$, $v_c = 3$ km s^{-1}, and $Q = 0.1$ (since Q can be no larger than the ratio of the shower duration to the interval between showers), after $t = 4.5 \times 10^9$ yr, we find $f = 0.025$. Thus a much larger mass, at least 30 to 50 $M_\oplus$, would be needed to clear out a well-defined gap. A mass this large for planet X is quite unacceptable, both because the planet would be so bright that it would probably have been discovered, and because it would cause large systematic residuals in the positions of the outer planets (see, e.g., Seidelmann 1971; Reynolds et al. 1980).

More distant encounters, with $p \gtrsim p_c$, help to clear a gap by causing gradual diffusion of comet orbits out of the region $r_p < r < r_a$. The mean square velocity change due to encounters with $p \gtrsim p_c$ is of order $\Delta v^2 \approx f v_c^2$ ln Λ, where $\Lambda = p_{max}/p_c$ and $p_{max} \approx 0.5\, r$. Thus, distant encounters with planet X could perhaps clear a gap with values of f as small as $(\ln \Lambda)^{-1} \approx 0.1$. However, the gap would not have sharp edges, since comets outside the gap are affected by distant encounters as well, and diffuse at a rate which is at least $\Delta v^2 \approx f v_c^2$. Thus, for $f \gtrsim 0.025$, $\Delta v/v_c \approx \Delta r/r \gtrsim 0.16$, so that the gap edges can be no sharper than $\Delta r \approx 11$ AU and brief showers cannot be produced (Whitmire and Matese [1985] estimate that $\Delta r \lesssim 0.3$ AU gives an acceptable shower duration). In other words, the requirements of a clean gap and sharp edges appear to be mutually contradictory. However, Whitmire and Matese have informed me that they now have a revised version of their theory which may avoid these problems.

This brings me to the final and most promising theory, namely that the periodicity might be caused by a distant unseen companion to the Sun (Nemesis), on an eccentric orbit with semimajor axis $\sim 90{,}000$ AU and period $(90{,}000)^{3/2} = 27$ Myr (Whitmire and Jackson 1984; Davis et al. 1984). At its assumed perihelion distance of around 10,000 to 20,000 AU, Nemesis would pass close enough to the Sun to rattle the dense inner Oort cloud of comets, filling the loss cone which is normally swept clean by Jupiter and leading to a brief, intense shower of comets. In my view the only substantial objection to the Nemesis hypothesis is that the required orbit is an unlikely one for a solar companion, because the escape time from the present orbit is much less than the present age of the solar system. Monte Carlo calculations by Hut (1984) which include both the galactic tidal field and what Thaddeus calls "the pitter-patter of passing stars" yield a half-life of about 1000 Myr, starting from the present orbit. This half-life must be reduced somewhat to include the effects of molecular clouds; estimates for comets at $a = 25{,}000$ AU (Hut and Tremaine 1985) scaled to Nemesis at $a = 90{,}000$ AU yield a somewhat shorter half-life of about 400 Myr, but with large uncertainties. (Clube and Napier [1984c] estimate that the half-life is 50 Myr, but this is much too extreme.) Thus Nemesis would escape from the solar system at a future time which is only around 10% larger than its present age, and we are forced to the improbable conclusion that Nemesis, if it exists, has been discovered just as it is about to disappear.

In my opinion the solar companion star Nemesis remains, so far, the best hypothesis to explain the periodicity in the craters and extinctions. It has been exposed to intense examination over the last year and I am not aware that any fatal flaws have shown up. Nevertheless, the Nemesis explanation is entirely *ad hoc*. There is no evidence for Nemesis except the periodicities, and Nemesis would have to be in an unusual orbit capable of surviving from now onward for only a fraction of the age of the solar system.

Terrestrial Evidence for Periodicity

The absence of a compelling astronomical explanation for the periodicity leads us to ask whether the terrestrial evidence for the periodicity is itself very strong. A number of authors have claimed to find statistical evidence for periodicities of between 20 and 40 Myr in the biological and geological record; the literature in 1984 alone contains claims of significant periodicities at 20, 21, 22, 23, 26, 28, 30, 31, 32, 33, 34, 35, 36 and 38 Myr. Unfortunately, there has been a long and dismal history in astronomy of spurious periodicities which have been claimed in many types of data; Feller (1971, p. 76) writes that "hidden periodicities used to be discovered as easily as witches in medieval times, but even strong faith must be fortified by a statistical test." Thus I have attempted to fortify my faith by reanalyzing some of the papers in which evidence for periodicity is presented, in collaboration with J. Heisler. We concentrated on two investigations: Raup and Sepkoski's (1984) analysis of the extinction record, and Alvarez and Muller's (1984) analysis of the crater record.

The Raup-Sepkoski analysis is based primarily on 12 extinction peaks which occurred in the last 250 Myr. They test for periodicity by comparing the times of these peaks to a set of markers with a given period P and arbitrary phase, assigning each peak to the nearest marker, finding the mean difference between the peaks and the markers and then shifting the phase of the markers to reduce the mean deviation to zero. This shift may give one or more peaks an error exceeding half a period, in which case they reassign the peak to the nearest marker, recompute the mean difference, and shift the phase of the markers once again. They iterate this procedure to convergence, and then compute the standard deviation of the differences $\sigma(P)$. For randomly spaced events the root mean square (rms) value of $\sigma(P)/P$ is 0.268 independent of P (except for edge effects which arise because P does not divide evenly into 250 Myr; for $P \leq 60$ Myr, edge effects change the rms value of $\sigma(P)/P$ by < 1%), and a value significantly below this is evidence that a periodicity is present. Raup and Sepkoski examine periods of $12, 13, \cdots, 60$ Myr and find a significant dip at 26 Myr, where $\sigma(P)/P = 0.148$. With 12 randomly spaced events, for a given P, $\sigma(P)/P$ would be larger than 0.148 in 99.74% of all samples. Raup and Sepkoski obtain significance levels of 99.55% to 99.99% using similar tests.

These estimates of significance are misleading, however, because the period of 26 Myr was not picked out in advance. Instead, it was chosen as giving the most significant result from many possible periods, i.e., from many possible statistical tests (Fisher 1929). In a time series of length T, Fourier components of frequencies $\nu = k/T$ are independent for $k = 0, 1, \cdots$. Therefore a search for periodicities between 12 Myr and 60 Myr in a time series of length 250 Myr involves frequencies with $k = 4, 5, \cdots, 21$, or 18 independent tests. Thus the significance level must be reduced from 0.9974 to $(0.9974)^{18}$ = 0.954.

This effect was already recognized by Raup and Sepkoski and its importance was estimated by them, using Monte Carlo tests. However, the validity of their tests was severely limited by two restrictions. First, they looked only for periods < 30 Myr, a restriction for which there is no *a priori* justification (indeed, in principle one should probably look for all periods up to 250 Myr rather than just 60 Myr). Second, they required that the dip in $\sigma(P)/P$ be significant over a 3 Myr interval in P. The phase shift over an interval T caused by a period change ΔP is $2\pi T \Delta P/P^2$, and a dip will persist only as long as the phase shift is $\lesssim 2\pi$. To have a significant dip persist for $\Delta P = 3$ Myr requires $P \gtrsim \sqrt{T \Delta P} = 27$ Myr. Thus the Raup-Sepkoski tests were blind to all periods except those in the range ~ 25 to 30 Myr.

To provide an independent estimate of the significance of the periodicity, Heisler and I have carried out our own Monte Carlo tests. We placed 12 extinction events randomly between 0 and 250 Myr and searched for values of $\sigma(P)/P$ below 0.148 in the interval 12 Myr $\leq P \leq 60$ Myr. We used the Raup-Sepkoski algorithm except that we searched in frequency rather than in period and used a much finer mesh with step size $\Delta \nu << 1/T$. Out of 1000 simulations, we found that 115 had periods between 12 and 60 Myr for which $\sigma(P)/P$ was < 0.148. Therefore, according to this test, the significance level of the Raup-Sepkoski periodicity is < 90%. We have carried out the same test with Sepkoski and Raup's (1986) revised list of eight extinction events, and find an even smaller significance level of only 50%. These results suggest that there is no strong evidence for periodicity in the extinction record.

The Alvarez-Muller analysis is based on 11 impact craters with ages between 5 and 250 Myr, with age uncertainties ≤ 20 Myr and diameters > 10 km. They describe a power-spectrum analysis which yields a period of 28.4

Myr, and ascribe a significance level of 97 to 99.5% to this periodicity. In the one test which they report in detail, they represent each crater by a Gaussian of unit area and rms width equal to the age uncertainty. They compute the power spectrum of the set of Gaussians, and then generate Monte Carlo simulations of 11 craters with random ages and the same set of age errors as in the real data. They then search for peaks as high as the peak seen in the real data at frequencies higher than 28.4 Myr^{-1}. They find higher peaks in fewer than 1% of the simulations and conclude that the peak is significant at the 99% level.

This significance level is misleading for the following reason. Convolution with a Gaussian is equivalent to multiplication of the power spectrum by a Gaussian. This process suppresses high-frequency peaks, and by searching only for peaks at frequencies higher than the observed frequency, Alvarez and Muller have looked only in the region where the peaks in the Monte Carlo simulations have been suppressed. Heisler and I have carried out an alternative analysis in which we calculated the power spectrum which results from representing the crater data by equally weighted delta functions rather than Gaussians. We then constructed simulations using 11 craters with random ages, and searched for peaks in the power spectra of the simulations with periods between 10 and 60 Myr. We found higher peaks in 4.3% of the simulations, implying that the periodicity in the crater data is significant at only about the 96% level, slightly lower than Alvarez and Muller's lower limit of 97%. Including periods longer than 60 Myr in the search would lower the significance level still further.

It might be argued that the significance of our result has been degraded because all craters are given equal weight, even if their age uncertainties are large. We therefore carried out an additional analysis in which each delta function was weighted by the inverse of the age uncertainty. This yielded a slightly lower significance level, 92.5%. To sum up, the evidence for a periodicity is stronger in the crater record than in the extinction record, but it is still rather weak.

It has sometimes been suggested that the evidence for the periodicity is stronger than these arguments suggest, because the extinction and crater records yield periodicities which agree in period and phase. However, this argument is misleading. The Alvarez group has accumulated very strong evidence that mass extinctions are associated with iridium layers and thus with impacts. With this strong correlation between extinctions and impact craters well established, if a spurious periodicity is found in one record, we should expect it also in the other.

One interesting point remains to be addressed. We have learned from W. Alvarez that there is now preliminary evidence that multiple iridium layers and multiple tektite layers are associated with at least some extinction events. This result suggests that major impacts are not random, i.e., that they tend to occur in bunches or showers of duration ~ 1 Myr, even though the showers may not be periodic. An additional advantage of a shower model is that some paleontologists claim that mass extinctions were not sudden but spread over 1 Myr or so, which is easier to reconcile with a shower than with a single major impact (Muller 1985). If impacts are bunched, then the statical analysis of the previous paragraphs should be modified. Rather than testing the null hypothesis that all craters occur at random times, we should test the hypothesis that they occur in random bunches. There are two clear bunches of craters in the Alvarez-Muller sample: two craters at 38 and 39 Myr and three at 95, 100 and 100 Myr. If we replace these bunches by single events at 38.5 Myr and 98 Myr, thus reducing the sample from 11 events to 8 events, the significance level drops from 96% to 75%, which is too low to provide any evidence of periodicity.

If there is no periodicity, and hence no evidence for Nemesis, then what caused the showers? Probably the answer is that they are comet showers which are set off by close stellar encounters, in precisely the manner envisaged by Hills (1981; see also his chapter). Hills showed that stellar encounters are frequent enough and strong enough to produce a steady flux of comets into the planetary system, but only for cometary semimajor axes $a > a_c = 2 \times 10^4$ AU. Comets with semimajor axes $< a_c$ enter the planetary system only in brief showers which occur when a passing star makes a close approach with impact parameter $p \lesssim a_c$. According to Hills, the mean interval between encounters with $p < a_c$ is about 10 Myr; for $p < 0.5\ a_c$ the interval is 40 Myr. Given the uncertainties, these intervals are consistent with the mean interval of 25 to 30 Myr between major impact showers. It would be very worthwhile to compare the statistics from a detailed Monte Carlo model of Hills' comet showers with the statistics of the cratering and extinction records.

Conclusions

To summarize, there is no very appealing astronomical model that would explain a 25 to 30 Myr periodicity in the cratering or extinction record, although the only argument against Nemesis is that the required orbit has a low *a priori* probability. However, the statistical evidence for a periodicity is weak, particularly if we accept the fact that extinctions are correlated with crater impacts and that the impacts occur in showers or bunches. Most likely, the extinctions are due to intense comet showers which occur at random times, as a result of occasional close encounters with passing field stars.

Thus the observational search for Nemesis is similar to the search for Pluto half a century ago, in at least two respects: first, it probably has no sound theoretical basis; and second, it is likely to yield extremely interesting results whether or not the theory is correct.

Acknowledgments. This work was supported by the National Science Foundation and by an Alfred P. Sloan Fellowship. I am grateful to W. Alvarez, J. Bahcall, J. Heisler, P. Hut, J. Matese, R. Muller, P. Thaddeus and A. Toomre for discussions, and to A. Toomre for a careful and constructive reading of the manuscript.

Color Section

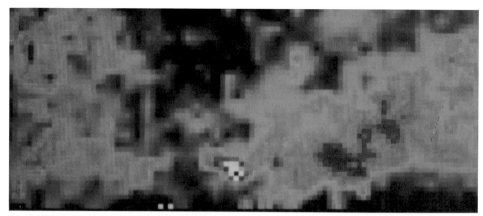

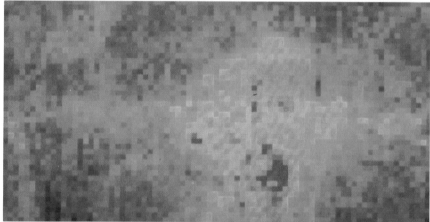

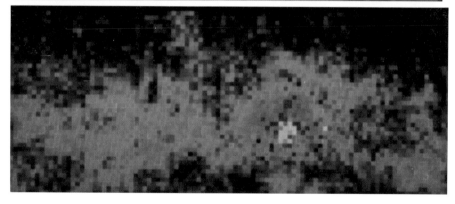

Color Plates 1–3. False color images of the CO emission are shown for the giant molecular cloud complexes associated with M17, W44 and W51. In each case, the strip extends vertically from $b = -1°\!.05$ to $1°\!.0$, and increasing longitude is to the left. The pixel size is $3'$. The M17 complex includes all the emission in Plate 1 from the bright (white) area just below center extending to the upper right approximately $2°\!.5$. The H II region, M17, coincides with the bright white emission region. The W44 complex shown in Plate 2 includes the entire $1°\!.5 \times 2°$ area on the right side of the center longitude. The W51 cloud encompasses the bright emission to the right of center in Plate 3. The CO data are from the Massachusetts-Stony Brook CO survey (Sanders et al. 1985b). (See the chapter by Scoville and Sanders.)

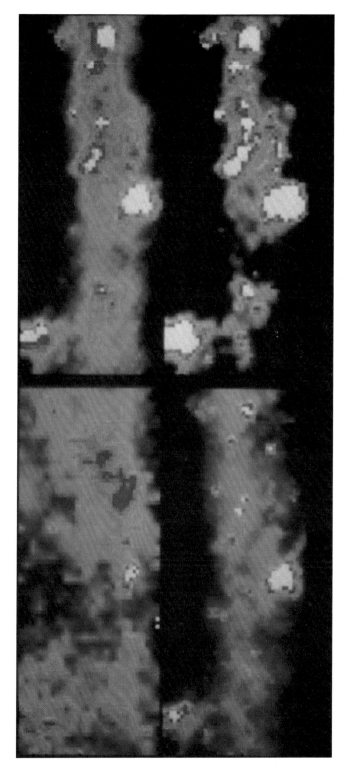

Color Plate 4. False color images of the CO (upper left) and IRAS 12 μm (lower left), 60 μm (lower right) and 100 μm (upper right) emission in the vicinity of M17 are shown. All images are to the same scale with 3' pixels (see Plate 1). The M17 H II region just below center is the brightest infrared luminosity source, but 60 and 100 μm emission is also seen from the extension of the cloud running to the upper right. These images were constructed at the Infrared Processing and Analysis Center with the assistance of J. Good. (See the chapter by Scoville and Sanders.)

Color Plate 5. A model of randomly coagulated core-mantle dust grains. Shown is an ensemble of 91 particles incorporated within their outer icy mantles the very small particles of the interstellar dust population accreted in the final stages of condensation. The degree of volumetric packing is 0.4 (60% open space) and the overall dimension is scaled to about 4 μm. (See the chapter by Greenberg.)

Glossary

GLOSSARY*

Compiled by Melanie Magisos

absolute magnitude	a measure of the intrinsic brightness of a star given by the apparent magnitude it would have if it were moved to a distance of 10 parsecs from the observer
albedo	geometric albedo: the ratio of planet brightness at zero phase angle to the brightness of a perfectly diffusing disk with the same position and apparent size as the planet; bond albedo: fraction of the total incident light reflected by a spherical body
Alpha Centauri	the nearest known star system to the Sun (distance 1.3 pc). Together with Proxima Centauri, it forms a triple-star system
antapex	the direction in the sky away from which the Sun seems to be moving (at a speed of $\sim$ 19 km s^{-1}) relative to the general field of nearby stars in the Galaxy
aphelion	the greatest separation between two bodies orbiting in an eccentric orbit
apogalacticon	the point of the orbit of a star which is farthest from the center of the Galaxy
apparent magnitude	a measure of the observed brightness of a celestial object as seen from the Earth. It is a function of the star's intrinsic brightness, its distance from the observer, and the amount of absorption by interstellar matter between the star and the observer. A sixth magnitude star is just visible to the naked eye

*We have used various definitions from *Glossary of Astronomy and Astrophysics* by J. Hopkins (by permission of the University of Chicago Press, copyright 1980 by the University of Chicago), from *Astrophysical Quantities* by W. W. Allen (London: Athlone Press, 1973) and from *Dictionary of Astronomy, Space, and Atmospheric Phenomena* by D. F. Tver (New York: Van Nostrand Reinhold Co., 1979).

asymmetric drift	the linear relation between the mean motion of stars relative to the Sun in the direction of the galactic rotation $\bar{v}$ and their radial component of the velocity dispersion σ_u^2: for small σ_u^2, $\bar{v}$ is also small
astrometric companion	the unseen companion of a star detected by observing perturbations of the proper motion of the primary
AU	astronomical unit = 1.496×10^{11}m $\cong$ the semimajor axis of the Earth's orbit around the Sun
Barnard's star	the second nearest star system and star with the largest known proper motion $10''.3$ yr^{-1} detected by E. Barnard in 1916
barycenter	the center of gravity of any two-body system such as the Earth and Moon, Earth and Sun, or a double star
basic solar motion	the Sun's velocity with respect to nearby A stars, giant K stars and M stars. These stars dominate the sample of nearby stars and provide similar (to each other) values of the solar motion
bolide	a term used to describe a "shooting star," the streak of light in the sky produced by the transit of a meteoroid through the Earth's atmosphere, that approaches the brightness of the full Moon
brown dwarf	a star that is not massive enough to burn hydrogen and which is therefore very faint (mass < 0.1 M$_\odot$). Its energy originates in gravitational contraction
Brownlee particles	interplanetary dust particles collected by aircraft in the Earth's stratosphere at an altitude of 20 km. Presumably they are of cometary origin
carbonatite	a carbonatite rock of magmatic origin composed mostly of calcium carbonate and magnesium carbonate. Carbonatites are commonly associated with kimberlites and other alkalic igneous rocks
Cas A phase	the phase at some epoch in the life of a supernova remnant when the mass of interstellar material with which the remnant interacts is comparable to the mass original-

GLOSSARY

ly ejected by the explosion. This phase is inherently complicated owing to instabilities in the interaction, even if the explosion and surrounding medium were isotropic. Cassiopeia A is a remnant which seems to be in this phase, and it is often cited as the prototype

closed-box limit
: the closed-box limit is the grammage traversed by a cosmic ray before being lost by a nuclear interaction. In the closed-box model, cosmic rays remain in the Galaxy until they are lost by interactions, principally nuclear

comet shower
: a large increase in the number of comets crossing the orbit of Jupiter produced by a perturbation of comets in the Oort cloud

craton
: part of the Earth's crust that has attained stability and has been little deformed for a prolonged period

deviation of the vertex
: the deviation of the principal axis of the velocity ellipsoid from the direction to the galactic center. Young stars display a deviation of the vertex of about 20°

dynamical standard of rest
: the velocity of an object at the Sun's distance from the galactic center, moving only in the galactic plane on a circular orbit

erg
: sometimes called dyne cm. The cgs unit of energy is the work done by a force of 1 dyne acting over a distance of 1 cm

Fokker-Planck equation
: a modified form of the Boltzmann equation allowing for collision terms in an approximate way. It is used, e.g., in the problem of charged-particle transport in fluctuating electromagnetic fields and in describing the stellar dynamics of star clusters

Forbush decrease
: a decrease in cosmic-ray intensity with an increase in solar activity (and vice versa). This phenomenon was first noted by Forbush in 1954

F region
: two layers in the Earth's ionosphere (F_1 and F_2 at about 150 and 300 km, respectively) immediately above the E region

FWHM
: full width at half maximum. The full width of a spectral line at half-maximum intensity

Gaussian distribution	a statistical distribution defined by the equation $p(x) = c \exp(-k^2 x^2)$, in which x is the statistical variable and c and k are constants. It yields the familiar bell-shaped curve. Accidental errors of measurement and similar phenomena follow this law
geologic time scale	*see table on facing page*
GeV	giga electron volt = 10^9 eV
GCR	galactic cosmic rays
GMC	giant molecular cloud, an ensemble of dust, H_2 and other molecules, especially CO. Their masses range up to 10^6 $M_\odot$ and diameters can exceed 100 pc
Gould's belt	a group of bright, local B stars which is apparently inclined about 16° to the galactic plane
grammage (cosmic ray grammage)	the mass per unit area through which cosmic rays propagate (g cm^{-2})
Gyr	giga year = 10^9 yr
Haas-Bottlinger diagram	a plot of the velocities of stars near the Sun with respect to the circular velocity of the Sun. The abscissa is the peculiar velocity component in the direction tangent to a galactocentric circle passing through the Sun while the ordinate is the radial component
halo population	old stars typical of those found in the halo of our Galaxy; also called population II
HEAO-3	High Energy Astronomical Observatory
Hill surface	*see* zero-velocity surface
Holmes tectonic cycle	apparent periodicities (e.g., ~ 30 Myr, ~ 200 Myr) associated with orogenic episodes and sea-level fluctuations, which were pointed out earlier this century by Holmes (1927) and others, and which have long motivated the search for an underlying galactic mechanism

Geologic Time Scale[a]

Era	Period		Epoch	Myr ago
CENOZOIC	QUATERNARY (Q)		Holocene	0.01
			Pleistocene	1.6
	TERTIARY (T)	NEOGENE	Pliocene	5.3
			Miocene	23.7
		PALEOGENE	Oligocene	36.6
			Eocene	57.8
			Paleocene	66.4
MESOZOIC	CRETACEOUS (K)		Upper	97.5
			Lower	144
	JURASSIC (J)		Upper	163
			Middle	187
			Lower	208
	TRIASSIC (Tr)		Upper	230
			Middle	240
			Lower	245
PALEOZOIC	PERMIAN (P)		Upper	258
			Lower	286
	PENNSYLVANIAN (P)			320
	MISSISSIPPIAN (M)			360
	DEVONIAN (D)		Upper	374
			Middle	387
			Lower	408
	SILURIAN (S)		Upper	421
			Lower	438
	ORDOVICIAN (O)		Upper	458
			Middle	478
			Lower	505
	CAMBRIAN		Upper	523
			Middle	540
			Lower	570
PRECAMBRIAN				

[a] Adapted from Palmer (1983).
[b] From Newell (1982).

horizon (in geology)	an interface indicative of a particular position in a sequence of sedimentary rocks
HR diagram (Hertzsprung-Russell diagram)	in present usage, a plot of bolometric magnitude of stars against their effective temperature. Related plots are the color-magnitude plot (absolute or apparent visual magnitude against color index) and the spectrum-magnitude plot (visual magnitude versus spectral type), the original form of the HR diagram
impulse approximation	an approach to modeling the orbital influence on a binary system due to a gravitational encounter with a third body that assumes that the relative velocity can be considered changed by an instantaneous impulse delivered at one point in the orbit
inverse Fermi mechanism	a mechanism which operates when a comet "rebounds" from the walls of an expanding zero-velocity surface around the Sun, as would happen for example during recession from a molecular cloud. A better known nongravitational analogue of this process arises when cosmic rays are accelerated between the jaws of a closing magnetic mirror. Essentially the reflected cosmic rays exchange no energy in the frame of reference of the moving mirrors but gain energy in the stationary frame of reference
IRAS	Infrared Astronomical Satellite
Kapteyn's two star stream hypothesis	on the basis of observed proper motions, an hypothesis proposed in 1904 that galactic disk stars near the sun are a mixture of two stellar populations which have different mean motions with respect to the Sun
Keplerian orbit	the orbit of a spherical particle or point interacting gravitationally with another spherical particle or point
kimberlite	an alkalic igneous rock composed largely of olivine with a number of accessory minerals, sometimes including diamond. Kimberlite intrusions commonly take the form of vertical pipes, apparently extending from the upper mantle
kpc	kiloparsec = 10^3 pc = 3.086×10^{21} cm
KT layer/boundary	Cretaceous-Tertiary boundary

ℓ, b	galactic longitude and latitude
leaky-box model	a model of space in which it is assumed that cosmic rays diffuse about, the heavier ones producing spallation by striking interstellar material, and all are assumed to have an escape probability $1/\tau_E$ s^{-1}. The cosmic-ray energy densities and the various spectra of nuclei are determined by balance among source, rate of spallation, loss by interaction and loss by escape. In the closed-box limit, τ_E approaches infinity
LHS	Luyten Half Second, a catalogue of stars with proper motions exceeding 0″.5 yr^{-1}
LISW	local interstellar wind
local fluff	a small cloud of density about 0.1 to 0.3 cm^{-3} surrounding the solar system and of about 1 pc radial extent around the Sun
local bubble	a cavity which seems to appear beyond the nearest 1 to 10 pc from the Sun (local bubble boundary). In different directions the cavity probably extends out to distances ranging from $\gtrsim 20$ pc to $\gtrsim 100$ pc. Hot gas ($\sim 10^6$ K) filling this cavity ($n \sim 4 \times 10^{-3}$ cm^{-3}) may provide much of the observed soft X-ray background. Other islands like the local fluff may be present in the cavity.
loss cone	a cone in the phase-space distribution of comets at large distances from the Sun resulting from the absence of comets that have perihelia inside the orbits of Jupiter and Saturn
LSR	local standard of rest. A frame of reference in which the mean motion of stars in the immediate neighborhood is zero. In such a reference system, the motions of stars in the solar neighborhood (volume of space about 100 pc in diameter) average out to zero. It can be used to define a coordinate system in which the origin is a point in the galactic plane moving in a circular orbit around the galactic center with a period equal to that appropriate to circular motion at the given galactocentric distance, and in which the three velocity components are: Π in the direction from the galactic center to the origin; θ, in the

direction of galactic rotation; and Z, perpendicular to the galactic plane. At $b = 0°$, v_{LSR} relative to the centroid of the Local Group is 300 km s^{-1} toward $\ell = 107°$, $b = -8°$

luminosity function
a number distribution of stars or galaxies with respect to their absolute magnitudes. The luminosity function shows the number of stars of a given luminosity (or the number of galaxies per integrated magnitude band) in a given volume of space

Lutz-Kelker effect
a systematic error caused by distance-limited stellar samples

$M_\odot$
solar mass = 1.989×10^{33} g

$M_\oplus$
mass of the Earth = 5.976×10^{27} g

main sequence
the principal band of stars on the HR diagram, containing more than 90% of the stars we observe, that runs diagonally from the upper left (high temperature, high luminosity) to the lower right (low temperature, low luminosity). A star appears on the main sequence after it has started to burn hydrogen in its core, and is estimated to stay on the main sequence until it has used up about 12% of its hydrogen (for a 1 $M_\odot$ star, about 10^{10} yr)

Malmquist correction
a statistical correction introduced into star counts distributed by apparent magnitude to remove sampling bias

Maxwell distribution
an expression for the statistical distribution of velocities among the molecules of a gas at a given temperature in equilibrium

microtektite
a small glassy object, less than one millimeter in diameter and usually spherical, found in some deep-sea sediments, similar to and possibly related to tektites in outward form and composition

middle comet zone
the part of the inner comet cloud lying between about 500 and 20,000 AU from the Sun. The comets in this zone are sufficiently far from the Sun that the loss cone generated by the perturbations of Jupiter and Saturn has been repopulated at times by the perturbations produced by close stellar passages (see Oort cloud, inner)

Milankovič cycles	the climatic changes putatively attributed to periodic cycles (20,000 to 100,000 yr) in the terrestrial position and inclination relative to the Sun, caused by orbital variations and precession, respectively (see, e.g., Steiner and Grillmair 1973)
Milky Way Galaxy	the collection of stars, gases, and clouds to which our Sun belongs. General characteristics are given in Bahcall's chapter, Table I
MKK system	a classification of stellar spectra according to spectral type and luminosity, devised by Morgan, Keenan and Kellman in 1943. The MKK system uses two parameters (spectral type and luminosity class) to describe the morphology of a stellar spectrum with respect to fundamental standards that define the framework of stellar types
Monte Carlo method	a statistical sampling technique used on computers to solve certain classes of problems not amenable to other forms of solution
M_{pg}	apparent photographic magnitude
m_R	apparent red magnitude in Luyten's (1979a,b, 1980) catalogues
M_V	absolute visual magnitude; the magnitude of a star if it were at a distance of 10 pc
Myr	10^6 yr
Nemesis theory	the assumption of the existence of a solar companion star to explain putative periodic cometary showers and extinction of species on the Earth
NLTT	New Luyten Two-Tenths, a catalogue in four volumes of stars with proper motions exceeding $0''.175$ yr^{-1}
Nyquist frequency	the highest frequency that can be determined by Fourier analysis of a time series. This limiting frequency is proportional to the number of observations and occurs because a sinusoidal waveform cannot be defined with fewer than three points
OB associations	associations of stars of spectral types OB-2. In general they are known only within about 3 kpc of the Sun
O-C plot	a plot of residuals, i.e., observed minus calculated positions

Oort cloud, classical	a spherical cloud of comets having semimajor axes $\geq$ 20,000 AU found by J. H. Oort in his empirical study of the orbits of long-period comets. Comets in this shell can be sufficiently perturbed by passing stars or GMCs so that a fraction of them acquire orbits that take them within the orbits of Jupiter and Saturn
Oort cloud, inner	the part of the solar comet cloud lying closer to the Sun than the classical Oort cloud. These comets have semimajor axes $<$ 20,000 AU. Normally, in this region there is an absence of comets with perihelia inside the orbits of Jupiter and Saturn. The comets in this zone are not likely to be perturbed by close stellar passages
Oort cloud, outer boundary	the boundary with a maximum semimajor axis observed for Oort cloud comets ranging roughly between 40,000 and 50,000 AU. This value is roughly one-half of the limit imposed by passing stars or galactic tides
orogenic tectonism	a large-scale disturbance of the continental crust resulting in the formation of mountains
pc	parsec: 1 parsec = the distance where 1 AU subtends 1 arcsec = 206,265 AU = 3.26 lightyear = 3.086×10^{18} cm
perigalacticon	the point of the orbit of a star which is nearest the center of the Galaxy
perihelion	the point in a solar system orbit closest to the Sun
Poisson distribution	an approximation to the binomial distribution used when the probability of success in a single trial is very small and the number of trials is very large
Poisson-Vlasov equation	the combined equation for the gravitational potential (Poisson's equation) and the distribution function for the stars (Vlasov's equation)
polarizing dust	interstellar dust grains aligned by weak interstellar magnetic fields which polarize starlight. The exact alignment mechanism of the dust is unknown
proper motion	the apparent angular rate of motion of a star across the line of sight on the celestial sphere. Proper motion was

GLOSSARY 433

	discovered by Halley (ca. 1715) from a comparison of the location of Arcturus with ancient star maps
Rayleigh number (*Ra*)	a dimensionless parameter involving the temperature gradient and the coefficients of thermal conductivity and kinematic viscosity, which determines when a fluid, under specified geometrical conditions, will become convectively unstable
Rayleigh-Taylor instability	a type of hydrodynamic instability for static fluids (e.g., cold dense gas above hot rarefied gas)
rms	root mean square, the square root of the mean square value of a set of numbers
r-process	the capture of neutrons on a very rapid time scale (i.e., one in which a nucleus can absorb neutrons in rapid succession so that regions of great nuclear instability are bridged), a theory advanced to account for the existence of all elements heavier than bismuth as well as the neutron-rich isotopes heavier than iron. The essential feature of the *r*-process is the release of great numbers of neutrons in a very short time (< 100 s). The presumed source of such a large flux of neutrons is a supernova, at the boundary between the neutron star and the ejected material
Sc spiral	a type of spiral galaxy in a classification scheme devised by E. Hubble which is characterized by open spiral arms and a small nucleus compared to the size of the galaxy
spallation	the chipping, fracturing or fragmentation and the upward and outward heaving of rock caused by interaction of a shock (compressional) wave with a free surface. Also, ejection of atoms from a solid by incident radiation
spectroscopic binaries (SB)	stars whose binary nature can be detected from the periodic Doppler shifts of their spectra, owing to their varying velocities in the line of sight. Spectroscopic binaries are typically of spectral type B, with nearly circular orbits whereas long-period M-type spectroscopic binaries have highly eccentric orbits
s-process	a process in which heavy, stable, neutron-rich nuclei are synthesized from iron-peak elements by successive captures of free neutrons in a weak neutron flux, so there is

	time for β decay before another neutron is captured (cf., *r*-process)
standard solar motion	the Sun's velocity with respect to the set of nearby stars whose radial velocities and proper motions are known
superbubble	a giant shell, ring or "loop" of interstellar gas and dust (diameter $\gtrsim$ 100 pc) around an association of massive O-B stars. Superbubbles appear to be formed by successive supernova explosions in these rapidly evolving associations and/or by the combined stellar winds of the massive association stars. A superbubble shell may be threaded by a magnetic field and coincide with a region of enhanced nonthermal radio emission. Superbubbles are not necessarily either complete (i.e., closed) or round
supercluster	a cluster of clusters of galaxies. The scale of superclusters is $\sim$ 50 Mpc, the scale of clusters $\sim$ 1 Mpc
supernova	a giant stellar explosion in which the star's luminosity suddenly increases by as much as 20 mag. Most of the star's mass is blown off, leaving behind, at least in some cases, an extremely dense core which (as in the Crab Nebula) may be a neutron star
tidal radius	the upper limit on orbital semimajor axis for comets orbiting the Sun imposed by the galactic tide
tidal stripping	the removal of objects in loosely bound orbits due to the tidal field of a passing massive object
Tisserand's approximation	a method of comparison of orbital elements of comets used for identification
unseen matter	the matter whose existence is inferred from studies of stellar (or gas) distributions and motion but is not observed directly
velocity ellipsoid	a convenient way of describing the velocities of stars near the Sun, introduced by K. Schwarzschild. The velocity locations of the center measures the solar motion with respect to the stars in the sample. The length of the axes of the ellipsoid measure the velocity dispersion. The largest velocity dispersion is found to lie in the ga-

lactic plane and points approximately at the galactic center

Walsh spectrum analysis
a method of spectrally analyzing a time series of pulse-like telegraph signals. Walsh functions are rectangular waveforms taking only $+1$ and -1 and amplitude values

Wolf-Rayet star
one of a class of very luminous, very hot (as high as 50,000 K) stars whose spectra have broad emission lines (mainly He I and He II), which are presumed to originate from material continually ejected from the star at very high velocities (about 2000 km s^{-1}) by stellar winds. They may be the exposed helium cores of stars that were at one time on the H-burning main sequence. Some Wolf-Rayet spectra show dominantly emission lines from ions of carbon (WC stars); other show dominantly emission lines from the ions of nitrogen (WN stars)

zero-velocity surface
a three-dimensional constant energy surface surrounding a binary system at which a third orbiting massless particle encounters the extremum of its motion (zero velocity) and is thus unable to cross. The shapes and sizes of these surfaces (also called Hill surfaces) depend principally on the energy of the massless particle

Bibliography

BIBLIOGRAPHY
Compiled by Melanie Magisos

Abell, G. O. 1982. *Exploration of the Universe* (Philadelphia: Saunders College Pub.).
Abt, H. A. 1979. The frequencies of binaries on the main sequence. *Astron. J.* 84:1591–1597.
Abt, H. A. 1983. Normal and abnormal frequencies. *Ann. Rev. Astron. Astrophys.* 21:343–372.
Abt, H. A., and Levy, S. G. 1976. Multiplicity among solar-type stars. *Astrophys. J. Suppl.* 30:273–306.
Adams, T. F., and Frisch, P. C. 1977. High-resolution observations of the Lyman alpha sky background. *Astrophys. J.* 212:300–308.
Agarwal, V. K., Schutte, W., Greenberg, J. M., Ferris, J. P., Briggs, R., Connor, S., Van de Bult, C. E. P. M., and Baas, F. 1985. Photochemical reactions in interstellar grains, photolysis of CO, NH_3 and H_2O. *Origins of Life* 16:21–40.
A'Hearn, M. F., and Feldman, P. D. 1985. S_2: A clue to the origin of cometary ice? In *Ices in the Solar System*, eds. J. Klinger, D. Benest, A. Dollfus, and R. Smoluchowski (Dordrecht: D. Reidel), pp. 463–471.
A'Hearn, M. F., Feldman, P. D., and Schleicher, D. G. 1983. The discovery of S_2 in comet IRAS-Aracki-Alcock 1983d. *Astrophys. J.* 274:L99-L103.
Ajello, J. M., and Thomas, G. E. 1985. Predicted interplanetary distribution of Lyman-α intensity and polarization. *Icarus* 61:163–170.
Ajello, J. M., Witt, N., and Blum, P. W. 1979. Four UV observations of the interstellar wind by Mariner 10: Analysis with spherically symmetric solar radiation models. *Astron. Astrophys.* 73:260–271.
Allan, R. R. 1970. The critical inclination problem. *Celestial Mech.* 2:121–122.
Allen, C. 1973. *Astrophysical Quantities* (London: Athlone Press).
Alvarez, L. W. 1983. Experimental evidence that an asteroid impact led to the extinction of many species 65 Myr ago. *Proc. Natl. Acad. Sci. U.S.A.* 80:627–642.
Alvarez, L. W., Alvarez, W., Asaro, F., and Michel, H. V. 1980. Extraterrestrial cause for the Cretaceous-Tertiary extinction: Experimental results and theoretical interpretation. *Science* 208:1095–1108.
Alvarez, W., and Muller, R. A. 1984. Evidence from crater ages for periodic impacts on the Earth. *Nature* 308:718–720.
Alvarez, W., Alvarez, L. W., Asaro, F., and Michel, H. V. 1982a. Current status of the impact theory for terminal Cretaceous extinction. *Geol. Soc. Amer. Spec. Paper 190*, pp. 305–315.
Alvarez, W., Asaro, F., Michel, H. V., and Alvarez, L. W. 1982b. Iridium anomaly approximately synchronous with terminal Eocene extinctions. *Science* 216:886–888.
Anderson, J. D. 1974. Lectures on physical and technical problems posed by precision radio tracking. In *Travitazione Sperimentale*, ed. B. Bertotti (New York: Academic Press).
Anderson, J. D., and Mashhoon, B. 1985. Pioneer 10 search for gravitational waves: Limits on a possible isotropic cosmic background of radiation in the microhertz region. *Astrophys. J.* 290:445–448.
Anderson, J. D., Levy, G. S., and Renzetti, N. A. 1985. Application of the Deep Space Network (DSN) to the testing of general relativity. In *Relativity in Celestial Mechanics and Astrometry*, IAU Symp. 114, ed. J. Kovalevsky (Dordrecht: D. Reidel), pp. 329–344.
Antonov, V. A., and Latyshev, I. N. 1972. Determination of the form of the Oort cometary cloud as the Hill surface in the galactic field. In *The Motion, Evolution of Orbits, and Origin of Comets*, IAU Symp. 45, eds. G. A. Chebotarev and E. I. Kazimirchak-Polonskaya (Dordrecht: D. Reidel), pp. 341–345.
Applegate, J. H., Douglas, M. R., Gürsel, Y., Hunter, P., Seitz, C. L., and Sussman, G. J. 1984. A digital orrery. Submitted to *IEEE Trans. on Computers*.

Arnold, J. R. 1965. The origin of meteorites as small bodies. II. The model. *Astrophys J.* 141:1536–1547.
Ash, M. E., Shapiro, I. I., and Smith, W. B. 1971. The system of planetary masses. *Science* 174:551–556.
Aumann, H. H. 1984. IRAS observations of nearby main-sequence dwarfs. *Bull. Amer. Astron. Soc.* 16:483 (abstract).
Aumann, H. H., Gillett, F. C., Beichman, C. A., de Jong, T., Houck, J. R., Low, F., Neugebauer, G., Walker, R. G., and Wesselius, P. 1984. Discovery of a shell around Alpha Lyrae. *Astrophys. J.* 278:L23-L27.
Axford, W. I. 1965. The modulation of galactic cosmic rays in the interplanetary medium. *Planet. Space Sci.* 13:115–130.
Axford, W. I. 1981. Acceleration of cosmic rays by shock waves. In *Proc. 17th Internatl. Cosmic Ray Conf.* 12:155–203.
Axford, W. I., Leer, E., and Skadron, G. 1977. Acceleration of cosmic rays at shock fronts. In *Proc. 15th Internatl. Cosmic Ray Conf.* 11:132 (abstract).
Bahcall, J. N. 1984a. Self-consistent determination of the total amount of matter near the sun. *Astrophys. J.* 276:169–181.
Bahcall, J. N. 1984b. K giants and the total amount of matter near the Sun. *Astrophys. J.* 287:926–944.
Bahcall, J. N. 1986. Brown dwarfs: Conference summary. In *Proc. Workshop on the Astrophysics of Brown Dwarfs*, ed. H. Kafatos (Cambridge: Cambridge Univ. Press). In press.
Bahcall, J. N., and Bahcall, S. 1985. The Sun's motion perpendicular to the galactic plane. *Nature* 316:706–708.
Bahcall, J. N., and Soneira, R. M. 1981. The distribution of stars to V = 16th magnitude near the north galactic pole: Normalization, clustering properties, and counts in various bands. *Astrophys. J.* 246:122–135.
Bahcall, J., Schmidt, M., and Soneira, R. M. 1983. The galactic spheroid. *Astrophys. J.* 265:730–747.
Bahcall, J. N., Hut, P., and Tremaine, S. 1985. Maximum mass of objects that constitute unseen disk material. *Astrophys. J.* 290:15–20.
Bailey, M. E. 1977. Some comments on the Oort cloud. *Astrophys. Space Sci.* 50:3–22.
Bailey, M. E. 1983a. The structure and evolution of the solar system comet cloud. *Mon. Not. Roy. Astron. Soc.* 204:603–633.
Bailey, M. E. 1983b. Comets, planet X and the orbit of Neptune. *Nature* 302:399–400.
Bailey, M. E. 1983c. Theories of cometary origin and the brightness of the infrared sky. *Mon. Not. Roy. Astron. Soc.* 205:47P-52P.
Bailey, M. E. 1984. The steady-state 1/a distribution and the problem of cometary fading. *Mon. Not. Roy. Astron. Soc.* 211:347–368.
Bailey, M. E. 1986. The mean energy transfer rate to comets in the Oort cloud and implications for cometary origins. *Mon. Not. Roy. Astron. Soc.* 218:1–30.
Baldwin, R. B. 1985. Relative and absolute ages of individual craters and the rate of infalls on the Moon in the post-Imbrium period. *Icarus* 61:63–91.
Barbanis, B., and Woltjer, L. 1967. Orbits in spiral galaxies and the velocity dispersion of population I stars. *Astrophys. J.* 150:461–468.
Bash, F. 1979. Density wave induced star formation: The optical surface brightness of galaxies. *Astrophys. J.* 233:524–538.
Basham, A. L. 1967. *The Wonder that was India* (London: Sidgwick Jackson).
Begelman, M. C., and Rees, M. J. 1976. Can cosmic clouds cause climatic catastrophes? *Nature* 261:298–299.
Bell, A. R. 1978. The acceleration of cosmic rays in shock fronts. I. *Mon. Not. Roy. Astron. Soc.* 182:147–156.
Benson, R. H., Chapman, R. E., and Deck, L. T. 1984. Paleoceanographic events and deep-sea ostracodes. *Science* 224:1334–1336.
Benton, M. J. 1985. Interpretations of mass extinction. *Nature* 314:496–497.
Berger, A. 1977. Long-term variations of the Earth's orbital elements. *Celestial Mech.* 15:53–74.
Berkhuijsen, E. M. 1972. A survey of linear polarization at 1415 MHz. I. Method of reduction and results for the North Polar Spur. *Astron. Astrophys. Suppl.* 5:205–238.

Bertaux, J. L. 1984. Helium and hydrogen of the local interstellar medium observed in the vicinity of the Sun. *Local Interstellar Medium, IAU Coll. 81,* eds. Y. Kondo, F. C. Bruhweiler, and B. D. Savage, NASA CP-2345, pp. 3–23.
Bertaux, J. L., Ammar, A., and Blamont, J. E. 1972. OGO 5 determination of the local interstellar wind parameters. *Space Res.* 12:1559.
Bhat, C. L., Issa, M. R., Mayer, C. J., and Wolfendale, A. W. 1985. Acceleration of cosmic rays in the Loop I "Supernova remnant"? *Nature* 314:515–517.
Bhatt, H. C., Rouse, D. P., and Williams, I. P. 1985. Mass distribution of interstellar clouds. *The Galaxy and the Solar System Abstract Booklet* (Tucson: Lunar and Planet. Lab.), p. 33.
Bidelman, W. P. 1985. G. P. Kuiper's spectral classifications of proper motion stars. *Astrophys. J. Suppl.* 59:197–227.
Biermann, L. 1978, Dense interstellar clouds and comets. In *Astronomical Papers Dedicated to Bengt Strömgren,* eds. A. Reiz and T. Anderson (Copenhagen: Copenhagen Obs.), pp. 327–335.
Biermann, L. 1981. The smaller bodies of the solar system. *Phil. Trans. Roy. Soc. London* A303:351–352.
Biermann, L., and Lüst, R. 1978. *Sitz. ber. Bayer. Akad. Wiss. Mat.-Naturw. Kl.*
Biermann, L., and Michel, K. W. 1978. On the origin of cometary nuclei in the presolar nebula. *Moon and Planets* 18:447–464.
Biermann, L., Huebner, W. F., and Lüst, R. 1983. Aphelion clustering of "new" comets: Star tracks through Oort's cloud. *Proc. Natl. Acad. Sci. U.S.A.* 80:5151–5155.
Binns, W. R., Fickle, F. K., Garrard, T. L., Israel, M. H., Klarmann, J., Stone, E. C., and Waddington, C. J. 1982. The abundance of the actinides in the cosmic radiation as measured on HEAO 3. *Astrophys. J.* 261:L117–L120.
Black, D. C. 1980. In search of other planetary systems. *Space Sci. Rev.* 25:35–81.
Blades, J. C., Wynne-Jones, I., and Wayte, R. C. 1980. Very high resolution spectroscopy of interstellar NaI. *Mon. Not. Roy. Astron. Soc.* 193:849–866.
Blandford, R. D., and Ostriker, J. P. 1978, Particle acceleration by astrophysical shocks. *Astrophys. J.* 221:L29–L32.
Blitz, L. 1979. A Study of the Molecular Complexes Accompanying Mon OB1, Mon OB2 and CMa OB1. Ph.D. Thesis, Columbia Univ., New York.
Blitz, L. 1982. Giant molecular cloud complexes in the Galaxy. *Sci. Amer.* 246:72–80.
Blitz, L. 1985. Molecular clouds in the solar vicinity. In *The Galaxy and the Solar System Abstract Booklet* (Tucson: Lunar Planet. Lab.), p. 34.
Blitz, L., Fich, M., and Kulkarni, S. 1983. The new Milky Way. *Science* 220:1233–1240.
Bloemen, J. B. G. M., Blitz, L., and Hermsen, W. 1984. The radial distribution of galactic gamma rays. *Astrophys. J.* 279:136–143.
Bloemen, J. B. G. M., Caraveo, P. A., Hermsen, W., Lebrun, F., Maddalena, R. J., Strong, A. W., and Thaddeus, P. 1985. Gamma rays from atomic and molecular gas in the large complex of clouds in Orion and Monoceros. *Astron. Astrophys.* In press.
Blum, P. W., Pfleiderer, J., and Wulf-Mathies, C. 1975. Neutral gases of interstellar origin in interplanetary space. *Planet. Space Sci.* 23:93–105.
Boeshaar, P. C. 1976. The Spectral Classification of M Dwarf Stars. Ph.D. Thesis, Ohio State Univ., Columbus.
Bogart, R. S., and Noerdlinger, P. D. 1982. On the distribution of orbits among long-period comets. *Astron. J.* 87:911–917.
Bohlin, R. C., Savage, B. D., and Drake, J. F. 1978, A survey of interstellar HI from L-alpha absorption. *Astrophys. J.* 224:132–142.
Bohor, B. F., Foord, E. E., Modreski, P. J., and Triplehorn, D. M. 1984. Mineralogic evidence for an impact event at the Cretaceous-Tertiary boundary. *Science* 224:867–869.
Bottomley, R. J. 1982. ^{40}Ar-^{39}Ar Dating of Melt Rock from Impact Craters. Ph.D. Thesis, Univ. of Toronto, Ontario.
Bottomley, R. J., York, D., and Grieve, R. A. F. 1978. ^{40}Ar-^{39}Ar ages of Scandinavian impact structures. *Contrib. Mineral. Petrol.* 68:79–84.
Bottomley, R. J., York, D., and Grieve, R. A. F. 1979. Possible source craters for the North American tektites: A geochronological investigation. *EOS Trans. AGU* 60:309 (abstract).
Bowen, E. G. 1956. The influence of meteoritic dust on rainfall. *J. Meteorol.* 13:142–152.

Brecher, K. 1984. The Canterbury Swarm. *Bull. Amer. Astron. Soc.* 16:476 (abstract).
Brin, D. 1984. The deadly thing at 2.4 kilo-parsecs. *Analog* 104(5):66–73.
Brouwer, D., and Clemence, G. M. 1961. *Methods of Celestial Mechanics* (New York: Academic Press).
Brownlee, D. E. 1978. Microparticle studies by sampling techniques. In *Cosmic Dust*, ed. J. A. M. McDonnell (New York: J. Wiley), pp. 295–336.
Brownlee, D. E. 1985. Cosmic dust: Collection and research. *Ann. Rev. Earth Planet. Sci.* 13:147–173.
Bruhweiler, F. C. 1982. The distribution of interstellar gas within 50 pc of the Sun. *Advances in Ultraviolet Astronomy: Four Years of IUE Research*, eds. Y. Kondo, J. M. Mead, R. D. Chapman, NASA CP-2238, p. 125.
Bruhweiler, F. C. 1984. Absorption line studies and the distribution of neutral gas in the local interstellar medium. *Local Interstellar Medium, IAU Coll. 81*, eds. Y. Kondo, F. C. Bruhweiler, and B. D. Savage, NASA CP-2345, pp. 39–50.
Bruhweiler, F. C., Gull, T. R., Kafatos, M., Sofia, S. 1980. Stellar winds, supernovae, and the origin of HI supershells. *Astrophys. J.* 238:L27–L30.
Bubeníček, J., Palouš, J., and Piskunov, A. E. 1985. Hyades moving group: New members, mass and age. *Sov. Astron. J.* 62:1073–1076.
Burstein, P., Barker, R. J., Kraushar, W. L., and Sanders, W. T. 1977. Three-band observations of the soft x-ray background and some implications of thermal emission. *Astrophys. J.* 213:405–420.
Burton, W. B., Gordon, M. A., Bania, T. M., and Lockman, F. J. 1975. The overall distribution of carbon monoxide in the plane of the Galaxy. *Astrophys. J.* 202:30–49.
Butler, D. M., Newman, M. J., and Talbot, R. J. 1978. Interstellar cloud material: Contribution to planetary atmospheres. *Science* 201:522–525.
Byl, J. 1983. Galactic perturbations on nearly parabolic cometary orbits. *Moon and Planets* 29:121–137.
Cameron, A. G. W. 1962. The formation of the Sun and the planets. *Icarus* 1:13–69.
Cameron, A. G. W. 1973. Accumulation processes in the primitive solar nebula. *Icarus* 18:407–450.
Cameron, A. G. W. 1978a. The primitive solar accretion disc and the formation of the planets. In *The Origin of the Solar System*, ed. S. F. Dermott (New York: J. Wiley and Sons), pp. 49–75.
Cameron, A. G. W. 1978b. Physics of the primitive solar accretion disk. *Moon and Planets* 18:5–40.
Carlberg, R. G., and Sellwood, J. A. 1985. Dynamical evolution in galactic disks. *Astrophys. J.* 292:79–89.
Carlberg, R. G., Dawson, P. C., Hsu, T., and Van den Berg, D. A. 1985. The age-velocity-dispersion relation in the solar neighborhood. *Astrophys. J.* 294:674–681.
Carrigy, M. A., and Short, N. M. 1968. Evidence of shock metamorphism in rocks from the Steen River structure, Alberta. In *Shock Metamorphism in Natural Materials*, eds. B. M. French and N. M. Short (Baltimore: Mono Book Corp.), pp. 367–378. In French.
Cash, W., Charles, P., Bowyer, S., Walter, F., Garmire, G., and Riegler, G. 1980. The x-ray superbubble in Cygnus. *Astrophys. J.* 238:L71–L76.
Cassé, M., and Paul, J. A. 1982. On the stellar origin of the ^{22}Ne excess in cosmic rays. *Astrophys. J.* 258:860–863.
Ceplecha, Z., and McCrosky, R. E. 1976. Fireball end heights: A diagnostic for the structure of meteoritic material. *J. Geophys. Res.* 81:6257–6275.
Chandrasekhar, S. 1942. *Principles of Stellar Dynamics* (Chicago: Univ. of Chicago Press).
Chang, S. 1979. Comets: Cosmic connections with carbonaceous meteorites, interstellar molecules and the origin of life. In *Space Missions to Comets*, eds. M. Neugebauer, D. K. Yeomans, J. C. Brandt and R. W. Hobbs, NASA Publ. 2089, pp. 59–112.
Chebotarev, G. A. 1965. On the dynamical limits of the solar system. *Sov. Astron. J.* 8:787–792.
Chebotarev, G. A. 1966. Cometary motion in the outer solar system. *Sov. Astron. J.* 10:341–344.
Christie-Blick, N. 1982. Pre-Pleistocene glaciation on Earth: Implications for climatic history of Mars. *Icarus* 50:423–443.
Christy, J. W., and Harrington, R. S. 1978. The satellite of Pluto. *Astron. J.* 83:1005–1008.
Clarke, J. T., Bowyer, S., Fahr, H. J., and Lay, G. 1984. IUE high resolution spectrophotometry of H Ly alpha emission from the local interstellar medium. *Astron. Astrophys.* 139:389–393.

Clayton, D. D. 1982. Cosmic radioactivity: A gamma ray search for the origins of atomic nuclei. In *Essays in Nuclear Astrophysics*, eds. E. A. Barnes, D. D. Clayton, and D. N. Schramm (Chicago: Univ. of Chicago Press), pp. 401–426.

Clayton, D. D. 1984. ^{26}Al in the interstellar medium. *Astrophys. J.* 280:144–149.

Clayton, D. D., Cox, D. P., Michel, F. C., and Szentgyorgi, A. 1985. The soft x-ray background versus cosmic rays: Could the solar system be inside a supernova remnant? Submitted to *Nature*.

Clemens, D. P. 1985. The Massachusetts-Stony Brook galactic plane CO survey: The galactic disk rotation curve. *Astrophys. J.* 295:422–436.

Clube, S. V. M. 1967. The kinematics of Gould's Belt. *Mon. Not. Roy. Astron. Soc.* 137:189–203.

Clube, S. V. M. 1978. Does our Galaxy have a violent history? *Vistas in Astron.* 22:77–118.

Clube, S. V. M. 1983a. Drift analysis: A forgotten technique. *Proc. Statistical Methods in Astronomy Symp.*, ESA SP-201, pp. 173–176.

Clube, S. V. M. 1983b. The origin and evolution of comets. In *Asteroids, Comets, Meteors*, eds. C.-I. Lagerkvist and H. Rickman (Uppsala: Uppsala Univ. Press), pp. 369–374.

Clube, S. V. M. 1984. Molecular clouds: Comet factories? In *Dynamics of Comets: Their Origin and Evolution, IAU Coll. 83*, eds. A. Carusi and G. B. Valsecchi (Dordrecht: D. Reidel), pp. 19–30.

Clube, S. V. M. 1985. The kinematics of nearby stars and large-scale radial motion in the galaxy. In *The Milky Way Galaxy, IAU Symp. 106*, eds. H. Van Woerden, R. J. Allen and W. B. Burton (Dordrecht: D. Reidel), pp. 145–147.

Clube, S. V. M., and Napier, W. M. 1982a. The role of episodic bombardment in geophysics. *Earth Planet. Sci. Lett.* 57:251–262.

Clube, S. V. M., and Napier, W. M. 1982b. Spiral arms, comets and terrestrial catastrophism. *Q. J. Roy. Astron. Soc.* 23:45–66.

Clube, S. V. M., and Napier, W. M. 1983. Some considerations relating to an interstellar origin of comets. *Highlights in Astron.* 6:355–362.

Clube, S. V. M., and Napier, V. M. 1984a. Comet capture from molecular clouds: A dynamical constraint on star and planet formation. *Mon. Not. Roy. Astron. Soc.* 208:575–588.

Clube, S. V. M., and Napier, W. M. 1984b. The microstructure of terrestrial catastrophism. *Mon. Not. Roy. Astron. Soc.* 211:953–968.

Clube, S. V. M., and Napier, W. M. 1984c. Terrestrial catastrophism: Nemesis or Galaxy? *Nature* 311:635–636.

Clube, S. V. M., and Napier, W. M. 1985a. Terrestrial catastrophism: Matters arising. *Nature* 313:503.

Clube, S. V. M., and Napier, W. M. 1985b. Comet formation in molecular clouds. *Icarus* 62:384–388.

Clube, S. V. M., and Pan, R. 1985. The kinematic centre of the Galaxy. *Mon. Not. Roy. Astron. Soc.* 216:511–519.

Cohen, C. J., Hubbard, E. C., and Oesterwinter, C. 1973. Elements of the outer planets for one million years. *Astron. Papers Amer. Ephemeris and Nautical Almanac*, 22:Part 1.

Cohen, R. S., Cong, H. I., Dame, T. M., and Thaddeus, P. 1980. Molecular clouds and galactic spiral structure. *Astrophys. J.* 239:L53–L56.

Cohen, R. S., Grabelsky, D. A., May, J., Bronfman, L., Alvarez, H., and Thaddeus, P. 1985. Molecular clouds in the Carina Arm. *Astrophys. J.* 290:L15–L20.

Cohen, R. S., Dame, T. M., and Thaddeus, P. 1986. The Columbia CO survey of molecular clouds in the First Galactic Quadrant. *Astrophys. J. Suppl.* 60:695–818.

Cohn, H., and Kulsrud, R. M. 1978. The stellar distribution around a black hole: Numerical integration of the Fokker-Planck equation. *Astrophys. J.* 226:1087–1108.

Colgate, S. A., and Johnson, M. H. 1960. Hydrodynamic origin of cosmic rays. *Phys. Rev. Lett.* 5:235–238.

Corliss, B. H., Aubry, M. P., Berggren, W. A., Fenner, J. M., Keigwin, L. D., and Keller, G. 1984. The Eocene/Oligocene boundary event in the deep sea. *Science* 226:806–810.

Cowie, L. L., Songaila, A., and York, D. G. 1979. A rapidly expanding shell of gas centered on the Orion OBI association. *Astrophys. J.* 230:469–484.

Cowie, L. L., McKee, C. F., and Ostriker, J. P. 1981a. Supernova remnant revolution in an inhomogeneous medium I. Numeric models. *Astrophys. J.* 247:908–924.

Cowie, L. L., Hu, E. M., Taylor, W., and York, D. G. 1981b. A search for expanding supershells of gas around OB associations. *Astrophys. J.* 250:L25–L29.
Cox, D. P., and Anderson, P. R. 1982. Extended adiabatic blast waves and a model of the soft x-ray background. *Astrophys. J.* 253:268–289.
Cox, D. P., and Smith, B. W. 1974. Large-scale effects of supernova remnants on the Galaxy: Generation and maintenance of a hot network of tunnels. *Astrophys. J.* 189:L105–L108.
Crain, I. K., and Crain, P. L. 1970. New stochastic model for geomagnetic reversals. *Nature* 228:39–41.
Crain, I. K., Crain, P. L., and Plaut, M. G. 1969. Long period Fourier spectrum of geomagnetic reversals. *Nature* 223:283.
Crochet, J. H., and Kuo, H. Y. 1979. Sources for gold, palladium and iridium in deep-sea sediments. *Geochim. Cosmochim. Acta* 43:831–842.
Crutcher, R. M. 1982. The local interstellar medium. *Astrophys. J.* 254:82–87.
Currie, K. L. 1969. Geological notes on the Carswell circular structure, Saskatchewan (74K). *Geol. Surv. Canada Paper* 67–32.
Dalaudier, F., Bertaux, J. L., Kurt, V. G., Mironova, E. N. 1983. Characteristics of interstellar helium observed with Prognoz-6 58.4 nm photometers. *Astron. Astrophys.* 134:171–184.
Dame, T. M. 1983. Molecular Clouds and Galactic Spiral Structure. Ph.D. Thesis, Columbia Univ., New York.
Dame, T. M., and Thaddeus, P. 1985. A wide-latitude CO survey of molecular clouds in the northern Milky Way. *Astrophys. J.* 297:751–765.
Damon, P. E. 1971. The relationship between late Cenozoic volcanism and tectonism and orogenic-epeirogenic periodicity. In *The Late Cenozoic Glacial Ages*, ed. K. K. Turekian (New Haven: Yale Univ. Press), pp. 15–35.
Daniels, P. A., and Hughes, D. W. 1981. The accretion of cosmic dust: A computer simulation. *Mon. Not. Roy. Astron. Soc.* 195:1001–1009.
Davidson, K. 1975. Does the solar system include distant but discoverable infrared dwarfs? *Icarus* 26:99–101.
Davis, M., Hut, P., and Muller, R. A. 1984. Extinction of species by periodic comet showers. *Nature* 308:715–717.
De Laubenfels, M. W. 1956. Dinosaur extinction: One more hypothesis. *J. Paleontol.* 30:207–218.
Delhaye, J. 1965. Solar motion and velocity distribution of common stars. In *Stars and Stellar Systems*, vol. 5, eds. A. Blaauw and M. Schmidt (Chicago: Univ. of Chicago Press), pp. 61–84.
Delsemme, A. H. 1973. Origin of the short-period comets. *Astron. Astrophys.* 29:377–381.
Delsemme, A. H. 1977. The origin of comets. In *Comets, Asteroids, Meteorites: Interrelations, Evolution and Origins*, ed. A. H. Delsemme (Toledo, OH: Univ. of Toledo), pp. 453–467.
Delsemme, A. H. 1985a. Empirical data from the Oort's cloud. In *Dynamics of Comets: Their Origin and Evolution, IAU Coll. 83*, eds. A. Carusi and G. B. Valsecchi (Dordrecht: D. Reidel), pp. 71–85.
Delsemme, A. H. 1985b. The nature of the cometary nucleus. *Publ. Astron. Soc. Pacific.* 97:861–869.
Dent, B. 1973. Glacial exhumation of impact craters on the Canadian Shield. *Geol. Soc. Amer. Bull.* 84:1667–1672.
de Vaucouleurs, G. 1983. The galaxy as fundamental calibrator of the extragalactic distance scale. I. *Astrophys. J.* 268:451–467.
d'Hendencourt, L., Allamandola, L. J., Baas, F., and Greenberg, J. M. 1982. Interstellar grain explosions: Molecule cycling between gas and dust. *Astron. Astrophys.* 109:L12–L14.
d'Hendencourt, L., Allamandola, L., Baas, F., and Greenberg, J. M. 1985. Time dependent chemistry in dense molecular clouds. *Astron. Astrophys.* 152:130–150.
D'Hondt, S. L., and Keller, G. 1985. Late-Cretaceous stepwise mass extinction of planktonic foraminifera. *Abstracts with Program 1985, Geol. Soc. Amer.*, pp. 557–558 (abstract).
D'Hondt, S., Keller, G., and Stallard, R. 1986. Major element compositional variation within and between microtektites of different Late Eocene strewn fields. Submitted to *Earth Planet. Sci. Lett.*
Donn, B. 1963. The origin and structure of icy cometary nuclei. *Icarus* 2:396–402.
Donn, B. 1976. The nucleus: Panel discussion. In *The Study of Comets, Part 2*, eds. B. Donn, M. Mumma, W. Jackson, M. A'Hearn and R. Harrington, NASA SP-373, pp. 611–621.

Dorman, J., Evans, S., Nakamura, Y., and Latham, G. 1978. On the time-varying properties of the lunar seismic meteoroid population. *Proc. Lunar Planet. Sci. Conf.* 9:3615–3626.
Drapatz, S., and Zinnecker, H. 1984. The size and mass distribution of galactic molecular clouds. *Mon. Not. Roy. Astron. Soc.* 210:11P–14P.
Drury, L. O'C. 1983. An introduction to the theory of diffusive shock acceleration of energetic particles in tenuous plasmas. *Rep. Prog. Phys.* 46:973–1027.
Duncombe, R. L., and Seidelmann, P. K. 1980. A history of the determination of Pluto's mass. *Icarus* 44:12–18.
Duncombe, R. L., Klepczynski, W. J., and Seidelmann, P. K. 1968a. Mass of Pluto. *Science* 162:800–802.
Duncombe, R. L., Klepczynski, W. J., and Seidelmann, P. K. 1968b. Orbit of Neptune and the mass of Pluto. *Astron. J.* 73:830–835.
Eddington, A. S. 1914. *Stellar Movements and the Structure of the Universe* (London: MacMillan and Co.).
Edgar, R. J., and Cox, D. P. 1984. A model of the soft X-ray background as a blastwave when viewed from inside. *Local Interstellar Medium, IAU Coll. 81,* eds. Y. Kondo, F. C. Bruhweiler, and B. D. Savage, NASA CP-2345, pp. 297–300.
Eggen, O. J. 1970. Stellar kinematics and parameters of the Sirius and Hyades superclusters. *Vistas in Astron.* 12:367–414.
Eggen, O. J. 1982. The Hyades main sequence. *Astrophys. J. Suppl.* 50:221–240.
Eggen, O. J. 1983a. Concentrations in the local association I. The southern concentrations NGC 2516, IC 22602, Centaurus-Lupus and Upper Scorpius. *Mon. Not. Roy. Astron. Soc.* 204:377–390.
Eggen, O. J. 1983b. Concentrations in the local association II. The northern concentrations including the α Persei, Pleiades, M34 and δ Lyrae clusters. *Mon. Not. Roy. Astron. Soc.* 204:391–403.
Eggen, O. J. 1983c. Concentrations in the local association III. Late-type bright giants, ages and abundances. *Mon. Not. Roy. Astron. Soc.* 204:405–414.
Eggen, O. J. 1984. The astrometric and kinematic parameters of the Sirius and Hyades superclusters. *Astron. J.* 89:1350–1357.
Elder, W. P. 1985. Biotic patterns across the Cenomanian-Turonian extinction boundary near Pueblo, Colorado: SEPM fieldtrip guidebook no. 4. 1985 Midyear Meeting, Golden, Colorado, *Soc. Econ. Paleontol. Mineral.*, pp. 157–169.
Elmegreen, B. G., and Lada, C. J. 1977. Sequential formation of subgroups in OB associations. *Astrophys. J.* 214:725–741.
Elmegreen, B. G., Lada, C. J., and Dickinson, D. F. 1979. The structure and extent of the giant molecular cloud near M17. *Astrophys. J.* 230:415–427.
Evans, J. C., Reeves, J. H., Rancitelli, L. A., and Bogard, D. 1982. Cosmogenic nuclides in recently fallen meteorites: Evidence for galactic cosmic ray variations during the period 1967–1978. *J. Geophys. Res.* 87:5577–5591.
Everhart, E. 1967a. Intrinsic distributions of cometary perihelia and magnitudes. *Astron. J.* 72:1002–1011.
Everhart, E. 1967b. Comet discoveries and observation selection. *Astron. J.* 72:716–726.
Everhart, E. 1968. Change in total energy of comets passing through the solar system. *Astron. J.* 73:1039–1052.
Everhart, E. 1969. Close encounters of comets and planets. *Astron. J.* 74:735–750.
Everhart, E. 1972. The origin of short-period comets. *Astrophys. Lett.* 10:131–135.
Everhart, E. 1973. Examination of several ideas of comet origins. *Astron. J.* 78:329–337.
Everhart, E. 1977. The evolution of comet orbits as perturbed by Uranus and Neptune. In *Comets, Asteroids, Meteorites: Interrelations, Evolution and Origins*, ed. A. H. Delsemme (Toledo: Univ. of Toledo), pp. 99–104.
Everhart, E. 1982. Evolution of long- and short-period orbits. In *Comets*, ed. L. L. Wilkening (Tucson: Univ. of Arizona Press), pp. 659–664.
Everhart, E., and Marsden, B. G. 1983. New original and future cometary orbits. *Astron. J.* 88:135–137.
Fahr, H. J. 1974. The extraterrestrial UV-background and the nearby interstellar medium. *Space Sci. Rev.* 15:483–540.
Faintich, M. B. 1971. Interstellar Gravitational Perturbations of Cometary Orbits. Ph.D. Thesis, Univ. of Illinois, Urbana.

Feller, W. 1971. *An Introduction to Probability Theory and its Applications*, vol. 2, 2nd ed. (New York: J. Wiley and Sons), p. 76.
Fernández, J. A. 1980. On the existence of a comet belt beyond Neptune. *Mon. Not. Roy. Astron. Soc.* 192:481–491.
Fernández, J. A. 1981. New and evolved comets in the solar system. *Astron. Astrophys.* 96:26–35.
Fernández, J. A. 1982. Dynamical aspects of the origin of comets. *Astron. J.* 87:1318–1332.
Fernández, J. A. 1984a. The formation and dynamical survival of the comet cloud. Preprint.
Fernández, J. A. 1984b. The distribution of perihelion distances of short period comets. *Astron. Astrophys.* 135:129–134.
Fernández, J. A., and Ip, W. H. 1981. Dynamical evolution of a cometary swarm in the outer planetary region. *Icarus* 47:470–479.
Fernández, J. A., and Ip, W. H. 1984. Some dynamical aspects of the accretion of Uranus and Neptune: The exchange of orbital angular momentum with planetesimals. *Icarus* 58:109–120.
Fernández, J. A., and Jockers, K. 1983. Nature and origin of comets. *Rep. Prog. Phys.* 46:665–772.
Firsov, L. F. 1965. Meteoritic origin of the Puchezh-Katunki crater. *Geotektonika* 2:106–118. In Russian.
Fischer, A. G. 1979. Rhythmic changes in the outer Earth. *Geol. Soc. of London Newsletter* 8(6):2–3.
Fischer, A. G., and Arthur, M. A. 1977. Secular variations in the pelagic realm. *Soc. Econ. Paleon. Mineral. Spec. Publ.* 25:19–50.
Fisher, R. A. 1929. Tests of significance in harmonic analysis. *Proc. Roy Soc.* 125:54–59.
Fisk, L. A. 1979. The interactions of energetic particles with the solar wind. In *Space Plasma Physics*, vol. 2 (Washington: Natl. Acad. Sci.), pp. 127–256.
Fisk, L. A., Kozlovsky, B., and Ramaty, R. 1974. An interpretation of the observed oxygen and nitrogen enhancements in low-energy cosmic rays. *Astrophys. J.* 190:L35–L37.
FitzGerald, M. P. 1970. The intrinsic colours of stars and two-colour reddening lines. *Astron. Astrophys.* 4:234–243.
Forbes, W. T. M. 1931. The great glacial cycle. *Science* 74:294–295.
Forman, M. A., and Schaeffer, O. A. 1979. Cosmic ray intensity over long time scales. *Rev. Geophys. Space Phys.* 17:552–560.
Fox, K., Williams, I. P., and Hughes, D. W. 1984. The Geminid asteroid (1983 TB) and its orbital evolution. *Mon. Not. Roy. Astron. Soc.* 208:11P–15P.
Freeman, J., Paresce, F., Bowyer, S., and Lampton, L. 1980. Observations of interstellar helium with a gas absorption cell: Limits on the bulk velocity of the interstellar medium. *Astron. Astrophys.* 83:58–64.
Frerking, M. A., Langer, W., and Wilson, R. W. 1982. The relationship between carbon monoxide abundance and visual extinction in interstellar clouds. *Astrophys. J.* 274:231–236.
Frisch, P. C. 1981. The nearby interstellar medium. *Nature* 293:377–379.
Frisch, P. C., and York, D. G. 1983. Synthesis maps of ultraviolet observations of neutral interstellar gas. *Astrophys. J.* 271:L59–L63.
Frisch, P. C., and York, D. G. 1984. Optical observations of nearby interstellar gas. *Local Interstellar Medium, IAU Coll. 81*, eds. Y. Kondo, F. C. Bruhweiler, and B. D. Savage, NASA CP-2345, pp. 113–115.
Frogel, J. A., and Stothers, R. 1977. The local complex of O and B stars. II. Kinematics. *Astron. J.* 82:890–901.
Ganapathy, R. 1982. Evidence for a major meteorite impact on the Earth 34 million years ago: Implications for Eocene extinctions. *Science* 216:885–886.
Ganapathy, R. 1983. The Tunguska explosion of 1908: Discovery of meteoritic debris at the explosion site and at the South Pole. *Science* 220:1158–1161.
Garcia-Munoz, M., Guzik, T. G., Simpson, J. A., and Wefel, J. P. 1984. The path-length distribution for galactic cosmic-ray propagation: An energy-dependent depletion of short path lengths. *Astrophys. J.* 280:L13–L17.
Garfinkel, B. 1960. On the motion of a satellite in the vicinity of the critical inclination. *Astron. J.* 65:624–627.
Gautier, T. N., Hauser, M. G., and Low, F. L. 1984. Parallel measurements of the zodiacal dust bands with the IRAS survey. *Bull. Amer. Astron. Soc.* 16:442–448.

Geballe, T., Baas, F., Greenberg, J. M., and Schutte, W. 1985. New infrared absorption features due to solid phase molecules containing sulfur in W33A. *Astron. Astrophys.* 146:L6–L8.
Gentner, W., and Wagner, G. A. 1969. Alterbestimmungen an Riesglasern und Moldaviten. *Geologica Bavarica* 61:296–303. In German.
Gentner, W., Lippolt, H. J., and Muller, O. 1964. Das Kalium-Argon-Alter das Bosumtwi-Kraters in Ghana und die chemische Beschaffenheit seiner Glaser. *Zeit. f. Natur.* 19a:150–153. In German.
Gentner, W., Kirsten, T., Storzer, D., and Wagner, G. A. 1973. K-Ar and fission track dating of Darwin crater glass. *Earth Planet. Sci. Lett.* 20:204–210.
Georgelin, Y. M., and Georgelin, Y. P. 1976. The spiral structure of our Galaxy determined from HII regions. *Astron. Astrophys.* 49:57–59.
Gehy, M. A., and Jakel, D. 1974. Late glacial and Holocene climatic history of the Sahara desert derived from a statistical assay of ^{14}C dates. *Paleogeogr., Paleoclim. and Paleoecol.* 15:205–208.
Gidon, P. 1970. Glaciations majeures et révolution galactique du système solaire. *Comptes Rendus Acad. Sci. Paris Série D* 271:385–387.
Glass, B. P. 1982. Possible correlations between tektite events and climatic changes? *Geol. Soc. Amer. Spec. Paper 190*, pp. 251–256.
Glass, B. P. 1986. No evidence for a 0.8–0.9 m.y. old micro-australite layer in deep-sea sediments. *Lunar Planet. Sci.* XVII:262 (abstract).
Glass, B. P., Swincki, M. B., and Zwart, P. A. 1979. Australasian, Ivory Coast and North American tektite strewn field: Size, mass, and correlation with geomagnetic reversals and other Earth events. *Proc. Lunar Planet. Sci. Conf.* 10:2535–2545.
Glass, B. P., DuBois, D. L., and Ganapathy, R. 1982. Relationship between an iridium anomaly and the North American microtektite layer in core RC9-58 from the Caribbean Sea. *J. Geophys. Res.* 87:425–428.
Glass, B. P., Burns, C. A., Crosbie, J. R., and DuBois, D. L. 1985. Late Eocene North American microtektites and clinopyroxene-bearing spherules. *J. Geophys. Res.*
Gliese, W. 1969. Catalogue of nearby stars. *Veroff. Astron. Rechen-Inst. Heidelberg* 22.
Gliese, W. 1981. Ein Stern auf Kollisionskurs zur Sonne? *Sterne und Weltraum* 20:445.
Gliese, W. 1982. The nearest stars. In Landolt-Bornstein, *Numerical Data and Functional Relationships in Science and Technology*, vol. 2, *Astronomy and Astrophysics*, subvol. c, *Interstellar Matter, Galaxy, Universe* (Berlin: Springer-Verlag), pp. 168–174.
Gliese, W., and Jahreiss, H. 1979. Nearby star data published 1969–1978. *Astron. Astrophys. Suppl.* 38:423–448.
Gloecker, G. 1979. Composition of energetic particle populations in interplanetary space. *Rev. Geophys. Space Sci.* 17:569–582.
Golay, M. 1973. Applications of the $UB_1B_2V_1G$ photometric system. *Spectral Classification and Multicolour Photometry*, eds. C. H. Fehrenbach and B. E. Westerlund (Dordrecht: D. Reidel), pp. 145–151.
Goldreich, P., and Ward, W. R. 1973. The formation of planetesimals. *Astrophys. J.* 183:1051–1061.
Goldsmith, D. 1985. *Nemesis: The Death-Star and Other Theories of Mass Extinction* (New York: Walker), pp. 128–129.
Golenetskii, S. P., Stepanok, V. V., and Murashov, D. A. 1981. Estimation of the pre-catastrophic composition of the Tunguska cosmic body. *Solar System Res.* 15:122–127. Trans. from *Astron. Vestn.* 15:167–173, 1981.
Gould, S. J. 1874. On the number and distribution of the bright fixed stars. *Proc. Amer. Assoc. Adv. Sci.*, pp. 115–120.
Gould, S. J. 1984. The cosmic dance of Siva. *Natural Hist.* 93(8):14–19.
Goulet, T. 1984. The Kinematics of Gas and Stars in the Solar Neighborhood. MS. Thesis, University of British Columbia, Vancouver.
Goulet, T., and Shuter, W. L. H. 1984. Kinematics of nearby gas and stars. In *Local Interstellar Medium, IAU Coll. 81*, eds. Y. Kondo, F. C. Bruhweiler, and B. D. Savage, NASA CP-2345, pp. 319–325.
Grabau, A. W. 1936. Oscillation or pulsation. *16th Internatl. Geol. Congress Rept.* 1:539–553.
Green, R. F. 1980. The luminosity function of hot white dwarfs. *Astrophys. J.* 238:685–698.
Greenberg, J. M. 1968. Interstellar grains. In *Stars and Stellar Systems*, vol. 7, *Nebulae and*

Interstellar Matter, ed. B. M. Middlehurst and L. H. Allen (Chicago: Univ. of Chicago Press), pp. 221–364.

Greenberg, J. M. 1979. Grain mantle photolysis: A connection between the grain size distribution function and the abundance of complex interstellar molecules. In *Stars and Star Systems*, ed. B. E. Westerlund (Dordrecht: D. Reidel), pp. 173–193.

Greenberg, J. M. 1982*a*. Dust in dense clouds: One stage in a cycle. In *Submillimetre Wave Astronomy*, eds. J. E. Beckmann and J. P. Phillips (Cambridge: Cambridge Univ. Press), pp. 261–306.

Greenberg, J. M. 1982*b*. What are comets made of? A model based on interstellar dust. In *Comets*, ed. L. L. Wilkening (Tucson: Univ. of Arizona Press), pp. 131–163.

Greenberg, J. M. 1983. Laboratory dust experiments: Tracing the composition of cometary dust. In *Cometary Exploration*, ed. T. I. Gombosi (Budapest: Hungarian Academy of Sci.), pp. 23–54.

Greenberg, J. M., and Chlewicki, G. 1983. A far ultraviolet extinction law: What does it mean? *Astrophys. J.* 272:563–578.

Greenberg, J. M., and d'Hendencourt, L. 1985. Evolution of ices from interstellar space to the solar system. In *Ices in the Solar System*, ed. J. Klinger, D. Benest, A. Dollfus and R. Smoluchowski (Dordrecht: D. Reidel), pp. 185–204.

Greenberg, J. M., and van de Bult, C.E.P.M. 1984. The 3 μm ice band. In *Proc. Hilo Workshop on Laboratory and Observational Infrared Spectra of Interstellar Dust*, eds. R. D. Wolstencroft and J. M. Greenberg, ROE Occasional Repts., pp. 70–82.

Greenberg, R., Weidenschilling, S. J., Chapman, C. R., and Davis, D. R. 1984. From icy planetesimals to outer planets and comets. *Icarus* 59:87–113.

Grieve, R. A. F. 1982. The record of impact on Earth: Implications for a major Cretaceous/Tertiary impact event. *Geol. Soc. Amer. Spec. Paper 190*, pp. 25–37.

Grieve, R. A. F. 1984. The impact cratering rate in recent time. *Proc. Lunar Planet. Sci. Conf. 14* in *J. Geophys. Res. Suppl.* 89:B403–B408.

Grieve, R. A. F., and Dence, M. R. 1979. The terrestrial cratering record. *Icarus* 38:230–242.

Grieve, R. A. F., Sharpton, V. L., Goodacre, A. K., and Garvin, J. B. 1985. A perspective on the evidence for periodic cometary impacts on Earth. *Earth Planet. Sci. Lett.* 76:1–9.

Grim, R., van Ijzedoorn, L., and Greenberg, J. M. 1986. Ultraviolet photolysis of H_2S in interstellar grain mantles as an explanation for S_2 in comets and implications for the thermal history of comets. To be submitted to *Astrophys. J. Lett.*

Grün, E. 1981. Physikalische und Chemische Eigenschaften des Interplanetaren Staubes-Messingen en Mikrometeoritenexperimentes auf Helios. Habilitationschrift Universitat, Heidelberg.

Grün, E., Pailer, N., Fechtig, H., and Kissel, J. 1980. Orbital and physical characteristics of micrometeorites in the inner solar system as observed by Helios I. *Planet. Space Sci.* 28:333–341.

Grün, E., Zook, H. A., Fechtig, H., and Giese, R. H. 1985. Collisional balance of the meteoritic complex. *Icarus* 62:244–272.

Gurov, Ye. P., Val'ter, A. A., Gurova, Ye. P., and Serebrennikov, A. I. 1978. The Elgygytgyn meteorite explosion crater in Chukotka. *Doklady Acad. Nauk SSSR* 240:1407–1410. In Russian.

Halbwachs, J. L. 1985. Binaries spectroscopiques dans le voisinage solaire. *CDS Inf. Bull.* 28:25–27.

Hallam, A. 1976. Stratigraphic distribution and ecology of European Jurassic bivalves. *Lethaia* 9:245–259.

Hallam, A. 1977. Jurassic bivalve biogeography. *Paleobiology* 3:58–73.

Hallam, A. 1984*a*. Pre-Quaternary sea-level changes. *Ann. Rev. Earth Planet. Sci.* 12:205–243.

Hallam, A. 1984*b*. The causes of mass extinctions. *Nature* 308:686–687.

Hamid, S. E., Marsden, B. G., and Whipple, F. L. 1968. Influence of a comet belt beyond Neptune on the motions of periodic comets. *Astron. J.* 73:727–729.

Harland, W. B., Cox, A. V., Llewellyn, P. G., Pickton, C. A. B., Smith, A. B., and Walters, R. 1982. *A Geologic Time Scale* (Cambridge: Cambridge Univ. Press).

Harms, J. E., Miton, D. J., Ferguson, J., Gilbert, D. J., Harris, W. K., and Goleby, B. 1980. Goat Paddock cryptoexplosion crater, Western Australia. *Nature* 286:704–706.

Harper, D. A., Lowenstein, R. F., and Davidson, J. A. 1984. On the nature of the material surrounding Vega. *Astrophys. J.* 285:808–812.

Harrington, R. S. 1985a. Implications of the observed distributions of very long period comet orbits. *Icarus* 61:60–62.

Harrington, R. S. 1985b. Quoted in Did comets kill the dinosaurs? *Time*, 6 May 1985, p. 81.

Harrington, R. A., and Van Flandern, T. C. 1979. The satellites of Neptune and the origin of Pluto. *Icarus* 39:131–136.

Hartmann, W. K. 1975. Cratering in the solar system. *Sci. Amer.* 236:84.

Hartung, J. B., Izett, G. A., Naeser, G. W., Kunk, M. V., and Sutter, J. F. 1986. The Manson, Iowa, impact structure and the Cretaceous-Tertiary boundary event. *Lunar Planet. Sci.* XVII:313–314 (abstract).

Hatfield, C. B., and Camp, M. J. 1970. Mass extinctions correlated with periodic galactic events. *Geol. Soc. Amer. Bull.* 81:911–914.

Hauck, B., and Mermilliod, M. 1983. Photometry of nearby stars. In *The Nearby Stars and the Stellar Luminosity Function, IAU Symp. 76*, eds. A. G. D. Philip and A. R. Upgren (Schenectady: L. Davis Press), pp. 341–343.

Hays, J. D. 1971. Faunal extinctions and reversals of the Earth's magnetic field. *Geol. Soc. Amer. Bull.* 82:2433–2447.

Hays, J. D., and Opdyke, N. D. 1967. Antarctic radiolaria, magnetic reversals, and climatic change. *Science* 158:1001–1011.

Hays, J. D., Imbrie, J., and Shackleton, N.J. 1976. Variations in the Earth's orbit: Pacemaker of the ice ages. *Science* 194:1121–1132.

Heckmann, O., and Strassl, H. 1934. Zur Dynamik des Sternsystems. *Veröff. U-Sternw. Göttingen*, No. 41.

Heggie, D. C. 1975. Binary evolution in stellar dynamics. *Mon. Not. Roy. Astron. Soc.* 173:729–787.

Heiles, C. 1979. HI shells and supershells. *Astrophys. J.* 229:533–544.

Heisler, J., and Tremaine, S. 1985. The influence of the galactic tidal field on the Oort comet cloud. *Icarus* 65:13–26.

Henry, R. C. 1977. Far ultraviolet studies. I. Predicted far-ultraviolet interstellar radiation field. *Astrophys. J. Suppl.* 33:451–458.

Herczeg, T. 1984. Duplicity on the main sequence. *Astron. Space Sci.* 99:29–39.

Heylmun, E. B. 1969. Geologic significance of the passage of the Earth through the galactic plane. *Geol. Soc. of Amer. Abstracts with Programs* 1:36.

Hill, E. R. 1960. The component of the galactic gravitational field perpendicular to the galactic plane, K_z. *Bull. Astron. Inst. Netherlands* 15:1–44.

Hill, G., Hilditch, R. W., and Barnes, J. V. 1979. Studies of A and F stars in the region of the north galactic pole. *Mon. Not. Roy. Astron. Soc.* 186:813–829.

Hills, J. G. 1975. Encounters between binary and single stars and their effect on the dynamical evolution of stellar star systems. *Astron. J.* 80:809–825.

Hills, J. G. 1981. Comet showers and the steady-state infall of comets from the Oort cloud. *Astron. J.* 86:1730–1740.

Hills, J. G. 1982. The formation of comets by radiation pressure in the outer protosun. *Astron. J.* 87:906–910.

Hills, J. G. 1983. The effect of low-velocity, low-mass intruders (collisionless gas) on the dynamical evolution of a binary system. *Astron. J.* 88:1269–1283.

Hills, J. G. 1984a. Dynamical constraints on the mass and perihelion distance of Nemesis and the stability of its orbit. *Nature* 311:636–638.

Hills, J. G. 1984b. Close encounters between a star-planet system and a stellar intruder. *Astron. J.* 89:1559–1564.

Hills, J. G. 1985. The passage of a "Nemesis-like" object through the planetary system. *Astron. J.* 90:1876–1882.

Hilton, J. L., and Bash, F. 1982. The ballistic particle model and the vertex deviation of young stars near the Sun. *Astrophys. J.* 255:217–226.

Hobbs, L. M. 1969. The profiles of the interstellar sodium D-lines. *Astrophys. J.* 157:165–173.

Hobbs, L. M. 1974. A comparison of interstellar NaI, CaII, and KI absorption. *Astrophys. J.* 191:381–393.

Hobbs, L. M. 1978. Optical interstellar lines toward 18 stars of low reddening. III. *Astrophys. J.* 222:491–507.
Hoffman, A. 1985. Patterns of family extinction depend on definition and geological timescale. *Nature* 315:659–662.
Hoffman, A., and Ghiold, J. 1985. Randomness in the pattern of "mass extinctions" and "waves of origination." *Geol. Mag.* 122:1–4.
Holmes, A. 1927. *The Age of the Earth: An Introduction to Geological Ideas* (London: Benn).
Holzer, T. E. 1977. Neutral hydrogen in interplanetary space. *Rev. Geophys. Space Phys.* 15:467–490.
Honig, R. E., and Hook, H. O. 1960. Vapor pressure data for some common gases. *R.C.A. Rev.* 21(3):360–368.
Hopkins, A. G., and Brown, C. W. 1975. Additional evidence for the Raman band of S_2. *J. Chem. Phys.* 62:1598.
Hörz, F. 1982. Ejecta of the Ries Crater, Germany. *Geol. Soc. Amer. Spec. Paper 190*, pp. 39–55.
House, F. C., and Innanen, K. A. 1975. A numerical study of local stellar motions. *Astrophys. Space Sci.* 32:139–151.
Hoyle, F. 1984. On the causes of ice ages. *Earth, Moon, and Planets* 31:229–248.
Hoyle, F., and Lyttleton, R. A. 1939. The effect of interstellar matter on climatic variation. *Proc. Cambridge Phil. Soc. Math. Phys. Sci.* 35:405–415.
Hoyt, W. G. 1980. *Planets X and Pluto* (Tucson: Univ. of Arizona Press).
Hughes, D. W. 1983. Cometary dust: Its source and characteristics. In *Asteroids, Comets, Meteors*, eds. C.-I. Lagerkvist and H. Rickman (Uppsala: Uppsala Univ. Press), pp. 379–381.
Hughes, D. W., and Daniels, P. A. 1982. Temporal variations in the cometary mass distribution. *Mon. Not. Roy. Astron. Soc.* 198:573–582.
Hut, P. 1983*a*. Binary-single star scattering. II. Analytical approximations for high velocity. *Astrophys. J.* 268:342–355.
Hut, P. 1983*b*. The topology of three-body scattering. *Astron. J.* 88:1549–1559.
Hut, P. 1983*c*. Binaries as a heat source in stellar dynamics: Release of binding energy. *Astrophys. J.* 272:L29–L33.
Hut, P. 1984*a*. How stable is an astronomical clock that can trigger mass extinctions on Earth? *Nature* 311:638–641.
Hut, P. 1984*b*. Hard binary-single star scattering cross sections for equal mass. *Astrophys. J. Suppl.* 55:301–317.
Hut, P. 1985. Binary formation and interaction with field stars. In *Dynamics of Star Clusters, IAU Symp. 113*, eds. J. J. Goodman and P. Hut (Dordrecht: D. Reidel), 103–106.
Hut, P. and Bahcall, J. N. 1983. Binary-single star scattering. I. Numerical experiments for equal masses. *Astrophys. J.* 268:319–341.
Hut, P., and Tremaine, S. 1985. Have interstellar clouds disrupted the Oort comet cloud? *Astron. J.* 90:1548–1557.
Hut, P., and Weissman, P. 1985. Dynamical evolution of cometary showers. *Bull. Amer. Astron. Soc.* 17:690 (abstract).
Hut, P., Alvarez, W., Elder, W. P., Hanson, T., Kauffman, E. G., Keller, G., Shoemaker, E. M., and Weissman, P. R. 1986. Comet showers as a possible cause of stepwise mass extinctions. Submitted to *Science*.
Innanen, K. A. 1979. The limiting radii of direct and retrograde satellite orbits, with applications to the solar system and stellar systems. *Astron. J.* 84:960–963.
Innanen, K. A. 1980. The coriolis asymmetry in the classical restricted 3-body problem and the Jacobian integral. *Astron. J.* 85:81–85.
Innanen, K. A., Patrick, A. T., and Duley, W. W. 1978. The interaction of the spiral density wave and the Sun's galactic orbit. *Astrophys. Space Sci.* 57:511–515.
Innes, D. E., and Hartquist, T. W. 1984. Are we in an old superbubble? *Mon. Not. Roy. Astron. Soc.* 209:7–13.
Irving, E., and Pullaiah, G. 1976. Reversals of the geomagnetic field, magnetostratigraphy, and relative magnitude of paleosecular variation in the Phanerozoic. *Earth-Science Rev.* 12:35–64.
Israel, F. P. 1985. Problems of galactic and extragalactic CO photometry. In *New Aspects of Galaxy Photometry*, ed. J. L. Nieto (Berlin: Springer-Verlag), pp. 101–110.

Iwan, D. 1980. X-ray observations of the North Polar Spur. *Astrophys. J.* 239:316–327.
Jablonski, D. 1984. Keeping time with mass extinctions. *Paleobiology* 10:139–145.
Jahn, B.-M., Floran, R. J., and Simonds, C. H. 1978. Rb-Sr isochron age of the Manicougan melt sheet, Quebec, Canada. *J. Geophys. Res.* 83:2799–2803.
Jahreiss, H. 1974. Die raumliche Verteilung, Kinematik und Alter der sonnennahen Sterne. Ph.D. Thesis, Heidelberg Univ.
Jahreiss, H., and Gliese, W. 1985. Radial velocities of nearby stars. *CDS Inf. Bull.* 28:19–23.
Jahreiss, H., and Weilen, R. 1975. The local mass of the galactic halo. Unpublished.
Jahreiss, H., and Wielen, R. 1983. Kinematics and ages of nearby stars. In *The Nearby Stars and the Stellar Luminosity Function, IAU Symp. 76*, eds. A. G. Davis Philip and A. R. Upgren (Schenectady: L. Davis Press), pp. 277–286.
Jeans, J. H. 1919. *Problems of Cosmology and Stellar Dynamics* (Cambridge: Cambridge Univ. Press).
Jenkins, E. B. 1984. Observations of absorption lines from highly ionized atoms. In *Local Interstellar Medium, IAU Coll. 81*, eds. Y. Kondo, F. C. Bruhweiler, and B. D. Savage, NASA CP-2345, pp. 155–168.
Jessberger, E. K., and Reimold, W. V. 1980. A late Cretaceous ^{40}Ar-^{39}Ar age for the Lappajarvi impact crater, Finland. *J. Geophys.* 48:57–59.
Jokipii, J. R. 1971. Propagation of cosmic rays in the solar wind. *Rev. Geophys. Space Phys.* 9:27–87.
Joly, J. 1924. *Radioactivity and the Surface History of the Earth* (Oxford: Clarendon Press).
Joss, P. C. 1973. On the origin of short-period comets. *Astron. Astrophys.* 25:271–273.
Joy, A. H., and Abt, H. A. 1974. Spectral types of M dwarf stars. *Astrophys. J. Suppl.* 28:1–18.
Kafatos, M., Sofia, S., Bruhweiler, F., and Gull, T. 1980. The evolution of supernova remnants in different galactic environments, and its affects on supernova statistics. *Astrophys. J.* 242:294–305.
Kahane, C., Guillotean, S., and Lucas, R. 1985. A multiline study of a typical giant molecular cloud: S147/S153. *Astron. Astrophys.* 146:325–336.
Kapteyn, J. C. 1905. *Brit. Assoc. Adv. Sci.*, Report 257.
Kapteyn, J. C. 1922. First attempt at a theory of the arrangement and motion of the sidereal system. *Astrophys. J.* 55:302–328.
Kauffman, E. G. 1984. The fabric of Cretaceous marine extinctions. In *Catastrophes and Earth History*, eds. W. A. Berggren and J. A. Van Couvering (Princeton: Princeton Univ. Press), pp. 151–246.
Keller, G. 1983. Biochronology and paleoclimatic implications of Middle Eocene to Oligocene planktonic foraminiferal faunas. *Marine Micropaleontology* 7:463–486.
Keller, G. 1986. Stepwise mass extinctions and impact events: Late Eocene to early Oligocene. In *Geological Events at the Eocene-Oligocene Boundary*, ed. L. Pomeizol (Amsterdam: Elsevier Scientific Publ.). In press.
Keller, G., D'Hondt, S., and Vallier, T. L. 1983. Multiple microtektite horizons in upper Eocene marine sediments: No evidence for mass extinctions. *Science* 221:150–152.
Keller, G., D'Hondt, S., Onstott, T., Orth, C. J., Gilmore, G. S., Shoemaker, E., and Keigwin, L. D. 1985. Multiple late Eocene impact events: Stratigraphic, isotopic and geochemical data. *Geol. Soc. Amer. Abst. Prog. 98th Ann. Meeting*, p. 626.
Keller, G., D'Hondt, S., Gilmore, J. S., Keigwin, L. D. Jr., Molina, E., Onstott, T., Orth, C. J., and Shoemaker, E. M. 1986. Late Eocene impact events: Stratigraphy, age, and geochemistry. Submitted to *Earth Planet. Sci. Lett.*
Keller, H. U., Richter, K., and Thomas, G. E. 1981. Multiple scattering of solar resonance radiation in the nearby interstellar medium II. *Astron. Astrophys.* 102:415–423.
Kendall, D. G. 1961. Some problems in the theory of comets, I and II. In *Proc. Fourth Berkeley Symp. on Mathematical Statistics and Probability* 3:99–148.
Kennett, J. P., and Thunnell, R. C. 1975. Global increase in explosive vulcanism. *Science* 187:497–503.
Kent, D. V., and Gradstein, F. M. 1985. A Jurassic to Recent chronology. In *The Western Atlantic Region*, vol. M, *The Geology of North America*, eds. B. E. Tucholke and P. R. Vogt (Boulder: Geol. Soc. Amer.). In press.
King, I. R. 1962. The structure of star clusters. I. An empirical density law. *Astron. J.* 67:471–485.

Kitchell, J. A., and Pena, D. 1984. Periodicity of extinctions in the geological past: Deterministic versus stochastic explanations. *Science* 226:689–692.
Kondo, Y., Bruhweiler, F. C., and Savage, B. D., eds. 1984. *Local Interstellar Medium, IAU Coll. 81,* NASA CP-2345.
Kota, J., and Jokipii, J. R. 1983. Effects of drift on the transport of cosmic rays. VI. A three-dimensional model including diffusion. *Astrophys. J.* 265:573–581.
Kowal, C. T. 1979. Chiron. In *Asteroids,* ed. T. Gehrels (Tucson: Univ. of Arizona Press), pp. 436–439.
Kowal, C. T., and Drake, S. 1980. Galileo's observations of Neptune. *Nature* 297:311–313.
Kozai, Y. 1959. The motion of a close Earth satellite. *Astron. J.* 64:367–377.
Kresák, L. 1977. On the differences between new and old comets. *Bull. Astron. Inst. Czech.* 28:346–355.
Kresák, L. 1978. The Tunguska object: A fragment of comet Encke? *Bull. Astron. Inst. Czech.* 29:129–134.
Kresák, L. 1981. The lifetimes and disappearance of periodic comets. *Bull. Astron. Inst. Czech.* 32:321–339.
Krisciunas, K. 1977. Toward the resolution of the local missing mass problem. *Astrophys. J.* 82:195–197.
Krogh, F. T., Ng, E. W., and Snyder, W. V. 1982. The gravitational field of a disk. *Celestial Mech.* 26:395–405.
Kropotkin, P. N. 1970. The possible role of cosmic factors in geotectonics. *Geotectonics* 2:80–88.
Kuhn, T. S. 1970. *The Structure of Scientific Revolutions* (Chicago: Univ. of Chicago Press).
Kuiper, G. P. 1951. On the origin of the solar system. In *Astrophysics,* ed. J. A. Hynek (New York: McGraw-Hill), pp. 357–424.
Kurfess, J. D., Johnson, W. N., Kinzer, R. L., Share, G. H., Strickman, M. S., Ulmer, M. P., Clayton, D. D., and Dyner, C. S. 1983. The oriented scintillation spectrometer experiment for the gamma ray observatory. *Adv. Space Res.* 3:109–112.
Kyte, F. T. 1984. Iridium sedimentation in the Denozoic: No evidence for a death star. Meteoritical Soc. Meeting (abstract), Albuquerque, NM, R-2.
Kyte, F. T., and Brownlee, D. E. 1985. Unmelted meteoritic debris in the Late Pliocene iridium anomaly: Evidence for the ocean impact of a nonchrondritic asteroid. *Geochim. Cosmochim. Acta* 49:1095–1108.
Kyte, F. T., and Wasson, J. T. 1986. Accretion rate of extraterrestrial matter: Iridium in marine sediments deposited 33-67 Ma ago. *Science* 231:1225–1229.
Kyte, F. T., Zhou, Z., and Wasson, J. T. 1981. High noble metal concentrations in a late Pliocene sediment. *Nature* 292:417–420.
Lâcarrieu, C. T. 1971. Moment equations in the study of the total mass density in the neighbourhood of the Sun and of the galactic force law K_2. *Astron. Astrophys.* 14:95–102.
Lacey, C. G. 1984. The influence of massive gas clouds on stellar velocity dispersions in galactic disks. *Mon. Not. Roy. Astron. Soc.* 208:687–707.
Lacy, J. H., Baas, F., Allamandola, L. J., Persson, S. E., McGregor, P. J., Lonsdale, C. J., Geballe, T. R., and van de Bult, C. E. P. M. 1984. 4.6 micron absorption features due to solid phase CO and cyano group molecules toward compact infrared sources. *Astrophys. J.* 276:533–543.
Lagage, P. O., and Cesarsky, C. J. 1983. The maximum-energy of cosmic rays accelerated by supernova shocks. *Astron. Astrophys.* 125:249–257.
Lallement, R., Bertaux, J. L., Kurt, V. G., and Mironova, E. N. 1984. Observed perturbations of the velocity distribution of interstellar hydrogen atoms in the solar system with Prognoz Lyman-Alpha measurements. *Astron. Astrophys.* 140:243–250.
Lambert, P. 1974. La structure d'impact de météorite géante de Rochechouart. Thèse Doc. Spec., Univ. Paris-Sud, Paris.
Landsman, W. B., Henry, R. C., Moos, H. W., and Linsky, J. L. 1984. Observations of interstellar hydrogen and deuterium toward Alpha Centauri A. *Astrophys. J.* 285:801–807.
Landsman, W. B., Murthy, J., Henry, R. C., Moos, H. W., Linsky, J. L., and Russell, J. L. 1986. IUE observations of interstellar hydrogen and deuterium toward Alpha Cen B. *Astrophys. J.* 303:791–796.
Laplace, P. S. 1816. Sur les cometes. *Addition à la Connaissance des Temps,* pp. 213–220.

Latham, D. W., Tonry, J., Bahcall, J. N., Soneira, R. M., and Schechter, P. 1984. Detection of binaries with projected separations as large as 0.1 parsec. *Astrophys. J.* 281:L41–L45.

Lavielle, B., Marti, K., and Regnier, S. 1985. Exposure ages of iron meteorites: Complex histories and the constancy of galactic cosmic rays. In *Proc. Conf. on Isotopic Anomalies in the Solar System* (Paris), Appendix 6-B.

La Violette, P. 1983. Galactic Explosions, Cosmic Dust Invasions and Climate Change. Ph.D. Thesis, Portland State University, Oregon.

Lebrun, F., Bennett, K., Bignami, G. F., Bloemen, J. B. G. M., Buccheri, R., Caraveo, P. A., Gottwald, M., Hermsen, W., Kanbach, G., Mayer-Hasselwander, H. A., Montemerle, T., Paul, J. A., Sacco, B., Strong, A. W., and Wills, R. D. 1983. Gamma-rays from atomic and molecular gas in the first galactic quadrant. *Astrophys. J.* 274:231–236.

Leggett, J. K., McKerrow, W. S., Cocks, L. R. M., and Rickard, R. B. 1981. Periodicity in the Lower Paleozoic marine realm. *J. Geol. Soc.* 138:167–176.

Leising, M. D., and Clayton, D. D. 1985. Angular distribution of interstellar ^{26}Al. *Astrophys. J.* 294:591–598.

Lesh, J. R. 1968. The kinematics of the Gould Belt: An expanding group? *Astrophys. J. Suppl.* 17:371–444.

Liebert, J., Dahn, C. C., Gresham, M., and Strittmatter, P. A. 1979. New results from a survey of faint proper-motion stars: A probable deficiency of very low luminosity degenerates. *Astrophys. J.* 233:226–238.

Lin, C. C., Yuan, C., and Shu, F. H. 1969. On the spiral structure of disk galaxies. III. Comparison with observations. *Astrophys. J.* 155:721–746.

Lin, S. C. 1966. Cometary impact and the origin of tektites. *J. Geophys. Res.* 71:2427–2437.

Lindblad, B. 1925. On the cause of star-streaming. *Astrophys. J.* 62:191–197.

Lindblad, B. 1927*a*. Cosmogonic consequences of a theory of the stellar system. *Arkiv für Mathematik, Astronomi och Fysik* 19A(35): 1–15.

Lindblad, B. 1927*b*. On the cause of the ellipsoidal distribution of stellar velocities. *Upsala Medd.* 26.

Lindblad, P. O. 1980. On the relation between local kinematics and galactic structure. *Mitt. Astro. Gesellshaft* 48:151–159.

Lindblad, P. O., Westin, T., Zentelis, N., and Loden, K. 1984. Local galactic kinematics. In *Symposium on Star Clusters and Associations and Their Relations to the Evolution of the Galaxy* (Prague: Czechoslovak Academy of Science). In press.

Lippincott, S. L. 1978. Astrometric search for unseen stellar and substellar companions to nearby stars and the possibility of their detection. *Space Sci. Rev.* 22:153–189.

Low, F. J., Beintema, D. A., Gautier, T. N., Gillett, F. C., Beichman, C. A., Neugebauer, G., Young, E., Aumann, H. H., Boggess, N., Emerson, J. P., Habing, H. J., Hauser, M. G., Houck, J. R., Rowan-Robinson, M., Soifer, B. T., Walker, R. G., and Wesselius, P. R. 1984. Infrared cirrus: New components of the extended IR emission. *Astrophys. J.* 278:L19–L22.

Lowell, P. 1915. Memoir on a trans-Neptunian planet. *Memoirs Lowell Obs.* 1(1).

Luck, J. M., and Turekian, K. K. 1983. Osmium-187/Osmium-186 in manganese modules and the Cretaceous-Tertiary boundary. *Science* 222:613–615.

Lucke, P. B. 1978. The distribution of color excesses and interstellar reddening material in the solar neighborhood. *Astron. Astrophys.* 64:367–377.

Lungershausen, G. F. 1957. Periodic changes in climate and the major glaciations of the globe (some problems of historical palaeogeography and absolute chronology). *Sovetskaya Geologiya* 59:88–115.

Lüst, R. 1984. The distribution of the aphelion directions of long-period comets. *Astron. Astrophys.* 141:94–100.

Lutz, T. E., and Kelker, D. H. 1973. On the use of trigonometric parallaxes for the calibration of luminosity systems: Theory. *Publ. Astron. Soc. Pacific* 85:573–578.

Lutz, T. M. 1985. The magnetic reversal record is not periodic. *Nature* 317:404–407.

Luyten, W. J. 1968. A new determination of the luminosity function. *Mon. Not. Roy. Astron. Soc.* 139:221–224.

Luyten, W. J. 1979*a*. *LHS Catalogue*, 2nd ed. (Minneapolis: Univ. of Minnesota).

Luyten, W. J. 1979*b*. *NLTT Catalogue*, vols. 1 and 2 (Minneapolis: Univ. of Minnesota).

Luyten, W. J. 1980. *NLTT Catalogue*, vols. 3 and 4 (Minneapolis: Univ. of Minnesota).

Lynds, B. J. 1962. Catalogue of dark nebulae. *Astrophys. J. Suppl.* 7:1–52.
Lyttleton, R. A. 1948. On the origin of comets. *Mon. Not. Roy. Astron. Soc.* 108:465–475.
Lyttleton, R. A. 1974. On the non-existence of the Oort cloud. *Astrophys. Space Sci.* 31:395–401.
Mackinnon, I. D. R., and Rietmeijer, F. J. M. 1984. Bismuth in interplanetary dust. *Nature* 311:135–138.
Maddalena, R. J., Morris, M., Moscowitz, J., and Thaddeus, P. 1986. The large system of molecular clouds in Orion and Monoceros. *Astrophys. J.* 303:375–391.
Magnani, L., Blitz, L., and Mundy, L. 1985. Molecular gas at high galactic latitudes. *Astrophys. J.* 295:402–421.
Mahoney, W. A., Ling, J. C., Jacobson, A. S., and Lingenfelter, R. E. 1982. Diffuse galactic gamma-ray line emission from nucleosynthetic ^{60}Fe, ^{26}Al, and ^{22}Ne: Preliminary limits from HEAO 3. *Astrophys. J.* 262:742–748.
Mahoney, W. A., Ling, J. C., Jacobson, A. S., and Lingenfelter, R. E. 1984. HEAO 3 discovery of ^{26}Al in the interstellar medium. *Astrophys. J.* 286:578–585.
Mak, E. K., York, D., Grieve, R. A. F., and Dence, M. R. 1976. The age of the Mistastin Lake crater, Labrador, Canada. *Earth Planet. Sci. Lett.* 31:345–357.
Malmquist, K. G. 1936. The effect of an absorption of light in space upon some relations in stellar statistics. *Stockholm Obs. Medd.* 26.
Marsden, B. G. 1983. *Catalog of Cometary Orbits* (Hillside: Enslow Publ.)
Marsden, B. G., and Roemer, E. 1982. Basic information and references. In *Comets*, ed. L. L. Wilkening (Tucson: Univ. of Arizona Press), pp. 707–733.
Marsden, B. G., Sekanina, Z., and Everhart, E. 1978. New osculating orbits for 110 comets and the analysis of the original orbits of 200 comets. *Astron. J.* 83:64–71.
Marti, K. 1985. Live ^{129}I- ^{129}Xe dating. In *Proc. Workshop on Cosmogenic Nuclides* (Houston: Lunar Planet. Sci. Inst.). In press.
Marti, K., Lavielle, B., and Regnier, S. 1984. Cosmic ray exposure ages of iron meteorites, complex irradiation and the constancy of cosmic ray flux in the past. *Lunar Planet. Sci.* XV:511–512 (abstract).
Masaitis, V. L., Mikhailov, M. V., and Solivanovskaia, T. V. 1975. *The Popigai Meteorite Crater* (Moscow: Nauka Press). In Russian. Trans. NASA Tech. Trans. 22F16900, 1976.
Masaitis, V. L., Danilin, A. N., Mashchak, M. S., Raikhlin, A. I., Solivanovskaia, T. V., and Shadenkov, E. M. 1980. *The Geology of Astroproblemes* (Leningrad: Nedra Press). In Russian. Trans. for the U.S. Geological Survey by D. B. Vitaliano, 1984.
Mason, B. 1962. *Meteorites* (New York: J. Wiley and Sons).
Matese, J. J. 1985. Perturbation of comet orbits by the galactic disk. In preparation.
Matese, J. J., and Whitmire, D. P. 1986. Planet X and the origins of the shower and steady state flux of short period comets. *Icarus* 65:37–50.
Mathewson, D. S., and Ford, V. L. 1970. Polarization observations of 1800 stars. *Mem. Roy. Astron. Soc.* 74:139–182.
Mauvais, F. V. 1847. On two observations of Neptune May 8 and 10, 1795, in the Histoire Celeste. *Astron. Nach.* 607.
Mayor, M. 1970. Possible influence of the spiral galactic structure on the local distributions of residual stellar velocities. *Astron. Astrophys.* 6:60–66.
Mayor, M. 1972. On the vertex deviation. *Astron. Astrophys.* 18:97–105.
Mayor, M. 1974. Kinematics and age of stars. *Astron. Astrophys.* 32:321–327.
Mayor, M. 1985. Programme en cours avec CORAVEL. Personal communication.
McCammon, D. 1984. The soft x-ray diffuse background: Implications for the nature of the local interstellar medium. In *Local Interstellar Medium, IAU Coll. 81*, eds. Y. Kondo, F. C. Bruhweiler, and B. D. Savage, NASA CP-2345, pp. 195–203.
McCammon, D., Burrows, D. N., Sanders, W. T., and Kraushar, W. L. 1983. The soft x-ray diffuse background. *Astrophys. J.* 269:107–135.
McCarthy, D. W., Probst, R. G., and Low, F. J. 1985. Infrared detection of a close cool companion to Van Biesbroeck 8. *Astrophys. J.* 290:L9–L13.
McCray, R., and Kafatos, M. 1986. Supershells and propagating star formation. Submitted to *Astrophys. J.*
McCrea, W. H. 1975. Ice ages and the Galaxy. *Nature* 255:607–609.
McCrea, W. H. 1981. Long time-scale fluctuations in the evolution of the Earth. *Proc. Roy. Soc. London* A375:1–41.

McDonald, F. B., Lal, N., Trainor, J. H., Van Hollebeke, M. A. I., and Webber, W. R. 1977. Observations of galactic cosmic ray energy spectra between 1 and 9 AU. *Astrophys. J.* 216:930–939.
McDougall, I., and Lovering, J. F. 1969. Apparent K-Ar dates on cores and excess Ar in flanges of australites. *Geochim. Cosmochim. Acta* 33:1057–1070.
McElhinny, M. W. 1971. Geomagnetic reversals during the Phanerozoic. *Science* 172:157–159.
McKay, C. P. 1985. Notilucent cloud formation and the effects of water vapor variability on temperatures in the middle atmosphere. *Planet. Space Sci.* 33:761–771.
McKay, C. P., and Thomas, G. E. 1978. Consequences of a past encounter of the Earth with an interstellar cloud. *Geophys. Res. Lett.* 5:215–218.
McKee, C., and Ostriker, J. P. 1977. A theory of the interstellar medium: Three components regulated by supernova explosions in an inhomogeneous substrate. *Astrophys. J.* 218:148–169.
Meier, R. R. 1977. Some optical and kinetic properties of the nearby interstellar gas. *Astron. Astrophys.* 55:211–219.
Meier, R. R. 1981. A parametric study of interstellar helium atoms incident upon the Earth. *NRL Memorandum 4423*.
Mewaldt, R. A. 1983. The elemental and isotopic composition of galactic cosmic ray nuclei. *Rev. Geophys. Space Sci.* 21:295–305.
Mewaldt, R. A., Spalding, J. D., and Stone, E. C. 1984. The isotopic composition of the anomalous low-energy cosmic rays. *Astrophys. J.* 283:450–456.
Meyer, P. 1969. Cosmic rays in the galaxy. *Ann. Rev. Astron. Astrophys.* 7:1–38.
Meyerhoff, A. A. 1973. Mass biotal extinctions, world climate changes, and galactic motions: Possible interrelations. *Canadian Soc. Petrol. Geol. Memoir* 2:745–758.
Mezzetti, M., Giuricin, G., and Mardirossian, F. 1983. Binarity and the local stellar mass density. *Astron. Astrophys.* 122:333–334.
Michel, F. C. 1981. The power law spectrum of shock-accelerated relativistic particles. *Astrophys. J.* 247:664–670.
Mihalas, D., and Binney, J. 1981. *Galactic Astronomy*, 2nd ed. (San Francisco: W. H. Freeman).
Mihalas, D., and Routley, P. 1968. *Galactic Astronomy* (San Francisco: W. H. Freeman).
Miller, G. E., and Scalo, J. M. 1979. The initial mass function and stellar birthrate in the solar neighborhood. *Astrophys. J. Suppl.* 41:513–547.
Milne, E. A. 1935. Stellar kinematics and the K-effect. *Mon. Not. Roy. Astron. Soc.* 95:560–573.
Milton, D. J., Barlow, B. C., Brett, R., Brown, A. R., Glickson, A. Y., Manwaring, E. A., Moss, F. J., Sedmik, E. C. E., Van Son, J., and Young, G. A. 1972. Gosses Bluff impact structure, Australia. *Science* 175:1199–1206.
Mohr, J. M. 1931. Sur le courant d'étoiles Ursa Major. *Publ. Inst. Astron. Univ. Charles Prague* 11.
Moniot, R. K., Kruse, T. H., Tuniz, C., Savin, W., Hall, G. A., Milazzo, T., Pal, D., and Herzog, G. F. 1983. The ^{21}Ne production rate in stony meteorites estimated from the ^{10}Be and other radionuclides. *Geochim. Cosmochim. Acta* 47:1887–1895.
Morfill, G. E., and Hartquist, T. W. 1985. Constraints on local supernovae and ^{26}Al in the interstellar medium from determinations of cosmic rays. *Astron. J.* 297:194–198.
Morfill, G. E., and Völk, H. J. 1984. Transport of dust and vapor and chemical fractionation in the early protosolar cloud. *Astrophys. J.* 287:371–395.
Morris, D. E., and Muller, R. A. 1986. Tidal gravitational forces: The infall of "new" comets and comet showers. *Icarus* 65:1–12.
Morton, D. C., and Purcell, J. D. 1962. Observations of the extreme ultraviolet radiation in the night sky using an atomic hydrogen filter. *Planet. Space Sci.* 9:455–458.
Muller, O., Hampel, W., Kirsten, T., and Gerzog, G. F. 1981. Cosmic-ray constancy and cosmogenic production rates in short-lived chondrites. *Geochim. Cosmochim. Acta* 45:447–460.
Montanari, A., Drake, R., Bice, D. M., Alvarez, W., Curtis, G. H., Turrin, B. D., and DePaolo, D. J. 1985. Radiometric time scale for the upper Eocene and Oligocene based on K/Ar and Rb/Sr dating of volcanic biotites from the pelagic sequence of Gubbio, Italy. *Geology* 13:596–599.
Muller, R. A. 1985. Evidence for a solar companion star. In *The Search for Extraterrestrial Life: Recent Developments, IAU Symp. 112*, ed. M. D. Papagiannis (Dordrecht: D. Reidel), pp. 233–244.

Muller, R. A., and Morris, D. E. 1985. Geomagnetic reversals driven by abrupt sea level changes. Submitted to *Nature*.
Mundy, L. G. 1984. The Density and Molecular Column Density Structure of Three Molecular Cloud Cores. Ph.D. Thesis, University of Texas, Austin.
Murray, C. A., and Argyle, R. W. 1984. Kinematics of middle main-sequence stars in the South Galactic cap. Herstmonceaux Workshop.
Napier, W. M. 1985. Dynamical interactions of the solar system with massive nebula. In *Dynamics of Comets: Their Origin and Evolution, IAU Coll. 83*, eds. A. Carusi and G. B. Valsecchi (Dordrecht: D. Reidel), pp. 31–41.
Napier, W. M., and Clube, S. V. M. 1979. A theory of terrestrial catastrophism. *Nature* 282:455–459.
Napier, W. M., and Staniucha, M. 1982. Interstellar planetesimals. I. Dissipation of a primordial cloud of comets by tidal encounters with massive nebulae. *Mon. Not. Roy. Astron. Soc.* 198:723–735.
Neftel, A., Oeschger, H., and Suess, H. E. 1981. Secular non-random variations of cosmogenic carbon-14 in the terrestrial atmosphere. *Earth Planet. Sci. Lett.* 56:127–147.
Negi, J. G., and Tiwari, R. K. 1983. Matching long term periodicities of geomagnetic reversals and galactic motions of the solar system. *Geophys. Res. Lett.* 10:713–716.
Neukum, G., Konig, B., and Fechtig, H. 1975. Cratering in the Earth-Moon system: Consequences for age determination by crater counting. *Proc. Lunar Planet. Sci. Conf.* 6:2597–2620.
Newell, N. D. 1967. Revolution in the history of life. In *Uniformity and Simplicity: A Symposium on the Principle of the Uniformity of Nature*, ed. C. Albritton, *Geol. Soc. Amer. Spec. Paper 89*, pp. 63–91.
Newell, N. D. 1982. Mass extinctions: Illusions or realities. *Geol. Soc. Amer. Spec. Paper 190*, pp. 257–263.
Newhall, X. X., Standish, E. M. Jr., and Williams, J. G. 1983. DE102:A numerically integrated ephemeris of the Moon and planets spanning forty-four centuries. *Astron. Astrophys.* 125:150–167.
Nichols, H. 1967. Central Canadian palynology and its relevance to North Western Europe in the late Quaternary period. *Rev. Paleobot., Palynol.* 3:231–243.
Nishiizumi, K., Regnier, S., and Marti, K. 1980. Cosmic ray exposure ages of chondrites, preirradiation and constancy of cosmic ray flux in the past. *Earth Planet. Sci. Lett.* 50:156–170.
Nishiizumi, K., Elmore, D., Honda, M., Arnold, J. R., and Gove, H. E. 1983. Measurements of ^{129}I meteorites and lunar rock by tandem accelerator mass spectrometry. *Nature* 305:611–612.
Nishiizumi, K., Murly, S. V. S., Marti, K., and Arnold, J. R. 1985. When did the average cosmic ray flux increase? Paper SH7.1-4. In *Proc. 19th Internatl. Cosmic Ray Conf.*, La Jolla, CA, pp. 379–381.
Nölke, F. 1909. Die Entstehung der Eiszeiten. *Deutsche Geographische Blätter* 32:1–30.
Nousek, J. A., Fried, P. M., Sanders, W. T., and Kraushaar, W. L. 1982. On the origin of the 1 keV diffuse x-ray background. *Astrophys. J.* 258:83–95.
Ogorodnikov, K. F. 1965. *Dynamics of Stellar Systems* (Oxford: Pergamon Press).
Oikawa, S., and Everhart, E. 1979. Past and future orbit of 1977 UB, object Chiron. *Astron. J.* 84:134–139.
Olano, C. A. 1982. On a model of local gas related to Gould's Belt. *Astron. Astrophys.* 112:195–208.
Olausson, E., and Svenonius, B. 1973. The relation between glacial ages and terrestrial magnetism. *Boreas* 2:109–115.
Oort, J. H. 1927. Observational evidence confirming Lindblad's hypothesis of a rotation of the galactic system. *Bull. Astron. Inst. Netherlands* 3:275–282.
Oort, J. H. 1928. Dynamics of the galactic system in the vicinity of the Sun. *Bull. Astron. Inst. Netherlands* 4:269–287.
Oort, J. H. 1932. The force exerted by the stellar system in the direction perpendicular to the galactic plane and some related problems. *Bull. Astron. Inst. Netherlands* 6:249–287.
Oort, J. H. 1950. The structure of the cloud of comets surrounding the solar system and a hypothesis concerning its origin. *Bull. Astron. Inst. Netherlands* 11:91–110.

Oort, J. H. 1960. Notes on the determination of K_z and on the mass density near the Sun. *Bull. Astron. Inst. Netherlands* 15:45–53.
Oort, J. H., and Schmidt, M. 1951. Differences between new and old comets. *Bull. Astron. Inst. Netherlands* 11:259–269.
Öpik, E. J. 1932. Note on stellar perturbations of nearly parabolic orbits. *Proc. Amer. Acad. Arts Sci.* 67:169–183.
Öpik, E. J. 1951. Collision probabilities with the planets and the distribution of interplanetary matter. *Proc. Roy. Irish Acad.* 54A:165–199.
Öpik, E. J. 1958. On the catastrophic effects of collisions with celestial bodies. *Irish Astron. J.* 5:34–36.
Öpik, E. J. 1963. Stray bodies in the solar system. Part 1. Survival of cometary nuclei and the asteroids. *Adv. Astron. Astrophys.* 2:219–262.
Öpik, E. J. 1966. Sun-grazing comets and tidal disruption. *Irish Astron. J.* 7:141–161.
Öpik, E. 1976. *Interplanetary Encounters: Close-Range Gravitational Interactions* (Amsterdam: Elsevier Sci. Publ.).
Oró, J. 1961. Comets and the formation of biochemical compounds in the primitive earth. *Nature* 190:384–390.
Ostriker, J. P., Spitzer, L. Jr., and Chevalier, R. A. 1972. On the evolution of globular clusters. *Astrophys. J.* 176:L51–L56.
Ovenden, M. W., and Byle, J. 1978. Comets and the missing planet. In *Dynamics of Planets and Satellites and Theories of their Motion*, ed. V. Szebenely (Dordrecht: D. Reidel), pp. 101–107.
Owens, A. J. 1973. Cosmic-Ray Scintillations. Ph.D. Thesis, California Institute of Technology, Pasadena.
Palme, H. 1982. Identification of projectiles of large terrestrial impact craters and some implications for the interpretation of Ir-rich Cretaceous-Tertiary boundary layers. *Geol. Soc. Amer. Spec. Paper 190*, pp. 223–233.
Palmer, A. R. 1983. The Decade of North American Geology 1983 Geologic Time Scale. *Geology* 11:503–504.
Palouš, J. 1983. Kinematics of B and A stars. I. Positions and space velocities. *Bull. Astron. Inst. Czech.* 34:286–302.
Palouš, J. 1985. Kinematics of B and A stars. II. Local velocity field derived from proper motions and radial velocities. *Bull. Astron. Inst. Czech.* 36:261–267.
Palouš, J., and Hauck, B. 1986. The Sirius super-cluster. *Astron. Astrophys.* In press.
Palouš, J., and Piskunov, A. E. 1985. The velocity-age relation for B and A stars. *Astron. Astrophys.* 143:102–107.
Parenago, P. P. 1950. Study of space velocities of stars. *Astron. J. USSR* 27:150–168. In Russian.
Paresce, F. On the distribution of interstellar material around the Sun. *Astron. J.* 89:1022–1027.
Paresce, F., Bowyer, S., and Kumar, S. 1974. Further evidence for an interstellar source of nighttime HeI 584 Å radiation. *Astrophys. J.* 188:L71–L73.
Parker, E. N. 1965. The passage of energetic charged particles through interplanetary space. *Planet. Space Sci.* 13:9–49.
Peltier, R. 1982. Dynamics of the ice-age Earth. *Adv. Geophys.* 24:1–146.
Pendleton, Y. J., and Black, D. C. 1983. Further studies on criteria for the onset of dynamical instability in general three-body systems. *Astron. J.* 88:1415–1419.
Perek, L. 1962. Distribution of mass in oblate stellar systems. In *Advances in Astronomy and Astrophysics*, vol. 1, ed. Z. Kopal (New York: Academic Press), pp. 165–287.
Perek, L. 1966. A dynamical model of the Galaxy. *Bull. Astron. Inst. Czech.* 17:333–341.
Perry, C. L., and Johnston, L. 1982. A photometric map of interstellar reddening within 300 parsecs. *Astrophys. J. Suppl.* 50:451–516.
Perry, C. L., Johnston, L., and Crawford, P. L. 1983. A photometric map of interstellar reddening within 100 pc. *Astron. J.* 87:1751–1774.
Pickering, W. H. 1909. A search for a planet beyond Neptune. *Ann. Astron. Obs. Harvard College* 61:113–162.
Playford, P. E., McLaren, D. J., Orth, C. J., Gilmore, J. S., and Goodfellow, W. D. 1984. Iridium anomaly in the upper Devonian of the Canning Basin, Western Australia. *Science* 226:437–439.

Poveda, A., and Allen, C. 1985. Does the maximum of the stellar luminosity function correspond to a maximum of the mass spectrum? Submitted to *Astrophys. J.*
Proctor, R. A. 1869. Preliminary paper on certain drifting motions of the stars. *Proc. Roy. Soc. London* 17:169–171.
Rampino, M. R., and Stothers, R. B. 1984a. Terrestrial mass extinctions, cometary impacts and the Sun's motion perpendicular to the galactic plane. *Nature* 308:709–712.
Rampino, M. R., and Stothers, R. B. 1984b. Geological rhythms and cometary impacts. *Science* 226:1427–1431.
Rampino, M. R., and Stothers, R. B. 1985. Terrestrial mass extinctions and galactic plane crossings. Reply. *Nature* 313:159–160.
Rampino, M. R., Self, S., and Fairbridge, R. W. 1979. Can rapid climatic change cause volcanic eruptions? *Science* 206:826–829.
Raup, D. M. 1985. Magnetic reversals and mass extinctions. *Nature* 314:341–343.
Raup, D. M. 1986. Biological extinction in Earth history. *Science* 231:1528–1533.
Raup, D. M., and Sepkoski, J. J. Jr. 1982. Mass extinctions in the marine fossil record. *Science* 215:1501–1503.
Raup, D. M., and Sepkoski, J. J. Jr. 1984. Periodicity of extinctions in the geologic past. *Proc. Natl. Acad. Sci. U.S.A.* 81:801–805.
Raup, D. M., and Sepkoski, J. J. Jr. 1986. Periodic extinctions of families and genera. *Science* 231:833–836.
Rawlins, D. 1970. The great unexplained residual in the orbit of Neptune. *Astron. J.* 75:856–857.
Rawlins, D. 1981. Galileo's observation of Neptune. *Nature* 290:164–165.
Rawlins, D., and Hammerton, M. 1973. Mass and position limits for an hypothetical tenth planet of the solar system. *Mon. Not. Roy. Astron. Soc.* 162:261–270.
Reedy, R. C., Arnold, J. R., and Lal, D. 1983. Cosmic ray record in solar system matter. *Ann. Rev. Nucl. Part. Sci.* 33:505–537.
Remy, F., and Mignard, F. 1985. Dynamical evolution of the Oort cloud. I. A Monte Carlo simulation. *Icarus* 63:1–19.
Renzetti, N. A., Jordan, J. F., Berman, A. L., Wackley, J. A., and Yunck, T. P. 1982. The Deep Space Network: An instrument for radio navigation of deep space probes. JPL Publ. 82-102.
Retterer, G. M., and King, I. R. 1982. Wide binaries in the solar neighborhood. *Astrophys. J.* 254:214–220.
Reynolds, R. J., and Ogden, P. M. 1979. Optical evidence for a very large expanding shell associated with the I Orion OB association, Barnard's Loop, and the high galactic latitude H Alpha filaments in Eridanus. *Astrophys. J.* 229:942–953.
Reynolds, R. T., Tarter, J. C., and Walker, R. G. 1980. A proposed search of the solar neighborhood for substellar objects. *Icarus* 44:772–779.
Rholfs, K. 1972. The local linearized velocity fields in the presence of a spiral density wave. *Astron. Astrophys.* 17:246–252.
Rickman, H. 1976. Stellar perturbations of the orbits of long-period comets and their significance for cometary capture. *Bull. Astron. Inst. Czech.* 27:92–105.
Rickman, H., and Froeschlé, C. 1980. A Monte Carlo estimate of the fraction of comets developing into sizeable asteroidal bodies. *Moon and Planets* 22:125–128.
Ripken, H. W., and Fahr, H. J. 1983. Modification of the local interstellar gas properties in the heliospheric interface. *Astron. Astrophys.* 122:181–192.
Robertson, P. B., Ostertag, R., Bischoff, L., Oskierski, W., Hickey, L. J., and Dawson, M. R. 1985. First results of a multidisciplinary analysis of the Haughton impact crater, Devon Island, Canada. *Lunar Planet. Sci.* XVI:702–703 (abstract).
Safronov, V. S. 1972. *Evolution of the Protoplanetary Cloud and Formation of the Earth and Planets.* NASA TT-F-677. Trans. from Russian (Moscow: Nauka Press, 1969).
Sanders, D. B., Solomon, P. M., and Scoville, N. Z. 1984. Giant molecular clouds in the Galaxy. I. The axisymmetric distribution of H_2. *Astrophys. J.* 276:182–203.
Sanders, D. B., Scoville, N. Z., and Solomon, P. M. 1985a. Giant molecular clouds in the Galaxy. II. Characteristics of discrete features. *Astrophys. J.* 289:373–387.
Sanders, D. B., Clemens, D. P., Scoville, N. Z., and Solomon, P. M. 1985b. The Massachusetts-Stony Brook galactic plane CO survey: b,V maps of the first galactic quadrant. *Astrophys. J. Suppl.* In press.

Sanders, W. T., Kraushaar, W. L., Nousek, J. A., and Fried, P. M. 1977. Soft diffuse x-rays in the Southern Galactic Hemisphere. *Astrophys. J.* 217:L87–L91.
Sanfilippo, A., Riedel, W. R., Glass, B. P., and Kyte, F. T. 1985. Late Eocene microtektites and radiolarian extinctions on Barbados. *Nature* 314:613–615.
Scalo, J. M. 1985. Fragmentation and hierarchical structure in the interstellar medium. In *Protostars & Planets II*, eds. D. Black and M. S. Matthews (Tucson: Univ. of Arizona Press), pp. 201–296.
Scalo, J. M., and Smoluchowski, R. 1984. Galactic gravitational shock and the extinction of species. *Bull. Amer. Astron. Soc.* 16:493–494 (abstract).
Schmidt, M. 1975. The mass of the galactic halo derived from the luminosity function of high-velocity stars. *Astrophys. J.* 202:22–29.
Scholl, H., Casenave, A., and Brahic, A. 1982. The effect of star passages on cometary orbits in the Oort cloud. *Astron. Astrophys.* 112:157–166.
Schreur, B. 1979. On the Production of Long-Period Comets by Stellar Perturbations of the Oort Cloud. Ph.D. Thesis, Florida State Univ., Tallahassee.
Schwartz, R. D., and James, P. B. 1984. Periodic mass extinctions and the Sun's oscillation about the Galactic plane. *Nature* 308:712–713.
Schwarzchild, K. 1907. Ueber die Eigenbewegungen der Fixsterne. *Nachrichten von der Königlichen Gesellschaft der Wissenschaften zu Göttingen*, pp. 614–632. In German.
Schwarzschild, K. 1908. Determination of vertex and apex according to the ellipsoid hypothesis using a small number of observed proper motions. *Gott. Nachr.*, pp. 191–200. In German.
Scott, J. S., and Chevalier, R. A. 1975. Cosmic ray production in the Cassiopeia: A supernova remnant. *Astrophys. J.* 197:L5–L8.
Scoville, N. Z., and Solomon, P. M. 1975. Molecular clouds in the galaxy. *Astrophys. J.* 199:L105–L110.
Seidelmann, P. K. 1971. A dynamical search for a transplutonian planet. *Astron. J.* 76:740–742.
Seidelmann, P. K., Klepczinski, W. J., Duncombe, R. L., and Jackson, E. S. 1971. The mass of Pluto. *Astron. J.* 76:488.
Seidelmann, P. K., Kaplan, G. H., Pulkkinen, K. F., Santoro, E. J., and Van Flandern, T. C. 1980. Ephemeris of Pluto. *Icarus* 44:19–28.
Seidelmann, P. K., Santoro, E. J., and Pulkkinen, K. F. 1985. Systematic differences between planetary observations and ephemerides. Preprint.
Sekanina, Z. 1976. A probability of encounter with interstellar comets and the likelihood of their existence. *Icarus* 27:123–133.
Sekanina, Z. 1982. The problem of split comets in review. In *Comets*, ed. L. L. Wilkening (Tucson: Univ. of Arizona Press), pp. 251–287.
Seligman, G. 1936. *Snow Structure and Ski Fields* (London: Macmillan).
Sellwood, J. A., and Carlberg, R. G. 1984. Spiral instabilities provoked by accretion and star formation. *Astrophys. J.* 282:61–74.
Sepkoski, J. J. Jr. 1982a. A compendium of fossil marine families. *Milwaukee Public Museum Contrib. in Biology and Geology* 51:1–125.
Sepkoski, J. J. Jr. 1982b. Mass extinctions in Phanerozoic oceans. A review. *Geol. Soc. Amer. Spec. Paper 190*, pp. 283–289.
Sepkoski, J. J. Jr., and Raup, D. M. 1986. Periodicity in marine mass extinctions. In *Dynamics of Extinction*, ed. D. Elliott (New York: J. Wiley and Sons), pp. 3–36.
Seyfert, C. K., and Sirkin, L. A. 1979. *Earth History and Plate Tectonics* (New York: Harper & Row).
Shapley, H. 1918. The distances, distribution in space, and dimensions of 69 globular clusters. *Astrophys. J.* 48:154–181.
Shapley, H. 1921. Note on a possible factor in changes of geological climate. *J. Geol.* 29:502–504.
Shapley, H. 1949. Galactic rotation and cosmic seasons. *Sky and Telescope* 9:36–37.
Share, G. H., Kinzer, R. L., Kurfess, J. D., Forest, D. J., Chupp, E. L., and Rieger, E. 1985. Detection of galactic ^{26}Al gamma radiation by the SMM spectrometer. *Astrophys. J.* 292:L61–L65.
Shemansky, D. E., Judge, D. L., and Jessen, J. M. 1984. Pioneer 10 observations of the interstellar medium in scattered emission of the He 584A and H Lyα 1216 Å lines. In *Local Interstellar Medium, IAU Coll. 81*, eds. Y. Kondo, F. C. Bruhweiler, and B. D. Savage, NASA CP-2345, pp. 24–27.

Shoemaker, E. M. 1983. Asteroid and comet bombardment of the Earth. *Ann. Rev. Earth Planet. Sci.* 11:461–494.
Shoemaker, E. M. 1984a. Cretaceous-Tertiary extinctions and asteroid impacts. *Bull. Amer. Astron. Soc.* 16:678 (abstract).
Shoemaker, E. M. 1984b. Large body impacts through geologic time. In *Patterns of Change in Earth Evolution, Dahlen Konferenzen*, eds. H. D. Holland and A. F. Trendall (Berlin: Springer-Verlag), pp. 15–40.
Shoemaker, E. M., and Wolfe, R. F. 1982. Cratering time scales for the Galilean satellites. In *Satellites of Jupiter*, ed. D. Morrison (Tucson: Univ. of Arizona Press), pp. 277–339.
Shoemaker, E. M., and Wolfe, R. F. 1984. Evolution of the Uranus-Neptune planetesimal swarm. *Lunar Planet. Sci.* XV:780–781 (abstract).
Shoemaker, E. M., Williams, J. G., Helin, E. F., and Wolfe, R. F. 1979. Earth-crossing asteroids: Orbital classes, collision rates with Earth and origin. In *Asteroids*, ed. T. Gehrels (Tucson: Univ. of Arizona Press), pp. 253–282.
Shu, F. H., Milione, V., Gebel, W., Yuan, C., Goldsmith, D. W., and Roberts, W. W. 1972. Galactic shocks in an interstellar medium with two stable phases. *Astrophys J.* 173:557–592.
Signor, P. W., and Lipps, J. H. 1982. Sampling bias, gradual extinction patterns, and catastrophes in the fossil record. *Geol. Soc. Amer. Spec. Paper 190*, pp. 281–291.
Silver, L. T., and Schultz, P. H., eds. 1982. *Geological Implications of Impacts of Large Asteroids and Comets on the Earth. Geol. Soc. Amer. Special Paper 190.*
Simonson, S. C. 1976. A density wave map of the galactic spiral structure. *Astron. Astrophys.* 46:261–268.
Singer, S. F., and Stanley, J. E. 1980. Submicron particles in meteor streams. In *Solid Particles in the Solar System, IAU Symp. 90*, ed. I. Halliday and B. A. McIntosh (Dordrecht: D. Reidel), pp. 329–332.
Smit, J., and Klaver, G. 1981. Sanidine spherules at the Cretaceous-Tertiary boundary indicate a large impact event. *Nature* 292:47–49.
Smit, J., and ten Kate, W. G. H. Z. 1982. Trace element patterns at the Cretaceous-Tertiary boundary: Consequences of a large impact. *Cretaceous Res.* 3:307–332.
Smith, B. A., and Terrile, R. J. 1984. A circumstellar disk around β Pictoris. *Science* 226:1421–1424.
Smoluchowski, R., and Torbett, M. 1984. The boundary of the solar system. *Nature* 311:38–39.
Solomon, P. M., and Sanders, D. B. 1980. Giant molecular clouds as the dominant component of interstellar matter in the Galaxy. In *Giant Molecular Clouds in the Galaxy*, ed. P. M. Solomon and M. G. Edmunds (New York: Pergamon Press), pp. 41–73.
Solomon, P. M., Sanders, D. B., and Rivolo, A. R. 1985. The Massachusetts-Stony Brook galactic plane survey: Disk and spiral arm molecular cloud populations. *Astrophys. J.* 292:L19–L24.
Sonett, C. P. 1985. Suess Wiggles: A comparison between radiocarbon records. *Meteoritics* 20:383–394.
Spitzer, L. 1958. Disruption of galactic clusters. *Astrophys. J.* 127:17–27.
Spitzer, L. 1978. *Physical Processes in the Interstellar Medium* (New York: Wiley).
Spitzer, L., and Schwarzschild, M. 1951. The possible influence of interstellar clouds on stellar velocities. I. *Astrophys. J.* 114:385–397.
Spitzer, L., and Schwarzschild, M. 1953. The possible influence of interstellar clouds on interstellar velocities. II. *Astrophys. J.* 118:106–112.
Standish, E. M. Jr. 1981. Galileo's observation of Neptune. *Nature* 290:164–165.
Stanley, S. M. 1984. Temperature and biotic crises in the marine realm. *Geology* 12:205–208.
Stanley, S. M., and Campbell, L. D. 1981. Neogene mass extinction of Western Atlantic molluscs. *Nature* 293:457–459.
Stark, A. A. 1979. Galactic Kinematics of Molecular Clouds. Ph.D. Thesis, Princeton Univ.
Stark, A. A. 1984. Kinematics of molecular clouds. I. Velocity dispersion in the solar neighborhood. *Astrophys. J.* 281:624–633.
Stark, A. A., and Blitz, L. 1978. On the masses of giant molecular cloud complexes. *Astrophys. J.* 225:L15–L19.
Stecker, F. W., Solomon, P. M., Scoville, N. Z., and Ryter, C. E. 1975. Molecular hydrogen in the galaxy and galactic gamma rays. *Astrophys. J.* 201:90–97.
Steiner, J. 1967. The sequence of geological events and the dynamics of the Milky Way Galaxy. *J. Geol. Soc. Australia* 14:99–131.

Steiner, J. 1973. Possible galactic causes for synchronous sedimentation sequences of the North American and Eastern European cratons. *Geology* 1:89–92.
Steiner, J. 1974. Possible galactic causes for geologic events. Reply. *Geology* 2:279–280.
Steiner, J. 1979. Regularities of the revised Phanerozoic time scale and the Precambrian time scale. *Geol. Rundschau* 68:825–831.
Steiner, J., and Grillmair, E. 1973. Possible galactic causes for periodic and episodic glaciations. *Geo. Soc. Amer. Bull.* 84:1003–1018.
Stigler, S. M. 1985. Terrestrial mass extinctions and galactic plane crossings. *Nature* 313:159.
Stohl, J. 1983. On the distribution of sporadic meteor orbits. In *Asteroids, Comets, Meteors*, eds. C. I. Lagerkvist and H. Rickman (Uppsala, Sweden: Uppsala Univ.), pp. 419–424.
Stokes, G. M. 1978. Interstellar titanium. *Astrophys. J. Suppl.* 36:115–141.
Storzer, D. 1985. The fission track age of high sodium/potassium australites revisited. *Meteoritics* 20:765–766 (abstract).
Storzer, D., and Wagner, G. D. 1977. Fission track dating of meteorite impacts. *Meteoritics* 12:368–369 (abstract).
Storzer, D., and Wagner, G. A. 1979. Fission track dating of Elgygytgyn, Popigai and Zhamanshin impact craters: No sources for Australasian or North-American tektites. *Meteoritics* 14:541–542 (abstract).
Storzer, D., and Wagner, G. A. 1980. Australites older than indochinites. *Naturwissenschaften* 67:90–91.
Storzer, D., Jessberger, E. K., Klay, N., and Wagner, G. A. 1984. ^{40}Ar-^{39}Ar evidence for two discrete tektite-forming events in the Australian-southeast Asian area. *Meteoritics* 19:317 (abstract).
Stothers, R. B. 1979. Solar activity cycle during classical antiquity. *Astron. Astrophys.* 77:121–127.
Stothers, R. B. 1984. Mass extinctions and missing matter. *Nature* 311:17.
Stothers, R. B. 1985. Terrestrial record of the Solar System's oscillation about the galactic plane. *Nature* 317:338–341.
Stothers, R. B., and Frogel, J. A. 1974. The local complex of O and B stars. I. Distribution of stars and interstellar dust. *Astron. J.* 79:456–471.
Streitmatter, R. E., Balasubrahmanyan, V. K., Protheroe, R. J., and Ormes, J. F. 1985. Cosmic ray propagation in the local superbubble. *Astron. Astrophys.* In press.
Strömberg, G. 1924. The asymmetry in stellar motions and the existence of a velocity-restriction in space. *Astrophys. J.* 59:228–251.
Strömgren, B. 1966. Spectral classification through photoelectric narrow-band photometry. *Ann. Rev. Astron. Astrophys.* 4:433–472.
Suess, H. E. 1980. The radiocarbon record in tree rings of the last 8000 years. *Radiocarbon* 20:200–207.
Talbot, R. J., and Newman, M. J. 1977. Encounters between stars and dense interstellar clouds. *Astrophys. J. Suppl.* 34:295–308.
Tamrazyan, G. P. 1957. Geotectonic hypothesis. *Izvestia Academii Nauk Azerbaidjanskoi SSR* 12:85–115.
Tamrazyan, G. P. 1967. The global historical and geological regularities of the Earth's development as a reflection of its cosmic origin (as a sequence of interaction in the course of galactic movement of the Solar System). *Ostrave Vysoke Skoly Banske Sbornik Ved. Praci Rada Horn. Geol.* 13:5–24.
Tedesco, E. F., and Gradie, J. 1986. 1986 DA and 1986 EB. *IAU, Central Bur. for Astron. Telegrams*, Circ. No. 4205.
Thaddeus, P., and Chanan, G. A. 1985. Cometary impacts, molecular clouds, and the motion of the Sun perpendicular to the galactic plane. *Nature* 314:73–75.
Thaddeus, P., and Dame, T. M. 1983. Number and distribution of molecular clouds in the galaxy. In *Proceedings of the Workshop on Star Formation*, ed. R. D. Wolstencroft (Edinburgh: Royal Observatory), pp. 15–26.
Thomas, G. E. 1971. Properties of nearby interstellar hydrogen deduced from Lyman Alpha sky background measurements. In *Solar Wind*, NASA SP-308, pp. 668–683.
Thomas, G. E., and Krassna, R. F. 1974. OGO-5 measurements of the Lyman Alpha sky background in 1970 and 1971. *Astron. Astrophys.* 30:223–232.
Tielens, A. G. G. M., and Hagen, W. 1982. Model calculations of the molecular composition of interstellar grain mantles. *Astron. Astrophys.* 114:245–260.

Tinbergen, J. 1982. Interstellar polarization in the immediate solar neighborhood. *Astron. Astrophys.* 105:53-64.
Tinsley, B. A. 1971. Extraterrestrial Lyman Alpha. *Rev. Geophys.* 9:89-102.
Tombaugh, C. 1961. The trans-Neptunian planet search. In *Planets and Satellites*, eds. G. P. Kuiper and B. M. Middlehurst (Chicago: Univ. of Chicago Press), pp. 12-30.
Torbett, M. V. 1986. Injection of Oort cloud comets to the inner solar system by galactic tidal fields. Submitted to *Icarus*.
Torbett, M. V., and Smoluchowski, R. 1984. Orbital stability of an unseen solar companion linked to periodic extinction events. *Nature* 311:641-642.
Trumpler, R., and Weaver, H. 1953. *Statistical Astronomy* (New York: Dover), pp. 586-615.
Turco, R. P., Toon, O. B., Ackerman, T. P., Pollack, J. B., and Sagan, C. 1983. Nuclear winter: Global consequences of multiple nuclear explosions. *Science* 222:1283-1292.
Ulrych, T. 1972. Maximum entropy power spectrum of long period geomagnetic reversals. *Nature* 235:218-219.
Umbgrove, J. H. F. 1947. *The Pulse of the Earth* (The Hague: Nijhoff).
Upgren, A. R. 1962. The space distribution of late-type stars in a north galactic pole region. *Astron. J.* 67:37-78.
Upgren, A. R. 1978. The motions of K and M dwarf stars of different ages. *Astron. J.* 83:626-635.
Upgren, A. R., and Armandroff, T. E. 1981. The degree of completeness of nearby stars and the stellar luminosity function. *Astron. J.* 86:1898-1908.
Upgren, A. R., and Chabotte, S. C. 1983. On the frequency of wide pairs among nearby dwarf stars. In *Statistical Methods of Astronomy*, International Colloquium (Strasbourg), pp. 231-232.
Urch, I. H., and Gleesen, L. J. 1973. Energy losses of galactic cosmic rays in the interplanetary medium. *Astrophys. Space Sci.* 20:177-185.
Urey, H. C. 1973. Cometary collisions and geological periods. *Nature* 242:32-33.
Vail, P. R., Mitchum, R. M., Todd, R. G., and Widmier, J. M. 1977. Seismic stratigraphy and global changes of sea level. In *Seismic Stratigraphy: Applications to Hydrocarbon Exploration*, ed. C. E. Payton (Tulsa, OK: Amer. Assoc. Petrol. Geol.), pp. 49-212.
Val'ter, A. A., Gurov, Ye. P., and Ryabenko, V. A. 1977. The Obolon' fossil meteorite crater (astroprobleme) on the northeast flank of the Ukranian shield. *Doklady Akad. Nauk S.S.S.R.* 232:170-173. Trans. *Doklady, Geol.* 232:37-40.
Valtonen, M. J. 1983. On the capture of comets into the solar system. *Observatory* 103:1-4.
Valtonen, M. J., and Innanen, K. A. 1982. The capture of interstellar comets. *Astrophys. J.* 255:307-315.
van de Bult, C. E. P. M., Greenberg, J. M., and Whittet, D. C. B. 1985. Ice in the Taurus molecular cloud: Modelling of the 3 μm profile. *Mon. Not. Roy. Astron. Soc.* 214:289-305.
van de Kamp, P. 1961. Double stars. *Publ. Astron. Soc. Pacific* 73:389-409.
van den Bergh, S. 1982. Giant molecular clouds and the Solar System comets. *J. Roy. Astron. Soc. Canada* 76:303-308.
Vandervoort, P. O. 1968. Moving pairs among the A-type stars within 20 pc. *Astrophys. J.* 152:895-904.
van der Zwet, G. P., Allamandola, L. J., Baas, F., and Greenberg, J. M. 1985. Laboratory identification of the emission features near 3.5 μm in the pre-main-sequence star HD97048. *Astron. Astrophys.* 145:262-268.
Van Flandern, T. A. 1978. A former asteroidal planet as the origin of comets. *Icarus* 36:51-74.
Van Rhijn, P. J. 1936. The absorption of light in interstellar galactic space and the galactic density distribution. *Publ. Kapteyn Astron. Lab., Groningen* 47.
Van Valen, L. M. 1984a. Catastrophies, expectations and the evidence. *Paleobiology* 10:121-137.
Van Valen, L. M. 1984b. A resetting of Phanerozoic community evolution. *Nature* 307:50-52.
Van Valen, L. M. 1985. Null hypotheses and prediction. *Nature* 314:230.
van Woerkom, A. F. F. 1948. On the origin of comets. *Bull. Astron. Inst. Netherlands* 10:445-472.
Vanýsek, V. 1983. Deuterium in comets. In *Asteroids, Comets, Meteors*, eds. C. I. Lagerkvist and H. Rickman (Uppsala: Uppsala Univ. Press), pp. 379-381.
Verniani, F. 1973. An analysis of the physical parameters of 5759 faint radio meteors. *J. Geophys. Res.* 78:8429-8462.

Villumsen, J. V. 1983. The vertical growth and structure of galactic disks. *Astrophys. J.* 274:632–645.
Völk, J. H. 1975. Cosmic ray propagation in interplanetary space. *Rev. Geophys. Space Phys.* 13:547–566.
Voshage, H. 1962. Eisenmeteorite als Raumsonden für die Untersuchung des Intensitatsverlaufes der Kosmischen Strahlung wahrend der Letzten Milliarden Jahre. *Z. Naturforsch* 17A:422–432.
Voshage, H. 1978. Investigations on cosmic-ray-produced nuclides in iron meteorites, 2. New results on $^{41}K/^{40}K/^{-40}He/^{21}Ne$ exposure ages and the interpretation of age distributions. *Earth Planet. Sci. Lett.* 40:83–90.
Vsekhsvyatskii, S. K. 1984. Physical characteristics of comets. NASA-TT F-80. Trans. from Russian.
Vyssotsky, A. N. 1963. Spectral surveys of K and M dwarfs. In *Basic Astronomical Data*, ed. K. A. Strand (Chicago: Univ. of Chicago Press), pp. 192–203.
Wallis, M. K., and Macpherson, A. K. 1981. On the outgassing and jet thrust of snowball comets. *Astron. Astrophys.* 98:45–49.
Weaver, H. 1974. Space distribution and motion of the local HI gas. In *Highlights of Astronomy*, vol. 3, ed. G. Contopoulos (Dordrecht: D. Reidel), pp. 423–440.
Weaver, H. 1979. Large supernova remnants as common features of the disk. In *Large-Scale Characteristics of the Galaxy, IAU Symp. 84*, ed. W. B. Burton (Dordrecht: D. Reidel), pp. 295–300.
Webber, W. R. 1985. Temporal variations of the anomalous oxygen component. In *Proc. Solar Wind V*, NASA. In press.
Weinberg, M. D., Shapiro, S. L., and Wasserman, I. 1986. The fate of wide binaries in the solar neighborhood: Encounters with stars and molecular clouds. *Icarus* 65:27–36.
Weis, E. W. 1984. Photometric parallaxes for selected stars of color class m from the NLTT catalogue. *Astrophys. J. Suppl.* 55:289–299.
Weissman, P. R. 1978. Physical and Dynamical Evolution of Long-Period Comets. Ph.D. Thesis, Univ. of California, Los Angeles.
Weissman, P. R. 1979. Physical and dynamical evolution of long-period comets. In *Dynamics of the Solar System*, ed. R. L. Duncombe (Dordrecht: D. Reidel), pp. 277–282.
Weissman, P. R. 1980*a*. Stellar perturbations of the cometary cloud. *Nature* 288:242–243.
Weissman, P. R. 1980*b*. Physical loss of long-period comets. *Astron. Astrophys.* 85:191–196.
Weissman, P. R. 1982*a*. Dynamical history of the Oort cloud. In *Comets*, ed. L. L. Wilkening (Tucson: Univ. of Arizona Press), pp. 637–658.
Weissman, P. R. 1982*b*. Terrestrial impact rates for long- and short-period comets. *Geol. Soc. Amer. Spec. Paper 190*, pp. 15–24.
Weissman, P. R. 1983*a*. Dynamical evolution of the Oort cometary cloud. *Highlights of Astron.* 6:363–370.
Weissman, P. R. 1983*b*. The mass of the Oort cloud. *Astron. Astrophys.* 118:90–94.
Weissman, P. R. 1984. The Vega particulate shell: Comets or asteroids? *Science* 224:987–989.
Weissman, P. R. 1985*a*. The origin of comets: Implications for planetary formation. In *Protostars & Planets II*, ed. D. C. Black and M. S. Matthews (Tucson: Univ. of Arizona Press), pp. 895–920.
Weissman, P. R. 1985*b*. Dynamical evolution of the Oort cloud. In *Dynamics of Comets*, ed. A. Carusi and G. B. Valsecchi (Dordrecht: D. Reidel), pp. 87–96.
Weissman, P. R. 1985*c*. Terrestrial impactors at geological boundary events: Comets or asteroids? *Nature* 314:517–518.
Weissman, P. R. 1985*d*. Cometary showers and unseen solar companions. *Nature* 312:380.
Weller, C. S., and Meier, R. R. 1981. Characteristics of the helium component of the local interstellar medium. *Astrophys. J.* 246:386–393.
Westin, T. N. G. 1985. The local system of early type stars: Spatial extent and kinematics. *Astron. Astrophys. Suppl.* 60:99–134.
Wetherill, G. W. 1975. Late heavy bombardment of the moon and terrestrial planets. *Proc. Lunar Sci. Conf.* 6:1539–1559.
Wetherill, G. W. 1980. Formation of the terrestrial planets. *Ann. Rev. Astron. Astrophys.* 18:77–113.
Wetherill, G. W., and ReVelle, D. O. 1981. Which fireballs are meteorites? A study of the Prairie Network photographic meteor data. *Icarus* 48:308–328.

Whipple, F. L. 1950. A comet model. I. Acceleration of comet Encke. *Astrophys. J.* 111:374–394.
Whipple, F. L. 1955. A comet model. III. The zodiacal light. *Astrophys. J.* 121:750–770.
Whipple, F. L. 1962. On the distribution of semimajor axes among comet orbits. *Astron. J.* 67:1–9.
Whipple, F. L. 1964. Evidence for a comet belt beyond Neptune. *Proc. Natl. Acad. Sci. U.S.A.* 51:711–718.
Whipple, F. L. 1972. The motion, evolution and orbits and origins of comets. In *The Motion, Evolution of Orbits, and Origin of Comets, IAU Symp. 45*, eds. G. A. Chebotarev, E. I. Kazimirchak-Polonskaya, and B. G. Marsden, pp. 401–408.
Whipple, F. L. 1975. Do comets play a role in galactic chemistry and γ-ray bursts? *Astron. J.* 80:525–531.
Whipple, F. L. 1976. The constitution of cometary nuclei. In *Comets, Asteroids, Meteorites*, ed. A. H. Delsemme (Toledo, OH: Univ. of Toledo), pp. 25–35.
Whipple, F. 1979. Scientific need for a cometary mission. In *Space Mission to Comets*, eds. M. Neugebauer, D. K. Yeomans, J. C. Brandt, and R. W. Hobbs. NASA Publ. 2089, pp. 1–32.
Whipple, F. L. 1982. The rotation of comet nuclei. In *Comets*, ed. L. L. Wilkening (Tucson: Univ. of Arizona Press), pp. 227–250.
Whipple, F. L., and Hamid, S. E. 1952. On the origin of the Taurid meteor streams. *Helwan Obs. Bull.* 41:224–252.
Whipple, F. L., and Hawkins, G. S. 1954. On meteors and rainfall. *J. Meteorol.* 13:236–240.
Whitmire, D. P., and Jackson, A. A. 1984. Are periodic mass extinctions driven by a distant solar companion? *Nature* 308:713–715.
Whitmire, D. P., and Matese, J. J. 1985. Periodic comet showers and Planet X. *Nature* 313:36–38.
Wiedenbeck, M. E., and Greiner, D. E. 1980. A cosmic ray age based on the abundance of ^{10}Be. *Astrophys. J.* 239:L139–L142.
Wielen, R. 1974. The kinematics and ages of stars in Gliese's catalogue. In *Highlights of Astronomy*, vol. 3, ed. G. Contopoulos (Dordrecht: D. Reidel), pp. 395–407.
Wielen, R. 1977. The diffusion of stellar orbits derived from the observed age-dependence of the velocity dispersions. *Astron. Astrophys.* 60:263–275.
Wielen, R. 1982. Kinematics and dynamics. In Landolt-Bornstein: *Numerical Data and Functional Relationships in Science and Technology*, vol. 2, *Astronomy and Astrophysics*, subvol. c, *Interstellar Matter, Galaxy, Universe* (Berlin: Springer-Verlag), pp. 208–231.
Wielen, R., Jahreiss, H., and Krueger, R. 1983. The determination of the luminosity function of nearby stars. In *The Nearby Stars and the Stellar Luminosity Function, IAU Symp. 76*, eds. A. G. Davis Philip and A. R. Upgren (Schenectady: L. Davis Press), pp. 163–169.
Williams, G. E. 1975. Possible relation between periodic glaciation and the flexure of the Galaxy. *Earth Planet. Sci. Lett.* 26:361–369.
Williams, G. E. 1981. *Megacycles* (Stroudsburg: Hutchinson Ross).
Williams, J. G. 1984. Determining asteroid masses from perturbations on Mars. *Icarus* 57:1–13.
Williams, J. G., and Benson, G. S. 1971. Resonances in the Neptune-Pluto system. *Astron. J.* 76:167–177.
Wilson, O. C., and Woolley, R. 1970. Calcium emission intensities as indicators of stellar age. *Mon. Not. Roy. Astron. Soc.* 148:463–475.
Wilson, R. E. 1953. *General Catalogue of Stellar Radial Velocities* (Washington: Carnegie Inst.), Publ. No. 601.
Wing, R. F., and Yorka, S. B. 1979. Comparisons of temperature classes obtained by spectroscopic and photometric techniques: M-type dwarfs, giants and supergiants. In *Spectral Classification of the Future, IAU Coll. 47*, eds. M. F. McCarthy, A. G. Davis Philip and G. V. Coyne (Ric. Astron. Specola Vaticana Vol. 9), pp. 519–534.
Winzer, S. R., Lum, R. K. L., and Shumann, S. 1976. Rb, Sr and strontium isotopic composition, K/Ar age and large ion lithophile trace element abundances in rocks and glasses from the Wanipitei Lake impact structure. *Geochim. Cosmochim. Acta* 40:51–57.
Wolbach, W. S., Lewis, R. S., and Anders, E. 1985. Cretaceous extinctions: Evidence for wildfires and search for meteoritic material. *Science* 230:167–170.
Wolfe, S. H. 1971. Potassium-argon ages of the Manicougan-Mushalagon lakes structure. *J. Geophys. Res.* 76:5424–5436.

Wolfendale, A. W. 1985. Enhanced cosmic ray intensity in the "Loop I" SNR. *The Galaxy and the Solar System Abstract Booklet* (Tucson: Lunar and Planet. Lab.), p. 32.
Wood, J. A. 1961. Stony meteorite orbits. *Mon. Not. Roy. Astron. Soc.* 122:79–88.
Woolley, R. 1970. Deviation of the vertex of the velocity ellipse of young stars and its connection with spiral structure. In *The Spiral Structure of Our Galaxy*, eds. W. Becker and G. Contopoulos (Dordrecht: D. Reidel), pp. 423–432.
Woolley, R., and Stewart, J. M. 1967. Motion of A stars perpendicular to the galactic plane. II. *Mon. Not. Roy. Astron. Soc.* 136:329–339.
Woolley, R., Epps, E. A., Penston, M. J., and Pocock, S. B. 1970. Catalogue of stars within twenty-five parsecs of the Sun. *Roy. Obs. Ann.* 5.
Woosley, S. E., and Weaver, T. A. 1980. Explosive neon burning and ^{26}Al gamma ray astronomy. *Astrophys. J.* 238:1017–1025.
Woosley, S. E., and Weaver, T. A. 1981. Anomalous isotopic composition of cosmic rays. *Astrophys. J.* 243:651–659.
Wright, T. 1755. *Original Theory of the Universe* (Philadelphia, 1837).
Wu, F. M., and Judge, D. L. 1980. A reanalysis of the observed interplanetary hydrogen Lα emission profiles and the derived local interstellar gas temperature and velocity. *Astrophys. J.* 239:389–394.
Yabushita, S. 1972. Stellar perturbations of orbits of long-period comets. *Astron. Astrophys.* 16:395–403.
Yabushita, S. 1979a. A statistical study of the evolution of orbits of long-period comets. *Mon. Not. Roy. Astron. Soc.* 187:445–462.
Yabushita, S. 1979b. Statistical tests of distribution of perihelion points of long-period comets. *Mon. Not. Roy. Astron. Soc.* 189:45–56.
Yabushita, S. 1983. Distribution of cometary binding energies based on the assumption of a steady state. *Mon. Not. Roy. Astron. Soc.* 204:1185–1191.
York, D. G. 1983. The interstellar medium near the sun. III. Detailed analysis of the line of sight to Lambda Scorpii. *Astrophys. J.* 264:172–195.
York, D. G., and Frisch, P. C. 1984. Synthesis of data on the local interstellar medium. *Local Interstellar Medium, IAU Coll. 81*, NASA CP-2345, pp. 51–59.
Young, J. S., and Scoville, N. Z. 1982. Extragalactic CO: Gas distribution which follow the light in IC342 and NGC6946. *Astrophys. J.* 258:467–489.
Yuan, C. 1969. Application of the density wave theory to the spiral structure of the Milky Way system. II. Migration of stars. *Astrophys. J.* 158:889–898.
Yuan, C. 1971. The kinematics of the nearby stars. I. The vertex deviation of the velocity ellipsoids. *Astron. J.* 76:664–669.
Zel'dovitch, Ya. B., and Raizer, Yu. P. 1967. *Physics of Shock Waves and High-Temperature Hydrodynamic Phenomena* (New York: Academic Press).
Zellner, B. 1979. Asteroid taxonomy and the distribution of the compositional types. In *Asteroids*, ed. T. Gehrels (Tucson: Univ. of Arizona Press), pp. 783–806.
Zimbelman, J. R. 1984. Planetary impact probabilities for long period comets. *Icarus* 57:48–54.
Zombeck, M. V. 1982. *Handbook of Space Astronomy and Astrophysics* (Cambridge: Cambridge Univ. Press).

List of Contributors

LIST OF CONTRIBUTORS WITH ACKNOWLEDGMENTS TO FUNDING AGENCIES

The following people helped to make this book possible, in organizing, writing, refereeing, or otherwise.

H. A. Abt, Kitt Peak National Observatory, Tucson, AZ
M. F. A'Hearn, Astronomy Program, University of Maryland, College Park, MD
K. Aksnes, Division for Electronics, Norwegian Defense Research Establishment, Kjeller, Norway
C. P. Allen, Instituto de Astronomia, Univ. Nacional Autonoma de Mexico, Mexico
L. W. Alvarez, Lawrence Radiation Laboratory, University of California, Berkeley, CA
W. Alvarez, Department of Geology and Geophysics, University of California, Berkeley, CA
M. H. Anders, Department of Geology and Geophysics, University of California, Berkeley, CA
J. D. Anderson, Jet Propulsion Laboratory, Pasadena, CA
G. Andlauer, Rue Castelnau 5, Mundolheim, France
J. N. Bahcall, Institute for Advanced Study, Princeton, NJ
M. E. Bailey, Department of Astronomy, University of Manchester, London, England
F. Bash, Astronomy Department, University of Texas, Austin, TX
G. Basri, Astronomy Department, University of California, Berkeley, CA
M. J. S. Belton, Kitt Peak National Observatory, Tucson, AZ
H. C. Bhatt, Physical Research Laboratory, Ahmedabad, India
J. W. Bieber, Bartol Research Foundation, University of Delaware, Newark, DE
J. J. Binney, Department of Theoretical Physics, Oxford University, Oxford, England
L. Blitz, Astronomy Department, University of Maryland, College Park, MD
F. Boley, Department of Physics and Astronomy, Dartmouth College, Hanover, NH
J. Bookbinder, Center for Astrophysics, Cambridge, MA
C. S. Bowyer, Astronomy Department, University of California, Berkeley, CA
A. Brahic, Observatoire de Paris, Meudon, France
K. Brecher, Department of Astronomy, Boston University, Boston, MA
V. A. Brumberg, Institute for Theoretical Astronomy, Leningrad, USSR
H. Campins, Planetary Science Institute, Tucson, AZ
R. G. Carlberg, Department of Physics, York University, Toronto, Canada
G. Chanan, Institute for Space Studies, New York, NY
D.-H. Chen, Purple Mountain Observatory, Academia Sinica, Nanking, China
D. D. Clayton, Department of Space Physics and Astronomy, Rice University, Houston, TX
S. V. M. Clube, Department of Astrophysics, Oxford University, Oxford, England
W. D. Cochran, Department of Astronomy, University of Texas, Austin, TX
G. Contopoulos, Department of Astronomy, Paneistimiopolis, Athens, Greece
D. P. Cox, Department of Physics, University of Wisconsin, Madison, WI

LIST OF CONTRIBUTORS

D. R. Criswell, University of California at San Diego, La Jolla, CA
C. C. Cunningham, Lunar and Planetary Laboratory, University of Arizona, Tucson, AZ
T. Dame, Institute for Space Studies, New York, NY
A. H. Delsemme, Department of Physics and Astronomy, University of Toledo, Toledo, OH
K. Denomy, Lunar and Planetary Laboratory, University of Arizona, Tucson, AZ
B. Donn, Astrochemistry Branch, NASA Goddard Space Flight Center, Greenbelt, MD
M. J. Duncan, Physical Sciences Division, Scarborough College, University of Toronto, West Hill, Ontario, Canada
R. L. Duncombe, Department of Aerospace Engineering and Engineering Mechanics, University of Texas, Austin, TX
E. Everhart, Physics Department, University of Denver, Denver, CO
A. Fahey, McDonnell Center for the Space Sciences, Washington University, St. Louis, MO
H. J. Fahr, Astronomisches Institut, University of Bonn, Bonn, West Germany
J. A. Fernández, Observatorio do Valongo, Universidade Federal do Rio de Janeiro, Rio de Janeiro, Brazil
P. C. Frisch, Astronomy and Astrophysics Center, University of Chicago, Chicago, IL
C. Froeschlé, Observatoire de Nice, Nice, France
J. A. Frogel, National Optical Astronomy Observatories, Tucson, AZ
C. G. Garmany, JILA, University of Colorado, Boulder, CO
D. Gautier, Observatoire de Paris-Meudon, Meudon, France
T. Gehrels, Lunar and Planetary Laboratory, University of Arizona, Tucson, AZ
D. L. Gilden, Institute for Advanced Study, Princeton, NJ
G. Gilmore, Royal Observatory, Edinburgh, Scotland
W. Gliese, Astronomisches Rechen-Institut, Heidelberg, West Germany
R. Gnedin, Pulkovo Observatory, Leningrad, USSR
J. M. Greenberg, Huygens Laboratory, University of Leiden, Leiden, The Netherlands
R. Greenberg, Lunar and Planetary Laboratory, University of Arizona, Tucson, AZ
B. W. Hapke, Department of Geology and Planetary Science, University of Pittsburgh, Pittsburgh, PA
R. S. Harrington, Astrometry Division, United States Naval Observatory, Washington, DC
W. K. Hartmann, Planetary Science Institute, Tucson, AZ
C. E. Heiles, Astronomy Department, University of California, Berkeley, CA
R. C. Henry, Johns Hopkins University, Baltimore, MD
J. Higdon, Jet Propulsion Laboratory, Pasadena, PA
A. Hildebrand, Lunar and Planetary Laboratory, University of Arizona, Tucson, AZ
J. G. Hills, Los Alamos National Laboratory, Los Alamos, NM
W. B. Hubbard, Lunar and Planetary Laboratory, University of Arizona, Tucson, AZ
W. F. Huebner, Theoretical Division, Los Alamos Scientific Laboratory, Los Alamos, NM
P. Hut, Astronomy Department, University of California, Berkeley, CA
K. A. Innanen, Physics Department, York University, Toronto, Canada
W.-H. Ip, Max-Planck-Institut für Aeronomie, Katlenburg, West Germany
A. A. Jackson, Computer Sciences Corporation, Houston, TX
H. Jahreiss, Astronomisches Rechen-Institut, Heidelberg, West Germany
E. B. Jenkins, Princeton University Observatory, Princeton University, Princeton, NJ
J. R. Jokipii, Lunar and Planetary Laboratory, University of Arizona, Tucson, AZ
C. Klatt, University of British Columbia, Vancouver, Canada
L. Kresák, Astronomical Institute, SAV, Bratislava, Czechoslovakia
M. Kutner, Department of Physics, Rensselaer Polytechnic Institute, Troy, NY
C. J. Lada, Steward Observatory, University of Arizona, Tucson, AZ
B. Lavielle, University of Bordeaux, Gradignan, France
P. A. LaViolette, Portland, OR
B. A. Lindblad, Lund Observatory, Lund, Sweden
A. Liu, Center for Radiophysics and Space Research, Cornell University, Ithaca, NY
L.-Z. Liu, Purple Mountain Observatory, Academia Sinica, Nanking, China
E. Lundelius, Jr., Department of Geological Sciences, University of Texas, Austin, TX
J. Lunine, Lunar and Planetary Laboratory, University of Arizona, Tucson, AZ
B. L. Lutz, Lowell Observatory, Flagstaff, AZ

LIST OF CONTRIBUTORS

B. T. Lynds, Kitt Peak National Observatory, Tucson, AZ
M. Magisos, Lunar and Planetary Laboratory, University of Arizona, Tucson, AZ
S. H. Margolis, McDonnell Center for Space Science, Washington University, St. Louis, MO
S. Marinus, Lunar and Planetary Laboratory, University of Arizona, Tucson, AZ
B. G. Marsden, Center for Astrophysics, Harvard and Smithsonian Observatories, Cambridge, MA
K. Marti, Department of Chemistry, University of California at San Diego, La Jolla, CA
K. Mason, Designs in Communication, Tucson, AZ
J. J. Matese, University of Southwestern Louisiana, Lafayette, LA
M. A. Matthews, Jurisprudence and Social Policies, University of California, Berkeley, CA
M. S. Matthews, Lunar and Planetary Laboratory, University of Arizona, Tucson, AZ
R. McCourt, National Public Radio, Tucson, AZ
K. D. McKeegan, McDonnell Center for the Space Sciences, Washington University, St. Louis, MO
G. McLaughlin, Lunar and Planetary Laboratory, University of Arizona, Tucson, AZ
R. S. McMillan, Lunar and Planetary Laboratory, University of Arizona, Tucson, AZ
E. E. Mendoza, V, Observatorio Astrofisico Nacional, Puebla, Mexico
F. C. Michel, Department of Space Physics and Astronomy, Rice University, Houston, TX
I. Mikulinsky, State University of Kazan, USSR
G. E. Morfill, Max-Planck-Institut für Extraterrestrische Physik, München, West Germany
D. Morris, Lawrence Berkeley Laboratory, University of California, Berkeley, CA
R. Muller, Lawrence Berkeley Laboratory, University of California, Berkeley, CA
C. A. Murray, Royal Greenwich Observatory, Hailsham, East Sussex, United Kingdom
W. M. Napier, Royal Observatory, Edinburgh, Scotland
M. J. Newman, Los Alamos National Laboratory, Los Alamos, NM
J. A. Nuth, III, Solar System Exploration Division, NASA Headquarters, Washington, DC
S. Ortolani, Asiago Astrophysical Observatory, University of Padova, Asiago, Italy
J. Palouš, Astronomical Institute of the Czechoslovak Academy of Sciences, Prague, Czechoslovakia
E. N. Parker, Laboratory for Astrophysics and Space Research, University of Chicago, Chicago, IL
M. Peimbert, Instituto de Astronomia, Universedad de Mexico, Mexico City, Mexico
A. E. Piskunov, Astronomical Council, USSR Academy of Science, Moscow, USSR
A. Poveda, Instituto de Astronomia, Universidad Nacional Autonoma de Mexico, Mexico City, Mexico
P. Raeburn, Associated Press, New York, NY
M. R. Rampino, NASA Goddard Institute for Space Studies, New York, NY
D. M. Raup, Department of Geophysical Sciences, University of Chicago, Chicago, IL
J. J. Rawal, Nehru Centre, Nehru Planetarium, Bombay, India
S. Regnier, University of Bordeaux, Gradignan, France
R. T. Reynolds, Theoretical and Planetary Studies Branch, NASA/Ames Research Center, Moffett Field, CA
H. Rickman, Astronomical Observatory, Uppsala, Sweden
E. Roemer, Lunar and Planetary Laboratory, University of Arizona, Tucson, AZ
E. E. Salpeter, Newman Laboratory, Cornell University, Ithaca, NY
D. H. Sampson, Department of Astronomy, Pennsylvania State University, University Park, PA
D. B. Sanders, Division of Math, Physics and Astronomy, California Institute of Technology, Pasadena, CA
S. A. Sandford, McDonnell Center for the Space Sciences, Washington University, St. Louis, MO
J. M. Scalo, Department of Astronomy, University of Texas, Austin, TX
P. L. Schecter, Mt. Wilson and Las Campanas Observatories, Pasadena, CA
T. Schmidt-Kaler, 581 Witten, Steinhugel, West Germany
N. Schneider, Lunar and Planetary Laboratory, University of Arizona, Tucson, AZ
R. D. Schwartz, Physics Department, University of Missouri, St. Louis, MO
N. Z. Scoville, Division of Math, Physics and Astronomy, California Institute of Technology, Pasadena, CA

P. K. Seidelmann, United States Naval Observatory, Washington, DC
H. E. Seuss, University of California at San Diego, La Jolla, CA
S. L. Shapiro, Center for Radiophysics and Space Research, Cornell University, Ithaca, NY
K. Shifrin, Institute of Oceanology, USSR
E. M. Shoemaker, United States Geological Survey, Flagstaff, AZ
W. L. H. Shuter, Department of Physics, University of British Columbia, Vancouver, Canada
B. A. Smith, Lunar and Planetary Laboratory, University of Arizona, Tucson, AZ
R. Smoluchowski, Department of Astronomy, University of Texas, Austin, TX
P. Solomon, Astronomy Program, State University of New York at Stony Brook, Stony Brook, NY
C. P. Sonett, Lunar and Planetary Laboratory, University of Arizona, Tucson, AZ
N. Sperling, 5248 Lawton Ave., Oakland, CA
E. M. Standish, Jr., Jet Propulsion Laboratory, Pasadena, CA
R. B. Stothers, NASA Goddard Institute for Space Studies, New York, NY
G. Strazzulla, Osservatorio Astrofisico, Catania, Italy
W. Sullivan, *New York Times*, New York, NY
A. Szentgyorgyi, Department of Space Physics and Astronomy, Rice University, Houston, TX
S. P. Tarafdar, Theoretical Astrophysics Section, Tata Institute of Fundamental Research, Bombay, India
P. Thaddeus, Institute for Space Studies, New York, NY
D. Thompson, *Science News*, Washington, DC
A. Toomre, Department of Mathematics, Massachusetts Institute of Technology, Cambridge, MA
M. V. Torbett, Department of Physics and Astronomy, University of Kentucky, Lexington, KY
S. Tremaine, CITA, McLennan Laboratory, University of Toronto, Toronto, Canada
A. Tyler, Lunar and Planetary Laboratory, University of Arizona, Tucson, AZ
A. R. Upgren, Van Vleck Observatory, Wesleyan University, Middletown, CT
P. O. Vandervoort, Astronomy and Astrophysics Center, University of Chicago, Chicago, IL
S. V. Vereshchagin, Astronomical Council, USSR Academy of Science, Moscow, USSR
J. V. Villumsen, Institute for Advanced Study, Princeton, NJ
R. M. Walker, McDonnell Center for the Space Sciences, Washington University, St. Louis, MO
I. M. Wasserman, Astronomy Department, Cornell University, Ithaca, NY
H. F. Weaver, 38 Highgate Rd., Berkeley, CA
W. R. Webber, Space Science Center, University of New Hampshire, Durham, NH
S. J. Weidenschilling, Planetary Science Institute, Tucson, AZ
J. L. Weinberg, Space Astronomy Laboratory, University of Florida, Gainesville, FL
M. D. Weinberg, Center for Radiophysics and Space Research, Cornell University, Ithaca, NY
P. R. Weissman, Jet Propulsion Laboratory, Pasadena, CA
T. Westin, Stockholm Observatory, Saltsjöbaden, Sweden
J. C. Wheeler, Astronomy Department, University of Texas, Austin, TX
F. L. Whipple, Planetary Science Division, Smithsonian Astrophysical Observatory, Cambridge, MA
S. White, Steward Observatory, University of Arizona, Tucson, AZ
D. P. Whitmire, Department of Physics, University of Southwestern Louisiana, Lafayette, LA
R. Wielen, Institute for Astronomy and Astrophysics, Technical University, Berlin, West Germany
I. P. Williams, Theoretical Astronomy Unit, Queen Mary College, London, England
J. Wisdom, Department of Earth, Atmospheric and Planetary Sciences, Cambridge, MA
R. F. Wolfe, United States Geological Survey, Flagstaff, AZ
A. W. Wolfendale, Physics Department, University of Durham, Durham, UK
B. Wopenka, McDonnell Center for the Space Sciences, Washington University, St. Louis, MO
S. Yabushita, Department of Applied Mathematics, Kyoto University, Kyoto, Japan
D. G. York, Astronomy and Astrophysics Center, University of Chicago, Chicago, IL
J.-Q. Zheng, Purple Mountain Observatory, Academia Sinica, Nanking, China
H. Zinnecker, Royal Observatory, Edinburgh, Scotland
E. Zinner, McDonnell Center for the Space Sciences, Washington University, St. Louis, MO

LIST OF CONTRIBUTORS

The editors acknowledge the support of NASA Grant NGL 03-002-081 and NSF Grant AST 8418962 for the preparation of the book. The following authors wish to acknowledge specific funds involved in supporting the preparation of their chapters.

Anderson, J. D.: Pioneer Project Office, NASA/Ames Research Center, Letter of Agreement ARC/PPO 17
Bahcall, J. N.: NSF Grants PHY-8217352 and NAS8-32902
Bash, F. N.: NSF Grant AST-8312332
Clayton, D. D.: NASA Grant NSG-7361 and NASA DPR S-10987C (as subcontract N 0001-83-C-2162 from the Naval Research Laboratory)
Cox, D. P.: NSF Grant AST-8415142
Delsemme, A. H.: NASA Grant NSG-7301 and NSF Grant 82-07435
Frisch, P. C.: NSF Grant AST-8317120 and NASA Grants NAGW-405, NAGW-671, and NAG5-704 to the University of Chicago
Greenberg, J. M.: NASA Grant NGR 33-018-148
Hut, P.: NSF Grant PHY-8440263
Jokipii, J. R.: NASA Grant NSG-7101 and NSF Grant ATM-220-18
Marti, K.: NASA Grant NAG9-41
Matese, J. J.: NSF Grant AST-8405077
Michel, F. C.: NASA Grant NGL-44-006-012
Muller, R. A.: Department of Energy Contract DE-AC03-76SF00098 and Research Corporation
Rampino, M. R.: NASA Cooperative Agreement NCC5-16, Task III, with Columbia University
Scoville, N.Z.: NSF Grant AST-8412473
Tremaine, S.: NSF Grant AST-8307654
Upgren, A. R.: NSF Grant AST-8318649
Whitmire, D. P.: NSF Grant AST-8405077
York, D. G.: NSF Grant AST-8317120 and NASA Grants NAGW-405, NAGW-671, and NAG5-704 to the University of Chicago

INDEX*

Absolute magnitude, *19*, 453
Accretion, planetary, 5, 381
^{26}Al, radioactive, 89
Albedo, 453
Alpha Centauri, 453
Antapex, 453
Antarctica, 356, 362
Apollo-Amor region, 268
Apparent magnitude, 453
Argon, 351
Asteroids
  Apollo, 260, 262, 272, 275
  Basaltic, 356
  Collisions in asteroid belt, 363, 368
  Earth-crossers, 247, 369–371
  General, 103, 368–369
  Impacts on Earth, 242, 315, 348, 368–369. *See also* Impact craters *under* Craters
  Main belt, 247
  Population, 271
Asymmetric drift, 454

B and A stars, *54*
β Pictoris, 222, 233, *234–235*, 295, 299
Baricenter, 454
Barnard's star, 29, 454
Binary stars
  Astrometric, 28, 454
  Disruption, 319, 321, 335
  Frequency, 28–30
  General, 161, 169, 270, 314, 317, 373, 375, 391
  Spectroscopic, *15–17*, 28–29, 463

Stability, 375. *See also* Orbit *under* Nemesis
Unseen companions, 14, 28, 225–227, 229, 412
Wide pairs, 28–29, *170*, 171, 217, 284, 313–314, 317, 320, 322, 335, 337
Black holes, 236, 258, 395
Bolide, 454
Bristlecone pine, *121*
Brownlee particles, 111, 274, 280, 454
Bruno, 273

Catastrophism, terrestrial, 260–261, 270, 275–280
Charon, 298, 308
Chiron, 260, 272
Chondrites, 274, *276*, 277
Circumstellar dust shells, *234*
Claystone, 358, 361
CO clouds, 97
Color-magnitude diagrams, *15–17*
Cometesimals
  Carbon grains, 110
  General, 110, 113, 234
  Morphological structure, 109
  Silicate grains, 110
Comets. *See also* Oort cloud
  Albedos, 324
  Angular momenta, 173, *175*
  Anisotropy, 174–182, *189*
  Antapex, 188, 192
  Aphelia, *186–187*, 188–193, *190–191*, 201, 453
  Bombardment episode, 263, 268, 271, 277

*Italicized page numbers indicate figures or tables

[475]

Comets (continued)
Cloud. *See* Oort cloud
Comet belt, 220
Composition
  D/H ratio, 113
  general, 103–115, *114*
  S_2, 109
Diffusion of orbits, 177, 210, 212–213, 215, 268, 282, 317, 412
Disk. *See* Comet disk *under* Oort cloud
Earth-crossers, 247, 253, 302, 329–331
Efficiency of injection, *157*
Fading, 261
Flux, 115, 339, 372, 385
Formation at low temperatures, 109
Fragmentation, 113
General, 260–285
Giant comets, 271, 273, 275, 277
Grains in, 112
Half-life, 217
Hyperbolic, 148, 326
Impact energy, 250
Impacts, 112–113, 115, 348
Impacts, terrestrial, 242, 287. *See also* Impact craters *under* Craters
Impulse approximation, 161, *162*, 166, 168, 206, 214–216, 321, 328, 458
Interstellar comets, 97, 148, 237, 255
Jupiter family, 302–303
Lifetime, 209
Long period, 369
Model
  dirty-snowball model, 223
  general, 103
  sandbank model, 223, 233
Neptune crossers, 221
"New" comets, 103, 166, *207*, 261
Nongravitational forces, 223, 295
Nuclei, 199, 370, 376
Orbital elements, 155
Orbits
  eccentricity, 148, 165
  general, *186–187*, 271
  Keplerian orbits, 325
  prograde, 150, 173, 175
  ratio, retrograde to prograde, *179*, *181*
  retrograde, 150, 173, 175, 218
Organic material, 113–114
Origins, 104, 148, 210–211
Perihelia, 155, 165, *212*
Periodic, 295
Perturbations. *See* Perturbations *under* Oort cloud
Perturbed comets, 182–185
Porosity, 111
Protocomets, 220–221, 233
Relation to the origin of life on Earth, 114
Semimajor axes, 165, *205*, *399*

Short period, 297–309, 369
Shower duration, 299
Showers, 103, 115, 222, 225–229, 231, 241, 247, 254, 295, 297, 299, 316, 324, 331, 336, 338, 363–371, 376, 380, 382–383, 385–386, 389–391, 396–397, 403–406, 409, 411, 416, 455
Snow, 112, 115
Sublimation lifetimes, 227
Sun-grazers, 111, 277
Tidal radius, 464
Tisserand's approximation, 464
Trans-Neptunian belt, 279, 294–295
Velocity impulse, 210
Young comets, *201–203*
Zero-velocity curve, *150–151*, 465
Coriolis forces, 151–152, 318, 320
Cosmic rays
  Anomalous component, 142, 144
  Closed-box limit, 455
  Diffusive shock acceleration, 127
  Energy density, 135
  Forbush decreases, 118, 124
  Galactic, 99, 117, 119, 121, 126, 255, 258
  General, 99, 116–128
  Grammage, 136–137, 142, 144, 456
  Inner solar system, 116
  Intensity variation, *126*
  Interstellar medium, 116
  Leaky-box model, 137, 459
  ^{22}Ne, 140–142
  Scale height, 139
  Solar, 117, 119
  Spallation grammage, 136
  Transport
    convection velocity, 124
    diffusion, 124–125
    general, 123–124, *125*
    magnetic field, 124–125
Craters
  Ages, 232, 348–363, *352*
  Asteroid impacts. *See* Impacts on Earth *under* Asteroids
  Chronostratigraphic position, 351
  Crater-age peak, 381
  Dating
    fission track, 350–351, 389
    isotopic techniques, 350
    potassium-argon, 351, 356, 389
  General, *228*, 241, 247, 250, 256, 287, 348–363, 415–416
  Impact craters, 241–242, *244*, 245, *246*, 247–248, 255, 257, 259–260, 387, 389, 414
  Impact metamorphosed rocks, 351
  Impact probability, 327, 329
  Impact pulses, 385

INDEX

Impacts, 61, 112–113, 115, 250, 253, 255, 266, 297, 301–302, 315, 323, 338, 361, 364, 367, 371, 393, 396
  Periodicity of impacts, 228–229, 241, 243, 245–246, 248, 251, 257, 299, 314–315, 337–338, 357–358, 361, 376, 383, 406
  Rate, 227, 247, 252, 263, 270, 278, 305, 314, 362–363, 372, 381, 383–385
  Record, 12, 270, 297, 308, 382–383, 385, 409, 415–416
  Surge, 363–364
Cratons, 348, 455
Cretaceous-Tertiary boundary, 193–194, 200, 242, 260, 265, 275–276, 339, 355, 358–362, *359*, 367–368, 371, 458
Cretaceous-Tertiary event, 225, 232

Dark nebulae
  Cepheus, 63
  Coal sack, 63
  General, 50
  Lupus, 63
  Ophiuchus, 63
  Scorpius, 63
  Vela, 63
  Visual extinctions by, 71
Darwin glass, 358, *359*
Death star, 314. *See also* Nemesis *and* Solar Companion
Deep Space Network (DSN), 293
Density waves, 54
Disk stars, velocity dispersions, 70
Distribution of stars, *50–51*
Doppler effect, 293–294
Dwarf stars
  Black dwarfs, 196
  Brown dwarfs, 11, 14, 34, 196, 200, 236, 395, 454
  F dwarfs, 4
  General, 14, 20, 26–27, 33–34
  GO dwarfs, 160
  Red dwarfs, 227, 395
Dynamical standard of rest, 455

Earth
  Albedo, 275
  Amino acids, 114
  Atmosphere, 84, 97–98, 100
  Biological extinctions. *See* Extinctions, biological
  Biomass, 114–115
  Climate, 98–100, 242, 253, 393
  Cometary impacts. *See* Impacts, terrestrial *under* Comets *and* Impacts, *under* Craters
  Comet bombardment, 242, 247–248, 263, 268, 271, 277, 327
  Cratering. *See* Craters
  Cratering rate. *See* Rate *under* Craters
  D/H ratio in oceans, 113. *See also* Composition *under* Comets
  Dust, 263, 273–274, 277, 280. *See also* Extinction, sunlight
  Earthquakes, 250
  Eccentricity, 411
  Evolution, biological, 104, 242, 263
  Evolution, geologic, 263
  Extinction of sunlight. *See* Extinction, sunlight
  Extinctions. *See* Extinctions, biological
  Forestation, 273, *274*
  F-region, 100, 455
  General, 139, 173, 178, 195, 198, 322
  Geologic record. *See* Geologic record
  Geomagnetic reversals, 250, 252, 254–255, 257, 277–278, 287, 393
  Glaciation, 99, 114, 147, 260, 267, 271–272, 274–275, 277
  Heat flux, 250
  Ice age, 99, 114
  Impact craters. *See* Impact craters *under* Craters
  Intelligent life, 84
  Magnetosphere, 99
  Mantle, 393
  Mantle plumes, 250
  Oceans, 113
  Orbit, 325, 401, 405
  Origin of life, 114
  Ozone, 100, 277
  Passage through cometary debris, 260
  Periodicity. *See* Periodicity, biological and geologic *and* Periodicity *under* Extinctions, biological
  Rainfall, 273, *274*
  Stratosphere, 100
  Tectonism. *See* Tectonism
  Troposphere, 100, 200
  Volcanism, 277
  Water, 100, 115
Enke, Comet, 260, 272, *273*
European Oak, *121*
Extinction, starlight, 105, 165
Extinction, sunlight, 114, 362
Extinctions, biological
  Families, 340
  Fossils, 339
  General, 5, 12, 114, 195, 199, 217, 222, 225, 228, 230–232, 241, 248–249, 252, 254–256, 258–259, 261–263, 266, 272, 275, *276*, 277–278, 286, 299, 307, 313–315, 323–324, 333, 338–348, *341*, *353*, 355, 357, 361–363, 385–386, 387, *388–389*, 392–394, 396, 409, 413
  Periodicity, 61, 160, 171, 173, 200, 231, 248–250, 335, 337–338, 340, 409, 414–415

Extinctions, biological (continued)
  Record, 409, 416
  Sea-level drops, 275, 393
  Vertebrate, 249

Fermi mechanism, inverse, 279, 458
Field stars, 319–320, 322–323
Fireballs, 370
Fokker-Planck equation, 455
Fomalhaut (α PsA), 233, *234*
Foraminifera, 362
Forbush decrease, 455

Galactic disk, Milky Way
  General, *76*, 256, 321, 382
  Missing matter, 382
  Radial forces, 155
  Surface density, 76
  Vertical forces, 155
Galactic plane
  General, 61, *63*, 65, *85–86*, 88, 262, 318–319, 323, 372, 382–383, 385, 387, 392, 394, 409–410
  Interstellar comets. *See* Interstellar comets *under* Comets
  Midplane, 374, 383, 385
Galactic rotation
  Density wave, 52, 54
  General, 196
  Oort-Lindblad model, 51
Galaxy, Milky Way
  Apogalacticon, 42–43
  Carina spiral arm, 244
  Cosmic rays. *See* Cosmic rays
  CO survey, *64*, 159
  Dark matter, 313–314
  Dynamics, 252
  Galactic center, 63, *85*, 87, 151, 275, 372
  General, 461
  H II region, *81*, 82
  Inner galaxy, 77, 81
  Magnetic field, 255, 258
  Models
    disk column densities, *12*
    general, 4–7
  Perigalacticon, 42–43
  Spiral arms, 35, 80, *81*, 82, 100, 127, 242, 255, 258–259, 263–264, 268–280, 371, 384–385
  Tides
    general, 320, 322–323, 372, 400, 410, 413
    gravitational potential, 149
    limit, 157
    radial forces, 155
    vertical forces, 155, *156*, 201
    zero-velocity surface, 149

Galileo Galilei, 292
Gaussian distribution, 456
Geologic epochs
  Eocene, 249, 275, 340, 342, 345, 355–356, 358, 361–362, 366–367, 371
  Holocene, 273
  Miocene, 249, 341, 345, 347, 356, 360
  Oligocene, 256, 345, 367
  Pleistocene, 249, 272, 274, 362, 367
  Pliocene, 249, 340, *353*, 356, 358, *359*, 362
Geologic eras
  Cenozoic, 245–246, 249, 251
  Mesozoic, 245–246, 249, 251, 342
Geologic periods
  Cretaceous, 249, *341*, 342, 387. *See also* Cretaceous-Tertiary boundary
  Devonian, 212
  Jurassic, 249, 340, *341*, 347, 351
  Permian, 340, 343, 357
  Quaternary, 348
  Tertiary, *341*, 342, 345
  Triassic, 343
Geologic record
  General, 271, 277, 392
  Periodicity, 257, 259
Geologic rhythms, 393
Geologic stages
  Albian, 342
  Bajocian, 340, 344, 357
  Barremian, 344
  Bathonian, *341*, 357
  Callovian, 344
  Cenomanian, 340, 344, 347, 357
  Coniacian, 344
  Djulfian, 340, 343, 347
  Guadalupian, 343, 347
  Hauterivian, 344, 347
  Kazanian, 343
  Maestrichtian, 340, 345, 361
  Norian, 340, 343, 347, 357
  Olenekian, 343
  Pliensbachian, 340, 344
  Rhaetian, 343, 347, 357
  Tithonian, 340, 344–345, 347
  Toarcian, 340
  Turonian, 344, 356
  Ufimian, 343
Geologic time scale, 231, 386
Geologic upheavals, 392
Geologic zones
  Globigerapsis semiinvoluta foraminiferal zone, *359*, 362, 366
  Globorotalia cerroazulensis foraminiferal zone, *359*, 361, 366
Giant molecular clouds. *See also* Molecular clouds

General, 35, 40, 42, 46, 52, 69–83, 149, 158, *164–165*, 172, 204, 216–217, 221, 225, 236, 255–257, 262, 264, 269, 282–283, 375, 456
  Properties, 159–160
Globular clusters, 48, 154
Gould's belt
  Density wave, 41
  General, 41, 49, 52–53, 84, 260, 264–265, 269, 456
  H I gas, 41
Great Rift, 63

H I regions, 264
H_2
  Distribution, 77
  Volume density, 71
Haas-Bottlinger diagram, 456
Halo population, 456
He, 279
HEAO-3, 127, 130, *131*, 456
Heliopause region, 93
Hephaistos, *273*
Hill surface, 218
Hipparcos, 11, 19
Holmes tectonic cycle, 265, 269–270, 456
Horizons, 361–362, 366–367
HR diagram, 18, 21, *22*, 23, 26, 54, *55*, 56, 458
Hugoniot curve, 112
Hyades supercluster, 55, *56*, 57

Impact glass, 348–351, 358, 360, 366
Impactors
  Achondrite, 370
  Chondrite, 370
  Iron, 370
Impact structures
  Boltysh, 356
  Bosumtwi, 362
  Carswell, 351
  General, *349–350*, 359
  Haughton, 356
  Kara, 356
  Manicouagan, 357
  Mien, 351
  Oboloń, 357
  Popigai, 351, 355–356
  Puchezh-Katunki, 357
  Ries, 356
  St. Martin, 357
  Steen River, 356–357
  Zhamanshin, 362
Impacts. *See also under* Comets *and under* Craters
  Impact melt sheets, 368
  Kamacite, 370

Siderophile trace elements, 368, 370
Taenite, 370
Impulse approximation. *See* Impulse approximation *under* Comets
Interplanetary dust, 229
Interstellar clouds
  Comets. *See* Interstellar comets *under* Comets
  Encounters with the Sun, 99
  General, 83–100, 163, 166, 252–254, 313–314, 319–322, 335–336, 410
  Gravitational field, 410
Interstellar dust
  Cold aggregation, 103
  Collisions with Earth, 258
  Extinction of starlight. *See* Extinction, starlight
  General, 85, 87, 105, *106–108*, 109–112, 115, 243, 272
  Magnetic fields, 87
Interstellar gas, 65, 87, 92–94, *96*, 98
Interstellar grains
  Constituents, 109
  General, 109–112, 144
  Photoprocessing, 109
  Porosity, 110
  Ultraviolet processing, 105, 109
Interstellar mantles, 105, *106–107*, 110–111
Interstellar material (ISM)
  General, 84, 86, 98, 99
  Gravitational energy, 99
Interstellar medium
  Cosmic radiation, *138*
  General, 104, 127
  Superbubbles. *See* Superbubble
Interstellar wind
  General, 83, 89–95, *90*, *92*, 97–98
  Neutral helium, 93–94
  Neutral hydrogen, 91–93
IRAS, 11, 204, *220*, 222, 233, *234*, 272, 287, 309, 401, 406, 458
Iridium, 225, 232, 242, 260, 275, 277, 316, 322, 348, 356, 358, *359*, 360, 366, 368, 371, 382, 415

Jarmillo magnetic event, 362
Jean's radius, 399
Jupiter, 262, 300, 302, 304, 326–327, 364, *378–379*, 380, 398, 402, 407, 412

Kapteyn's two star stream hypothesis, 458
K giants, 4, 7, *8*, 11

l Lac Association, 84
Laplace, 261, 263, 279
Libyan glass, 358, *359*
Local standard of rest, 217, 459–460
Loop I bubble, 88–89, 95–96, 98

Luminosity callibrations, 23–26
Luminosity function, 6, 13, 18, 21, *25*, 26, 31, *32*, 33–34, 460
Lunar laser ranges, 288
Luyten Catalogue, 20, 459, 461

Magnetic field, interplanetary, 123–124, 126
Main-sequence stars
  CM groups, 27
  General, 14, 18, 24, 29, 269
  HK groups, 27
Malmquist bias, 24, 460
Mass, interstellar, 237
Meteorites
  Carbonaceous, 370
  Chondritic, 116, 121–122
  General, 119–120, *122*, 123, 127, 229
  Iron, 116, 120, 122, 135
  Stony, 370
Meteors
  β Taurid, 274
  Geminid stream, 272
  General, 103, 111
  Stohl streams, 272
  Taurid-Arietid stream, 272–273
Microspherules, 358, *360*, 366, 382, 385
Microtektites, 225, 247, 249, 360–362, *360*, 366–367, 460
Milankovič cycles, 275, 461
MKK system, 461
Molecular clouds. *See also* Giant molecular clouds
  Clumpiness, 169
  CO, 69–71
  Density, 253
  Distribution, 70, 254
  Encounters with the Oort cloud, 78. *See also* Perturbations *under* Oort cloud
  Galactic plane, concentration in, 391
  General, 39, 42–43, 61–68, *63*, 104–105, 135, 173, 196, 265, 267, 269, 275, 277, 279–283, 285, 322, 371–373, *375*, 382–385, 410, 413
  Giant. *See* Giant molecular clouds
  H_2 volume density, 71
  Half-thickness, 77–78
  Impact parameters, 158
  Internal density, 75
  Interstellar molecular matter, 158
  Mass, 79
  Mass density, 70, 321
  Mass spectrum, 75
  Mean density, 78
  Mean free path, 78
  Metalicity gradients, 72
  Nemesis, effect on orbit, 391
  Nongravitational effects, 79
  Perturbation of Oort cloud. *See* Perturbations *under* Oort cloud
  Properties, 72, *73–76*, 77
  Ram pressure, 79–80
  Size distribution, 159
  Solar energy flux, 80
  Surface densities, 78
  Tidal stripping of Oort cloud. *See* Tidal stripping *under* Oort cloud
  Velocity dispersion, 78
  Z scale height, *77*
Molecular disks, 63, *67*
I Mon Association, 84
Monoceros, 63
Moon, 247, 252, 263, 272, 336, 369
Motion of stars. *See also* Velocity field
  General, 47–48, 53
  Proper motions, 47–48, 462–463

Nemesis. *See also* Solar Companion
  General, 61, 173–174, 229–230, 233, 270–271, 276, 287, 308, 313–315, 317–319, 326, 336, 387–396, 400, 402–403, 409, 412–413, 416, 461
  Half-life, 321–322
  History, 391
  Hypothesis, 193–194, 337, 396
  Lifetime, 319, 322
  Mass, 195–198
  Orbit
    general, 390
    Keplerian, 373, 458
    period, 322–323
    stability, 198
    tulip orbit, *394*, 395
  Search, 200, 395
  Sun-Nemesis system, 319, 322–323, 326–329
Neptune
  Calculated position, 292, 461
  General, 148, 234, 236–237, 286–287, 295, 297–300, 302–304, 306, 308, 330, 377–381, 397, 407
  Observed position, 292, 461
  Orbit, 288–289, *290*, 291–292
  Postdiscovery orbit, 291
Nereid, 307
Novae, *134*
Nr/Ir plane, 274
Nyquist frequency, 243, 461

OB associations, 70, 88, 97, 461
OB stars, 48, 52, 57
Oort cloud. *See also* Comets
  Boundary of stability, 151–152, *153*
  Comet collisions, 405
  Comet disk, 377

INDEX 481

Depletion, 270
Dissipation, 284
Distances, 287
Dynamical modeling, 209–214
End states, *210*
Formation, 317
Galactic tides. *See* Tides *under* Galaxy
Gap, 299, 376–377, 380
General, 153, 155–156, 158, 173–174, 176–178, *179*, 180, *181*, 182, *183*, 189, 193, 195–199, 204–237, 260, 262–263, 277–278, 280–281, 297–300, 313, 316, 326–327, 333–335, 367, 384, 393, 398, 462
Hypothesis, 204–209, *224*, 225
Lifetime, 283
Loss cone, 178, 180, 326, 329–330, 333–334, 401, 459
Mass, 207, 231, 237
Maxwellian velocity distribution, 181, 460
Mean free path, 78
Mean time, 78–79
Monte Carlo simulations, *208*, 209, 461
Number of comets, 115
Orbits, 271
Origin
  in situ formation, 148
  primordial origin, 166–167
Outer boundary, 149, 372
Perturbations. *See also* Molecular clouds *and* Binary stars
  galactic gravitational field, 214, 218–219, 230, 233
  general, 269–271, 274, 287, 394, 410
  interstellar clouds, 214, 216–218, 230, 233
  molecular clouds, 78–79, 182, 253, 267, 315, 321, 335, 364, 396–397
  Monte Carlo simulations, 169
  nongravitational forces, 206–207, 214–215, 221
  oscillation of solar system through galactic plane, 225–227, 231
  planetary, 205, 214–215
  spiral arms, 397
  stellar, 148, 157–158, 161, 199, 215–216, 233, 236, 319–320, 335, 338, 363, 383, 390–391, 396
Population, *211*, 214
Power-law distribution, 302
Precession, 300
Protosolar disk, 148
Rotation, primordial, 176
Size
  diameter, 78, 283
  limits, 158
  Monte Carlo simulations, 158

Solar velocity, 78
Tidal stripping by GMC's, 161–163, 464
Oort cloud, inner
  General, 115, 204, 219, *220*, 222, 226, 231, 233–237, 253, 317, 324, 328, 332, 363, 372, 384, 398–403, 407, 412, 462
  Mass, 222
Oort cloud, middle, 398, 403–407, 460
Oort cloud, outer, 115, 219, 221, 237, 398, 405–407, 462
Oort limit, 11
Orion, 63, 264
Orion Association, 50, 52, 84
Orion Molecular Cloud, 71
Orion Spiral Arm, 264–265, 269
Oscillation of the solar system through the galactic plane
  Amplitude, 63, 68, 80
  General, 12, 62, 65, 80, 241, 252–256, 263, 269, 275, 287, 315, 371, 383, 389, 391, 409, 411
  Model, 308
Osmium, 275

Palmer time scale, 249
Passing stars, 373, 377, 383, 397–408
Periodicity, biological and geologic. *See also* Periodicity *under* Extinctions, biological *and* Periodicity of impacts *under* Craters
  Astronomical clocks, 371–381
  General, *264*, 294, 307, 409–411, 414–416
  Monte Carlo simulations, 253–254, 256
II Perseus Association, 84
Perth Catalogue, 288
Phanerozoic, 245, 248, 252, 254, 257, 270, 356–357, 360
Pickering's planets O, P, and S, 291
Pioneers 10 and 11
  General, 286–288, 292–295, *296*, 299
  Radiation pressure, 292
Planetary companions, 14, 29
Planetesimals, 148, 236, 245, 260, 272, 307, 377
Planet X
  Comet shower mechanism, *300*
  General, 204, 225–228, 231, 233, 270–271, 276, 286–296, 297–309, 316, 371, 376, *378–379*, 380, 385, 387, 393, 411–412
  Mass, 292
  Model, 299–307
Planets
  Accretion, 234
  General, 403
Pleiades supercluster, 55, *56*

Pluto, 205, 236, 270–271, 298, 308, 376, 380–381, 407, 411, 416
Poisson-Vlasov equation, 4, 6–7, 9, 12, 462
Polarizing dust, 462
Power spectrum, 415
Protostars, 284
Protostellar gas clouds, 47, 52–53

Radio continuum bubble, 94
Radiolaria, 361
Rayleigh number (Ra), 463
Rayleigh-Taylor interaction, 140, 463
r-process, 128, 463

Satellite orbits, 218
Saturn, 300, 302, 304, 326–327, 333, *378–379*, 380, 407–408
Sco-Cen bubble, 95–96, 98
Scorpius-Centaurus Association, 49–50, 52, 84–85, 88, 98, 262, 265
Sc spiral, 463
Shot noise, 61, 65–66, 355, 363, 368
Sirius supercluster, 55, *56*, 57
Smithsonian Astrophysical Observatory, 288
Solar circle, molecular clouds, 77, 82
Solar companion. *See also* Nemesis
  Eccentricity, 329
  General, 172–204, 230, 313–337, 372–376, 384–386, 390, 409–416
  Impulse approximation, 373–374
  Mass, 328, 332–333
  Orbit, 328–329
  Orbit, stability, 372–373, *374–375*
  Orbital diffusion, 335
Solar motion, 27, 36–37, 43, 46, 464
Solar nebula, 148
Solar neighborhood
  Data, *4*
  Flux of stars, 386
  General, 13–34, 84, 129, 134–135
  Limit, 14
  Mass density, 7–9, 32–34, *33*
  Matter
    amount, 3–4
    distribution, 3–4, 11, 18
  Missing matter
    general, 4, 7–12, 236, 255, 335, 464
    dissipational matter, 5, 11
  Photometric data, 21
  Proper motion, 20–21, 33
  Spectral types, distance indicators, 20, 24
  Star-forming region, 264
  Trigonometric parallaxes
    general, *14*, 18–20, *22*, 24, 26, 33
    Lutz-Kelker effect, 19, 24, 460
Solar oscillation. *See* Oscillation of the solar system through the galactic plane
Solar surface, 97

Solar system
  Disk, 295–296, 303
  Encounters with molecular clouds, 160. *See also* Perturbations *under* Oort cloud
  Galactic environment, 3–12
  Gravitational instability, 220
  Orbit around galactic nucleus, 258
  Origin, accretion disk, 220, 233–236
  Outer solar system, gravitational forces, 286, 288
  Position of Sun, 12
  Protosolar disk, 148
  Velocity, 159–160
Solar wind
  General, 93, 97–99, 117–118, 123, *124*, *143*
  Velocity, 126
Spallation, 463
Spiral arms. *See* Spiral arms *under* Galaxy
s-process, 463–464
Star formation, 52
Star stream, 29, 38
Stars, space density, 53
Stellar cluster, 162, 166, 381
Stellar density, 384
Suess wiggle, 120
Sun. *See also* Solar system
  Cosmic rays, 117, 119
  Heliosphere, 120, 142
  Local bubble, 359
  Local fluff, *133*, 144, 459
  Magnetic fields, 117–118, 123, *124*
  Motion, 35, 225, 454. *See also* Oscillation of the solar system through the galactic plane
  -Nemesis system. *See under* Nemesis
  Orbit, 42–43, *44–45*, 46, 262, 265
  Period, 3
  Perturbation of, 269
  Protosun, 399
  Solar environment. *See* Solar neighborhood
  Solar flares, 117, 119
  Solar maxima, 120
  Solar modulation, 123
  Solar wind. *See* Solar wind
  Sunspots, 117, *118*, 119, 243, 259
  T Tauri, 220
  Vertical motion, 217
  Vertical oscillation through galactic plane. *See* Oscillation of the solar system through the galactic plane
  X-ray background, 130–131, *134*
  Z-velocity, 254, 256
Superbubble, 83, 98, 129–130, 132, *133–134*, 135–136, 139, 141–142, 464
Superclusters
  Field stars, *56*, 57
  General, 47, 55–57, 464

Hyades, 55, *56*, 57
Pleiades, 55, *56*
Sirius, 55, *56*, 57
Supernova
   ^{26}Al gamma rays, 130, *131*, *134*
   Cas A phase, 454–455
   General, 88, 97, 127, 129–144, 464
   Nucleosynthesis, 130, 144
   Presupernova, 140
   Remnant, 83, 87, 89, 116, 127, 130, 132, 135, 139
   X-ray background, 130–131, *134*

Tectonism
   Carbonatite intrusions, *151*, 454
   Correlation with impact cratering, *152*
   General, 156–158, 263, *267*
   Kimberlite intrusions, *151*, 458
   Orogeny, 263, *266*, 278, 462
   Periodicity, 150, *151*
   Sea-level, 251. *See also* Sea-level drops *under* Extinctions, biological
Tektites
   Australasia, 367, 369
   Australite, 272, 359, 367
   Barbados, 361
   Caribbean, 361, 366
   China, 367
   Clinopyroxene, 359
   General, 356, 358, *359*, *360*, 362, 415
   Gulf of Mexico, 361, 366
   Indian Ocean, 362, 366
   Indochinites, 359, 367
   Irgizites, 359
   Ivory Coast, 272, 362
   Moldavites, 272, 359
   North America, 359, 361, 366
   Pacific, 361, 366
   Philippinites, 359, 367
   South Australia, 359
   Southeast Asia, 362
   Tasmania, 367
   Tenth planet. *See* Planet X

Tinbergen dust patch, 87, 95, 97–98
Tracer stars, 4, 11
Tunguska event, *273*, 274

Uranus
   Declination residuals, 291
   Ephemeris, 288
   General, 148, 234, 236, 286–287, 294, 297–300, 302–303, 306, 330, 377, 380–381, 407
   Orbit, 288, *289*, 290–292
   Prediscovery data, 291
Ursae Major stream, 57
U.S. Naval Observatory, 288–289

van Biesbroeck 8, 29
Vega (α Lyra), 233, *234*, 295
Velocity field
   Distribution function, 53
   General, 35–57
   Mean velocities, 37
   Milne's model, 52
   Radial velocity, *23*, 48
   Space distribution, *30–31*
   Space velocities, 26
   Stellar velocities, 47, 53
   Velocity dispersions, 27, *28*, 35, 37–40, *54*, 160
   Velocity distribution, 26, 47, 57
   Velocity ellipsoid
     general, 35, 37–38, 48–49, 52, 54, 57, 464–465
     third integral of motion, 39
     vertex, 38–39
     vertex deviation, 39, 52–54, 56–57, 455, 465
Voyager 1 and 2, 288

Walsh spectrum analysis, 465
Wolf-Rayet ejecta, 141, 465
Wright of Durham, Thomas, 261

Zodiacal cloud, 221, 262, 268, 273, 277